# Organic Reactions

# Organic Reactions

## VOLUME 46

JOHN WILEY & SONS, INC.

New York   •   Chichester   •   Brisbane   •   Toronto   •   Singapore

This text is printed on acid-free paper.

Published by John Wiley & Sons, Inc.

Copyright © 1994 by Organic Reactions, Inc.

All rights reserved. Published simultaneously in Canada.

Reproduction or translation of any part of this work beyond
that permitted by Section 107 or 108 of the 1976 United
States Copyright Act without the permission of the copyright
owner is unlawful. Requests for permission or further
information should be addressed to the Permissions Department,
John Wiley & Sons, Inc., 605 Third Avenue, New York, NY
10158-0012

Library of Congress Catalog Card Number 42-20265

ISBN 0-471-08619-3

Printed in the United States of America

10  9  8  7  6  5  4  3  2

# PREFACE TO THE SERIES

In the course of nearly every program of research in organic chemistry the investigator finds it necessary to use several of the better-known synthetic reactions. To discover the optimum conditions for the application of even the most familiar one to a compound not previously subjected to the reaction often requires an extensive search of the literature; even then a series of experiments may be necessary. When the results of the investigation are published, the synthesis, which may have required months of work, is usually described without comment. The background of knowledge and experience gained in the literature search and experimentation is thus lost to those who subsequently have occasion to apply the general method. The student of preparative organic chemistry faces similar difficulties. The textbooks and laboratory manuals furnish numerous examples of the application of various syntheses, but only rarely do they convey an accurate conception of the scope and usefulness of the processes.

For many years American organic chemists have discussed these problems. The plan of compiling critical discussions of the more important reactions thus was evolved. The volumes of *Organic Reactions* are collections of chapters each devoted to a single reaction, or a definite phase of a reaction, of wide applicability. The authors have had experience with the processes surveyed. The subjects are presented from the preparative viewpoint, and particular attention is given to limitations, interfering influences, effects of structure, and the selection of experimental techniques. Each chapter includes several detailed procedures illustrating the significant modifications of the method. Most of these procedures have been found satisfactory by the author or one of the editors, but unlike those in *Organic Syntheses* they have not been subjected to careful testing in two or more laboratories.

Each chapter contains tables that include all the examples of the reaction under consideration that the author has been able to find. It is inevitable, however, that in the search of the literature some examples will be missed, especially when the reaction is used as one step in an extended synthesis. Nevertheless, the investigator will be able to use the tables and their accompanying bibliographies in place of most or all of the literature search so often required.

Because of the systematic arrangement of the material in the chapters and the entries in the tables, users of the books will be able to find information desired by reference to the table of contents of the appropriate chapter. In the interest of economy the entries in the indices have been kept to a minimum, and, in particular, the compounds listed in the tables are not repeated in the indices.

The success of this publication, which will appear periodically, depends upon the cooperation of organic chemists and their willingness to devote time and effort to the preparation of the chapters. They have manifested their interest already by the almost unanimous acceptance of invitations to contribute to the work. The editors will welcome their continued interest and their suggestions for improvements in *Organic Reactions.*

Chemists who are considering the preparation of a manuscript for submission to *Organic Reactions* are urged to write either secretary before they begin work.

## HAROLD RAY SNYDER
### May 21, 1910–March 8, 1994

Harold Ray Snyder had a long and distinguished career (1937–1976) in teaching, research, editing, and administration at the University of Illinois. He was born in Mt. Carmel, Illinois and was graduated from the University in 1931, where he did his senior research with Professor R.C. Fuson. He completed his Ph.D. at Cornell University in 1935, where he did his thesis research with Professor John R. Johnson, who had previously been on the staff at Illinois. After a year at the Solvay Process Company in Syracuse, N.Y., Dr. Snyder decided to give up industrial work to seek an academic position. Roger Adams obliged with an offer of a postdoctoral research assistantship for the academic year 1936–1937, and Harold Snyder joined the teaching staff at Illinois in 1937.

Roger Adams had initiated *Organic Syntheses* in 1920. In 1940, he conceived the idea of a companion series, *Organic Reactions*. Its first volume was published in 1942, with Roger as Editor-in-Chief and four other Editors: Professors Werner E. Bachmann of Michigan, Louis F. Fieser of Harvard, John R. Johnson of Cornell, and Associate Professor Harold R. Snyder. Harold served on the Editorial Board through the issuance of Volume 7 in 1953. Chapters were shorter in those days, and the first seven volumes of *Organic Reactions* contained a total of 66 chapters. Roger was away when the first four volumes were issued, and busy catching up when he returned. Thus the substantial task of editing those 66 chapters fell largely to Harold and the other Editors, and because some of the other Editors suffered poor health, Harold shouldered much of this burden for the first 3–4 volumes. This task appealed to Harold's devotion to organic chemistry, his thoroughness, and his language style, which included logical development of ideas, correct choice of words, and good grammar. His careful work on those early volumes set many of the standards of style and format that have persisted in *Organic Reactions* ever since. Although he never authored a chapter for *Organic Reactions,* one phase of his early research was covered in a chapter in Volume 7 by his Ph.D. students James H. Brewster and Ernest L. Eliel on "Carbon–Carbon Alkylations with Amines and Ammonium Salts."

Harold assembled and maintained a group of industrious, dedicated, and loyal research students that consisted at first of senior undergraduates, then of a mixture of seniors and graduate students, and finally of graduate students plus an occasional postdoctorate or visiting professor on sabbatical leave. He had a close and continuing relationship with his students, a reflection in part of the hospitality, warmth, and genuine interest of Harold and his wife, Mary Jane McIntosh Snyder, in the students. The major contribution to this relationship came from Harold's style of research direction. He inspired his students to follow their own ideas, stressed that research was a learning experience for them, and was always willing to take an extra step on their behalf. His excellent graduate course in classical organic synthesis was an avenue for recruitment of some of the best research students. His attitude toward the developing mechanistic theories was that they were "useful when they suggest new experiments, but dangerous when they discourage them" (James H. Brewster).

Toward his colleagues—one of us writes here as a one-time junior colleague—he was tolerant, generous, and helpful. He had a dry sense of humor that occasionally provoked a hilarious response. Perhaps unknown to his students, but appreciated by his staff colleagues and friends, was his propensity for engaging in practical jokes, often well-staged and rather elaborate.

When Harold was not at his office desk he was likely to be found in the library or in the laboratory across the hall from his office in Noyes Laboratory. There he tried out new reactions on a test-tube scale before he assigned them to students, especially undergraduate research students. He explained that it was wise to generate a bit of optimism at the outset of a research problem. He also encouraged his graduate students to try test-tube reactions initially. His early research involved the practical synthesis of amino acids, from which logical developments followed for the synthesis of unnatural amino acids and antagonists of the natural amino acids. It is pleasing to note that his synthetic methodology has been applied recently to the point mutation of peptides using biotechnology.

He and his students developed *C*-alkylation with quaternary ammonium salts and nucleophilic displacements on such salts, including the stereochemistry, as described in the *Organic Reactions* chapter. His name is associated with important innovations in the use of polyphosphoric acid for inter- and intramolecular condensations, cyclizations, and functional conversions in organic chemistry. He pioneered the use of boron trifluoride as an efficient catalyst in the Fischer indole synthesis and discovered new reactions of anils, including Diels-Alder reactions. He and his students delineated the requirements for disproportionation of tertiary amines. He developed the synthesis and chemistry of arylboronic acids. One of his fundamental ideas was to incorporate sufficient boron into organ-specific drugs that they could then be selectively neutron-irradiated at their in vivo locus. Chemists/pharmacologists are still trying to meet this challenge. In addition to Harold's many research publications and patents, he coauthored a textbook, *Organic Chemistry,* (John Wiley, 1st edition 1942) with R. C. Fuson and was Editor-in-Chief of Volume 28 of *Organic Syntheses* in 1948.

During the Second World War, Harold carried out work for the National Defense Research Committee, the War Production Board Rubber Research Program (with Marvel), and the Committee on Medical Research. Within the latter program, the team of Snyder, Price, and Leonard, together with their graduate students, helped develop the process for production of the antimalarial drug Chloroquine in time for its use in the Pacific. The drug is still in use today, although resistant strains of malarial parasites have become a problem in some areas of the world. Just before the War, Harold had been awarded a John Simon Guggenheim fellowship (1939), which represented an unusually early appreciation of his record in chemical research. His trip to Europe, however, which included a series of lectures in Italy, had to be postponed until 1952. A Professor of Chemistry at Illinois from 1945, Harold served as Associate Head of the Department during 1957–1960 and Associate Dean of the Graduate College during 1960–1975. In the latter role, he also served as Secretary of the Research Board, a very responsible position because the Board distributed all internal research grants and graduate fellowships. It is clear that Harold's objectivity and his modest nature contributed to his long and effective tenure in that office. Outside the University, Harold was a consultant to Merck and Company and to Phillips Petroleum Company. He was also an advisor to the Office of Naval Research in the evaluation of ONR research grant applications.

A Symposium in Honor of Harold Ray Snyder in 1976 was attended by friends, associates, and former students who gathered to pay tribute. The Symposium also marked his 65th year and his retirement from the University of Illinois. Further, his former students organized the Harold Snyder Endowment Fund, which supports undergraduate students who have an interest in organic chemistry to do research with a University of Illinois faculty member. Since 1990, two Snyder Scholars have been supported annually. Additional scholarships are probable with increases in the Endowment Fund.

Harold Snyder was predeceased by his first wife, Mary Jane Snyder, and by his second wife, her sister Bonnie McIntosh Snyder. He is survived by his three children, Dr. Jane Snyder of Columbus, Ohio, Dr. John Snyder of Basking Ridge, New Jersey, and Dr. Mary Ann Nirdlinger of Sylvania, Ohio; four grandchildren and several nieces and nephews; and a sister, Joanne Dorsch of Lake Kiowa, Texas. He will be long remembered by his own family and by many former students and colleagues who constitute his professional family.

NELSON J. LEONARD
ROBERT M. JOYCE

# CONTENTS

CHAPTER                     PAGE

1. TIN (II) ENOLATES IN THE ALDOL, MICHAEL, AND
   RELATED REACTIONS
   *Teruaki Mukaiyama and Shū Kobayashi*. . . . . . . . . . . . . . . . . .  1

2. THE [2,3]-WITTIG REACTION
   *Takeshi Nakai and Koichi Mikami*. . . . . . . . . . . . . . . . . . . . . 105

3. REDUCTIONS WITH SAMARIUM (II) IODIDE
   *Gary A. Molander*. . . . . . . . . . . . . . . . . . . . . . . . . . . . . . 211

CUMULATIVE CHAPTER TITLES BY VOLUME. . . . . . . . . . . . . . . . . . . . . 369

AUTHOR INDEX, VOLUMES 1–46. . . . . . . . . . . . . . . . . . . . . . . . . 381

CHAPTER AND TOPIC INDEX, VOLUMES 1–46. . . . . . . . . . . . . . . . . . . 385

# Organic Reactions

# CHAPTER 1

# TIN(II) ENOLATES IN THE ALDOL, MICHAEL, AND RELATED REACTIONS

Teruaki Mukaiyama and Shū Kobayashi

*Science University of Tokyo, Kagurazaka, Shinjuku-ku, Tokyo, Japan*

## CONTENTS

|  | PAGE |
|---|---|
| Introduction | 2 |
| Scope and Limitations | 4 |
| General Preparative Methods for Tin(II) Enolates | 5 |
| Simple Diastereoselective Reactions | 7 |
| Aldol Reactions | 7 |
| Michael Reactions | 17 |
| Asymmetric Reactions | 19 |
| Enantioselective Aldol Reactions | 19 |
| Enantioselective Michael Reactions | 21 |
| Asymmetric Sulfenylation | 23 |
| Diastereoselective Reactions | 23 |
| Comparison with Other Metal Enolates | 31 |
| Experimental Procedures | 36 |
| Tin Powder | 36 |
| Tin (II) Chloride | 36 |
| Tin(II) Triflate [Sn(OTf)$_2$] | 36 |
| *syn*-5-Hydroxy-4-methyl-3-octanone [Tin(II) Triflate-Promoted Aldol Reaction Between a Ketone and an Aldehyde] | 36 |
| *syn*-3-Hydroxy-2-methyl-1-phenyl-3-(*p*-tolyl)-1-propanone (Metallic Tin) | 37 |
| Ethyl 3-Hydroxy-3-phenylpropionate (SnCl$_2$-LAH) | 37 |
| *cis*-3,4-Epoxy-4-phenyl-2-butanone (Synthesis of an $\alpha,\beta$-Epoxyketone) | 38 |
| *anti*-3-Hydroxy-2-methyl-1,3-diphenylbutan-1-one (Cross Aldol Reaction Between Ketones) | 38 |
| *syn*-3-(3-Hydroxy-2-methyl-3-phenylpropanoyl)thiazolidine-2-thione (Reaction of 3-Acylthiazolidine-2-thione with an Aldehyde) | 39 |
| 3,4-Dihydroxy-6-phenyl-2-hexanones [Tin(II) Enediolate] | 40 |
| 3-Ethylthiomethyl-4-hydroxy-6-phenyl-2-hexanone [Preparation of $\beta$-Ethylthiomethyl Aldols with a Catalytic Amount of Tin(II) Enolate] | 41 |
| Ethyl 3-(*tert*-Butylthio)carbonyl-2-furfurylaminobutylate [Reaction of a Tin(II) Thioester Enolate with an $\alpha$-Iminoester] | 41 |

*Organic Reactions, Vol. 46*, Edited by Leo A. Paquette et al.
ISBN 0-471-08619-3 © 1994 Organic Reactions, Inc. Published by John Wiley & Sons, Inc.

4-(1-Benzoylethyl)phenol [Addition–Reduction Reaction of a Tin(II) Enolate with
1,4-Benzoquinone] . . . . . . . . . . . . . . . . . 42
3-(2,3-Dimethyl-5-oxo-5-phenylpentanoyl)-1,3-oxazolidin-2-one [Reaction of a
Tin(II) Enolate with an α,β-Unsaturated Ketone in the Presence of TMSCl] . . . 42
3-Hydroxy-2-methyl-1,3-diphenyl-1-propanone [Enantioselective Cross Aldol Reaction
of a Ketone and an Aldehyde] . . . . . . . . . . . . . . 43
Oxo-3-phenylhexanedithioate [Catalytic Asymmetric Michael Reaction
of a Tin(II) Enethiolate] . . . . . . . . . . . . . . . 44
3-[(5-Oxo-2(S)-pyrrolidinyl)acetyl]-4(S)-isopropyl-1,3-thiazolidine-2-thione [Alkylation
Reaction of a Chiral Tin(II) Enolate] . . . . . . . . . . . 44
TABULAR SURVEY . . . . . . . . . . . . . . . . . 45
Table I. Aldol Reaction with Achiral Substrates; No or Simple Diastereoselection . . 46
Table II. Michael Reaction with Achiral Substrates; No or
Simple Diastereoselection . . . . . . . . . . . . . . . 70
Table III. Enantioselective Aldol Reaction . . . . . . . . . . 74
Table IV. Enantioselective Michael Reaction . . . . . . . . . . 78
Table V. Enantioselective Sulfenylation . . . . . . . . . . . 80
Table VI. Diastereoselective Aldol Reaction . . . . . . . . . . 81
Table VII. Diastereoselective Aldol Alkylation . . . . . . . . . 93
REFERENCES . . . . . . . . . . . . . . . . . . 99

## INTRODUCTION

The element tin has played an increasingly important role in organic chemistry as well as organometallic chemistry, serving as a source of new reagents for selective transformations.[1–3] The main activity in these fields has been focused for a long time on tin(IV) compounds, and tin(II) compounds have been used primarily as reductants of aromatic nitro compounds to aromatic amines.[4]

During the last decade, generation and reactions of various metal enolates have been extensively studied, and successful applications to the controlled formation of carbon–carbon bonds have been realized under mild conditions.[5–16] The chemistry of tin(IV) enolates has also been studied and several interesting features of these enolates have been reported,[17,18] whereas tin(II) analogs were relatively unknown in synthetic organic chemistry, probably because of the lack of general methods for generating them.

In 1979, tin(II) fluoride was employed as a reductant of several α-halocarbonyl compounds and allylic halides to generate tin(IV) species, which subsequently reacted with aldehydes to form new carbon–carbon bonds (Eqs. 1–5).[19–23] When metallic tin was used instead of tin(II) fluoride, similar reactions proceeded smoothly. Because the reducing ability of metallic tin is superior to that of tin(II) fluoride, allyl bromide and α-bromoesters, which could not be reduced by tin(II) fluoride, were easily reduced by metallic tin to generate tin(II) enolates, which in turn reacted with aldehydes to yield homoallyl alcohols or β-hydroxyesters (Eqs. 6 and 7).[24,25] The latter reaction was the first example of the reaction of tin(II) enolates. More conveniently, tin(II) enolates could be generated

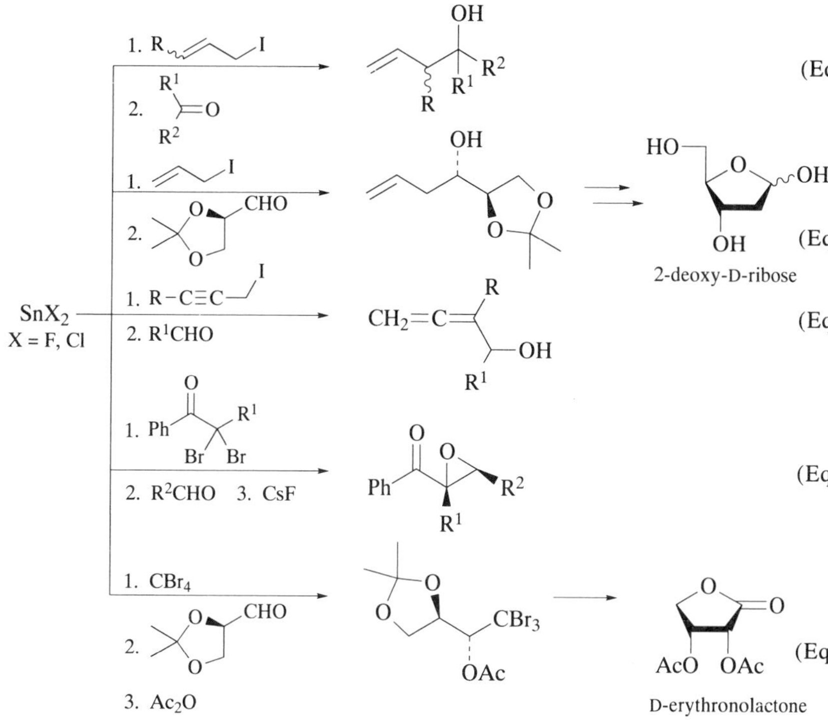

(Eq. 1)

(Eq. 2)

2-deoxy-D-ribose

(Eq. 3)

(Eq. 4)

(Eq. 5)

D-erythronolactone

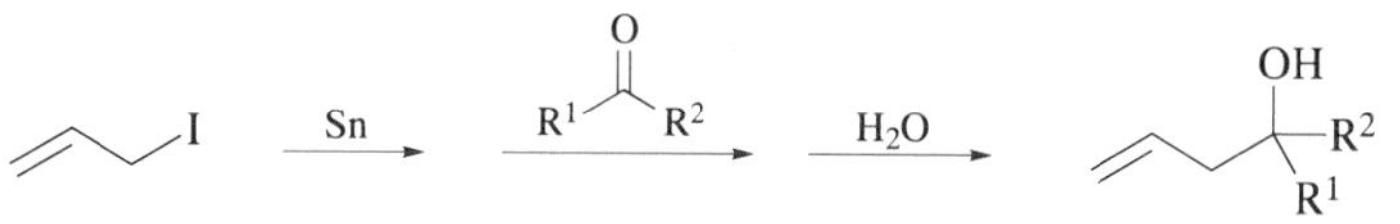

(Eq. 6)

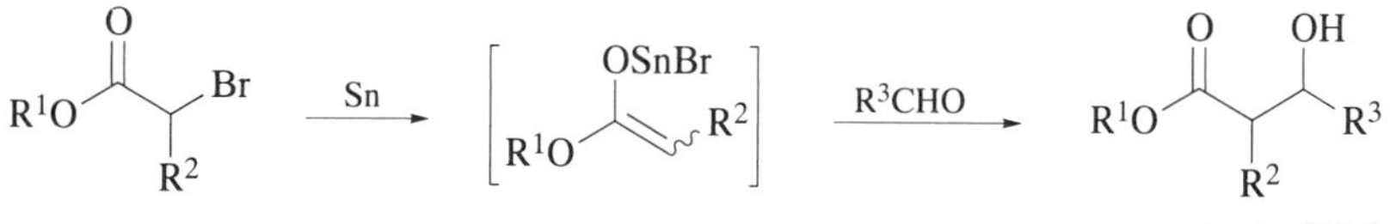

$syn/anti$ >90/10

(Eq. 7)

directly from ketones by using tin(II) triflate and a tertiary amine under neutral conditions.[26]

While strongly basic conditions are required to prepare lithium enolates, tin(II) enolates can be generated under extremely mild conditions. Tin(II) enolates can behave as interesting chemical species in synthetic reactions that cannot be realized with other metal enolates. One of the characteristic features of tin(II) enolate mediated reactions is a highly enantioselective version employing chiral diamines.

Recent developments in the field of stereoselective aldol reactions have resulted in exploitation of the asymmetric version of this reaction, and several successful methods have been reported using chiral carbonyl compounds and/or chiral enolates.[6,27,28] However, the efficiency of these reactions is greatly diminished by the tedious procedures for attachment and removal of the chiral auxiliaries. Thus development of a highly enantioselective aldol reaction between two achiral carbonyl compounds utilizing chiral chelating agents became desirable, though the use of chiral addends in the aldol reaction had not met with much success.[29-31] Chiral diamines derived from (S)-proline, which are postulated to form rigid cis-fused 5-membered bicyclic structures by chelation to a metal center, were found to be effective ligands for several highly enantioselective reactions.[32-35] Coordination of a chiral diamine to the metal center of the tin(II) enolate effected highly enantioselective cross aldol and Michael reactions between two prochiral reactants.

This review covers the literatures on tin(II) enolates to the middle of 1991.

## SCOPE AND LIMITATIONS

At the beginning of this section, general preparative methods for tin(II) enolates are briefly surveyed, followed by a detailed description of the reactions of tin(II) enolates. The reactions are roughly classified into two parts, simple diastereoselective reactions and asymmetric reactions. In the first part, the simple diastereoselective reaction between two achiral substrates is described; for example, the aldol reaction of enolates having prochiral faces with aldehydes gives four diastereomeric and enantiomeric aldols (Eq. 8). In an unselective reac-

$$R^1CHO \quad + \quad R^2 \overset{OM}{\underset{}{\diagdown}} R^3 \quad \longrightarrow$$

$$R^2 \overset{O}{\underset{R^3}{\diagup}} \overset{OH}{\diagup} R^1 \quad + \quad R^2 \overset{O}{\underset{R^3}{\diagup}} \overset{OH}{\diagup} R^1$$

$$+ \quad R^2 \overset{O}{\underset{R^3}{\diagup}} \overset{OH}{\diagup} R^1 \quad + \quad R^2 \overset{O}{\underset{R^3}{\diagup}} \overset{OH}{\diagup} R^1$$

(Eq. 8)

tion, two racemic diastereomers result. The simple diastereoselective reaction gives a surplus of one of these diastereomers. The second part, asymmetric reactions, is further grouped into enantioselective reactions and diastereoselective reactions. One characteristic feature of tin(II) enolates is the asymmetric version in which two achiral substrates combine to give a chiral, optically active product with the aid of a chiral ligand which is not covalently bonded to the substrates (enantioselective reaction). Conventional asymmetric reactions of achiral tin(II) enolates with chiral compounds, chiral tin(II) enolates with achiral compounds, and chiral tin(II) enolates with chiral compounds are discussed in the section on diastereoselective reactions.

## General Preparative Methods for Tin(II) Enolates

Tin(II) enolates are generated in situ according to the following procedures and then immediately reacted with electrophiles. Isolation of tin(II) enolates is generally difficult compared to that of silyl and tin(IV) enolates.

*The Tin(II) Triflate Method.* Most commonly and conveniently, tin(II) enolates can be generated by reaction of ketones and tin(II) triflate in the presence of a tertiary amine (Eq. 9).[26,36] The choice of the tertiary amine is crucial; for ex-

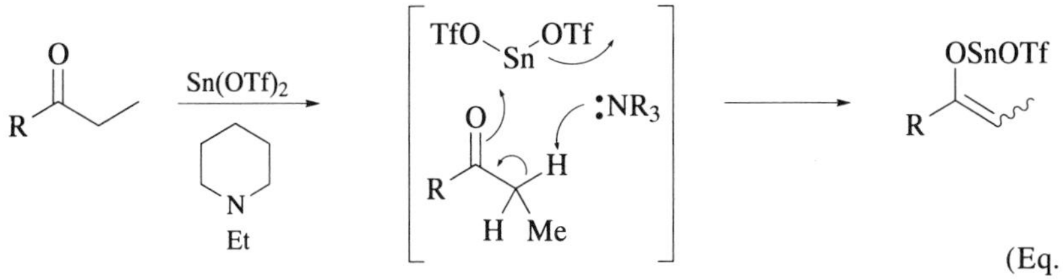

$$\text{(Eq. 9)}$$

ample, pyridine or 1,8-diazabicyclo[5,4,0]undec-7-ene (DBU) fail to promote the reaction because of the formation of a coordinated complex with divalent tin, while other amines give problems with self-aldol reactions. N-Ethylpiperidine gives excellent results in the cross-aldol reaction of propiophenone with 3-phenylpropanal (Eq. 10).

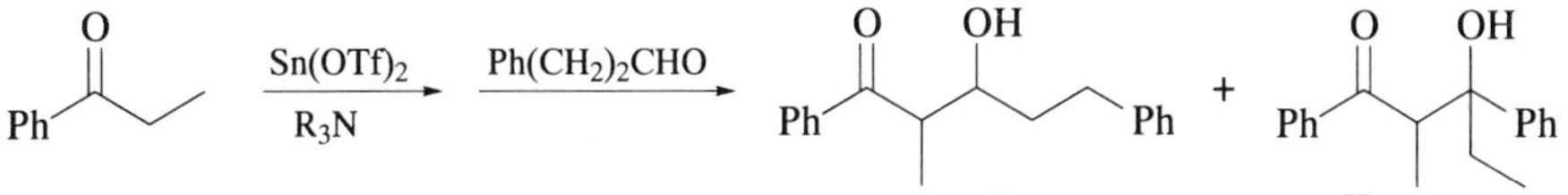

| $R_3N$ | I (%) | II (%) |
|---|---|---|
| pyridine | (0) | (0) |
| DBU | (0) | (0) |
| Et$_3$N | (50) | (15) |
| N-methylmorpholine | (22) | (65) |
| N-ethylpiperidine | (80) | (trace) |

$$\text{(Eq. 10)}$$

Tin(II) enolates of 3-acylthiazolidine-2-thiones,[37] 3-acyloxazolidine-2-thiones,[38] or 3-acyloxazolidinones[39,40] are readily generated by the procedure described above, while tin(II) enolates of esters or thioesters cannot be smoothly generated by the same procedure because of the relatively low acidity of their α-protons.

Tin(II) dienolates are also successfully generated by this procedure (Eq. 11).[41]

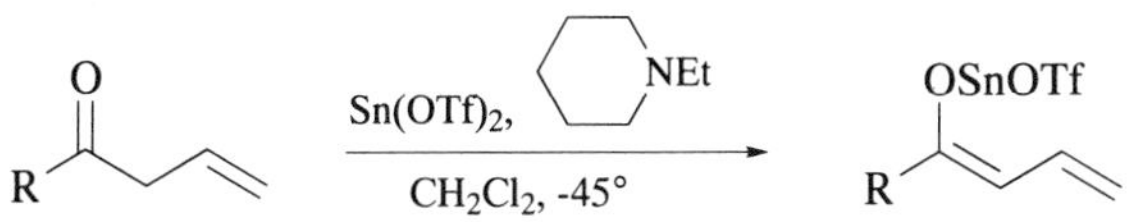

$$\text{(Eq. 11)}$$

*Reduction of α-bromocarbonyls.*  Tin(II) enolates can also be generated in situ by the reduction of α-bromoesters or α-bromoketones by metallic tin (Eq. 12).[42] Instead of metallic tin, the tin(II) chloride–lithium aluminum hydride (LAH) system is effective with some α-bromoesters.[25]

$$R^1 \overset{O}{\underset{R^2}{\diagup\!\!\!\diagdown}} Br \quad \xrightarrow[\text{DMF, rt}]{\text{Sn}} \quad R^1 \overset{OSnBr}{\diagup\!\!\!\diagdown} R^2 \qquad \text{(Eq. 12)}$$

*Addition to Ketenes.*  Tin(II) enolates of thioesters are conveniently generated by the addition of tin(II) thiolates (generated in situ from 1,1′-dimethylstannocene and a thiol) to ketenes (Eq. 13).[43]

$$\underset{H}{\overset{R^1}{\diagup\!\!\!\diagdown}}\!\!=\!\!C\!\!=\!\!O \quad \xrightarrow[\text{THF, rt}]{\text{Sn}(SR^2)_2} \quad R^2S \overset{OSnSR^2}{\diagup\!\!\!\diagdown} R^1 \qquad \text{(Eq. 13)}$$

*Metal Exchange Reaction of Lithium Enolates.*  Tin(II) enolates of thioesters are also generated by metal exchange of lithium enolates with tin(II) chloride or tin(II) triflate (Eq. 14).[44–46]

$$t\text{-BuS}\overset{O}{\diagup\!\!\!\diagdown}R \quad \xrightarrow[\text{2. SnCl}_2]{\text{1. LDA, Et}_2\text{O}} \quad t\text{-BuS}\overset{OSnCl}{\diagup\!\!\!\diagdown}R \qquad \text{(Eq. 14)}$$

*1,4-Addition of Tin(II) Triflate Sulfide to α,β-Unsaturated Ketones.*  Tin(II) enolates of β-thioketones are generated by the conjugate addition of tin(II) triflate sulfide, prepared in situ from tin(II) triflate and a lithium thiolate, to α,β-unsaturated ketones (Eq. 15).[47]

$$\overset{O}{\diagup\!\!\!\diagdown}\!\!\diagdown \quad \xrightarrow[\text{THF, -45°}]{\text{EtSSnOTf}} \quad \overset{OSnOTf}{\diagup\!\!\!\diagdown}\!\!\diagdown\!\!\diagup\text{SEt} \qquad \text{(Eq. 15)}$$

*Reduction of α-Dicarbonyl Compounds [Tin(II) Enediolates].*  Tin(II) enediolates are generated by reduction of α-dicarbonyl compounds with activated metallic tin prepared from tin(II) chloride and metallic potassium (Eq. 16).[48]

$$R^1\overset{O}{\underset{O}{\diagup\!\!\!\diagdown}}R^2 \quad \xrightarrow[\text{THF, rt}]{\text{SnCl}_2,\ K} \quad \overset{R^1 \quad R^2}{\underset{O\diagdown\underset{Sn}{}\diagup O}{\diagup\!\!=\!\!\diagdown}} \qquad \text{(Eq. 16)}$$

## Simple Diastereoselective Reactions

### Aldol Reaction

*Aldehydes and Ketones as Acceptors.* α-Haloesters react with metallic tin to generate tin(II) enolates, which readily add to carbonyl compounds to give β-hydroxyesters in high yields (Reformatsky-type reaction) (Eq. 17). Activated

Sn
DMF, rt
X = halogen


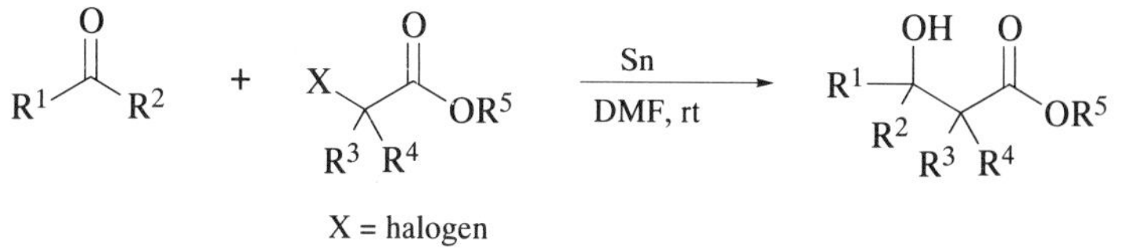

(Eq. 17)

tin, prepared by reduction of tin(II) chloride with lithium aluminum hydride (LAH) (2:1 molar ratio), is also effective in some of these reactions (Eq. 18).[25]

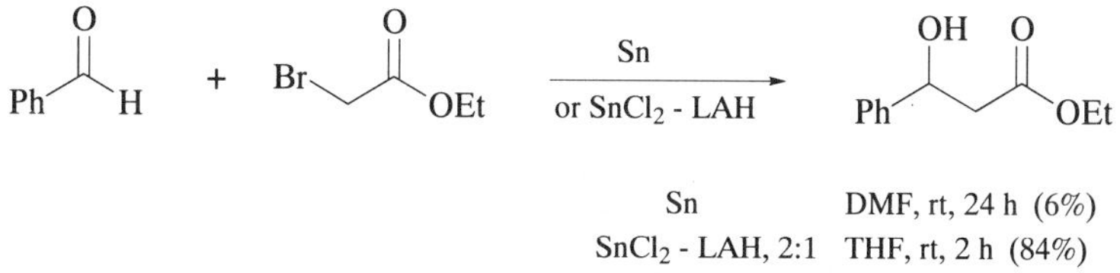

Sn
or $SnCl_2$ - LAH
Sn DMF, rt, 24 h (6%)
$SnCl_2$ - LAH, 2:1 THF, rt, 2 h (84%)


(Eq. 18)

The reduction of α-bromoketones by metallic tin also generates tin(II) enolates which react with aldehydes and ketones regiospecifically with high *syn* selectivity (Eq. 19).[42] Generation of tin(IV) species via reduction of α-bromoketones or α-bromoesters with initially generated tin(II) species does not take place under the reaction conditions.

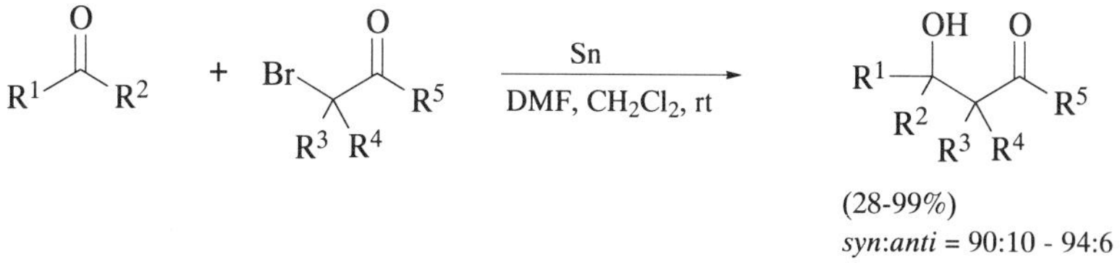

Sn
DMF, $CH_2Cl_2$, rt
(28-99%)
syn:anti = 90:10 - 94:6


(Eq. 19)

In the above described reaction, α-bromoketones are prepared by the bromination of ketones[49–51] and then are reduced by metallic tin to generate tin(II) enolates. Instead, more conveniently and efficiently, tin(II) enolates can be generated by reaction of ketones with tin(II) triflate in the presence of *N*-ethylpiperidine, and they undergo aldol reactions with aldehydes to give the corresponding β-hydroxyketones in good yields under extremely mild conditions with good to excellent *syn* selectivity (Eq. 20).[26]

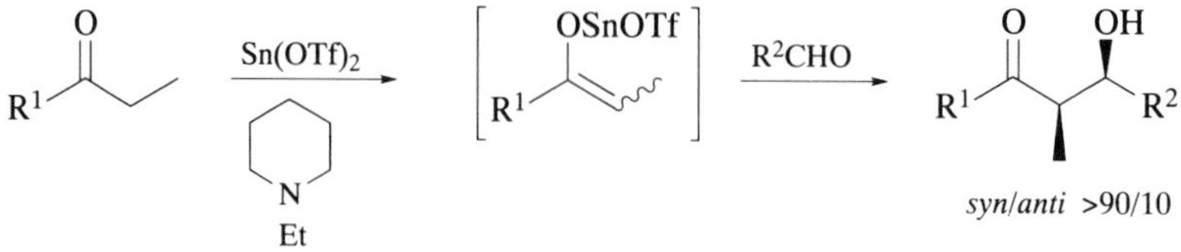

(Eq. 20)

Tin(II) enolates generated by the procedure described above are highly reactive with a ketone acceptor, and ketone–ketone cross-coupling products can be obtained in good yields (Eq. 21).[52] Enhanced *anti*-selectivity is observed when

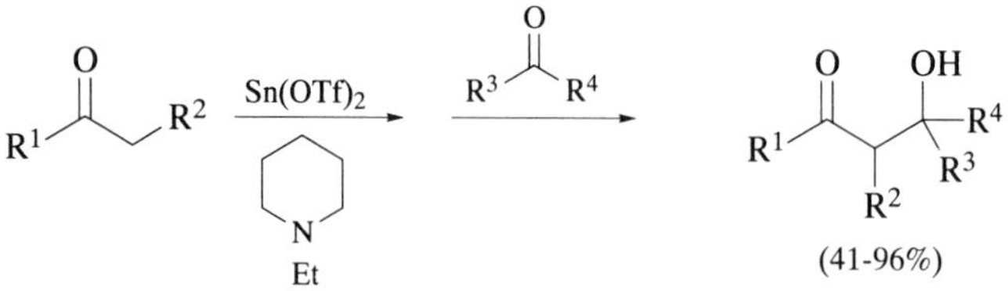

(Eq. 21)

(41-96%)

aromatic ketones are the acceptor carbonyl compounds. This high reactivity towards ketones is characteristic of tin(II) enolates. Versatile and frequently employed boron enolates display poor reactivity toward ketones,[53–56] and the even more nucleophilic lithium enolates react with only less hindered ketones in moderate yields.[57,57a]

A branched-chain sugar, ethyl 2-*C*-methyl-DL-lyxofuranoside, is synthesized stereoselectively by application of this ketone–ketone cross-coupling reaction starting from a 1,3-dihydroxy-2-propanone derivative and methyl pyruvate (Eq. 22).[58]

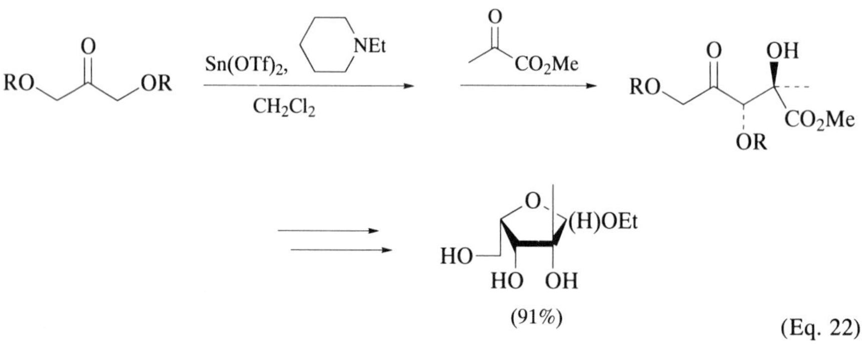

(91%)

(Eq. 22)

A convenient method for the stereoselective synthesis of *cis*-β-substituted-α,β-epoxyketones involves the tin(II) triflate mediated cross aldol reaction of α-bromoketones with aldehydes. The corresponding adducts, *syn*-α-bromo-β-hydroxyketones, are preferentially produced. In turn, these are converted to

$cis$-$\alpha,\beta$-epoxyketones with minimum isomerization to $trans$ isomers via intramolecular $S_N2$-type ring closure to oxiranes by the action of potassium fluoride-dicyclohexyl-18-crown-6 (Eq. 23).[59]

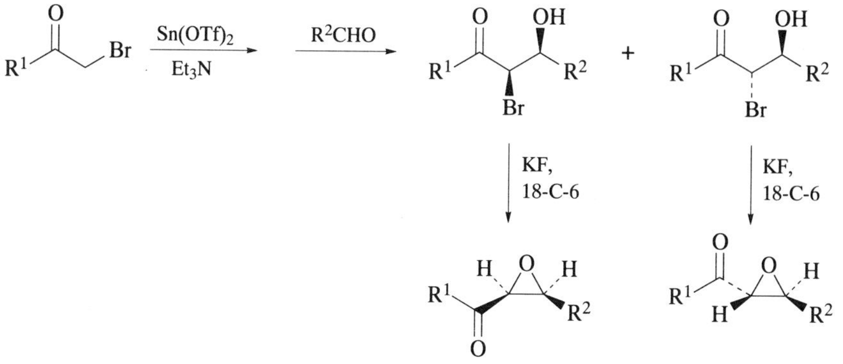

(Eq. 23)

Dialkyl-(2,3-epoxy-4-oxoalkyl)phosphonates are synthesized by this procedure starting from (1-formylalkyl)phosphonates and the tin(II) enolates of $\alpha$-bromoketones (Eq. 24).[60]

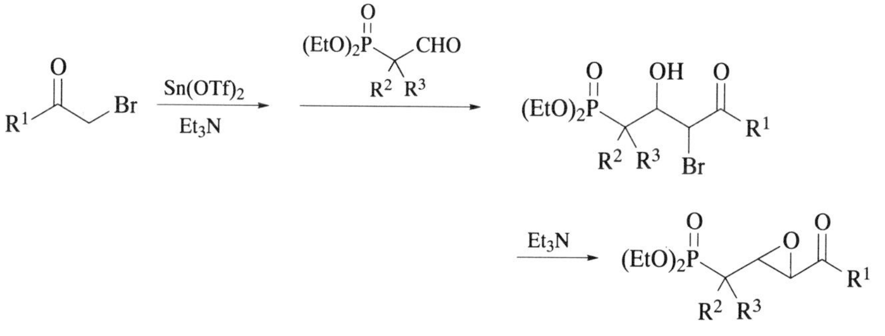

(Eq. 24)

$\alpha,\beta$-Epoxyesters or aldehydes are frequently prepared by the Darzens reaction[61-63] or by epoxidation of the corresponding $\alpha,\beta$-unsaturated ketones or esters.[64] However, the stereoselective synthesis of $cis$-$\beta$-substituted $\alpha,\beta$-epoxyesters or aldehydes by these methods is not easily achieved because of the low stereoselectivity of the Darzens reaction or the difficulty in obtaining the starting Z olefins. In an extension of the tin(II) enolate mediated stereoselective aldol reaction, various $cis$-$\alpha,\beta$-epoxy esters and $cis$-$\alpha,\beta$-epoxy aldehydes are prepared with high stereoselectivities by using an $\alpha$-bromo-$\alpha'$-siloxyketone as the starting carbonyl compound.[65] The cross-aldol product is obtained by treatment of tin(II) triflate with the $\alpha$-bromoketone in the presence of $N$-ethylpiperidine, followed by addition of the aldehyde. On treatment of the crude adduct with sodium carbonate, the *cis* oxirane ring is formed stereospecifically. Oxidative cleavage of the α-hydroxyketone affords an α,β-epoxycarboxylic acid, which is in turn converted to the methyl ester with diazomethane. The *cis*-α,β-epoxyketone is also easily converted to a *cis*-α,β-epoxyaldehyde by reducing the carbonyl group with sodium borohydride prior to oxidative cleavage (Eq. 25).[66]

(Eq. 25)

*cis:trans* 95:5

β-Hydroxy aldehydes or β-hydroxy carboxylic acid derivatives are useful synthetic building blocks. In particular, β-hydroxy aldehydes have been used for the construction of a variety of polyoxygenated natural products.[67,68] Direct generation of tin(II) enolates from esters by using the tin(II) triflate method is not generally successful. Though the reaction with benzaldehyde affords the aldol product in good yield, only products of polymerization of the aldehyde result from aldehydes containing an α hydrogen atom. Self-polymerization of the starting material takes place with aldehyde enolates.

3-Acylthiazolidine-2-thiones, prepared from acyl chlorides and thiazolidine-2-thione or from carboxylic acids and thiazolidine-2-thione with dicyclohexyl-carbodiimide (DCC) or pyridinium salts,[69] are aldehyde or ester equivalents because they can be cleanly converted to the corresponding aldehydes by the reduction with diisobutylaluminum hydride (DIBAL) or transformed into a variety of carboxylic acid derivatives under mild conditions.[70,71] They undergo a similar tin(II) triflate-mediated aldol-type reaction to give β-hydroxy carbonyl compounds in excellent yields with excellent *syn* selectivity (Eq. 26). These cross coupling products are versatile synthetic intermediates and can be transformed into esters, amides, aldehydes, and diols in good yields (Eqs. 27–29).[36,72]

A dramatic reversal is observed in the stereochemical course of the tin(II) triflate mediated aldol-type reaction of 3-(2-benzyloxyacetyl)thiazolidine-2-thione. The *syn* isomers are preferentially obtained by the reaction with aldehydes in the absence of tetramethylethylenediamine (TMEDA), while *anti* isomers are observed in the same reaction carried out in the presence of TMEDA (Eq. 30).[73]

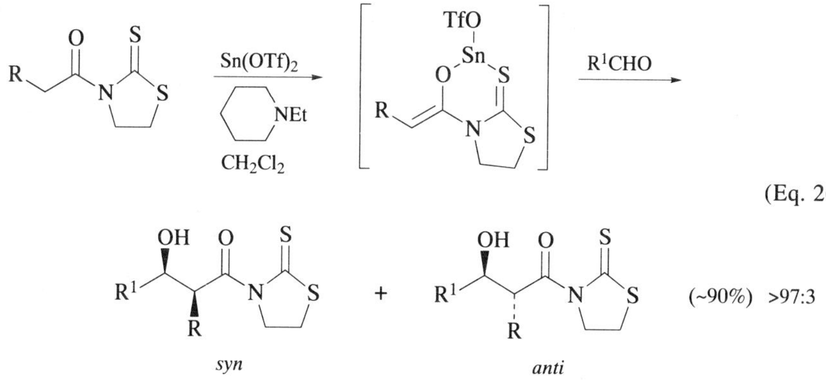
Sn(OTf)2
NEt
CH2Cl2
TfO
Sn
R1CHO
OH O S
syn
anti
(~90%) >97:3

(Eq. 26)

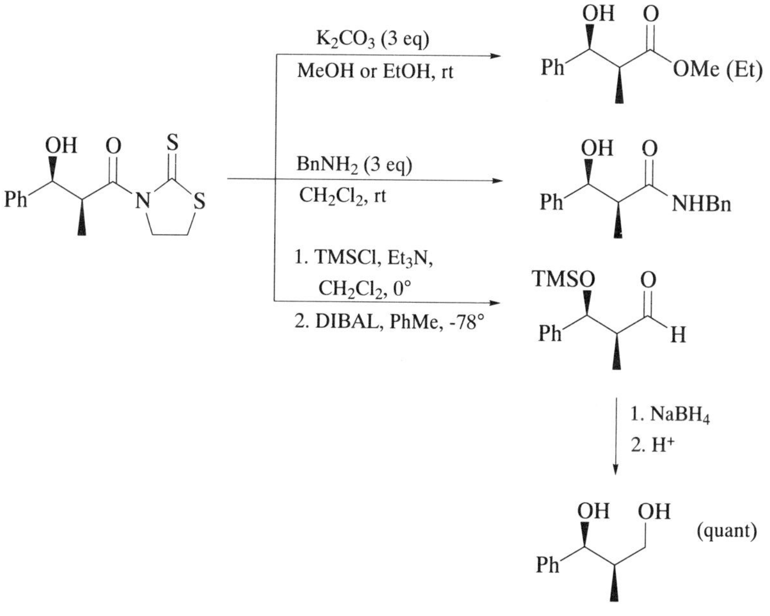
K2CO3 (3 eq)
MeOH or EtOH, rt
OMe (Et)
(~95%)
BnNH2 (3 eq)
CH2Cl2, rt
NHBn
(90%)
1. TMSCl, Et3N,
CH2Cl2, 0°
2. DIBAL, PhMe, -78°
TMSO
(65%)
1. NaBH4
2. H+
(quant)

(Eq. 27)

(Eq. 28)

(Eq. 29)

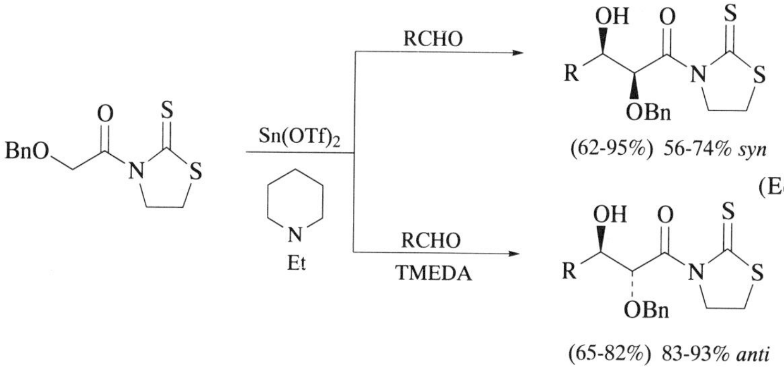
BnO
Sn(OTf)2
Et
RCHO
RCHO
TMEDA
OBn
(62-95%) 56-74% syn
(65-82%) 83-93% anti

(Eq. 30)

Furthermore, enolate formation from 3-acylthiazolidine-2-thiones bearing heteroatoms (X) on the $\beta$-carbon of the acyl group takes place without $\beta$ elimination of the heterosubstituent, and polyfunctionalized aldol adducts are obtained by subsequent treatment with aldehydes (Eq. 31). Adducts where X = nitrogen

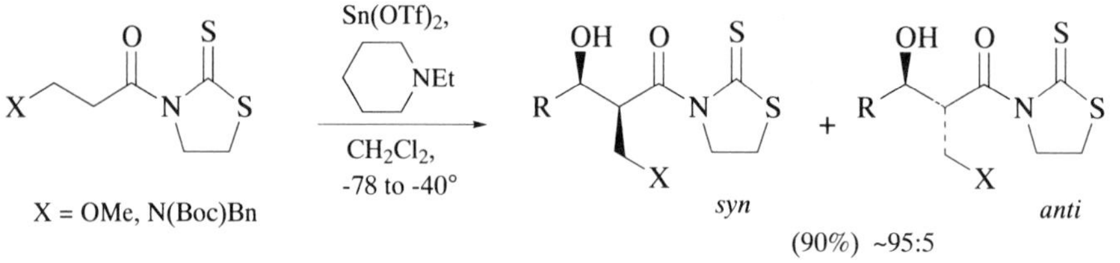

$$(\text{Eq. 31})$$

are cyclized to $\beta$-lactams on treatment of the corresponding acids with 2-chloro-1-methylpyridinium iodide (Eq. 32).[74]

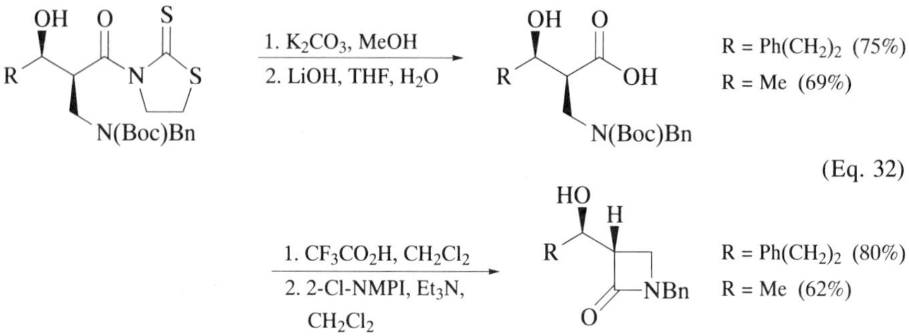

$$(\text{Eq. 32})$$

Tin(II) enolates of thioesters can be generated by the reaction of ketenes with tin(II) thiolates and they react smoothly with aldehydes to give syn-$\beta$-hydroxy thioesters in high yields with high diastereoselectivity (Eq. 33).[43]

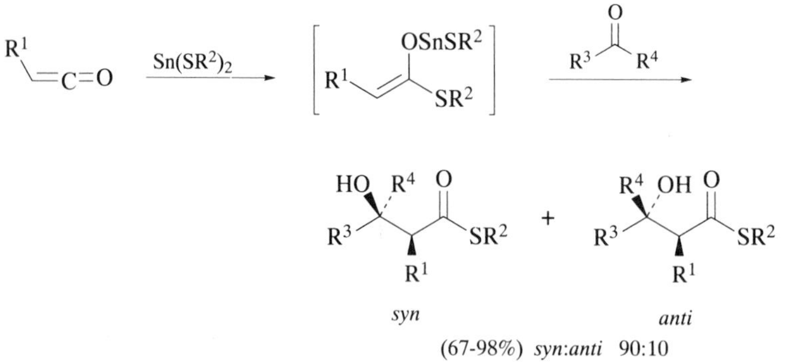

$$(\text{Eq. 33})$$

$\alpha,\beta$-Dihydroxyketones are obtained by the reaction of $\alpha$-dicarbonyl compounds with aldehydes in the presence of activated tin, prepared from tin(II) chloride and potassium. In this reaction, tin enediolates are formed as key intermediates, and improvement of diastereoselectivity is observed by the addition of hexafluorobenzene (Eq. 34).[48]

(68-95%)

THF

THF/$C_6F_6$

syn:anti ≈ 1:1

syn:anti ≈ 3:1

(Eq. 34)

Although methylglyoxal is a synthetically useful compound, it is difficult to handle because of its volatility and ease of polymerization. This unstable compound is successfully converted to the tin(II) enediolate on treatment with activated tin. The tin(II) enediolate reacts with aldehydes to give $\alpha,\beta$-dihydroxyketones in good yields (Eq. 35).[75]

(Eq. 35)

Activated tin is effective for the preparation of aldehyde enolates from $\alpha$-bromoaldehydes. The cross aldol reaction between two aldehydes is generally quite difficult because of competitive side reactions such as preferential self-condensation, or further reaction of the enolate with the initially produced $\beta$-hydroxy aldehydes. Tin(II) enolates prepared from 2-bromo-2-methylpropanal and the activated tin ($SnCl_2$-K) react smoothly with aldehydes to give $\beta$-hydroxy aldehydes, cross aldol adducts between two different aldehydes, in fairly good yields (Eq. 36).[76]

(64-74%)

(Eq. 36)

In most aldol reactions based on intermediate metal enolates, the starting carbonyl compound is quantitatively converted to the metal enolate prior to reaction with the second carbonyl compound. Therefore a stoichiometric amount of metal reagent is generally required. By use of only a catalytic amount of the tin(II) thiolate, the aldol reaction of tin(II) enolates with aldehydes takes place successfully to afford the cross aldol products in good yields with high stereoselectivity (Eq. 37).[77,78]

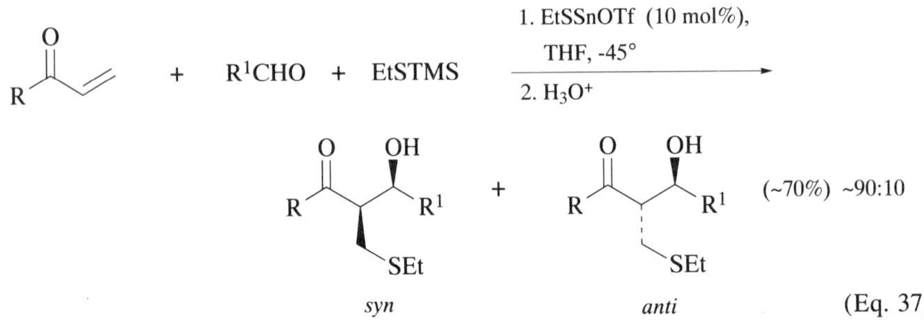

(Eq. 37)

Regeneration of the tin(II) enolate in the catalytic cycle is assumed as shown in Eq. 38. In the first step, conjugate addition of the tin(II) species **1** to vinyl

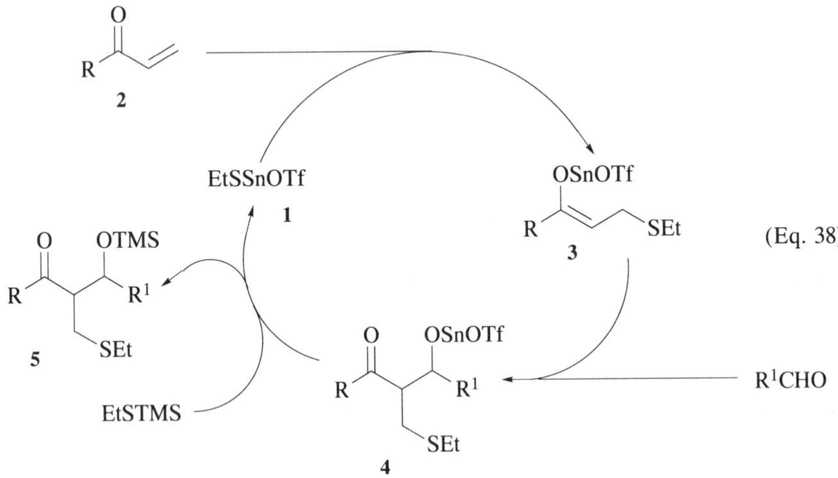

(Eq. 38)

ketone **2** produces tin(II) enolate **3**, which in turn reacts with an aldehyde to give the aldol product **4** as its tin(II) alkoxide. In the next step, **4** reacts with the alkylthiotrimethylsilane to regenerate **1** along with the aldol product in the form of its trimethylsilyl ether **5**. The driving force for this reaction may be attributed to the difference in the relative bond strengths of tin(II)–sulfur and silicon–sulfur bonds.

*Imines (Schiff bases) as Acceptors.* Imines (Schiff bases) are one of the most promising classes of compounds for the synthesis of nitrogen-containing molecules such as amino sugars, amino acids, and $\beta$-lactams.[79–82] In spite of efforts to employ imines for the preparation of these valuable natural products,[83–85] the low reactivity of imines compared to the corresponding carbonyl compounds and side reactions,[86] such as oligomerization or abstraction of the $\alpha$-proton, are still problems in attempts to use imines as electrophiles for carbon–carbon bond-forming reactions.

Tin(II) enolates of thioesters, formed in situ from tin(II) 2-methyl-2-propanethiolate and ketenes, react with imines in the presence of tin(II) triflate to give $\beta$-amino thioesters with high *anti* selectivity (Eq. 39). In the absence of

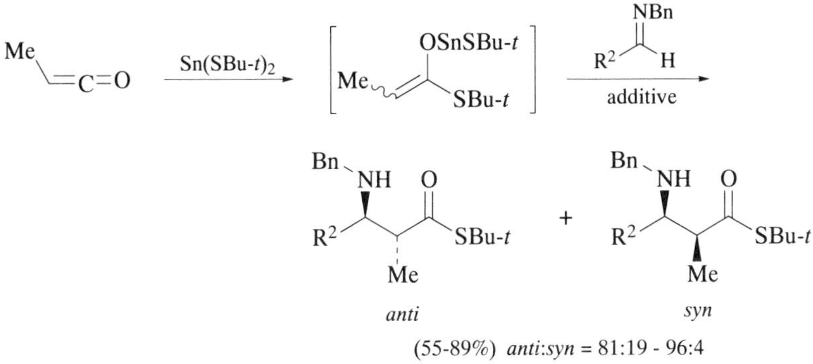

(Eq. 39)

(55-89%) *anti:syn* = 81:19 - 96:4

tin(II) triflate, this reaction is usually sluggish. Carbapenem antibiotic PS-5 is formally synthesized by this reaction (Eq. 40).[87]

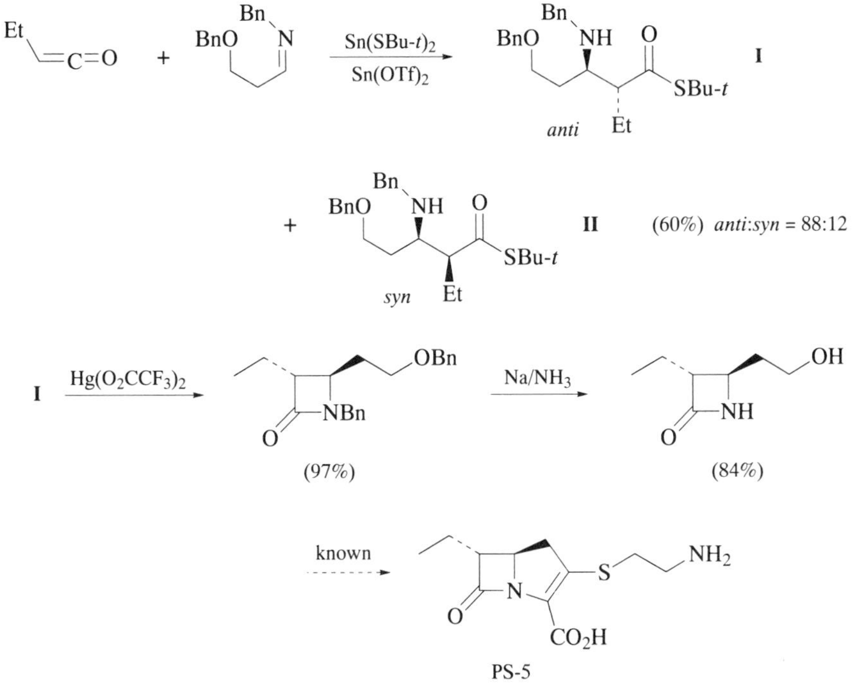

(Eq. 40)

Tin(II) enolates of thioesters, generated from lithium enolates and tin(II) chloride, react smoothly with $\alpha$-iminoesters to afford the corresponding $\beta$-amino thioesters in good yields with high diastereoselectivity. However, these enolates are less reactive than those prepared from lithium enolates and tin(II) triflate.[44]

The type of metal enolate has a dramatic effect on the addition to $\alpha$-iminoesters. No adduct is obtained with lithium and magnesium enolates, probably because of decomposition of the $\alpha$-iminoester under the strong basic conditions. Titanium and aluminum enolates react with $\alpha$-iminoesters to give adducts in rather low yields. Tin(II) enolates afford *syn* isomers of $\beta$-amino thioesters in good yields with high stereoselectivity, while preferential formation of *anti* isomers is observed with titanium enolates. The best yield and very high *syn* stereoselectivity are achieved with the tin(II) enolate prepared from a lithium enolate and tin(II) chloride (Eq. 41).

| $MX_n$ | Yield (%) | syn:anti |
|---|---|---|
| Li | (0) | — |
| $MgCl_2$ | (0) | — |
| $TiCl_2(OPr\text{-}i)_2$ | (30) | 29:71 |
| $Et_2AlCl$ | (25) | 50:50 |
| $SnCl_2$ | (72) | 95:5 |
| $Sn(OTf)_2$ | (60) | 80:20 |

*Other Electrophiles as Acceptors.* Tin(II) enolates formed from ketones and tin(II) triflate in the presence of N-ethylpiperidine react with dithioacetals in the presence of triphenylmethyl tetrafluoroborate (trityl tetrafluoroborate, $TrBF_4$) to afford $\gamma$-ketosulfides in good yields (Eq. 42).[88]

(Eq. 42)

The addition of organometallic reagents to a quinone often affords complicated products because of undesirable side reactions.[89-91] For example, when Grignard reagents are employed, electron transfer from the anionic species to the quinone often predominates and the self-coupling product of the nucleophile is the major product. Also, it is often difficult to achieve selective 1,2 or 1,4 additions and mono- or diadditions of the nucleophile, though several successful results have recently been reported by using alkyllithium reagents,[92-96] trialkylboranes,[97-102] π-allyl Ni complexes,[103,104] or allylstannanes,[105-109] which give either selective 1,2- or 1,4-addition products. However, addition of metal enolates derived from ketones or carboxylic acid derivatives has rarely been reported, probably because of competing side reactions.

Tin(II) enolates smoothly react with p-benzoquinone or its mono-N-tosylimino derivative to give 1,2-addition products in good yields. These can be reduced in situ to α-(p-hydroxy- or p-aminophenyl)carbonyl derivatives by addition of dichloromethylsilane and dimethylaminopyridine (Eq. 43).[110,111] This is a novel addition–reduction reaction providing a convenient route for the synthesis of α-arylcarbonyl compounds.

(Eq. 43)

**Michael Reactions.** The Michael reaction is one of the most fundamental carbon–carbon bond forming reactions and it is widely employed in organic synthesis.[112-114] The reaction in which metal enolates behave as nucleophiles is of particular interest with great potential as a versatile method for the stereoselective synthesis of both acyclic and cyclic systems. Fruitful results have been reported in the Michael reaction of silyl enol ethers with α,β-unsaturated carbonyl compounds by use of titanium tetrachloride,[115-119] trityl salts,[120-124] and other additives.[125-131] In the reaction between other metal enolates and α,β-unsaturated carbonyl compounds, however, the problems of competitive 1,2 addition along with some polymerization of enones limit the application of metal enolate nucleophiles in the Michael reaction.[132-152]

Though tin(II) enolates do not react with α,β-unsaturated ketones in the absence of an additive at −78°, the corresponding Michael adducts are obtained in good yields by using an equimolar amount of chlorotrimethylsilane as an activator (Eq. 44).[111,153] Other silicon compounds, such as chlorodimethylsilane, dichlorodimethylsilane, and trimethylsilyl triflate, are also effective.

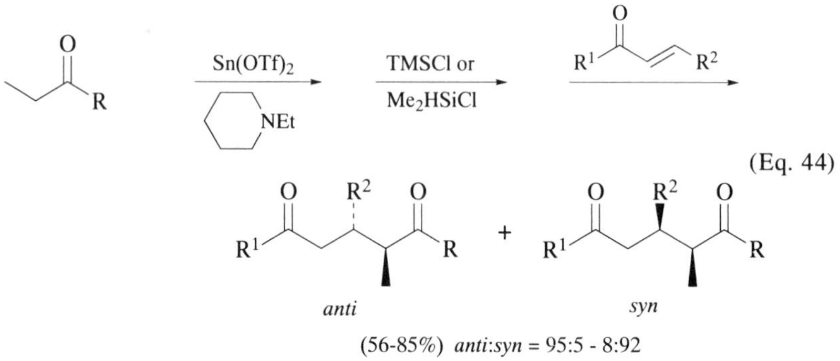

$$(Eq. 44)$$

$$(56-85\%) \; anti:syn = 95:5 - 8:92$$

The tin(II) enolates of cyclic ketones react with $\beta$-nitrostyrene to afford 4-nitroketones with unprecedented *anti* selectivity (Eq. 45).[154]

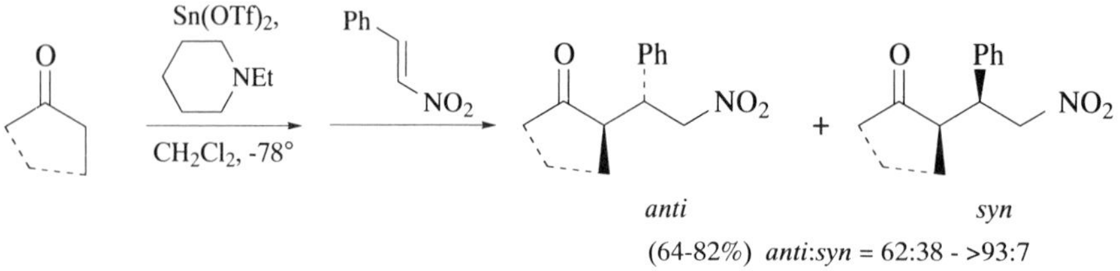

$$(64-82\%) \; anti:syn = 62:38 - >93:7$$

$$(Eq. 45)$$

Tin(II) dienolates react exclusively at the $\gamma$ carbon of acyclic $\alpha,\beta$-unsaturated ketones in the Michael sense (Eq. 46).[41] It has been well established that enolate

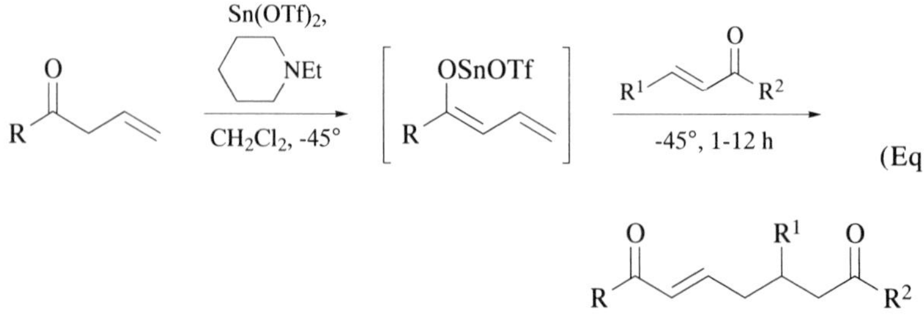

$$(Eq. 46)$$

anions, notably lithium enolates of $\alpha,\beta$-unsaturated carbonyl compounds, react predominantly at the $\alpha$ carbon. A series of silicon-directed $\gamma$-substitution reactions with a variety of electrophiles in the presence of a Lewis acid catalyst has been reported.[155-162]

In such reactions of dienolates, either acyclic 1,7-diketones or 2-cyclohexenol derivatives can be obtained in moderate to excellent yields by appropriate choice of reaction parameters. In the latter case, only a single cyclohexanol adduct is formed (Eq. 47).[41]

(Eq. 47)

### Asymmetric Reactions

**Enantioselective Aldol Reactions.** Divalent tin has vacant $d$ orbitals to which amines, especially diamines, can easily coordinate.[163] Using this property of tin(II), a highly enantioselective cross-aldol reaction between ketones and aldehydes is accomplished by using a chiral diamine derived from ($S$)-proline as a ligand to a tin(II) enolate (Eq. 48).[164] Thus the tin(II) enolate formed from propiophenone and tin(II) triflate in the presence of $N$-ethylpiperidine is treated with diamine **6** and then with benzaldehyde at $-78°$ to give the cross-aldol product in 65% yield with an optical purity of 60%. The optical purity can be improved to 65% by conducting the reaction at $-95°$. Screening of reaction conditions reveals that the chiral diamine forms a 1:1 complex with the tin(II) enolate. The structure of the chiral diamine strongly influences the enantioselectivity, and the aldol product is obtained in up to 80% ee when chiral diamine **7** is employed as a ligand.

In the cross-aldol reaction of aromatic ketones with aromatic aldehydes, good to high ee is achieved by employing diamine **7** as a chiral ligand. With aliphatic aldehydes, proper choice of the chiral ligand affords the cross-aldol products in high optical purity.[36]

The enantioselectivity achievable in the cross-aldol reaction of aliphatic ketones with aldehydes is rather low (up to 50% ee) when chiral diamine **6** or **10** is employed. However, the tridentate ligand **11** is most effective for this reaction, and 70–80% ee is realized in the reaction of aldehydes with the tin(II) enolate of *tert*-butyl ethyl ketone.[36]

The asymmetric aldol reaction of 3-acetylthiazolidine-2-thione with achiral aldehydes is also successfully carried out via tin(II) enolates by using chiral diamine **7** as a ligand,[165] and the aldol-type adducts are obtained in high yields (Eq. 49). The enantioselectivities obtained in this reaction are high because the tin(II) enolate of 3-acetylthiazolidine-2-thione is almost completely fixed by coordination of the thiazolidine–thione part to tin(II). As described in the previous

Chiral diamines

|                 |       |                  |                             |          |         |              |
|-----------------|-------|------------------|-----------------------------|----------|---------|--------------|
|       6         |       7          |       8          |       9          |

$$R^1\text{COCH}_2\text{CH}_3 \xrightarrow[\text{CH}_2\text{Cl}_2,\ -78°]{\text{Sn(OTf)}_2,\ \text{(piperidine)NEt}} \xrightarrow{\text{chiral diamine}} \xrightarrow{R^2\text{CHO}} R^1\text{CO}\ \overset{|}{\text{CH}}\text{CH}(\text{OH})R^2$$

| $R^1$ | $R^2$ | Chiral diamine | $R^2\text{CHO}$ added at | Yield (%) | syn:anti | ee (%) of syn |
|-------|-------|----------------|--------------------------|-----------|----------|---------------|
| Ph    | Ph    | 6              | -78°                     | (66)      | 6:1      | 60            |
| Ph    | Ph    | 6 (0.5 eq)     | -78°                     | (74)      | 6:1      | 30            |
| Ph    | Ph    | 6 (2.0 eq)     | -78°                     | (61)      | 6:1      | 60            |
| Ph    | Ph    | 6              | -95°                     | (66)      | 6:1      | 65            |
| Ph    | Ph    | 6              | -95° to 25°              | (75)      | 1:2      | 0             |
| Ph    | Ph    | 6              | -95°                     | (66)      | 6:1      | 65            |
| Ph    | Ph    | 7              | -95°                     | (74)      | 6:1      | 80            |
| Ph    | Ph    | 8              | -95°                     | (56)      | 6:1      | 50            |
| Ph    | Ph    | 9              | -95°                     | (72)      | 6:1      | 75            |
| Ph    | Ph    | 10             | -95°                     | (66)      | 20:1     | 20            |
| Ph    | Ph    | 7              | -78°                     | (66)      | 6:1      | 60            |
| Ph    | Ph    | 7 (0.5 eq)     | -78°                     | (74)      | 6:1      | 30            |
| Ph    | Ph    | 7 (2.0 eq)     | -78°                     | (61)      | 6:1      | 60            |
| Ph    | Ph    | 7              | -95°                     | (66)      | 6:1      | 65            |
| Ph    | Ph    | 7              | -95° to rt               | (75)      | 1:2      | 0             |
| Ph    | i-Pr  | 10             | -95°                     | (69)      | >20:1    | 75            |
| Ph    | t-Bu  | 7              | -95°                     | (57)      | 100:0    | 90            |
| t-Bu  | R     | 11             | -78°                     | (—)       | —        | 70-80         |

(Eq. 48)

$$\text{CH}_3\text{CO-thiazolidinethione} \xrightarrow{\text{Sn(OTf)}_2} \xrightarrow[\text{Me}\quad 7]{} \xrightarrow{\text{RCHO}} R\overset{\text{OH}\ \text{H}}{\text{CH}}\text{CH}_2\text{CO-thiazolidinethione}$$

~90% ee

(Eq. 49)

section, the adduct is easily converted to a $\beta$-hydroxy aldehyde or $\beta$-hydroxy carboxylic acid derivative, and thus this method is useful for the preparation of a variety of optically active compounds.

A highly enantioselective synthesis of 2-substituted malates is achieved by application of this reaction. The tin(II) enolate of 3-acetylthiazolidine-2-thione reacts with $\alpha$-ketoesters to afford the aldol-type products generally in greater than 95% ee (Eq. 50).[166]

(Eq. 50)

>95% ee

In the reaction of 3-(2-benzyloxyacetyl)thiazolidine-2-thione with aldehydes, up to 94% ee is observed in the *anti* adduct by employing chiral diamine **7** (Eq. 51).[73]

(Eq. 51)

87-94% ee

**Enantioselective Michael Reactions.** The asymmetric Michael reaction of tin(II) enolates is achieved in moderate to high enantioselectivities by using the coordination of chiral diamine ligands to the intermediate tin(II) enolate. Trimethylsilyl triflate (TMSOTf) is used as an activator of $\alpha,\beta$-unsaturated ketones. The reactions do not proceed in the absence of TMSOTf. The tin(II) enolate of 3-propanoyl-1,3-oxazolidin-2-one or methyl dithioacetate reacts with benzalacetone in the presence of a chiral diamine and TMSOTf to give the Michael adduct in good yield with good enantioselectivity (Eq. 52).[78,167]

The catalytic asymmetric Michael reaction of tin(II) enethiolates forms 5-oxodithioesters in high yields with moderate to good enantioselectivities by employing catalytic amounts of tin(II) triflate and chiral diamine **12** (Eq. 53).[78,168] The catalytic cycle of Eq. 54 is postulated.

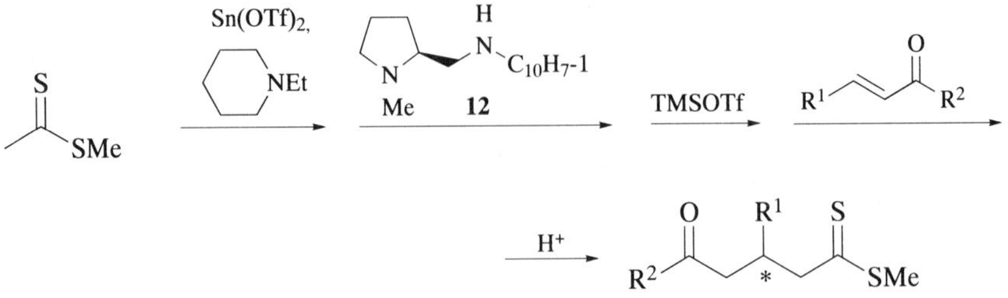

(44-82%)  15-70% ee

(Eq. 52)

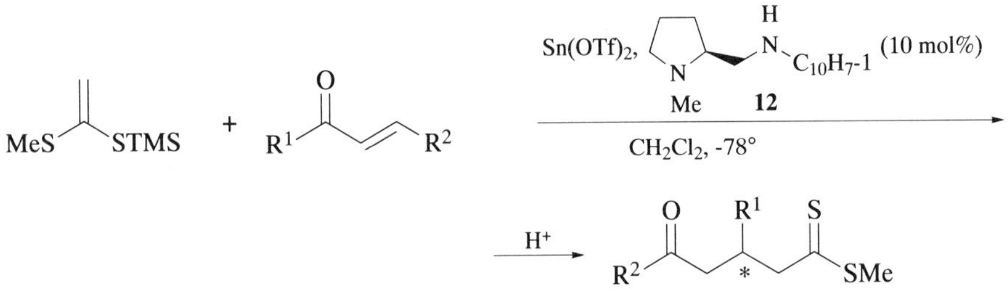

| R[1] | R[2] | Yield (%) | %ee | %ee (1 eq of **12**) |
|------|------|-----------|-----|----------------------|
| Me | Ph | (80) | 70 | 70 |
| Me | 1-furyl | (82) | 60 | 60 |
| Ph | Ph | (79) | 40 | 40 |

(Eq. 53)

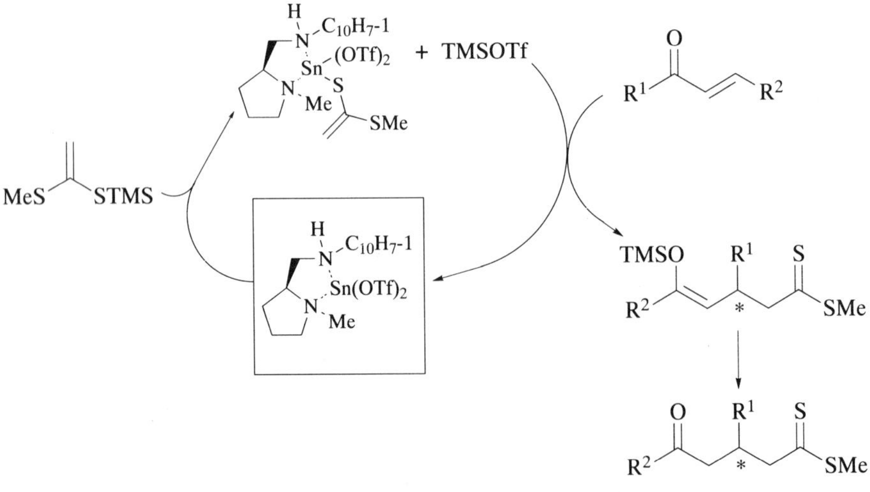

(Eq. 54)

**Asymmetric Sulfenylation.**   In the presence of a chiral diamine, the reaction of tin(II) enolates of ketones or 3-acyl-2-oxazolidones with thiosulfonates proceeds smoothly to give $\beta$-keto sulfides in high enantioselectivities. These products can be easily converted to optically active epoxides or allylic alcohols (Eq. 55).[169]

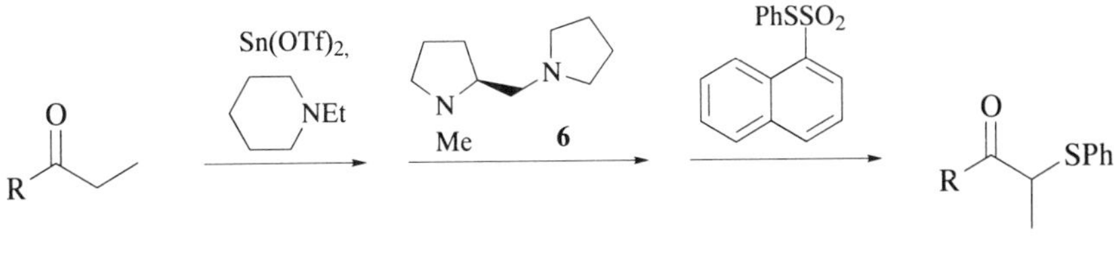

(Eq. 55)

### Diastereoselective Reactions

*Aldol Reaction of Chiral Aldehydes or Imines.*   There are several examples of the reaction of tin(II) enolates with 2,3-$O$-isopropylidene-$D$-glyceraldehyde, 4-$O$-benzyl-2,3-$O$-isopropylidene-$L$-threose, their derivatives. In most cases, selectivities are good to high, and they are clearly explained by the Felkin–Anh model.[170-172]   For example, 4-$O$-benzyl-2,3-$O$-isopropylidene-$L$-threose reacts with the tin(II) enolate prepared from ethyl bromoacetate and metallic tin to give predominantly the *anti* adduct (Eq. 56). This selectivity is reasonably explained

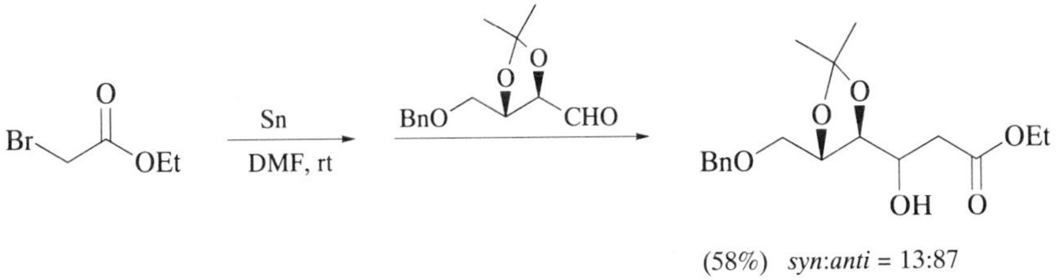

(58%)  *syn:anti* = 13:87

(Eq. 56)

by the model shown in Fig. 1; the $-CH_2OBn$ group, being away from the reaction site, has little effect on the reaction path.[173]

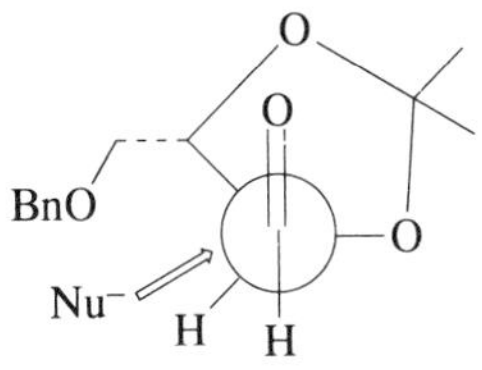

**FIG. 1**

This reaction provides a convenient method for the preparation of monosac-charides. Some biologically important sugars, such as methyl D-glucosaminate[75] and 2-amino-2-deoxy-D-arabinitol[64] are synthesized by this reaction (Eqs. 57 and 58).

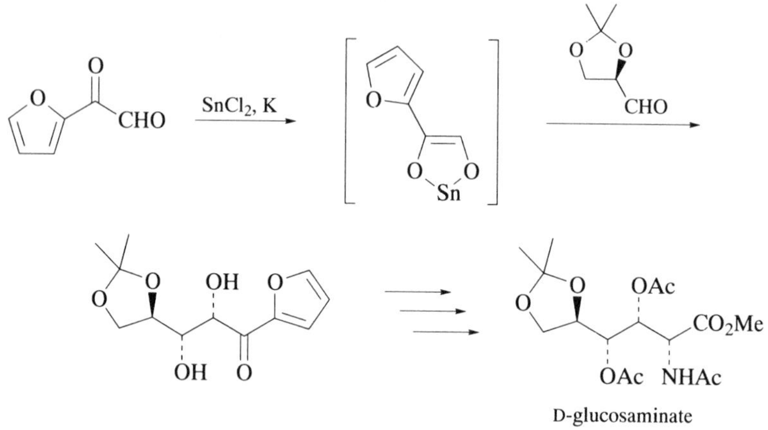

D-glucosaminate

(Eq. 57)

2-amino-2-deoxy-D-arabinitol pentaacetate

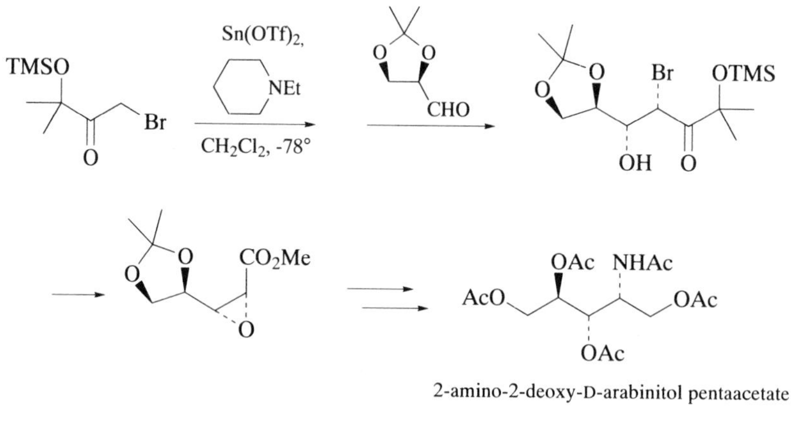

(Eq. 58)

The asymmetric addition of tin(II) enolates derived from thioesters to α-iminoesters having a chiral auxiliary on the nitrogen atom proceeds smoothly to afford *syn-β*-amino acid derivatives, which are in turn converted to optically active *cis*-substituted β-lactams (Eq. 59).[174]

*Aldol Reactions of Chiral Tin(II) Enolates.*    Highly efficient internal asymmetric induction is achieved in the aldol reaction of tin(II) enolates generated from 3-amino-substituted pentanoylthiazolidine-2-thiones. This reaction is successfully applied to the formal total synthesis of thienamycin (Eq. 60).[175]

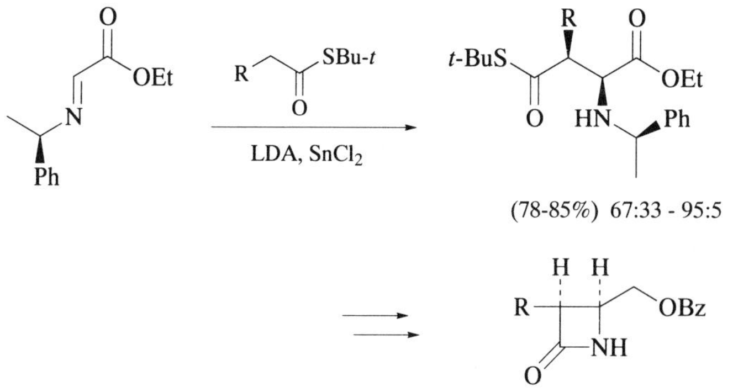

(Eq. 59)

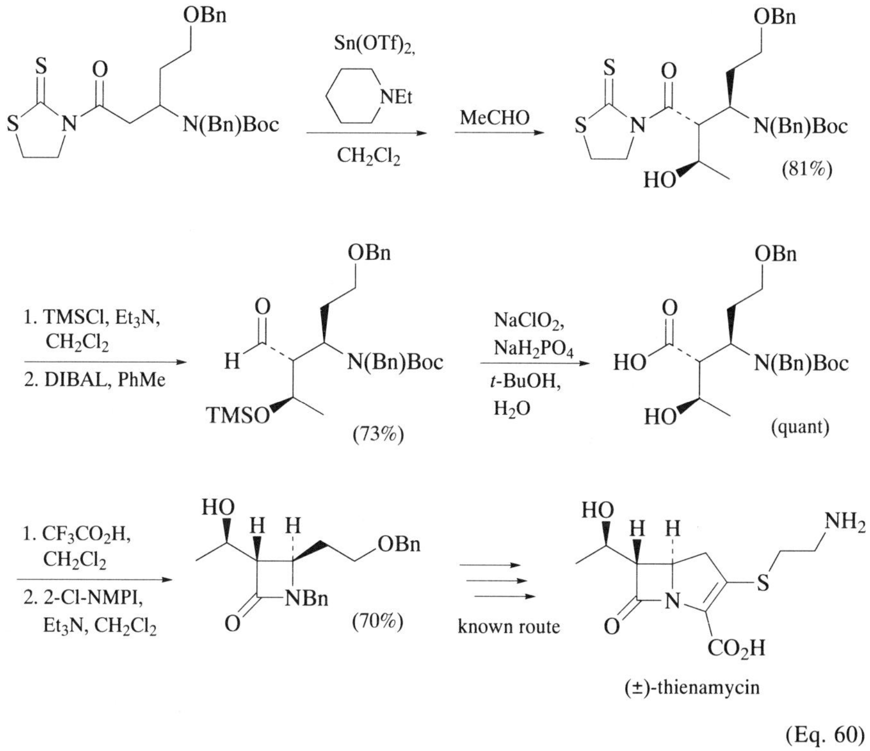

(78-85%)  67:33 - 95:5

(81%)

(73%)

(quant)

(70%)        known route

(±)-thienamycin

(Eq. 60)

Several examples of the reaction of tin(II) enolates of chiral amides or imides with aldehydes are reported. Though these reactions require a tedious procedure for the attachment and removal of chiral auxiliaries, they are synthetically useful because diastereoselectivities are generally high and chiral sources can be recovered.

Tin(II) enolates of chiral 3-acetyloxazolidine-2-thiones react stereoselectively with aldehydes to give the corresponding adducts in good yields (Eq. 61). High diastereoselectivities are explained by a cyclic transition state which involves

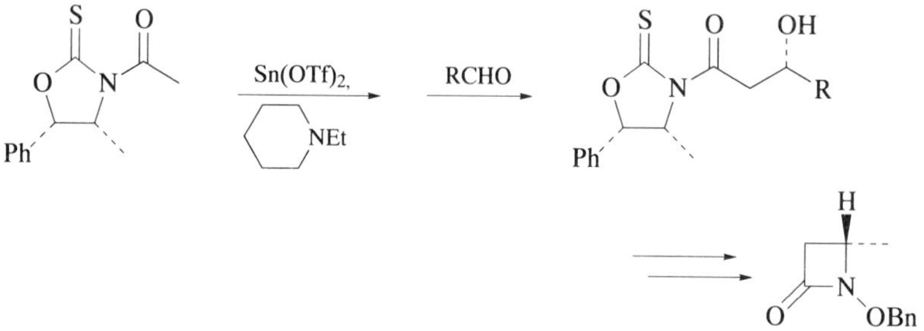

(Eq. 61)

chelation of the thione portion of the chiral auxiliary as shown in Fig. 2.[38] This transition state contrasts remarkably with that of the aldol reaction of boron enolates of chiral 3-acyloxazolidine-2-ones with aldehydes[176] (see the next section). This reaction is applied to the synthesis of a chiral azetidinone (Eq. 61).[38]

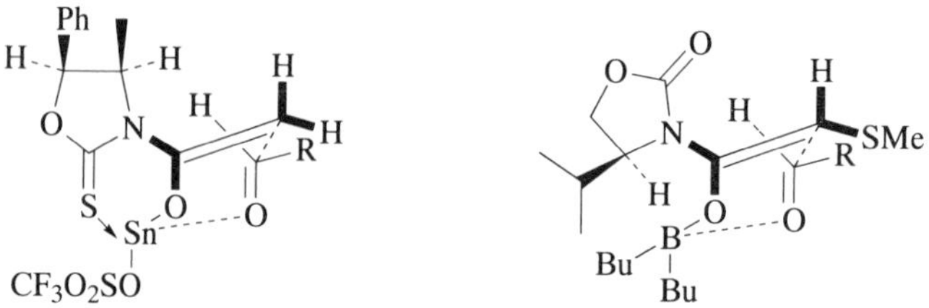

**FIG. 2**

When 3-acylthiazolidine-2-thiones are employed as enolate components instead of 3-acyloxazolidine-2-thiones, superior diastereoselectivities are obtained. Diastereocontrolled aldol reactions between tin(II) enolates, prepared from 3-acetylthiazolidine-2-thiones, tin(II) triflate, and $N$-ethylpiperidine, and $\alpha,\beta$-unsaturated aldehydes are successfully carried out to give the corresponding adducts in good yields (Eq. 62).[55] This reaction is applied to the synthesis of the chiral Geissman–Waiss lactones **13** (Eq. 63).[177]

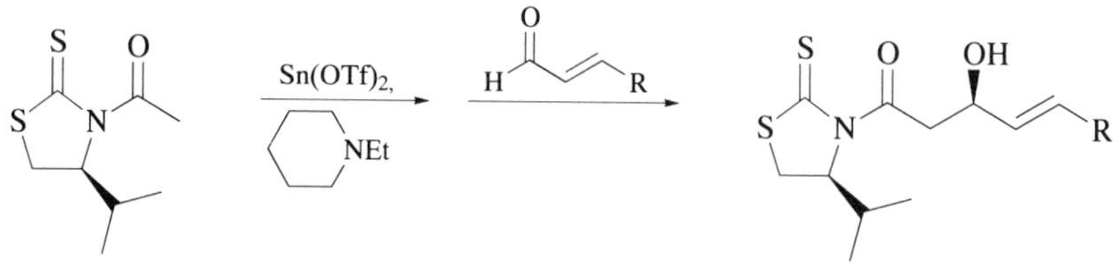

(Eq. 62)

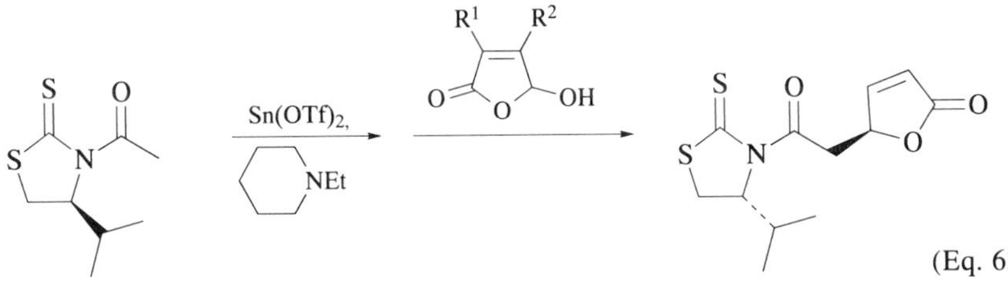

(Eq. 63)

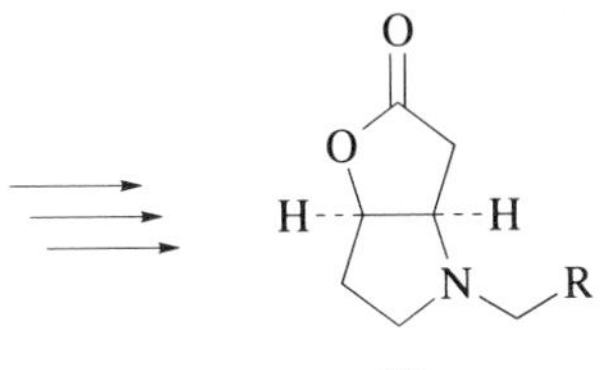

**13**

The aldol reaction of tin(II) enolates of chiral 3-acyloxazolidine-1-ones with aldehydes gives lower diastereoselectivities than those obtained in the reactions discussed above. Benzyl $cis$-$\alpha,\beta$-epoxycarboxylates are prepared by a modified Darzens procedure by using the aldol reaction of chiral $\alpha$-haloimidates with aldehydes (Eq. 64).[39] The selectivities of this reaction are improved when the tin(II)

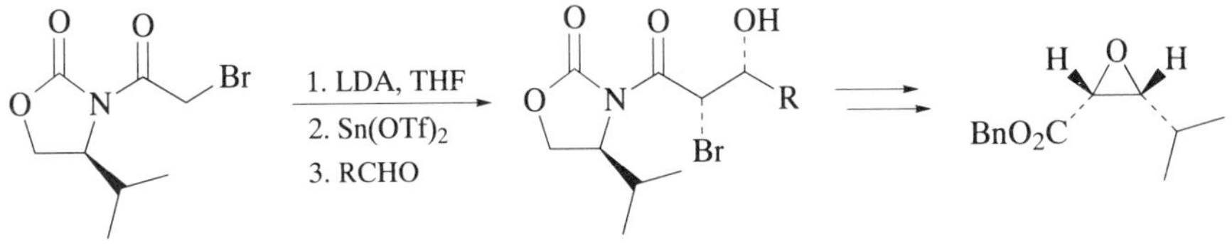

(Eq. 64)

enolates are prepared by the metal exchange reaction of lithium enolates with tin(II) triflate.

Cysteine-derived thiazolidinethiones serve as efficient chiral auxiliaries in tin(II)-mediated aldol reactions (Eq. 65). These chiral auxiliaries are easily re-

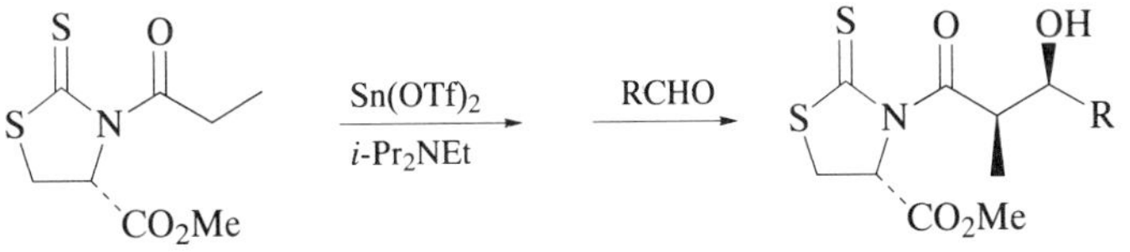

(Eq. 65)

moved and recovered by methanolysis or hydroxaminolysis.[178] It is remarkable that the diastereofacial preference of the chiral 3-acylthiazolidine-2-thione is opposite to that of 3-acyloxazolidine-2-thiones or 3-acylthiazolidine-2-thiones.

The sense of the selectivity in this reaction is the same as that obtained in the reaction of boron enolates, and higher selectivity is observed when using the boron enolate.

A chiral glycine synthon, as its derived tin(II) enolate, undergoes a highly *syn* diastereoselective aldol reaction with aldehydes to give the corresponding adducts in high yields. The utility of these intermediates is demonstrated by their subsequent transformation to enantiomerically pure *N*-methyl-β-hydroxy amino acids (Eq. 66).[179]

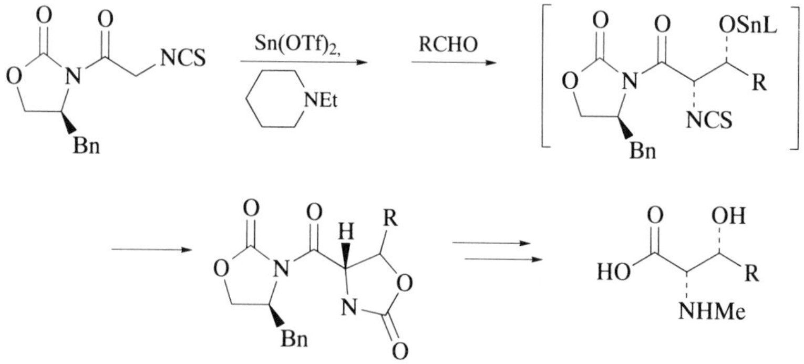

(Eq. 66)

The tin(II) enolate generated by the action of $SnCl_2$–$LiAlH_4$ on an (α-bromoisobutanoyl)oxazolidinone reacts with benzaldehyde or a conjugated (Z)-enal to yield oxazinedione derivatives with excellent diastereoselectivities (Eq. 67).[180] Neooxazolomycin, a novel oxazole polyene lactam–lactone antitumor antibiotic, is synthesized by use of this reaction as one of key steps (Eq. 68).[181]

Tin(II) azaenolates are prepared from chiral 1,3-oxazolidines and react with aldehydes to afford aldol-type adducts in high yields with high diastereoselectivities (Eq. 69).[46,182]

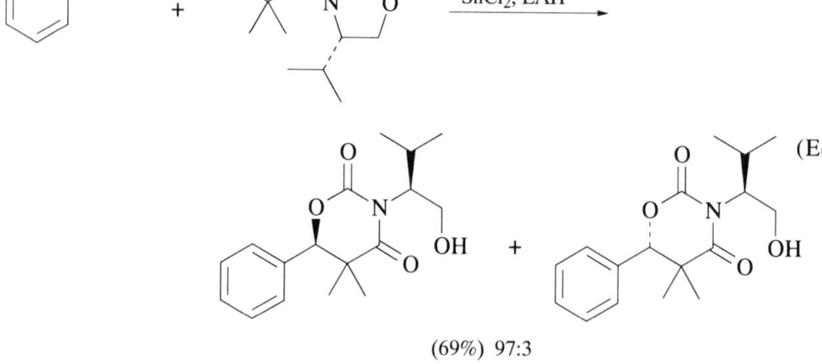

(Eq. 67)

(69%) 97:3

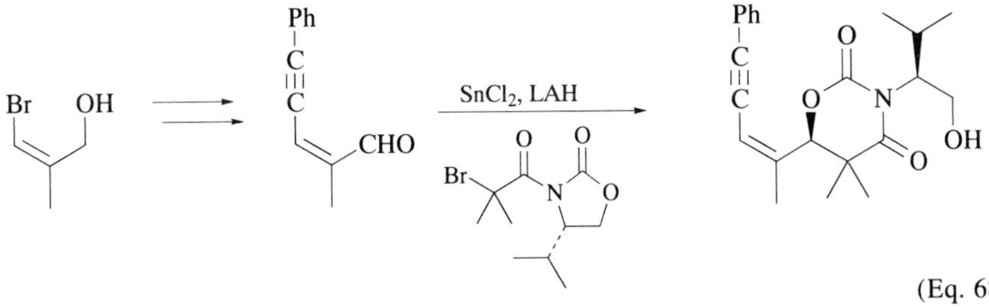

(Eq. 68)

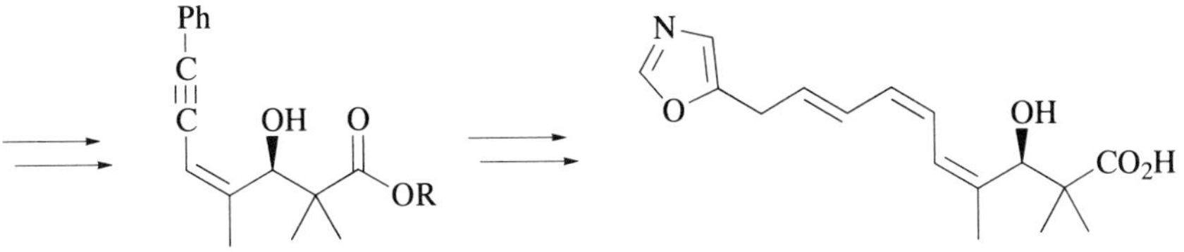

Neooxazolomycin

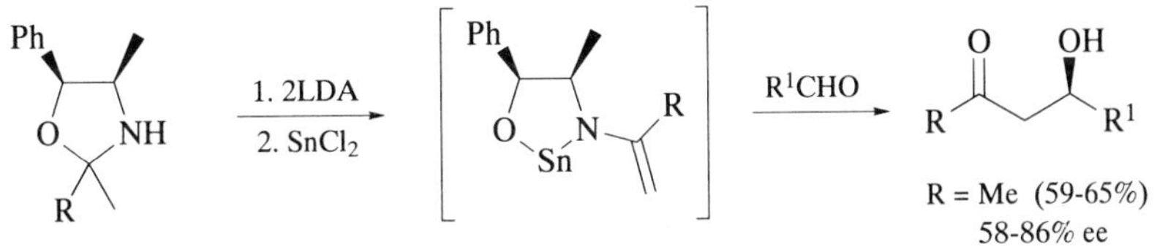

R = Me (59-65%)
58-86% ee

(Eq. 69)

Highly diastereoselective aldol-type alkylation is carried out by employing tin(II) enolates derived from chiral 3-acylthiazolidine-2-thiones and 4-acetoxy-2-azetidinones. These tin(II) enolates also react with optically active 2-azetidi-nones in a highly stereoselective manner to give the corresponding adducts in high yield (Eq. 70).[183]

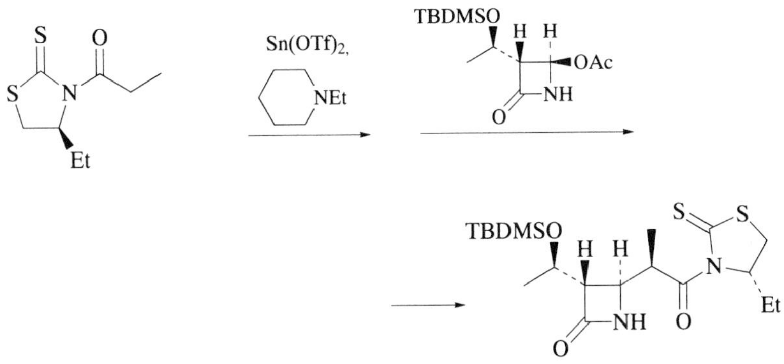

(Eq. 70)

The same level of diastereoselectivity is attained by using the tin(II) enolate of 3-acyloxazolidine-2-one in the presence of zinc bromide as an activator of the azetidinone (Eq. 71).[184] The corresponding boron enolate gives better selectivity in this reaction.

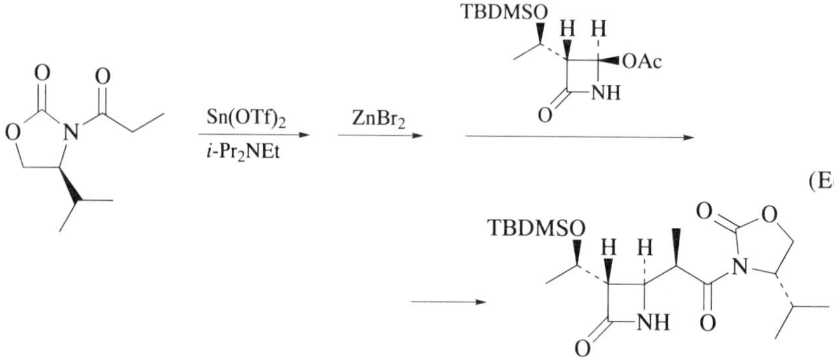

(Eq. 71)

These reactions provide a new approach to the construction of chiral carbapenem precursors.[185–187]

A general method for the enantioselective synthesis of bicyclic alkaloids with a nitrogen atom ring juncture involves the aldol-type alkylation of tin(II) enolates with cyclic acyl imines followed by reductive annulation of the resultant cyclic imines (Eq. 72). (+)-Epilupinine (**14**) is synthesized by this methodology.[188,189]

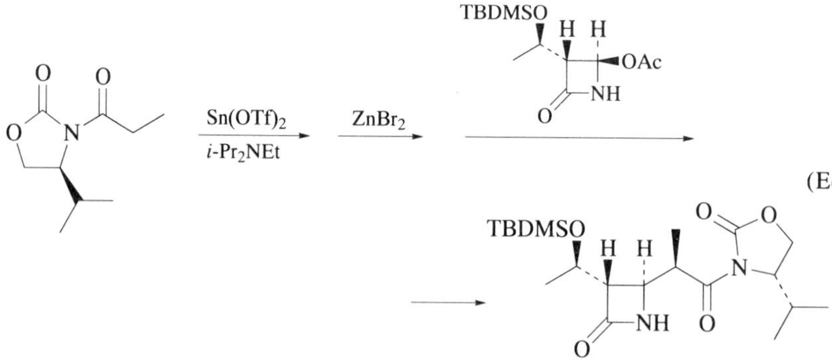

(Eq. 72)

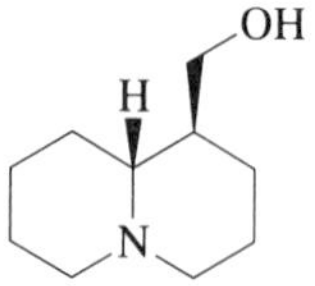

**14**

The $\beta$-methylcarbapenem key intermediate **15** is prepared by this aldol-type alkylation of the tin(II) enolate generated in situ from the bromoketone and metallic tin with the azetidinone (Eq. 73).[190]

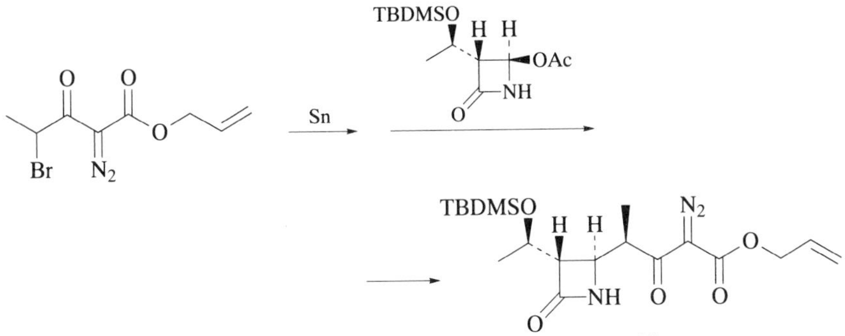

(Eq. 73)

## COMPARISON WITH OTHER METAL ENOLATES

Tin(II) enolates are easily prepared under extremely mild conditions compared to other metal enolates. Although lithium enolates are commonly employed, their use is restricted because of their strong basicity. For example, the formation of lithium enolates of $\beta$-alkoxy carbonyl compounds fails because $\beta$ eliminations rapidly follow deprotonations of the $\alpha$ protons under the strongly basic conditions. Enolate formation using the tin(II) triflate method is quite effective in these cases. The tin(II) enolates are smoothly generated from 3-acetylthiazolidine-2-thione with $\beta$-alkoxy functionality by treatment with tin(II) triflate in the presence of N-ethylpiperidine. These enolates react with aldehydes to afford aldol-type products in high yields (Eq. 74).[74]

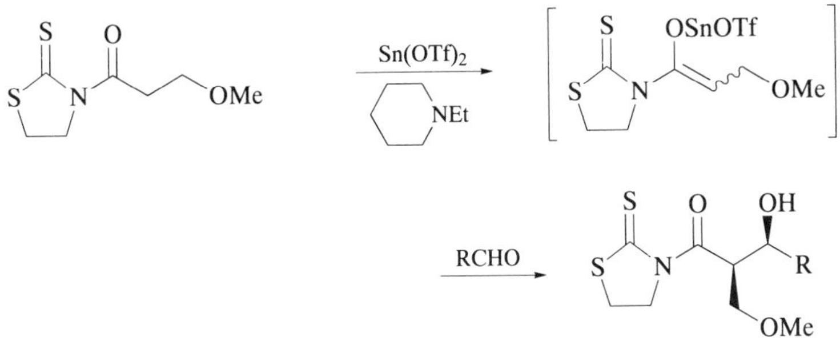

(Eq. 74)

Tin(II) enolates are also generated from 3-(3-aminopropanoyl)thiazolidine-2-thiones and they react smoothly with aldehydes to give aldol-type adducts with

high *syn* selectivity. These aldol-type adducts are stereospecifically converted to
β-lactam derivatives with hydroxy side chains (Eq. 75).[74, 191]

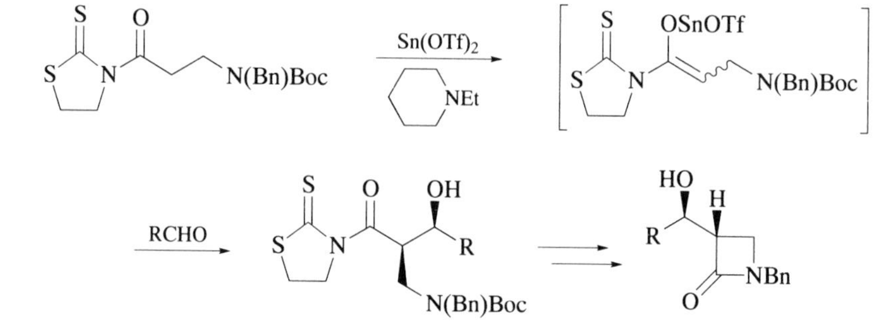

(Eq. 75)

Enolate formation from active esters sometimes gives disappointing results
owing to easy elimination of the ester groups to form ketenes. For example, attempts to generate the lithium enolate from 3-acetylthiazolidine-2-thione under
standard conditions fail because of ketene formation (Eq. 76). On the other hand,

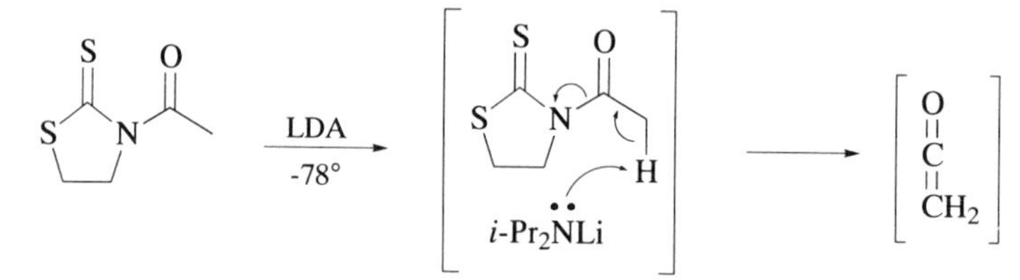

(Eq. 76)

the tin(II) enolates of 3-acylthiazolidine-2-thiones are easily prepared by treatment with tin(II) triflate and *N*-ethylpiperidine and react with electrophiles such
as aldehydes, ketones, and α,β-unsaturated carbonyl compounds (Eq. 77).[36,72]

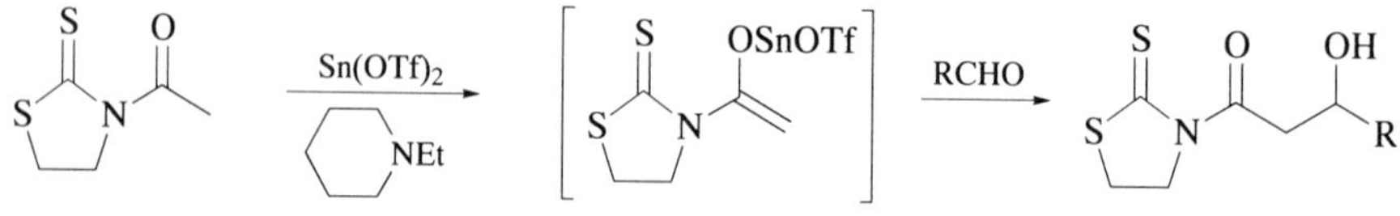

(Eq. 77)

It is interesting that tin(II) and tin(IV) enolates have quite different properties
in preparation, reactivity, and selectivity. Tin(IV) enolates sometimes exist as an
equilibrium mixture of *C*-stannyl and *O*-stannyl enolates, and their behavior is
rather similar to that of silicon enolates.

In the aldol reaction of tin(IV) enolates with aldehydes, the stereoselectivities
depend on the reaction temperature,[17,18,192,193] and *anti* aldol adducts predominate

at low temperature (Eq. 78). On the other hand, the reaction of tin(II) enolates proceeds with excellent *syn* selectivity at $-78°$ (Eq. 79).[26,36]

-78°, 20:80
45°, 77:23

(Eq. 78)

>95:5

(Eq. 79)

Boron enolates can also be prepared by using dialkylboryl triflate and a tertiary amine (usually Hünig's base) under mild conditions.[53,54] While tin(II) triflate is a white solid and can be stored for a long time under argon, dibutylboryl triflate, which is the most commonly used dialkylboryl triflate, is a liquid at room temperature and can be distilled. It is usually used as a stock solution in an appropriate solvent (e. g., dichloromethane).

One of the most striking differences between tin(II) and boron enolates is their reactivity toward ketones as acceptors. While the reaction of boron enolates with ketones proceeds sluggishly, tin(II) enolates generated from ketones react smoothly with another ketone to give ketone–ketone cross-coupling adducts in high yields.[52]

The higher reactivity of tin(II) enolates than boron enolates is also observed in the reaction of sterically hindered ketones with aldehydes. The boron enolate generated from the chiral *N*-isobutyryloxazolidinone reacts with benzaldehyde to give the oxazinedione as a 92:8 diastereomeric mixture in 23% yield. When the lithium enolate from the same oxazolidinone is treated with benzaldehyde, the opposite stereoselection is observed in better yield. On the other hand, the tin(II) enolate generated from α-bromoisobutyryloxazolidinone **16** reacts with benzaldehyde to give oxazinediones **17** and **18** (69%) in a 97:3 ratio (Eq. 80).[180,181]

Tin(II) and boron enolates generally give better selectivities than other metal enolates. The sense of the selectivities depends strongly on the combination of metals and substrates, and even opposite selectivity is sometimes observed between tin(II) and boron enolates. Thus the tin(II) enolate derived from the 3-acyloxazolidine-2-thione reacts smoothly with aldehydes to give adducts in high

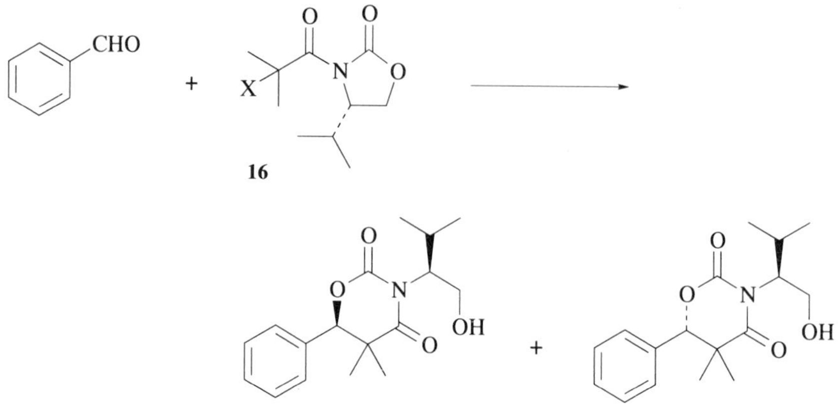


**16**

**17** (*R,S*) **18** (*S,S*)

| X | Reagent | Yield (%) | (*R,S*)/(*S,S*) |
|---|---|---|---|
| H | $Bu_2BOTf$, *i*-$Pr_2NEt$ | (23) | 92:8 |
| H | LDA | (64) | 13:87 |
| Br | $SnCl_2$, $LiAlH_4$ | (69) | 97:3 |

(Eq. 80)

yields (Eq. 81).[38,176,177] The high selectivities obtained in these cases are opposite to those obtained in the reaction of the boron enolates derived from 3-acyloxazolidine-2-ones (Eq. 82).[55] These selectivities can be explained by the chelation and nonchelation models (Fig. 3).

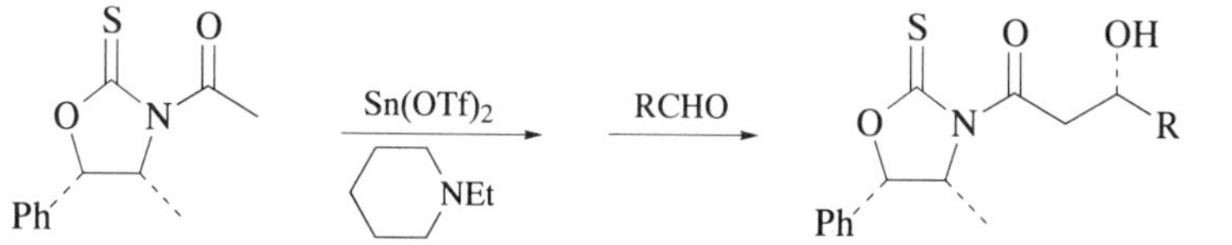


(Eq. 81)

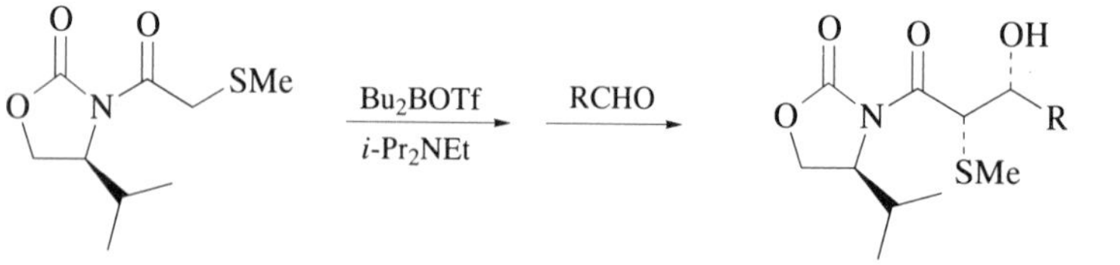


(Eq. 82)

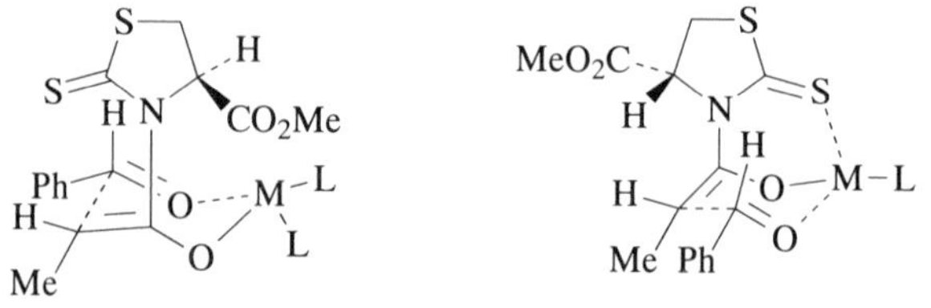


Nonchelation model Chelation model

**FIG. 3**

On the other hand, similar selectivities are observed with tin(II) and boron enolates in the reaction of cysteine-derived thiazolidinethione enolates with aldehydes (Eq. 83).[178] These selectivities are explained by the nonchelation model.

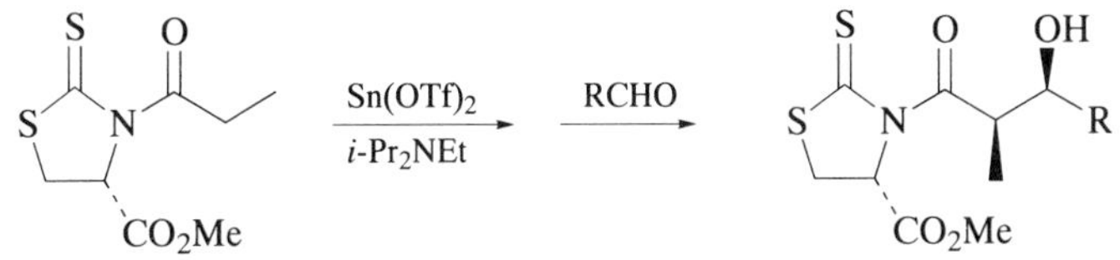

(Eq. 83)

A remarkable contrast in selectivities is observed between tin(II) and titanium(IV) $\beta$-keto imide-derived enolates with aldehydes (Eq. 84).[194]

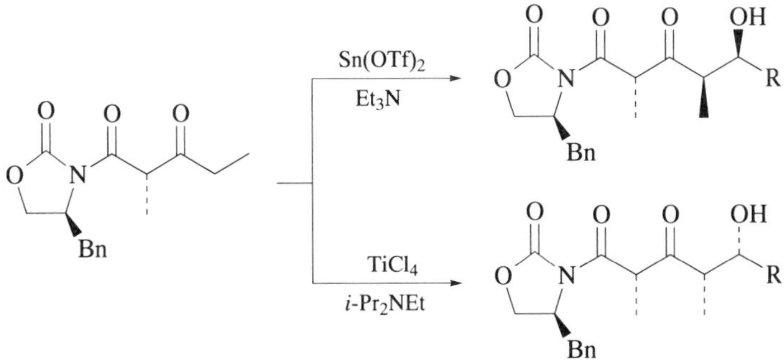

(Eq. 84)

Effects of different metal enolates are observed in the reaction of $\eta^5$-CpFe(PPh$_3$)COCOCH$_3$ with benzaldehyde (Eq. 85).[195]

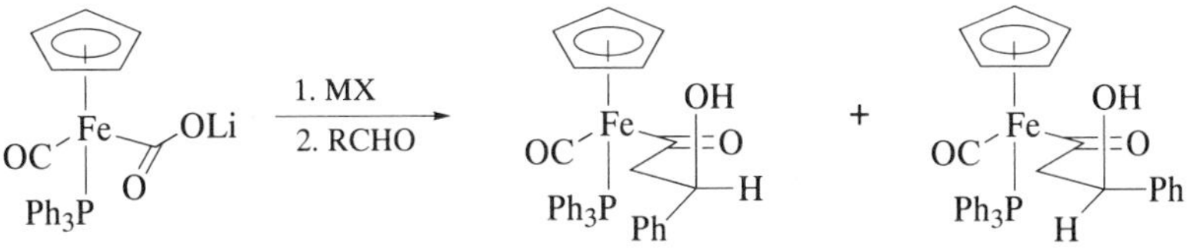

| MX | Yield (%) | I:II |
|---|---|---|
| Et$_2$AlCl | (81) | 3.5:1 |
| TiCl(OPr-$i$)$_3$ | (75) | 1:1 |
| MgBr$_2$ | (69) | 1.1:1 |
| SnCl$_2$ | (66) | 1:13.3 |
| Sn(OTf)$_2$ | (38) | 1:1.9 |
| Cp$_2$ZrCl$_2$ | (39) | 1:4.8 |

(Eq. 85)

Finally, it is noteworthy that tin(II) enolates can be coordinated by up to three ligands such as amines, thus differing from other metal enolates. Good results are obtained in the asymmetric aldol reaction of tin(II) enolate with chiral di-amine ligands derived from (S)-proline. Chiral tin(II) Lewis acids have recently been developed by using these properties of tin(II) metal, and several useful asymmetric reactions have been developed.[196-216]

## EXPERIMENTAL PROCEDURES

**Tin Powder.** Commercially available tin powder was added to about twice its weight of 10% aqueous sodium hydroxide solution and the mixture was shaken vigorously for 10 minutes. The powder was then washed with water until the washings showed no alkalinity to litmus, rinsed with methanol, and dried in vacuo at 130° for an hour.[42,217]

**Tin(II) Chloride.** Commercially available anhyd. tin(II) chloride was thoroughly dried before use by heating in vacuo at ca. 120° for an hour.[25]

**Tin(II) Triflate [Sn(OTf)₂].[36]** A modification of the procedure of Aubke et al.[218] was employed. Excess trifluoromethanesulfonic acid (40 g, 26.7 mmol) was added with vigorous stirring to anhydrous $SnCl_2$ (16 g, 8.4 mmol). Gaseous HCl was evolved in an exothermic reaction. To ensure complete reaction, the mixture was heated at 80° for 48 hours. After removal of essentially all volatile products in vacuo, the resultant white solid was washed with dry ether (50 mL × 3) to remove the last traces of acid. After drying in vacuo with warming (ca. 100°) for several hours, the powdery white solid tin(II) triflate was used as such without further purification. Tin(II) triflate is stored under Ar over $P_2O_5$. If the activity becomes low, the old tin(II) triflate must be rewashed with ether and dried in vacuo. *Note: Preparation and all handling of Sn(OTf)₂ should be carried out under an inert atmosphere in the strict absence of moisture.*

***syn*-5-Hydroxy-4-methyl-3-octanone [Tin(II) Triflate Promoted Aldol Reaction Between a Ketone and an Aldehyde].[26,36]** To a suspension of tin(II) triflate (0.458 g, 1.1 mmol) and N-ethylpiperidine (0.138 g, 1.2 mmol) in $CH_2Cl_2$ (2 mL) was added dropwise 3-pentanone (0.086 g, 1.0 mmol) in $CH_2Cl_2$ (2 mL) at −78° under argon with stirring. At this point, the suspension became a solution. After the mixture was stirred for 30 minutes, n-butyraldehyde (0.093 g, 1.3 mmol) in $CH_2Cl_2$ (2 mL) was added dropwise at −78°. The mixture was allowed to stand for 2.5 hours, then added to a vigorously stirred pH 7 phosphate buffer−$CH_2Cl_2$ mixture at 0°. After separation of the organic layers, the aqueous layer was extracted with $CH_2Cl_2$ (three times), then the combined organic layers were dried over $Na_2SO_4$. After concentration in vacuo, the resultant oil was purified by flash column chromatography (hexane–$Et_2O$, 4:1) to yield 5-hydroxy-4-methyl-3-octanone (0.136 g, 86%, *syn*:*anti*=>91:9). IR (neat) 3450, 1705 cm$^{-1}$; 60 MHz $^1$H NMR (CCl$_4$) $\delta$ 0.88–1.56 (m, 13H), 2.4–2.7 (m, 3H), 3.6 (m, 1H), 3.85 (m, 1H); 270 MHz $^1$H NMR (CDCl$_3$) $\delta$ 3.89 (m, $J$ = 3 Hz, *syn*) and 3.65 (m, $J$ = 7 Hz, *anti*).

*syn*-**3-Hydroxy-2-methyl-1-phenyl-3-(*p*-tolyl)-1-propanone (Metallic Tin).**[42] A solution of $\alpha$-bromopropiophenone (212 mg, 1.0 mmol) in $CH_2Cl_2$ (2 mL) was added dropwise to metallic tin (131 mg, 1.1 mmol) in DMF (2 mL) with stirring under Ar at 0°. The resulting mixture was stirred vigorously at this temperature for 35 minutes. At this point, most of the metallic tin had disappeared and a dark green slurry resulted. The mixture was cooled to −78° and a solution of *p*-tolualdehyde (96 mg, 0.8 mmol) in $CH_2Cl_2$ (2 mL) was added slowly over 20 minutes. The reaction mixture was stirred for 2 hours at −78°, then pH 7 phosphate buffer was added. After removal of the precipitate by filtration, the organic layer was extracted with ether, and the extract was dried over $MgSO_4$. 3-Hydroxy-2-methyl-1-phenyl-3-(*p*-tolyl)-1-propanone (201 mg, 99%, *syn*:*anti* = 92:8) was isolated by preparative TLC on silica gel (hexane:$Et_2O$ = 4:1); IR (neat) 3450, 1670, 1450, 1220, 970, 700, 680 cm$^{-1}$; $^1$H NMR (CDCl$_3$) $\delta$ 1.15 (d, 3H), 2.3 (s, 3H), 3.85 (m, 2H), 4.9 (d, trace, *anti*, $J$ = 7.5 Hz), 5.1 (d, 1H, *syn*, $J$ = 3 Hz), 7.55–7.0 (m, 7H), 8.0–7.75 (m, 2H).

**Ethyl 3-Hydroxy-3-phenylpropionate (SnCl$_2$–LAH).**[25] Lithium aluminum hydride (19 mg; 0.5 mmol) was added portionwise to anhydrous tin(II) chloride (190 mg; 1.0 mmol) suspended in THF (1 mL) under argon. A spontaneous exothermic reaction occurred and a dark gray material was deposited. To this suspension were added dropwise at room temperature ethyl bromoacetate (100 mg; 0.6 mmol) dissolved in THF (1 mL), followed by benzaldehyde (53 mg; 0.5 mmol) dissolved in THF (1 mL), and the mixture was stirred for 2 hours at this temperature. Water was added to the reaction mixture and then the THF was removed under reduced pressure. The organic materials were extracted with ether and the extract was dried over $MgSO_4$. Ethyl 3-hydroxy-3-phenylpropionate (81 mg, 83%) was isolated by preparative TLC on silica gel. NMR (CDCl$_3$) $\delta$ 1.17 (t, 3H, $J$ = 7 Hz), 2.55 (d, 2H, $J$ = 7 Hz), 3.23-3.70 (broad, 1H), 4.03 (q, 2H, $J$ = 7 Hz), 4.93 (t, 1H, $J$ = 7 Hz), 7.15 (s, 5H). IR (neat) 3450, 1720 cm$^{-1}$.

***cis*-3,4-Epoxy-4-phenyl-2-butanone (Synthesis of an *α,β*-Epoxyketone).[36,59]**
To a suspension of tin(II) triflate (355 mg, 0.85 mmol) and Et$_3$N (110 mg,
1.09 mmol) in 2 mL of THF was added dropwise bromoacetone (87 mg,
0.64 mmol) in THF (2 mL) at $-78°$ under Ar with stirring. After the mixture
was stirred for 30 minutes, benzaldehyde (108 mg, 1.02 mmol) in THF (2 mL)
was added dropwise and the mixture was stirred for another 30 minutes at $-78°$.
The reaction was quenched with 10% aqueous citric acid and the organic materials were extracted with ether (three times) and dried over Na$_2$SO$_4$. After evaporation of the solvent, the resultant crude adduct in DMF (2 mL) was added
dropwise to a suspension of KF (136 mg, 2.34 mmol) and dicyclohexyl-18-crown-6
(914 mg, 2.46 mmol) in DMF (2 mL) at room temperature under Ar with stirring.
After the mixture was stirred for 12 hours, the reaction was quenched with pH 7
phosphate buffer and the organic materials were extracted with ether (three
times). The combined organic extracts were washed with brine and dried over
Na$_2$SO$_4$. After evaporation of the solvent, the resultant oil was purified by silica
gel column chromatography (hexane–Et$_2$O $= 8:1$) to afford 3,4-epoxy-4-phenyl-
2-butanone in 72% yield (*cis*:*trans* $= 70:30$). IR (neat) 1715, 700 cm$^{-1}$; $^1$H NMR
(CDCl$_3$) $\delta$ for *cis* isomer, 1.77 (s, 3H), 3.72 (d, 1H, $J = 5$ Hz), 4.25 (d, 1H,
$J = 5$ Hz), 3.93 (d, 1H, $J = 2$ Hz), 7.26 (s, 5H); for *trans* isomer, 2.13 (s, 1H),
3.42 (d, 1H, $J = 2$ Hz), 3.93 (d, 1H, $J = 2$ Hz), 7.26 (s, 5H).

***anti*-3-Hydroxy-2-methyl-1,3-diphenylbutan-1-one (Cross Aldol Reaction
Between Ketones).[52]**    To a suspension of tin(II) triflate (1.458 g, 1.1 mmol) and
*N*-ethylpiperidine (0.138 g, 1.2 mmol) in CH$_2$Cl$_2$ (2 mL) was added dropwise
propiophenone (0.134 g, 1.0 mmol) in CH$_2$Cl$_2$ (2 mL) at $0°$ under Ar with stirring. After the mixture had been stirred for 15 minutes, acetophenone (0.156 g,
1.3 mmol) in CH$_2$Cl$_2$ (1 mL) was added dropwise at $0°$. The mixture was allowed
to stand for 1 hour, then pH 7 phosphate buffer was added. After separation of
the organic layer, the aqueous layer was extracted with CH$_2$Cl$_2$ (three times), and
the combined organic extracts were dried over Na$_2$SO$_4$. After evaporation under
reduced pressure, the resultant oil was purified by preparative TLC (hexane–
Et$_2$O $= 9:1$) to yield crystalline *anti*-3-hydroxy-2-methyl-1,3-diphenylbutan-1-
one (0.151 g, 60%). IR (neat) 3480, 1660 cm$^{-1}$; $^1$H NMR (CCl$_4$) $\delta$ 1.35 (d, $J = 7$ Hz,

3H), 1.45 (s, 3H), 4.0 (q, $J$ = 7 Hz,1H), 4.6 (s, 1H), 4.0–4.6 (m, 8H), 7.68–8.0 (m, 2H).

**syn-3-(3-Hydroxy-2-methyl-3-phenylpropanoyl)thiazolidine-2-thione (Reaction of 3-Acylthiazolidine-2-thione with an Aldehyde).[36,72]** To a $CH_2Cl_2$ suspension (2.0 mL) of tin(II) triflate (480 mg, 1.15 mmol) and N-ethylpiperidine (155 mg, 1.37 mmol) was added dropwise a $CH_2Cl_2$ solution (1.2 mL) of 3-propanoylthiazolidine-2-thione (163 mg, 0.93 mmol) at −78°. After further stirring at this temperature for 15 minutes, a $CH_2Cl_2$ solution (1.2 mL) of benzaldehyde (146 mg, 1.38 mmol) was added, and the mixture was further stirred for 20 minutes. The reaction was quenched with pH 7 phosphate buffer and the white precipitate was removed through Celite. The organic material was extracted with ether (three times), and the extracts were dried over $Na_2SO_4$ and evaporated in vacuo. The residual oil was purified by silica gel column chromatography to afford 3-(3-hydroxy-2-methyl-3-phenylpropanoyl) thiazolidine-2-thione in 94% yield (syn:anti = 97:3). IR (neat) 3450, 1690 cm$^{-1}$, $^1$H NMR (CDCl$_3$) $\delta$ 1.21 (d, 3H, $J$ = 7 Hz), 2.77–3.17 (m, 3H), 4.00–4.25 (m, 2H), 4.63–4.97 (m, 2H), 7.33 (s, 5H).

*Conversion of the Aldol Adduct into Ester.*[36,72] To a MeOH solution (2 mL) of syn-3-(3-hydroxy-2-methyl-3-phenylpropanoyl)thiazolidine-2-thione (86 mg, 0.31 mmol) was added powdered $K_2CO_3$ (100 mg, 0.72 mmol) and the mixture was stirred for several minutes until the yellow color disappeared completely. A pH 7 phosphate buffer was added and the organic materials were extracted with ether. The extracts were dried over $Na_2SO_4$ and then evaporated in vacuo. The crude product was purified by silica gel TLC to afford the corresponding methyl ester (56 mg, 95%).

*Conversion of the Aldol Adduct into Amide.*[36,72] To a $CH_2Cl_2$ solution (2 mL) of syn-3-(3-hydroxy-2-methyl-3-phenylpropanoyl)thiazolidine-2-thione (90 mg,

0.32 mmol) was added a $CH_2Cl_2$ solution of benzylamine (101 mg, 0.94 mmol) at room temperature. The yellow color disappeared immediately, and pH 7 phosphate buffer was added. The organic materials were extracted with AcOEt and the extracts were dried over $Na_2SO_4$. After evaporation of the solvent, the crude product was purified by silica gel TLC to afford the corresponding amide (77 mg, 90%).

*Conversion of the Aldol Adduct into Aldehyde.*[36,72] The hydroxy function of the adduct was protected by reaction with isopropyldimethylsilyl chloride (2 equiv), and $Et_3N$ (2 equiv) in $CH_2Cl_2$ at 0° overnight (87%).

To a toluene solution (1.0 mL) of this protected compound (82 mg, 0.22 mmol) was added a toluene solution (0.80 mL, 1.83 mL/mmol) of DIBAL at −78° and the mixture was further stirred for 10 minutes at −78°. The reaction was quenched with pH 7 phosphate buffer solution (0.2 mL) and $Na_2SO_4$ was added as a drying agent. After removal of the precipitate through Celite, the solvent was evaporated in vacuo and the residual oil was purified by silica gel column chromatography to afford *syn*-3-isopropyldimethylsiloxy-2-methyl-3-phenylpropanal in 75% yield.

**3,4-Dihydroxy-6-phenyl-2-hexanones [Tin(II) Enediolate].**[75] A THF suspension of tin(II) chloride (1.52 g, 8 mmol) and metallic potassium (313 mg, 8 mmol) was stirred at room temperature for 1 hour under Ar and was then refluxed carefully for another 30 minutes. After the suspension had been cooled to 0° in an ice bath, a THF solution (2 mL) of 3-phenylpropanal (67.1 mg, 0.5 mmol) was added and a THF solution (8 mL) of fresh methylglyoxal (0.24 g, 3.3 mmol) was added dropwise. The reaction mixture was stirred at 0° overnight and poured into a phosphate buffer solution (pH 7, 20 mL). After filtration of insoluble materials, the organic materials of the filtrate were extracted with AcOEt and the extracts were dried over anhydrous $MgSO_4$. After evaporation of the solvent, the residue was purified by preparative TLC on silica gel (AcOEt:petroleum ether, 1:1.5) to give 3,4-dihydroxy-6-phenyl-2-hexanones (73.1 mg, 70%, *syn*:*anti* = 40:60). *syn*: $^1H$ NMR ($CDCl_3$) $\delta$ 1.67–2.50 (m, 3H), 2.21 (s, 3H), 2.57–3.00 (m, 2H), 3.17–4.30 (m, 3H), 7.23 (s, 5H); IR (NaCl) 3430, 1710 cm$^{-1}$. *anti*: $^1H$ NMR ($CDCl_3$) 1.45–1.92 (m, 2H), 2.12 (s, 3H), 2.18–2.41 (m, 1H), 2.55–2.93 (m, 2H), 3.55 (d, 1H, $J = 5$ Hz), 3.68–4.02 (m, 1H), 4.24 (dd, 1H, $J = 3.3$ Hz), 7.22 (s, 5H); IR (NaCl) 3430, 1710 cm$^{-1}$.

**3-Ethylthiomethyl-4-hydroxy-6-phenyl-2-hexanone [Preparation of $\alpha,\beta$-Ethylthiomethyl Aldol with a Catalytic Amount of Tin(II) Enolate].**[77,78]   To a solution of ethanethiol (10 mg, 0.17 mmol) in THF (2 mL) was added n-butyllithium (1.54 M, 0.11 mL) in hexane at 0° under Ar. Tin(II) triflate (69 mg, 0.17 mmol) was added, and after 20 minutes, the mixture was cooled to −45°. Methyl vinyl ketone (118 mg, 1.68 mmol) in THF (1.5 mL) and 3-phenylpropanal (350 mg, 2.61 mmol) in THF (1.5 mL) were successively added to the mixture. The reaction mixture was stirred for 12 hours, then quenched with 10% aqueous citric acid, and the organic materials were extracted with $CH_2Cl_2$ (three times). To completely hydrolyze the trimethylsilyl ether group, the crude aldol product obtained after evaporation of the solvent was dissolved in methanol and to this solution was added citric acid. After stirring for 30 minutes, the reaction was quenched with pH 7 phosphate buffer. The organic layer was extracted with $CH_2Cl_2$ (three times) and the combined extracts were dried over anhydrous $Na_2SO_4$. After evaporation of the solvent, the crude product was purified by silica gel column chromatography to afford 3-ethylthiomethyl-4-hydroxy-6-phenyl-2-hexanone (336 mg, 75% yield, syn:anti = 90:10). IR (neat) 3450, 1750 cm$^{-1}$; $^1$H NMR (CCl$_4$) $\delta$ 1.1 (t, 3H, $J$ = 8 Hz), 1.3–1.7 (m, 2H), 1.9 (syn), 2.0 (anti) (s, 3H), 2.1–3.2 (m, 8H), 3.6 (brs, 1H), 7.0 (s, 5H). The diastereomeric ratio was determined by integration of the methyl signal. Relative stereochemistry was determined by $^{13}$C NMR spectra.[108] $^{13}$C NMR (CDCl$_3$) $\delta$ 70.97 (syn), 71.73 (anti).

**Ethyl 3-(tert-Butylthio)carbonyl-2-furfurylaminobutyrate [Reaction of a Tin(II) Thioester Enolate with an $\alpha$-Iminoester].**[44]   To a solution of lithium diisopropylamide (0.68 mmol) in dry ether (2 mL) under argon at −78° was added tert-butyl propanethioate (100 mg, 0.68 mmol) in dry ether (1.5 mL) over 5 minutes. After the mixture was stirred for 30 minutes, tin(II) chloride (156 mg, 0.82 mmol) was added as a powder and the mixture was stirred vigorously for another 30 minutes. To this mixture, ethyl N-furfuryliminoacetate (62 mg, 0.34 mmol) was added and the reaction mixture was stirred for 3 hours at −78°. The reaction was quenched by adding pH 7 phosphate buffer, the mixture was filtered through a Celite pad, and the product was extracted with AcOEt. The organic layer was washed with brine and dried over Na$_2$SO$_4$. After evaporation of the solvents, the crude product was purified by silica gel TLC (AcOEt–hexane) to afford ethyl 3-(*tert*-butylthio)carbonyl-2-furfurylaminobutyrate (92 mg, 82% yield, *syn/anti* = 95:5).

**4-(1-Benzoylethyl)phenol [Addition-reduction Reaction of a Tin(II) Enolate with 1,4-Benzoquinone].**[110,111] Propiophenone (0.44 mmol) in CH$_2$Cl$_2$ (1 mL) was added dropwise at −78° to a stirred mixture of tin(II) triflate (0.55 mmol) and *N*-ethylpiperidine (0.61 mmol) in CH$_2$Cl$_2$ (1 mL). The resulting mixture was stirred at −78° for 40 minutes. 1,4-Benzoquinone (0.19 mmol) in CH$_2$Cl$_2$ (2 mL) was added dropwise at −78° to the yellow-green suspension, and stirring was continued for 30 minutes, after which dichloromethylsilane (0.97 mmol in 2 mL of CH$_2$Cl$_2$) and dimethylaminopyridine (0.44 mmol in 2 mL of CH$_2$Cl$_2$) were added dropwise in quick succession. Stirring was continued at −78° until preparative TLC (50% ethyl acetate/hexane) showed consumption of all the intermediate (ca. 30 minutes). The reaction was then quenched with 10% aqueous citric acid solution (10 mL) and the resulting two-phase mixture was extracted with CH$_2$Cl$_2$ (3 × 10 mL). The combined extracts were dried (MgSO$_4$), filtered, and evaporated under reduced pressure to give an oily residue which was purified by preparative TLC on silica gel using 30% ethyl acetate/hexane as eluant to give the desired phenol (0.16 mmol, 83%) as a colorless oil, which slowly solidified on standing: mp. 89–90° (recrystallized from chloroform/hexane). IR (KBr) 3400, 1660, 1605, 1590, 1570, 1505, 1220, 730 cm$^{-1}$; $^1$H NMR (CDCl$_3$) δ 1.5 (d, 3H, *J* = 7 Hz), 4.6 (q, 1H, *J* = 7 Hz), 6.7 (d, 2H, *J* = 7 Hz), 7.1 (d, 2H, *J* = 7 Hz), 7.4 (m, 3H), 7.8–8.0 (m, 2H).

**3-(2,3-Dimethyl-5-oxo-5-phenylpentanoyl)-1,3-oxazolidin-2-one [Reaction of a Tin(II) Enolate with an α,β-Unsaturated Ketone in the Presence of TMSCl].**[111,153] Tin(II) triflate (343 mg, 0.82 mmol) was cooled to −78° and *N*-ethylpiperidine (116 mg, 1.03 mmol) in CH$_2$Cl$_2$ (2 mL) was added dropwise. After

10 minutes, a solution of 3-propanoyl-1,3-oxazolidin-2-one (97 mg, 0.67 mmol) in dichloromethane (1.5 mL) was added dropwise with stirring to the yellow-green suspension, and stirring was continued at −78° for 1 hour. Phenyl propenyl ketone (77 mg, 0.53 mmol) and TMSCl (107 mg, 0.99 mmol) were added successively to the mixture, and after stirring for 2 hours the reaction was quenched at −78° with 10% citric acid (10 mL). Dichloromethane (10 mL) was added and the organic and aqueous phases were separated. The aqueous phase was extracted with CH$_2$Cl$_2$ (10 mL × 3) and the combined organic layers were dried (MgSO$_4$), filtered, and evaporated under reduced pressure. To completely hydrolyze the trimethysilyl ether, the residue was dissolved in methanol and citric acid was added. After stirring for 1 hour the reaction was quenched with pH 7 phosphate buffer. The organic layer was extracted with dichloromethane (10 mL × 3) and the combined extracts were dried (MgSO$_4$). After evaporation of the solvent, the crude product was purified by preparative TLC to afford white crystalline 3-(2,3-dimethyl-5-oxo-5-phenylpentanoyl)-1,3-oxazolidin-2-one (120 mg, 79%). IR(KBr) 2970, 1760, 1680, 1380, 1240, 760, 690 cm$^{-1}$; $^1$H NMR(CDCl$_3$) $\delta$ 0.9–1.3 (m, 6H), 2.4–3.3 (m, 3H), 3.6–4.5 (m, 5H), 7.2–7.6 (m, 3H), 7.8–8.1 (m, 2H).

**3-Hydroxy-2-methyl-1,3-diphenyl-1-propanone [Enantioselective Cross Aldol Reaction of a Ketone and an Aldehyde].[36,164]** To a suspension of tin(II) triflate (296 mg, 0.71 mmol) and *N*-ethylpiperidine (95 mg, 0.84 mmol) in CH$_2$Cl$_2$ (2 mL) was added dropwise propiophenone (78 mg, 0.58 mmol) in CH$_2$Cl$_2$ (1.5 mL) at −78° under Ar. After the mixture was stirred for 30 minutes, (*S*)-1-methyl-2-[(piperidin-1-yl)methyl]pyrrolidine (158 mg, 0.87 mmol) in CH$_2$Cl$_2$ (1.5 mL) was added dropwise, and the mixture was stirred for 5 minutes at −78°. The mixture was cooled to −95°, and then benzaldehyde (91 mg, 0.86 mmol) in CH$_2$Cl$_2$ (1.5 mL) was added dropwise. The mixture was further stirred for 30 minutes at the same temperature, then quenched with pH 7 phosphate buffer. The organic layer was extracted with ether (three times) and the combined extracts were dried over Na$_2$SO$_4$. After evaporation of the solvent, the crude product was purified by silica gel TLC to afford 3-hydroxy-2-methyl-1,3-diphenyl-1-propanone (103 mg, 74%). The optical purity of the product was determined by measurement of the $^1$H and $^{19}$F NMR spectra of the MTPA ester (80% ee).[219]

**5-Oxo-3-phenylhexanedithioate [Catalytic Asymmetric Michael Reaction of a Tin(II) Enethiolate].**[78,168] To a $CH_2Cl_2$ suspension (2 mL) of tin(II) triflate (38 mg, 0.09 mmol), benzalacetone (0.93 mmol), and (S)-1-methyl-2-[(N-1-naphthylamino)methyl]pyrrolidine (23 mg, 0.10 mmol) under Ar was added a $CH_2Cl_2$ solution (1 mL) of 4-phenyl-3-buten-2-one (136 mg, 0.93 mmol) at −78°. To this mixture was slowly added a $CH_2Cl_2$ solution (2 mL) of vinylthiosilane (199 mg, 1.11 mmol) over 4 hours. The mixture was stirred for 1 hour at this temperature, 10% aqueous citric acid was added, and the organic materials were extracted with $CH_2Cl_2$ (three times). The combined extracts were washed with 4% aqueous $NaHCO_3$ (twice) and brine successively, and the organic phase was dried with anhydrous $MgSO_4$. After evaporation of the solvent, the residue was dissolved in methanol (10 mL), citric acid (800 mg, 4.16 mmol) was added, and the mixture was stirred for 1 hour at room temperature. To the mixture was added pH 7 phosphate buffer, the organic materials were extracted with $CH_2Cl_2$ (three times), and the combined extracts were dried over anhydrous $MgSO_4$. After evaporation of the solvent, the crude product was purified by silica gel TLC to afford methyl 5-oxo-3-phenylhexanedithioate (187 mg, 80% yield, 70% ee). IR (neat) 1675, 1255 $cm^{-1}$; [1]H NMR (CDCl$_3$) δ 1.9 (s, 3H), 2.4 (s, 3H), 2.7 (d, 2H, $J = 7$ Hz), 3.2 (d, 2H, $J = 7$ Hz), 3.6–4.1 (m, 1H), 7.1 (s, 5H).

**3-[(5-Oxo-2(S)-pyrrolidinyl)acetyl]-4(S)-isopropyl-1,3-thiazolidine-2-thione [Alkylation of a Chiral Tin(II) Enolate].**[188,189] Tin(II) triflate (1.08 g, 2.59 mmol) was dissolved in dry THF (6 mL) under Ar at room temperature. To the solution cooled to −50° in a dry ice–acetonitrile bath were added successively N-ethylpiperidine (0.41 ml, 2.99 mmol) and 3-acetyl-4(S)-isopropyl-1,3-dithiazolidine-2-thione (0.406 g, 2.0 mmol) in dry THF (1.8 mL), and the mixture was then stirred for 3 hours between −50 and −40° to form the tin(II) enolate. To this enolate was added a 1.0 M solution of 5-acetoxy-2-pyrrolidinone (3.0 mmol) in dry THF at −5°, and the mixture was then stirred for 2 hours between −5 and 0°. The reaction mixture was poured into a mixture of phosphate buffer solution (pH 7.0, 50 mL) and AcOEt (50 mL) with vigorous stirring. After the precipitate was filtered through Celite and washed with AcOEt (3 × 50 mL), the combined filtrate was washed with brine and then submitted to workup to provide a crude product. A sample of the crude product was submitted to HPLC analysis (column, Diasil 5C 184.6 mm i.d. × 25 cm; eluant, CH₃CN–H₂O, 9:1; flow rate, 1.0 mL/min; detection, UV 305 nm) to determine diastereomeric excess (94% de). Flash column chromatography of the crude product (elution with 67% AcOEt in CHCl₃) afforded the pure product, 3-([5-oxo-2(S)-pyrrolidinyl]acetyl)-4(S)-isopropyl-1,3-thiazolidine-2-thione (67%). IR (CH₃Cl) 3340, 1690 (br), 1255, 1175 cm⁻¹; ¹H NMR (100 MHz, CDCl₃) δ 0.98 and 1.06 (d, 6H, J = 6.8 Hz), 1.68–2.08 (m, 1H), 2.18–2.60 (m, 4H), 3.04 (dd, 1H, J = 11.5, 1.5 Hz), 3.18 (dd, 1H, J = 17.8, 9.8 Hz), 3.57 (dd, 1H, J = 11.5, 8.0 Hz), 3.78 (dd, 1H, J = 17.8, 3.5 Hz), 3.96–4.28 (m, 1H), 5.20 (ddd, 1H, J = 8.0, 6.0, 1.5 Hz), 6.28 (brs, 1H).

## TABULAR SURVEY

The seven tables include all examples of the reaction found in the literature through the middle of 1991. Entries in each table are arranged in order of increasing number of carbon atoms in the donor carbonyl compound.
The following abbreviations are used in the tables:

| | |
|---|---|
| Ac | acetyl |
| Bn | benzyl |
| Boc | *tert*-butoxycarbonyl |
| Bz | benzoyl |
| Cbz | benzyloxycarbonyl |
| de | diastereomeric excess |
| 18-C-6 | dicyclohexyl-18-crown-6 |
| DEIPS | diethylisopropylsilyl |
| ee | enantiomeric excess |
| IPDMS | isopropyldimethylsilyl |
| LAH | lithium aluminum hydride |
| LDA | lithium diisopropylamide |
| NEPIP | *N*-ethylpiperidine |
| NMMOR | *N*-methylmorpholine |
| Py | pyridine |
| TBDMS | *tert*-butyldimethylsilyl |
| Tf | trifluoromethanesulfonyl (triflyl) |
| TMS | trimethylsilyl |
| TrBF₄ | trityl tetrafluoroborate |
| Ts | *p*-toluenesulfonyl |

TABLE I. ALDOL REACTION WITH ACHIRAL SUBSTRATES; NO OR SIMPLE DIASTEREOSELECTION

| Starting Material | Enolization Reagent | Acceptor | Product(s) and Yield(s) (%), syn/anti | Refs. |
|---|---|---|---|---|
| **C2** | | | | |
| $CH_2=C=O$ | $Sn(SBu\text{-}t)_2$ | PhCHO | [t-BuS–C(O)–CH2–CH(OH)–Ph] (75) | 43 |
| **C3** | | | | |
| [1-bromopropan-2-one] | $Sn(OTf)_2$, NEPIP, KF, 18-C-6 | PhCHO | R = Ph, (72) 70/30 | 59 |
| | " | $Ph(CH_2)_2CHO$ | R = Ph(CH2)2, (65) 73/27 | 59 |
| | $Sn(OTf)_2$, $Et_3N$ | $(EtO)_2(O)PCH_2CHO$ | R = CH2P(O)(OEt)2, (63) 60/40 | 60 |
| | " | 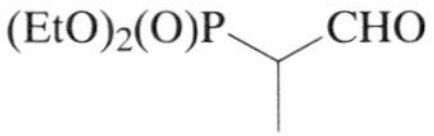 $(EtO)_2(O)P$—CHO | R = CH(Me)P(O)(OEt)2, (43) 67/33 | 60 |
| | " |  $(EtO)_2(O)P$—CHO | R = C(Me)2P(O)(OEt)2, (30) 100/0 | 60 |
| [methylglyoxal] | SnCl2, K | $Ph(CH_2)_2CHO$ | R = Ph(CH2)2, (70) 43/57 | 75 |
| | " | $BnO(CH_2)_3CHO$ | R = BnO(CH2)3, (80) 16/84 | |
| | " | $n\text{-}C_7H_{15}CHO$ | R = n-C7H15, (55) 33/67 | |
| | " | $p\text{-}ClC_6H_4CHO$ | R = p-ClC6H4, (54) 25/75 | |
| [Me(H)C=C=O] | | | [t-BuS–C(O)–CH(Me)–CH(OH)–R] | 43 |

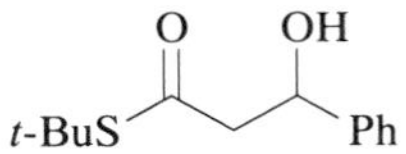

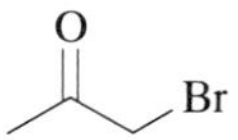

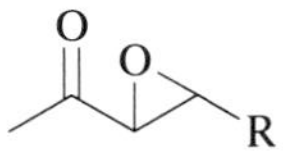

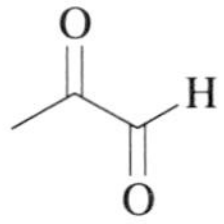

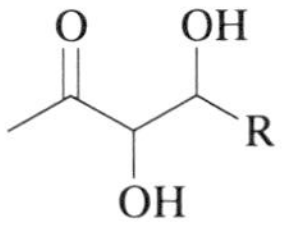

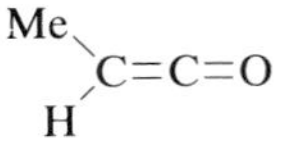

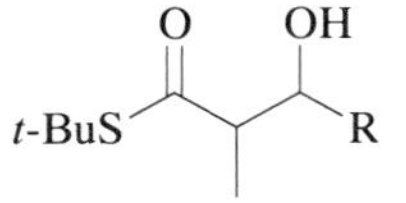

TABLE I. ALDOL REACTION WITH ACHIRAL SUBSTRATES; NO OR SIMPLE DIASTEREOSELECTION (*Continued*)

| Starting Material | Enolization Reagent | Acceptor | Product(s) and Yield(s) (%), *syn/anti* | Refs. |
|---|---|---|---|---|
| | $Sn(SBu\text{-}t)_2$ | PhCHO | R = Ph, (98) 90/10 | |
| | " | $Ph(CH_2)_2CHO$ | R = $Ph(CH_2)_2$, (89) 97/3 | |
| | " | *i*-PrCHO | R = *i*-Pr, (78) 93/7 | |
| | " | *t*-BuCHO | R = *t*-Bu, (72) 90/10 | 43 |
| | | | | |
| | $Sn(SBu\text{-}t)_2$ | 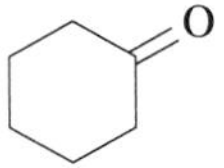 | $R^1, R^2 = \text{-}(CH_2)_6\text{-}$, (78) | |
| | " | PhCOMe | $R^1$ = Ph, $R^2$ = Me, (77) 70/30 | |
| | " | $Ph(CH_2)_2COMe$ | $R^1 = Ph(CH_2)_2$, $R^2$ = Me, (69) 58/42 | |
| | " | $EtO_2CCOMe$ | $R^1 = CO_2Et$, $R^2$ = Me, (96) 19/81 | |
| | " | $MeCO(CH_2)_8CHO$ | $R^1 = MeCO(CH_2)_8$, $R^2$ = H | |
| | | | -100°, (67) | |
| | | | -45°, (81) | |
| | | | | |
| | $Sn(OTf)_2$ | " | R = Ph, (89) 96/4 | |
| | " | " | R = PhCH=CH, (60) 81/19 | |
| | $SnBr_2$ | " | R = PhCH=CH, (55) 84/16 | 87 |

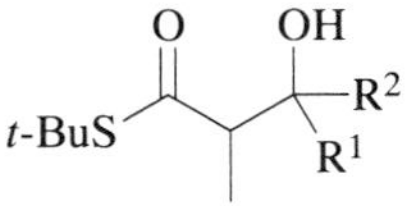

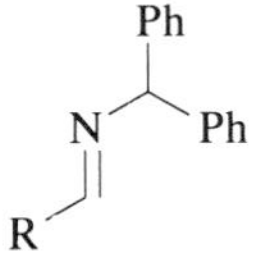

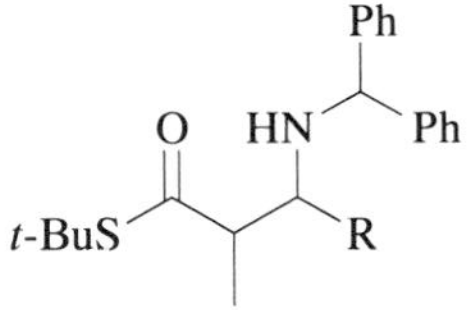

47

TABLE I. ALDOL REACTION WITH ACHIRAL SUBSTRATES; NO OR SIMPLE DIASTEREOSELECTION (*Continued*)

| Starting Material | Enolization Reagent | Acceptor | Product(s) and Yield(s) (%), *syn/anti* | Refs. |
|---|---|---|---|---|
| | Sn(OTf)$_2$ | " | R = *i*-Pr, (83) 92/8 | |
| | " | " | R = Ph(CH$_2$)$_2$, (65) 92/8 | |
| | Sn(SPy)$_2$ | PhCHO | 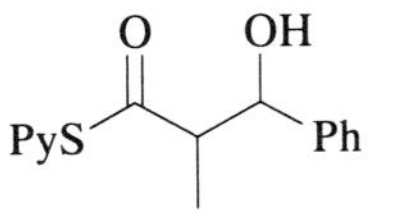 (89) 93/7 | 43 |
| C$_4$ Br$\diagup$CO$_2$Et | Sn | PhCHO | (84) | 25 |
| O=⟨⟩=O (pentane-2,4-dione type) | SnCl$_2$, K | Ph(CH$_2$)$_2$CHO | R = Ph(CH$_2$)$_2$, (95) 50/50 | 48 |
| | SnCl$_2$, K, C$_6$F$_6$ | " | R = Ph(CH$_2$)$_2$, (86) 75/25 | 48 |
| | SnCl$_2$, K | *n*-C$_8$H$_{17}$CHO | R = *n*-C$_8$H$_{17}$, (78) 50/50 | 48 |
| | SnCl$_2$, K, C$_6$F$_6$ | " | R = *n*-C$_8$H$_{17}$, (70) 75/25 | 48 |
| | SnCl$_2$, K | *i*-PrCHO | R = *i*-Pr, (83) 67/33 | 48 |
| | " | BnOCH$_2$CHO | R = CH$_2$OBn, (77) 67/33 | 48 |

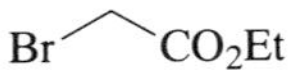

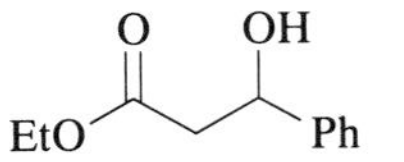

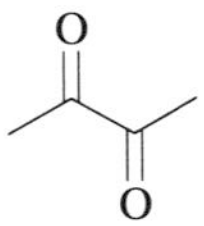

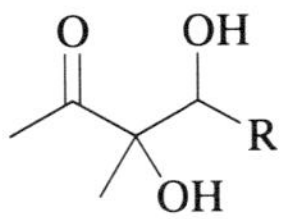

TABLE I. ALDOL REACTION WITH ACHIRAL SUBSTRATES; NO OR SIMPLE DIASTEREOSELECTION (*Continued*)

| Starting Material | Enolization Reagent | Acceptor | Product(s) and Yield(s) (%), *syn/anti* | Refs. |
|---|---|---|---|---|
| Br—⟨⟩—CHO | SnCl$_2$, K | PhCHO | R = Ph, (64) | 76 |
| | " | p-ClC$_6$H$_4$CHO | R = p-ClC$_6$H$_4$, (71) | 76 |
| | " | p-BrC$_6$H$_4$CHO | R = p-BrC$_6$H$_4$, (70) | 76 |
| | " | n-C$_{11}$H$_{23}$CHO | R = n-C$_{11}$H$_{23}$, (65) | 76 |
| | " | BnO(CH$_2$)$_3$CHO | R = (CH$_2$)$_3$OBn, (74) | 76 |
| | EtSSnOTf | PhCHO | R = Ph, (68) 95/5 | 77, 78 |
| | " | Ph(CH$_2$)$_2$CHO | R = Ph(CH$_2$)$_2$, (75) 90/10 | 77, 78 |
| 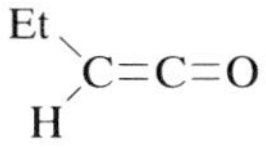 | Sn(SBu-t)$_2$ | PhCHO | 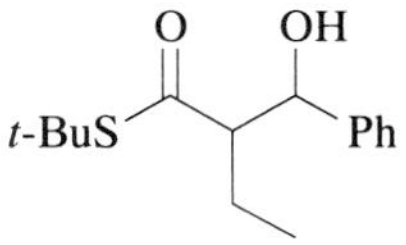 (90) >95/5 | 43 |

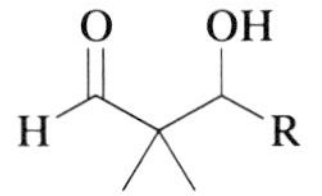

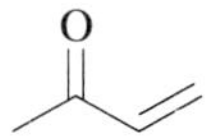

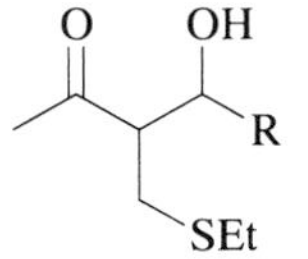

49

TABLE I. ALDOL REACTION WITH ACHIRAL SUBSTRATES: NO OR SIMPLE DIASTEREOSELECTION (*Continued*)

| Starting Material | Enolization Reagent | Acceptor | Product(s) and Yield(s) (%), *syn/anti* | Refs. |
|---|---|---|---|---|
| C5 | | | | |
| (ethyl 2-bromopropanoate) | " | 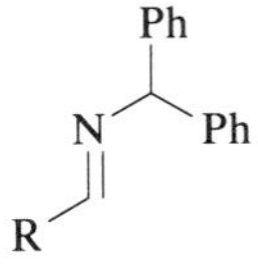 (R–CH=N–CH(Ph)Ph) | 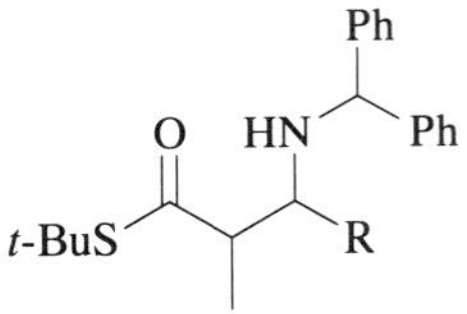 (60) 12/88 | 87 |
| (methyl 4-bromocrotonate) | Sn | PhCHO | 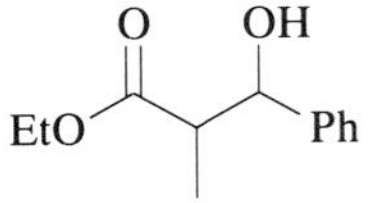 (81) 55/45 | 25 |
| 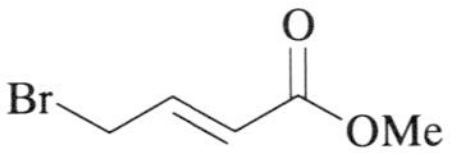 | Sn | PhCHO | 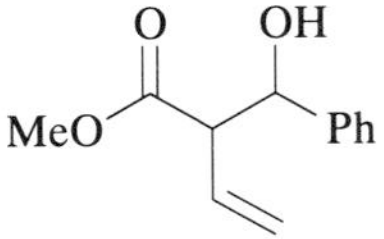 (85) 57/43 | 25 |
| (3-pentanone) | Sn(OTf)₂, NEPIP | PhCHO | R = Ph, (77) 87/13 | 26, 36 |
| | " | *i*-PrCHO | R = *i*-Pr, (73) 93/7 | 26, 36 |
| | " | *n*-PrCHO | R = *n*-Pr, (86) >91/9 | 26, 36 |

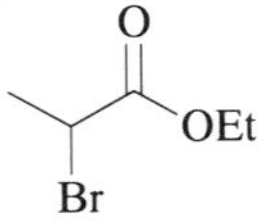

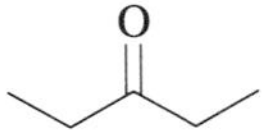

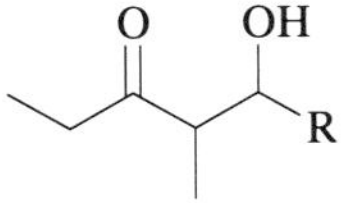

TABLE I. ALDOL REACTION WITH ACHIRAL SUBSTRATES; NO OR SIMPLE DIASTEREOSELECTION (*Continued*)

| Starting Material | Enolization Reagent | Acceptor | Product(s) and Yield(s) (%), *syn/anti* | Refs. |
|---|---|---|---|---|
| | Sn(OTf)$_2$, NEPIP | 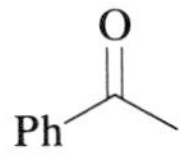 PhCOCH$_3$ | 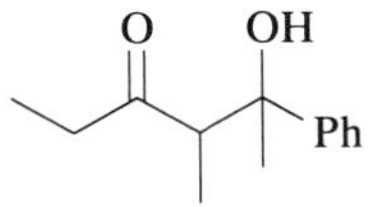 (45) 13/87 | 52 |
| | Sn(OTf)$_2$, NEPIP | 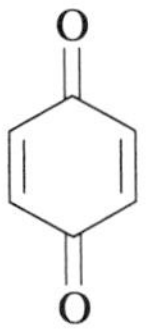 , MeHSiCl$_2$ (benzoquinone) | 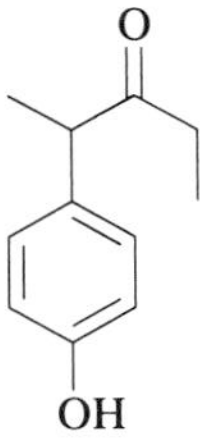 (68) | 110, 111 |
| | Sn(OTf)$_2$, NEPIP | 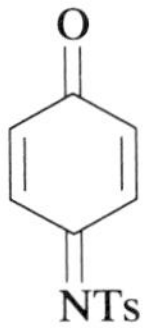 , MeHSiCl$_2$ (NTs quinone imine) | 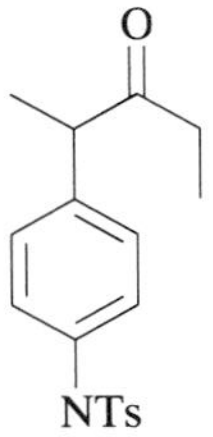 (68) | 110, 111 |
| (2-bromo-3-pentanone) | Sn | PhCHO | R = Ph, (70) 91/9 | 42 |
| | Sn | *i*-PrCHO | R = *i*-Pr, (65) 91/9 | 42 |
| | " | *n*-PrCHO | R = *n*-Pr, (77) 92/8 | 42 |

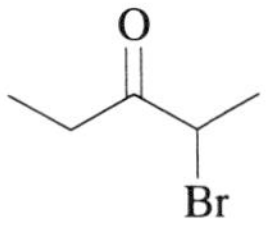

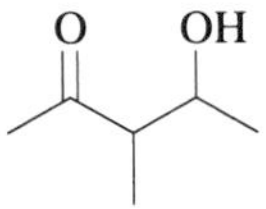

TABLE I. ALDOL REACTION WITH ACHIRAL SUBSTRATES; NO OR SIMPLE DIASTEREOSELECTION (*Continued*)

| Starting Material | Enolization Reagent | Acceptor | Product(s) and Yield(s) (%), *syn/anti* | Refs. |
|---|---|---|---|---|
| 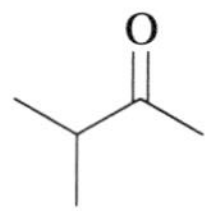 (3-methyl-2-butanone) | Sn(OTf)$_2$, NEPIP | 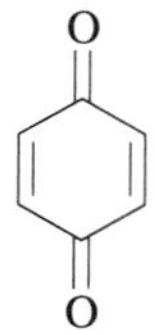 (1,4-benzoquinone), MeHSiCl$_2$ | 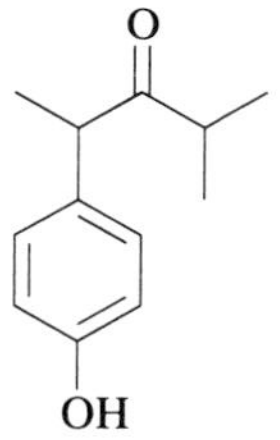 (65) | 110, 111 |
| 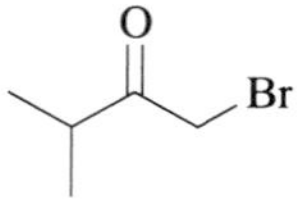 (1-bromo-3-methyl-2-butanone) | Sn(OTf)$_2$, Et$_3$N | (EtO)$_2$(O)PCH$_2$CHO | 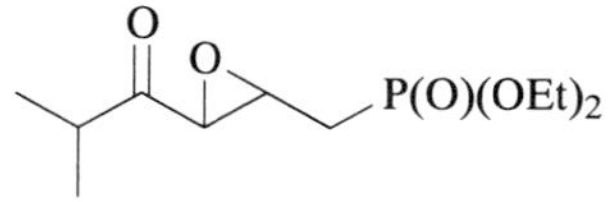 epoxide—P(O)(OEt)$_2$ (46) 75/25 | 60 |
| (N-acetyl-1,3-thiazolidine-2-thione) | Sn(OTf)$_2$, NEPIP | PhCHO | R = Ph, (90) | 36, 37 |
| | " | Ph(CH$_2$)$_2$CHO | R = Ph(CH$_2$)$_2$, (88) | 36, 37 |
| | " | *i*-PrCHO | R - *i*-Pr, (94) | 36, 37 |

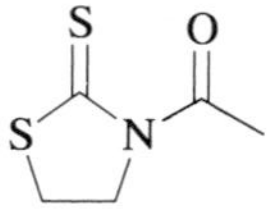

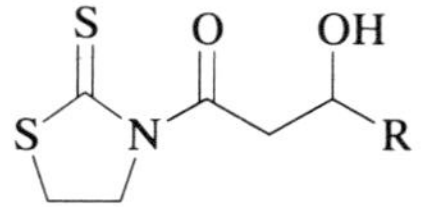

TABLE I. ALDOL REACTION WITH ACHIRAL SUBSTRATES; NO OR SIMPLE DIASTEREOSELECTION (*Continued*)

| Starting Material | Enolization Reagent | Acceptor | Product(s) and Yield(s) (%), *syn/anti* | Refs. |
|---|---|---|---|---|
| 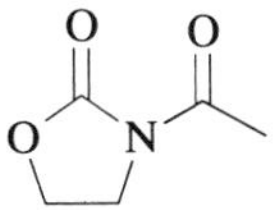 (3-acetyloxazolidin-2-one) | Sn(OTf)$_2$, NEPIP | 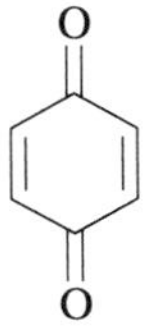 (1,4-benzoquinone), MeHSiCl$_2$ | 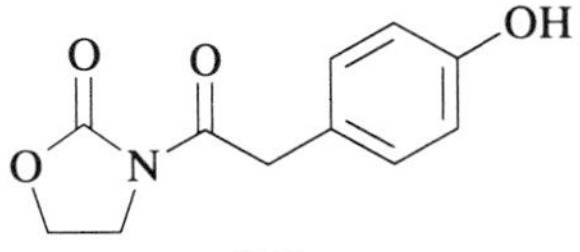 (48), phenol-OH product | 110, 111 |
| | Sn(OTf)$_2$, NEPIP | 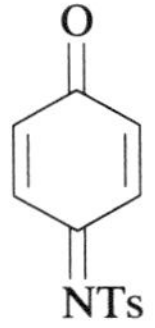 (N-tosyl quinone imine, NTs), MeHSiCl$_2$ | 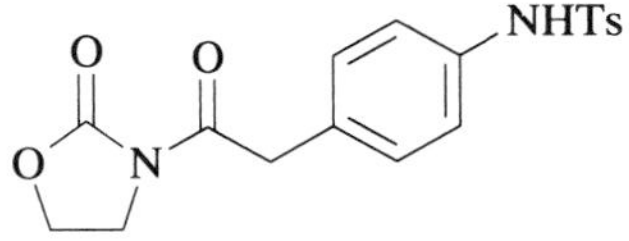 (67), NHTs product | 110, 111 |
| C$_6$ (ethyl 2-bromobutanoate) | Sn | PhCHO | EtO–C(O)–CH(Et)–CH(OH)Ph (81) 57/43 | 25 |
| (ethyl 2-bromo-2-methylpropanoate) | Sn | PhCHO | EtO–C(O)–C(CH$_3$)$_2$–CH(OH)Ph (95) | 25 |
| 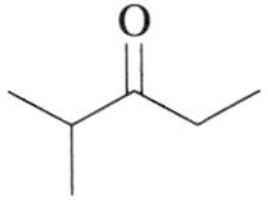 (2-methyl-3-pentanone) | Sn(OTf)$_2$, NEPIP | PhCHO | 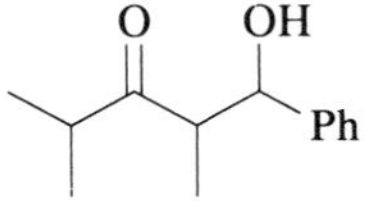 (O)C–CH(CH$_3$)–CH(OH)Ph (72) 91/9 | 26, 36 |

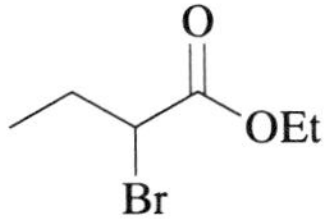

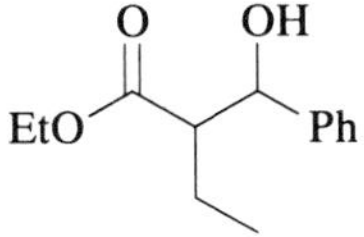

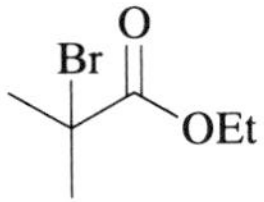

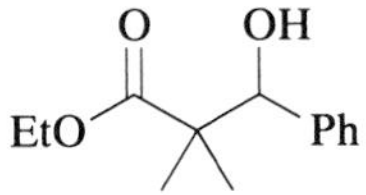

53

TABLE I. ALDOL REACTION WITH ACHIRAL SUBSTRATES; NO OR SIMPLE DIASTEREOSELECTION (*Continued*)

| Starting Material | Enolization Reagent | Acceptor | Product(s) and Yield(s) (%), *syn/anti* | Refs. |
|---|---|---|---|---|
| (cyclohexanone) | Sn(OTf)$_2$, NEPIP | 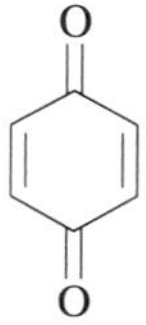 (benzoquinone), MeHSiCl$_2$ | 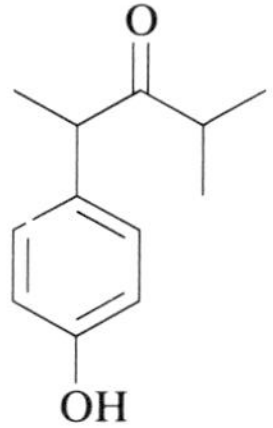 (70) | 110, 111 |
| 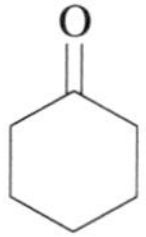 (cyclohexanone) | Sn(OTf)$_2$, NEPIP | PhCHO | 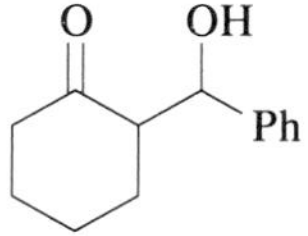 [OH, Ph product] (41) >95/5 | 26, 36 |
| (cyclohexanone) | Sn(OTf)$_2$, NEPIP | PhCH(SEt)$_2$ | 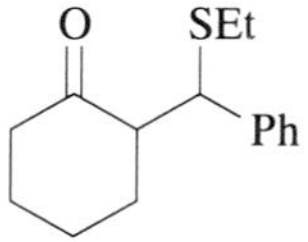 [SEt, Ph product] -45°, (78) 77/23; -78°, (87) 87/13 | 88 |
| 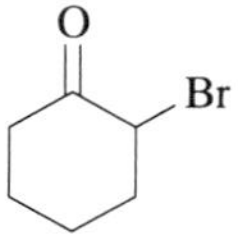 (2-bromocyclohexanone) | Sn | PhCHO | 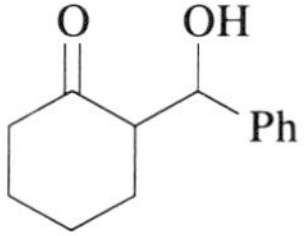 [OH, Ph product] (28) 94/6 | 42 |
| 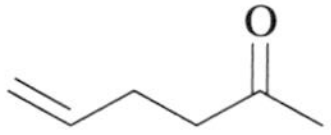 (hept-6-en-2-one derivative) | Sn(OTf)$_2$, NMMOR | 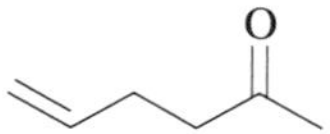 (hex-5-en-2-one) | 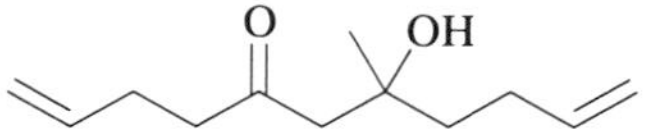 (84) | 52 |

54

TABLE I. ALDOL REACTION WITH ACHIRAL SUBSTRATES; NO OR SIMPLE DIASTEREOSELECTION (*Continued*)

| Starting Material | Enolization Reagent | Acceptor | Product(s) and Yield(s) (%), syn/anti | Refs. |
|---|---|---|---|---|
| 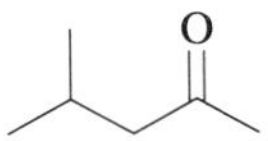 (4-methyl-2-pentanone) | Sn(OTf)$_2$, NMMOR | 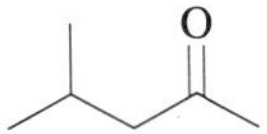 (4-methyl-2-pentanone) | 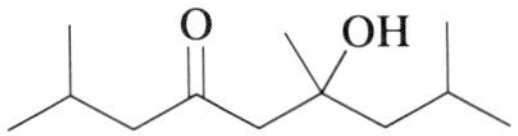 (structure with OH) (76) | 52 |
| 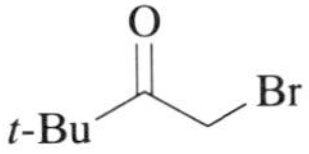 (t-Bu bromomethyl ketone, Br) | Sn(OTf)$_2$, NEPIP, KF, 18-C-6 | PhCHO | 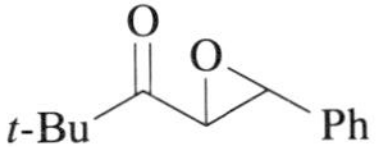 (epoxy ketone, t-Bu, Ph) (64) >95/5 | 59 |
| | Sn(OTf)$_2$, NEPIP, KF, 18-C-6 | Ph(CH$_2$)$_2$CHO | 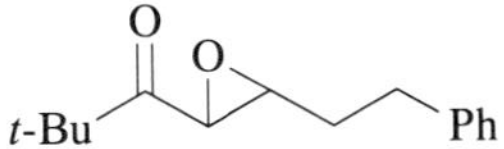 (epoxy ketone, t-Bu, Ph) (48) >95/5 | 59 |
| | Sn(OTf)$_2$, NEPIP, KF, 18-C-6 | i-PrCHO | 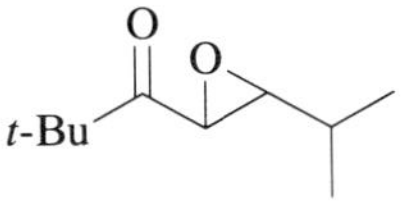 (epoxy ketone, t-Bu, i-Pr) (47) >95/5 | 59 |
| 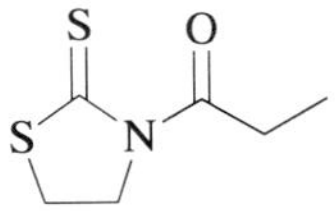 (N-propionyl thiazolidinethione) | Sn(OTf)$_2$, NEPIP | PhCHO | 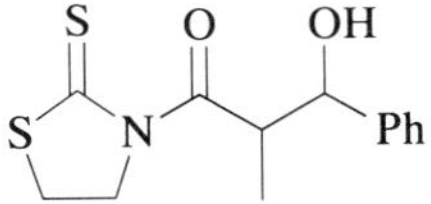 (OH, Ph, thiazolidinethione) (94) 97/3 | 36, 37 |
| | Sn(OTf)$_2$, NEPIP | Ph(CH$_2$)$_2$CHO | 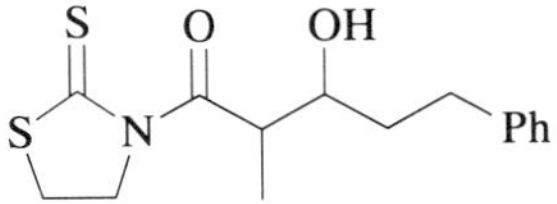 (OH, Ph, thiazolidinethione) (91) >97/3 | 36, 37 |

55

TABLE I. ALDOL REACTION WITH ACHIRAL SUBSTRATES; NO OR SIMPLE DIASTEREOSELECTION (*Continued*)

| Starting Material | Enolization Reagent | Acceptor | Product(s) and Yield(s) (%), *syn/anti* | Refs. |
|---|---|---|---|---|
| (oxazolidinone propanoyl) | Sn(OTf)$_2$, NEPIP | *i*-PrCHO | 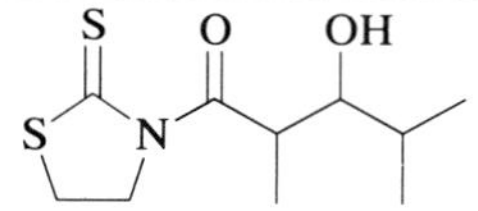 (95) >97/3 | 36, 37 |
| 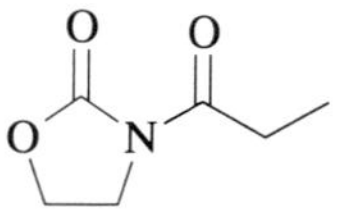 | Sn(OTf)$_2$, NEPIP | 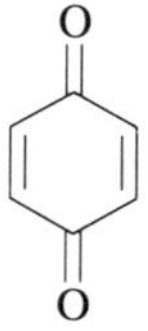 , MeHSiCl$_2$ | 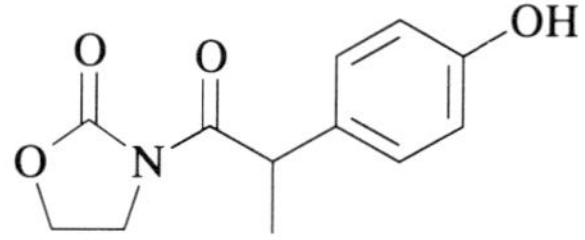 (71) | 110, 111 |
| (3,5-heptanedione derivative) | Sn | Ph(CH$_2$)$_2$CHO | R = Ph(CH$_2$)$_2$, (92) 67/33 | 48 |
| | Sn, C$_6$F$_6$ | " | R = Ph(CH$_2$)$_2$, (87) 80/20 | 48 |
| | Sn | *n*-C$_8$H$_{17}$CHO | R = *n*-C$_8$H$_{17}$, (86) 67/33 | 48 |
| | Sn, C$_6$F$_6$ | " | R = *n*-C$_8$H$_{17}$, (81) 75/25 | 48 |
| C$_7$ | Sn(OTf)$_2$, NMMOR | | (83) | 52 |

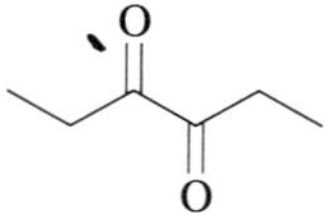

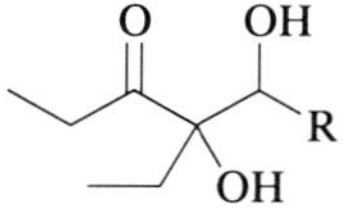

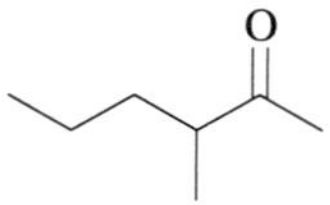

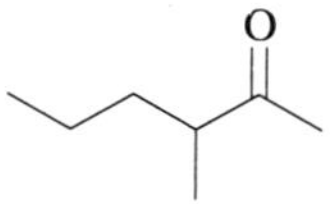

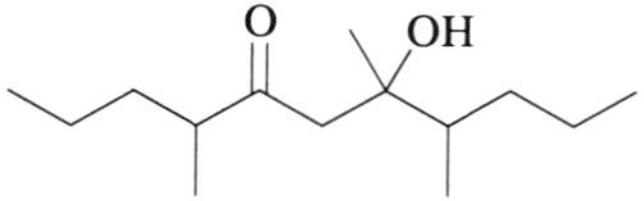

TABLE I. ALDOL REACTION WITH ACHIRAL SUBSTRATES; NO OR SIMPLE DIASTEREOSELECTION (Continued)

| Starting Material | Enolization Reagent | Acceptor | Product(s) and Yield(s) (%), syn/anti | Refs. |
|---|---|---|---|---|
| 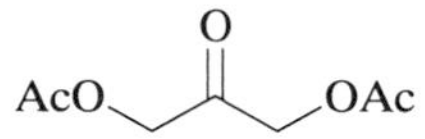 (AcO—CH2—CO—CH2—OAc) | Sn(OTf)$_2$, NEPIP | 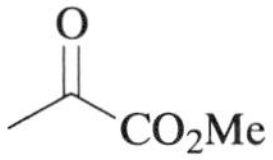 (O)CH$_3$COCO$_2$Me | 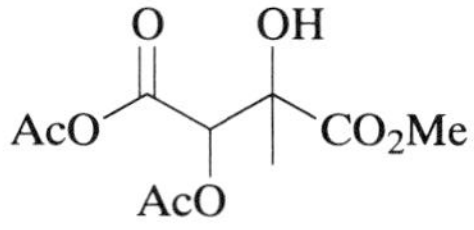 AcO / OH / CO$_2$Me / AcO structure; 0°, (45) 85/15; −78°, (37) 71/29 | 58 |
| 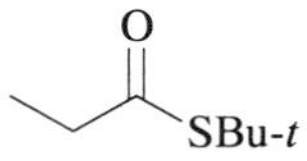 (thione, SBu-$t$) | LDA, SnCl$_2$ | 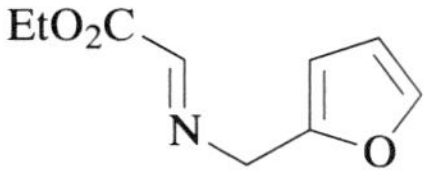 EtO$_2$C—CH=N—CH$_2$-furan | 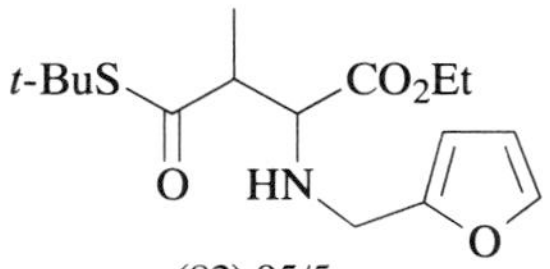 $t$-BuS, O, CO$_2$Et, HN—CH$_2$-furan; (82) 95/5 | 44 |
| (thiazolidine-2-thione N-propionyl) | Sn(OTf)$_2$, NEPIP | PhCHO | R = Ph, (95) 25/75 | 74 |
| | " | Ph(CH$_2$)$_2$CHO | R = Ph(CH$_2$)$_2$, (82) 84/16 | 74 |
| | " | $n$-C$_5$H$_{11}$CHO | R = $n$-C$_5$H$_{11}$, (77) 85/15 | 74 |
| | " | $i$-PrCHO | R = $i$-Pr, (68) 67/33 | 74 |
| | " | PhCH=CHCHO | R = PhCH=CH, (73) 50/50 | 74 |
| C$_8$ | | | | |
| ($n$-C$_6$H$_{13}$COCH$_3$) | Sn(OTf)$_2$, NMMOR | ($n$-C$_6$H$_{13}$COCH$_3$) | $n$-C$_6$H$_{13}$, OH, C$_6$H$_{13}$-$n$ structure; (82) | 52 |

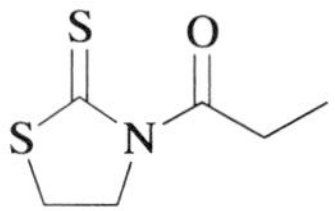

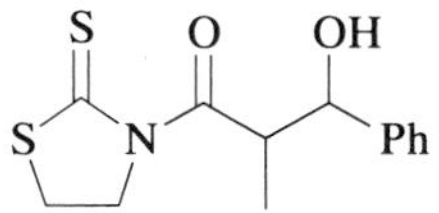

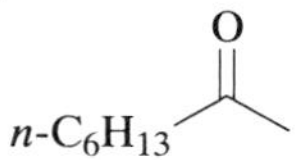

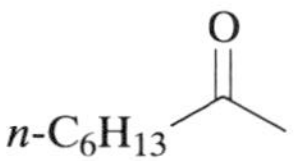

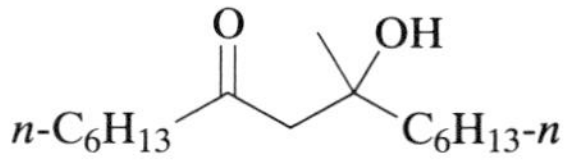

57

TABLE I. ALDOL REACTION WITH ACHIRAL SUBSTRATES; NO OR SIMPLE DIASTEREOSELECTION (*Continued*)

| Starting Material | Enolization Reagent | Acceptor | Product(s) and Yield(s) (%), *syn/anti* | Refs. |
|---|---|---|---|---|
| 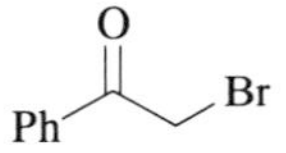 (Ph−CO−CH₂Br) | Sn | PhCHO | 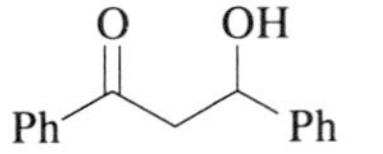 (63) | 42 |
| | Sn(OTf)₂, NEPIP, KF, 18-C-6 | Ph(CH₂)₂CHO | R = Ph(CH₂)₂, (80) 66/34 | 59 |
| | " | *i*-PrCHO | R = *i*-Pr, (80) 65/35 | 59 |
| 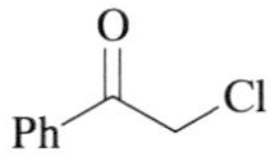 (Ph−CO−CH₂Cl) | Sn(OTf)₂, NEPIP | 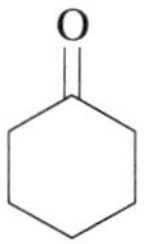 | 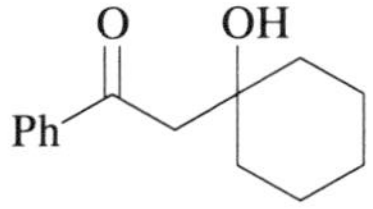 (87) | 52 |
| (TMS ketone with CH₂Br) | Sn(OTf)₂, NEPIP | PhCHO | R = Ph, (70) >95/5 | 66 |
| | " | Ph(CH₂)₂CHO | R = Ph(CH₂)₂, (83) >95/5 | 66 |
| | " | *n*-C₁₁H₂₃CHO | R = *n*-C₁₁H₂₃, (79) >95/5 | 66 |
| 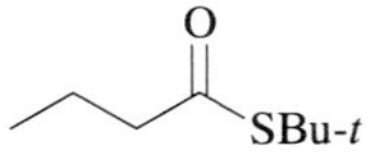 (O=C−SBu-*t*) | LDA, SnCl₂ | 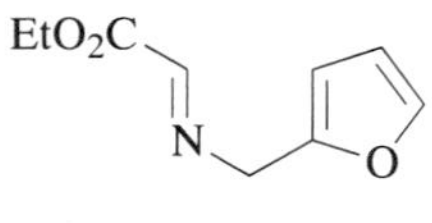 $EtO_2C$−CH=N−CH₂-furan | 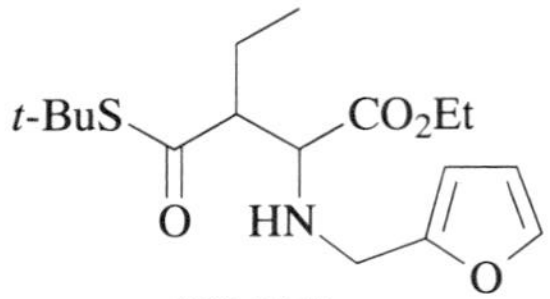 (61) 95/5 | 44 |

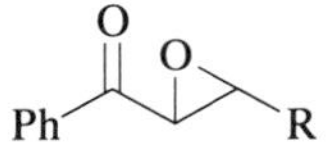

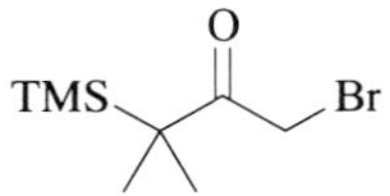

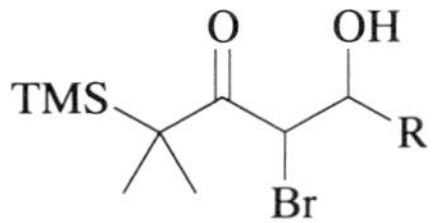

58

TABLE I. ALDOL REACTION WITH ACHIRAL SUBSTRATES; NO OR SIMPLE DIASTEREOSELECTION (*Continued*)

| Starting Material | Enolization Reagent | Acceptor | Product(s) and Yield(s) (%), *syn/anti* | Refs. |
|---|---|---|---|---|
| 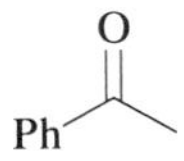 Ph—C(=O)—CH₃ | Sn(OTf)₂, NEPIP | 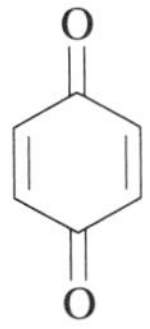 quinone , MeHSiCl₂ | 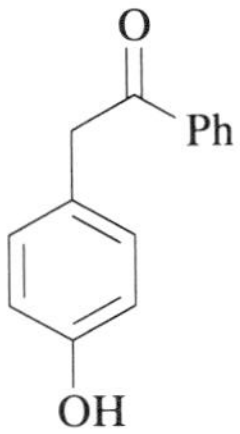 (65) | 110, 111 |
| Ph—C(=O)—CHO | SnCl₂, K / " | Ph(CH₂)₂CHO / n-C₈H₁₇CHO | R = Ph(CH₂)₂, (73) 50/50 / R = n-C₈H₁₇, (68) 67/33 | 76 / 76 |
| C₉ Ph—C(=O)—CH₂CH₃ | Sn(OTf)₂, NEPIP / " / " | PhCHO / i-PrCHO / n-PrCHO | R = Ph, (71) >95/5 / R = i-Pr, (80) 91/9 / R = n-Pr, (79) 86/14 | 26, 36 / 26, 36 / 26, 36 |

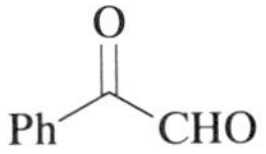

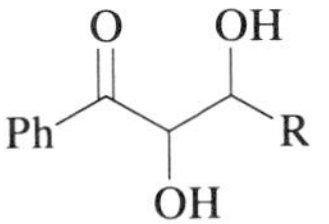

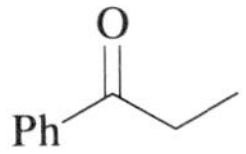

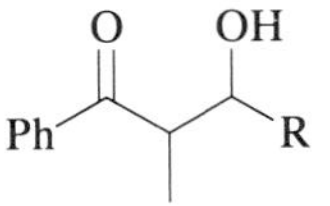

TABLE I. ALDOL REACTION WITH ACHIRAL SUBSTRATES; NO OR SIMPLE DIASTEREOSELECTION (*Continued*)

| Starting Material | Enolization Reagent | Acceptor | Product(s) and Yield(s) (%), *syn/anti* | Refs. |
|---|---|---|---|---|
| | Sn(OTf)$_2$, NEPIP | 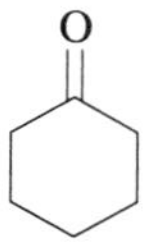 | R$^1$R$^2$ = (CH$_2$)$_5$, (83) | 52 |
| | " |  | R$^1$R$^2$ = (CH$_2$)$_4$, (75) | 52 |
| | " | EtCOEt | R$^1$ = R$^2$ = Et, (80) | 52 |
| | " | PhCOMe | R$^1$ = Ph, R$^2$ = Me, (60) 0/100 | 52 |
| | " | *i*-PrCOMe | R$^1$ = *i*-Pr, R$^2$ = Me, (69) 50/50 | 52 |
| | " | *i*-BuCOMe | R$^1$ = *i*-Bu, R$^2$ = Me, (69) 50/50 | 52 |
| | " | Ph(CH$_2$)$_2$COMe | R$^1$ = Ph(CH$_2$)$_2$, R$^2$ = Me, (78) 50/50 | 52 |
| | Sn(OTf)$_2$, NEPIP | 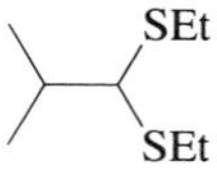 | 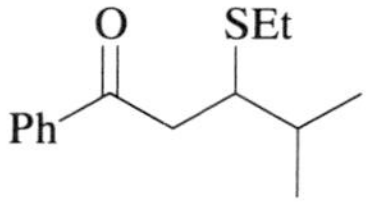 (64) 56/44 | 88 |
| | Sn(OTf)$_2$, NEPIP | CH(SEt)$_3$ | 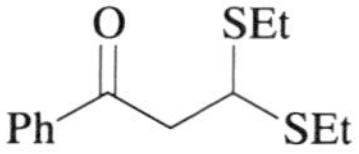 (80) | 88 |

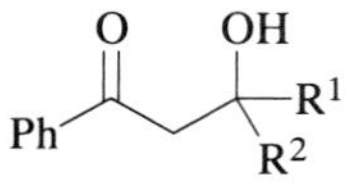

60

TABLE I. ALDOL REACTION WITH ACHIRAL SUBSTRATES; NO OR SIMPLE DIASTEREOSELECTION (*Continued*)

| Starting Material | Enolization Reagent | Acceptor | Product(s) and Yield(s) (%), *syn/anti* | Refs. |
|---|---|---|---|---|
| | Sn(OTf)$_2$, NEPIP | 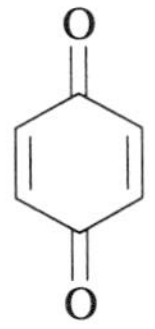 , MeHSiCl$_2$ | 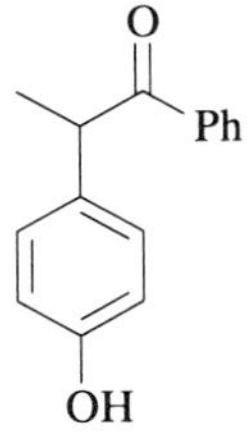 (83) | 110, 111 |
| | Sn(OTf)$_2$, NEPIP | 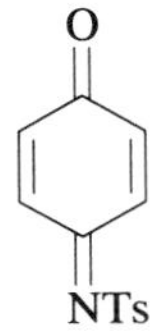 , MeHSiCl$_2$ | 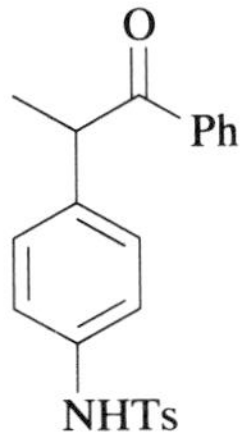 (77) | 110, 111 |
| | Sn(OTf)$_2$, NEPIP | 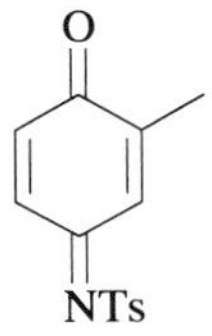 , MeHSiCl$_2$ | 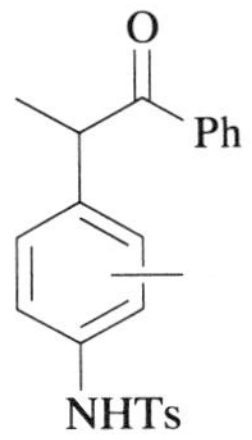 (59) | 110, 111 |

TABLE I. ALDOL REACTION WITH ACHIRAL SUBSTRATES; NO OR SIMPLE DIASTEREOSELECTION (*Continued*)

| Starting Material 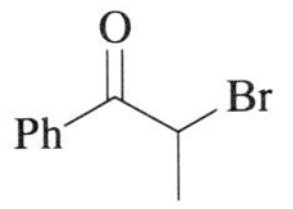 | Enolization Reagent | Acceptor | Product(s) and Yield(s) (%), *syn/anti* 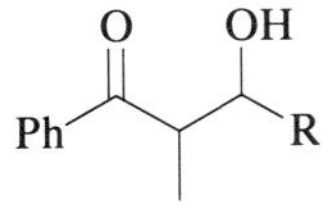 | Refs. |
|---|---|---|---|---|
| | Sn | PhCHO | R = Ph, (93) 93/7 | 42 |
| | " | p-MeC₆H₄CHO | R = p-MeC₆H₄, (99) 92/8 | 42 |
| | " | p-ClC₆H₄CHO | R = p-ClC₆H₄, (91) 92/8 | 42 |
| | " | n-PrCHO | R = n-Pr, (85) 91/9 | 42 |
| | " | i-BuCHO | R = i-Bu, (72) 92/8 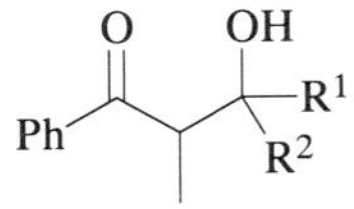 | 42 |
| | Sn | 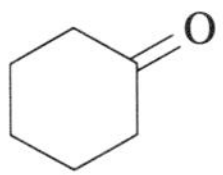 | R¹R² = (CH₂)₅, (88) | 42 |
| | " | BnCOBn | R¹ = R² = Bn, (44) | 42 |
| 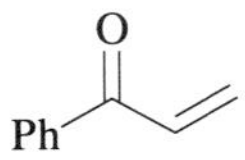 | EtSSnOTf | Ph(CH₂)₂CHO | 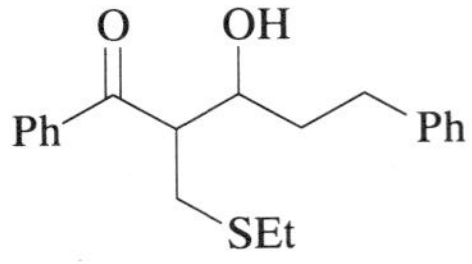 (73) 84/16 | 77, 78 |

62

TABLE I. ALDOL REACTION WITH ACHIRAL SUBSTRATES: NO OR SIMPLE DIASTEREOSELECTION (*Continued*)

| Starting Material | Enolization Reagent | Acceptor | Product(s) and Yield(s) (%), *syn/anti* | Refs. |
|---|---|---|---|---|
| 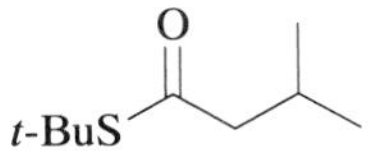 | LDA, SnCl$_2$ | 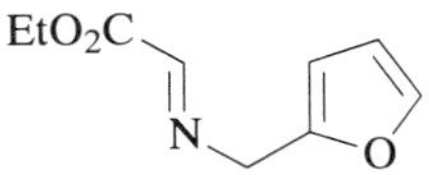 | 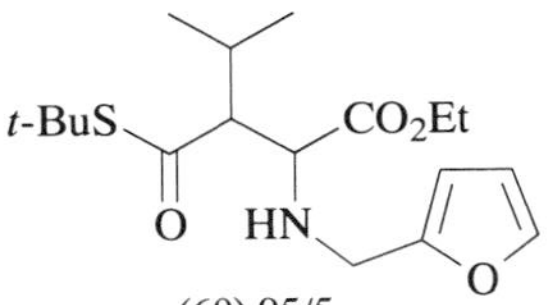 (60) 95/5 | 44 |
| 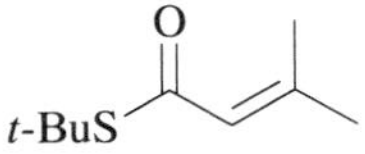 | LDA, SnCl$_2$ | 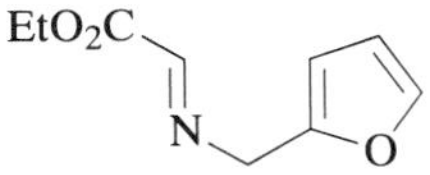 | 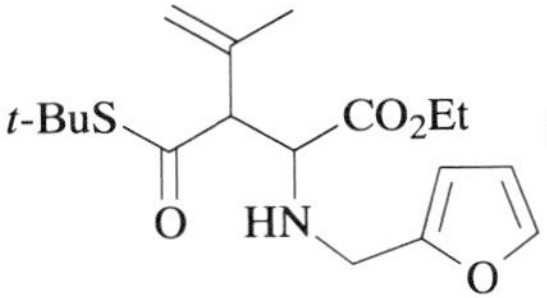 (60) | 44 |
| C$_{10}$ 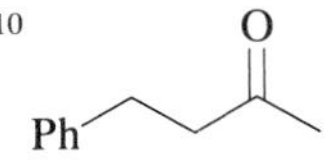 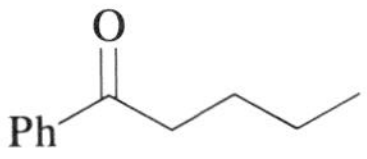 | Sn(OTf)$_2$, NMMOR | 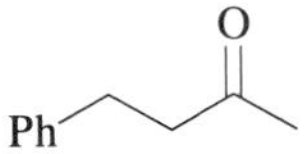 | 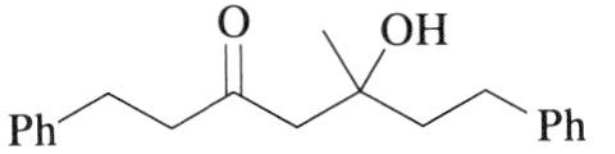 (76) 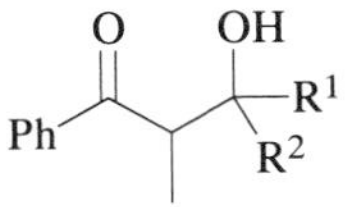 | 52 |
| | Sn(OTf)$_2$, NEPIP | 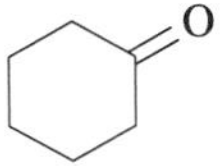 | R$^1$R$^2$ = (CH$_2$)$_5$, (87) | 52 |
| | " | 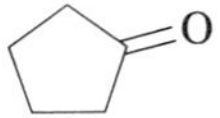 | R$^1$R$^2$ = (CH$_2$)$_4$, (76) | 52 |
| | " | PhCOMe | R$^1$ = Ph, R$^2$ = Me, (41) 0/100 | 52 |

TABLE I. ALDOL REACTION WITH ACHIRAL SUBSTRATES; NO OR SIMPLE DIASTEREOSELECTION (*Continued*)

| Starting Material | Enolization Reagent | Acceptor | Product(s) and Yield(s) (%), *syn/anti* | Refs. |
|---|---|---|---|---|
| (structure: EtO–C(=O)–CH(Br)–Ph) | Sn(OTf)$_2$, NEPIP | 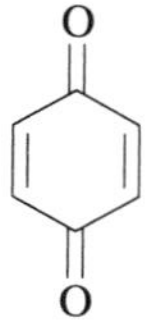 (structure: benzoquinone), MeHSiCl$_2$ | 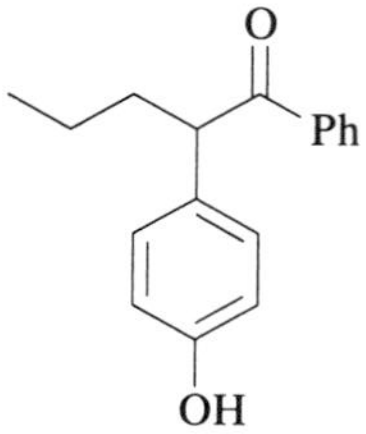 (structure with Ph, O, OH) (83) | 110, 111 |
| | Sn | *p*-ClC$_6$H$_4$CHO | (structure: EtO–C(=O), O, OH, R, Ph) R = *p*-ClC$_6$H$_4$, (88) 80/20 | 25 |
| | " | *p*-MeC$_6$H$_4$CHO | R = *p*-MeC$_6$H$_4$, (93) 77/23 | 25 |
| | " | 1-C$_{10}$H$_7$CHO | R = 1-C$_{10}$H$_7$, (93) 77/23 | 25 |
| | " | Ph(CH$_2$)$_2$CHO | R = Ph(CH$_2$)$_2$, (91) 71/29 | 25 |
| | " | PhCH=CHCHO | R = PhCH=CH, (83) 79/21 | 25 |
| | Sn | BnCOBn | (structure: Ph–C(=O), O, OH, R$^1$, R$^2$) R$^1$ = R$^2$ = Bn, (80) | 25 |
| | " | (structure: cyclohexanone) | R$^1$R$^2$ = (CH$_2$)$_5$, (96) | 25 |

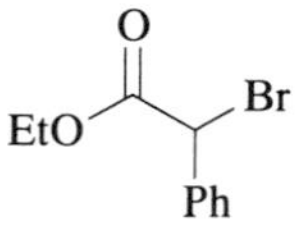

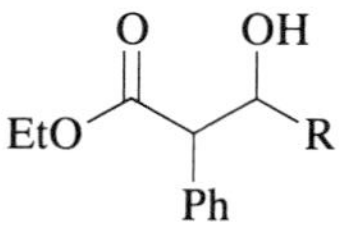

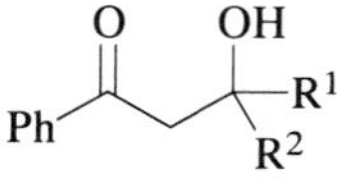

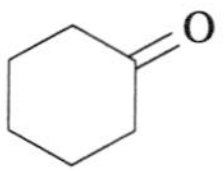

64

TABLE I. ALDOL REACTION WITH ACHIRAL SUBSTRATES; NO OR SIMPLE DIASTEREOSELECTION (*Continued*)

| Starting Material | Enolization Reagent | Acceptor | Product(s) and Yield(s) (%), *syn/anti* | Refs. |
|---|---|---|---|---|
| 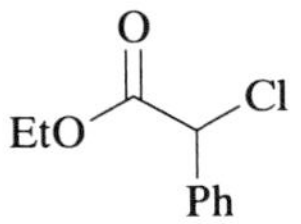 EtO–CO–CHCl–Ph | Sn | PhCHO | 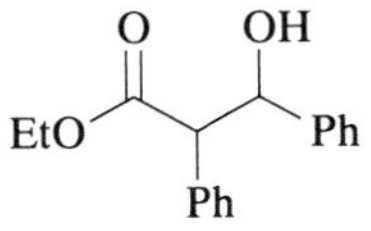 EtO–CO–CH(Ph)–CH(OH)–Ph  (88) 59/41 | 25 |
| Ph–CO–CHBr–CH$_2$CH$_3$ | Sn | PhCHO | Ph–CO–CH(CH$_2$CH$_3$)–CH(OH)–Ph  (96) | 42 |
| Ph–CO–C(CH$_3$)$_2$–Br | Sn | PhCHO | Ph–CO–C(CH$_3$)$_2$–CH(OH)–R,  R = Ph, (92) | 42 |
|  | " | Ph(CH$_2$)$_2$CHO |  R = Ph(CH$_2$)$_2$, (94) | 42 |
|  | " | PhCH=CHCHO |  R = PhCH=CH, (91) | 42 |
| 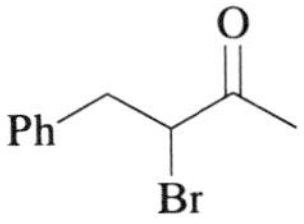 Ph–CO–CH$_2$–CHBr–CH$_3$ | Sn | PhCHO | 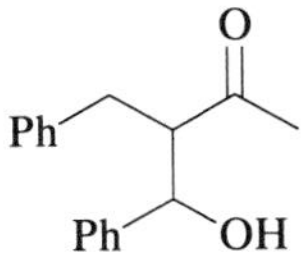 Ph–CH$_2$–CH(C(O)CH$_3$)–CH(OH)–Ph  (82) | 42 |
| C$_{11}$ 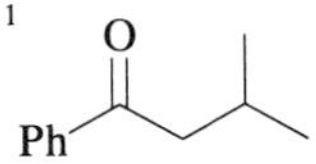  Ph–CO–CH$_2$–CH(CH$_3$)$_2$ | Sn(OTf)$_2$, NEPIP | 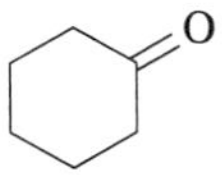 cyclohexanone | 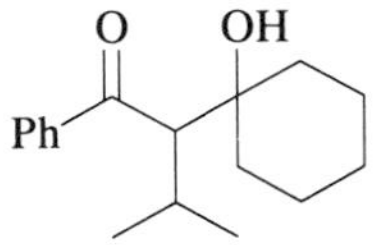 Ph–CO–CH(iPr)–C(OH)(cyclohexyl)  (48) | 52 |

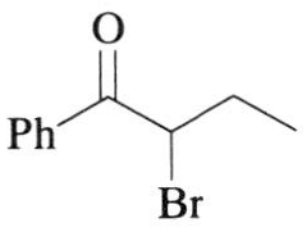

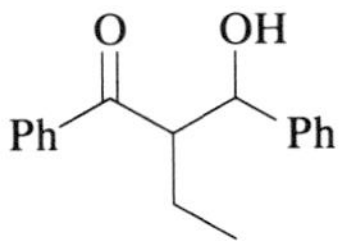

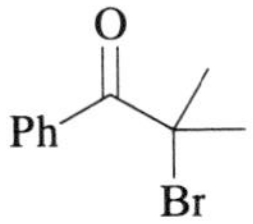

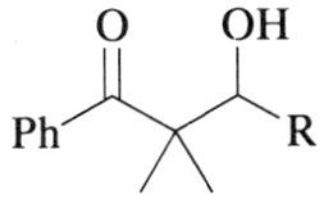

TABLE I. ALDOL REACTION WITH ACHIRAL SUBSTRATES; NO OR SIMPLE DIASTEREOSELECTION (*Continued*)

| Starting Material | Enolization Reagent | Acceptor | Product(s) and Yield(s) (%), *syn/anti* | Refs. |
|---|---|---|---|---|
| C$_{12}$ | Sn(OTf)$_2$, NEPIP | PhCHO | R = Ph, (88) >97/3 | 36, 37 |
| | " | Ph(CH$_2$)$_2$CHO | R = Ph(CH$_2$)$_2$, (95) >97/3 | 36, 37 |
| | | | | |
| 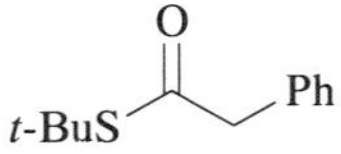 | LDA, SnCl$_2$ | 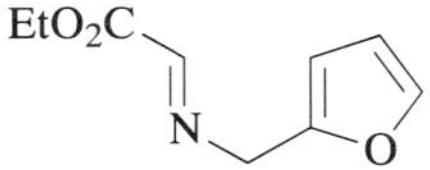 | 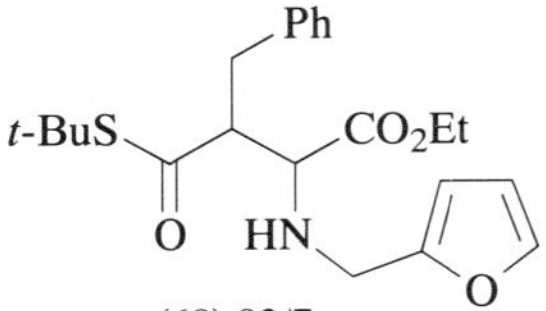 (68) 93/7 | 44 |
| 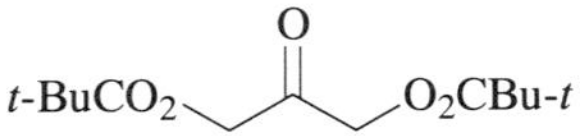 | Sn(OTf)$_2$, NEPIP | 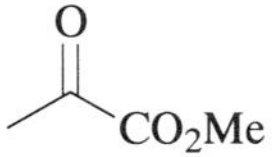 | 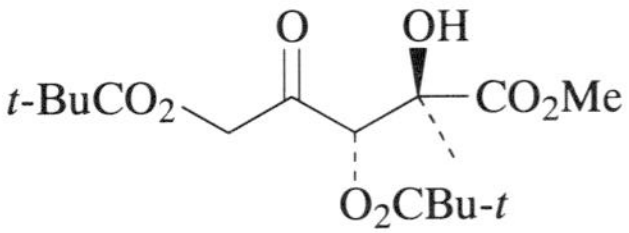 0°, (68) 64/36 −78°, (55) 82/18 | 58 |
| C$_{14}$ 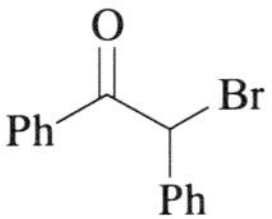 | Sn | PhCHO | 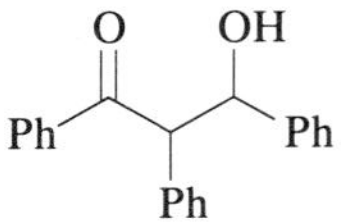 (82) | 42 |

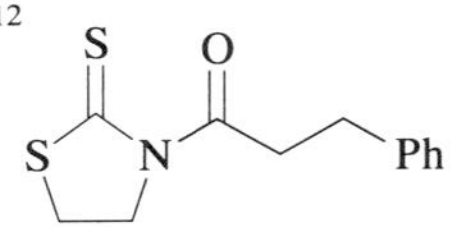

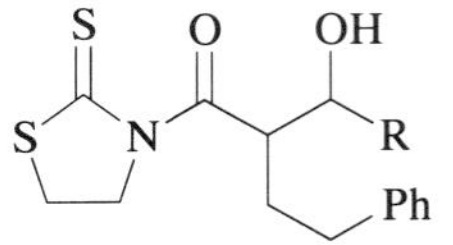

TABLE I. ALDOL REACTION WITH ACHIRAL SUBSTRATES; NO OR SIMPLE DIASTEREOSELECTION (*Continued*)

| Starting Material | Enolization Reagent | Acceptor | Product(s) and Yield(s) (%), *syn/anti* | Refs. |
|---|---|---|---|---|
| C₁₅ | Sn(OTf)₂, NEPIP | 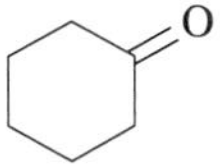 | <br>R¹R² = (CH₂)₅, (86) | 52 |
| | " | | R¹R² = (CH₂)₄, (78) | 52 |
| | " | PhCOMe | R¹ = Ph, R² = Me, 0°, (85) 30/70<br>-78°, (48) >95/5 | 52 |
| | Sn(OTf)₂, NEPIP | 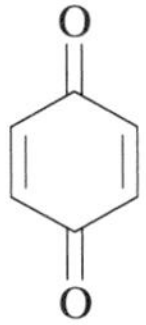 , MeHSiCl₂ | 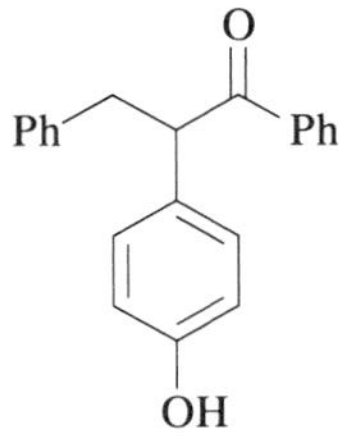 <br>(66) | 110, 111 |
| 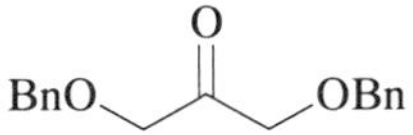 C₁₇ | Sn(OTf)₂, NEPIP | 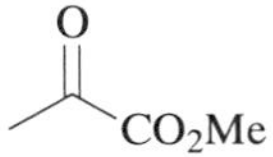 | 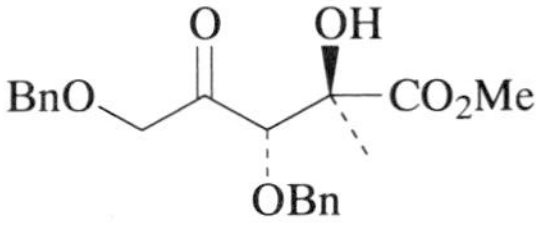 <br>(63) 74/26 | 58 |

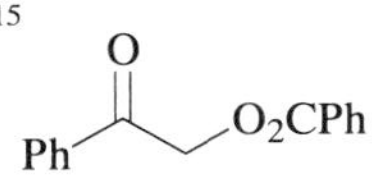

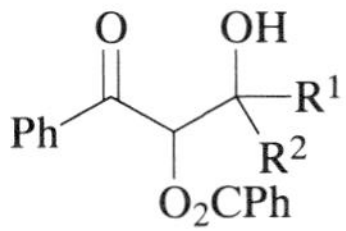

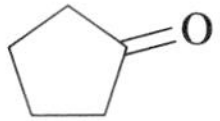

67

TABLE I. ALDOL REACTION WITH ACHIRAL SUBSTRATES; NO OR SIMPLE DIASTEREOSELECTION (*Continued*)

| Starting Material | Enolization Reagent | Acceptor | Product(s) and Yield(s) (%), *syn/anti* | Refs. |
|---|---|---|---|---|
| 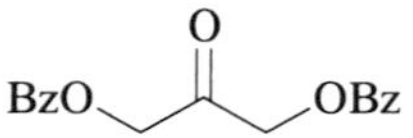 | Sn(OTf)₂, NEPIP | 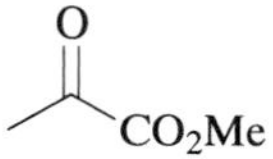 | 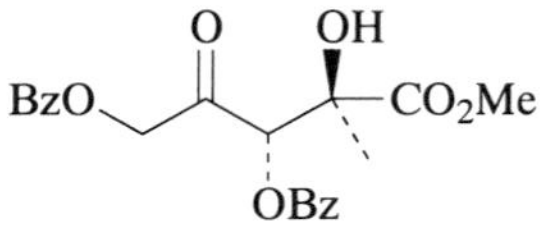 <br>(93) 85/15 | 58 |
| | Sn(OTf)₂, NEPIP | PhCHO | <br>R = Ph, (94) 90/10 | 74 |
| | " | Ph(CH₂)₂CHO | R = Ph(CH₂)₂, (96) 95/5 | 74 |
| | " | n-C₅H₁₁CHO | R = n-C₅H₁₁, (98) 95/5 | 74 |
| | " | i-PrCHO | R = i-Pr, (98) 95/5 | 74 |
| | " | MeCHO | R = Me, (99) 95/5 | 74 |

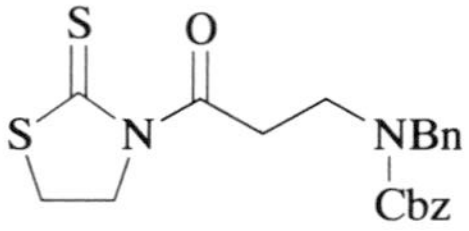

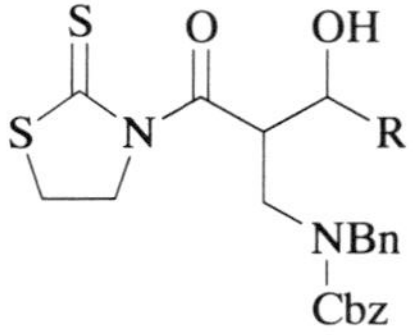

68

TABLE I. ALDOL REACTION WITH ACHIRAL SUBSTRATES; NO OR SIMPLE DIASTEREOSELECTION (*Continued*)

| Starting Material | Enolization Reagent | Acceptor | Product(s) and Yield(s) (%), *syn/anti* | Refs. |
|---|---|---|---|---|
| C$_{21}$ | Sn(OTf)$_2$, NEPIP | PhCHO | R = Ph, (83) 90/10 | 74 |
| | " | Ph(CH$_2$)$_2$CHO | R = Ph(CH$_2$)$_2$, (99) 95/5 | 74 |
| | " | *n*-C$_5$H$_{11}$CHO | R = *n*-C$_5$H$_{11}$, (99) 95/5 | 74 |
| | " | MeCHO | R = Me, (97) 95/5 | 74 |

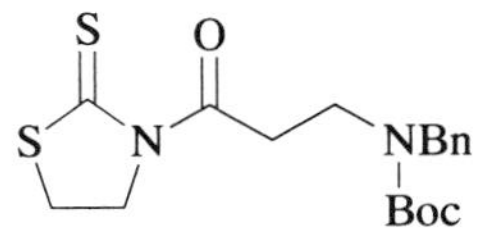

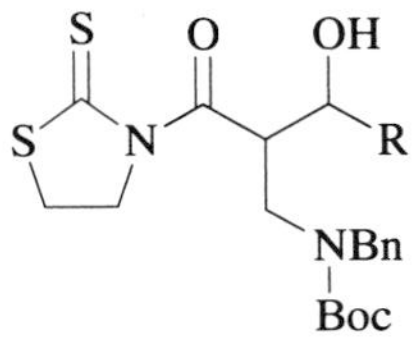

69

## TABLE II. MICHAEL REACTION WITH ACHIRAL SUBSTRATES; NO OR SIMPLE DIASTEREOSELECTION

| Starting Material | Enolization Reagent | Acceptor | Product(s) and Yield(s) (%), syn/anti | Refs. |
|---|---|---|---|---|
| C5 | | | | |
| 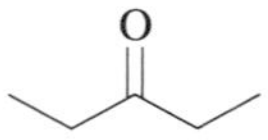 (3-pentanone) | Sn(OTf)$_2$, NEPIP | 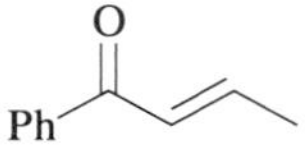 (Ph-CO-CH=CH-CH$_3$) (Me$_2$HSiCl) | 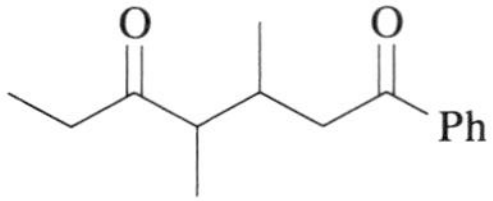 (72) | 110, 111 |
| 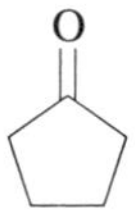 (cyclopentanone) | Sn(OTf)$_2$, NEPIP | 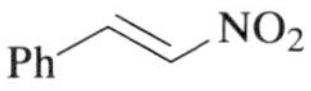 (Ph-CH=CH-NO$_2$) | 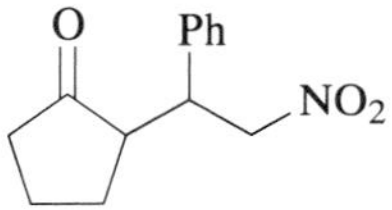 (82) 30/70 | 154 |
| C6 | | | | |
| 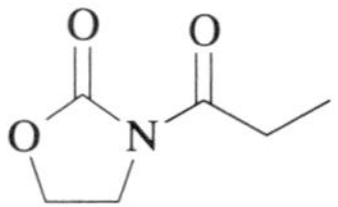 (N-propionyl oxazolidinone) | Sn(OTf)$_2$, NEPIP | 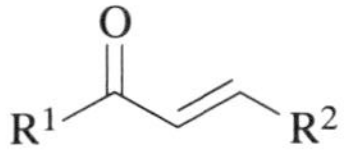 (R$^1$-CO-CH=CH-R$^2$) | 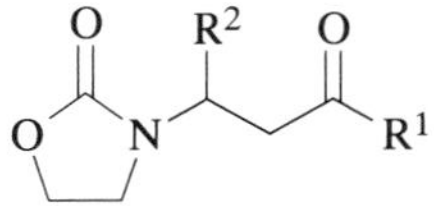 | 111, 153 |
| | | R$^1$ = Me, R$^2$ = Ph, Me$_2$HSiCl | (77) 92/8 | |
| | | Me$_3$SiCl | (78) 29/71 | |
| | | R$^1$ = Me, R$^2$ = Me, Me$_2$HSiCl | (56) 5/95 | |
| | | Me$_3$SiCl | (56) 29/71 | |
| | | R$^1$ = Ph, R$^2$ = Me, Me$_2$HSiCl | (85) 25/75 | |
| | | R$^1$R$^2$ = (CH$_2$)$_3$ Me$_3$SiCl | (84) 38/62 | |
| 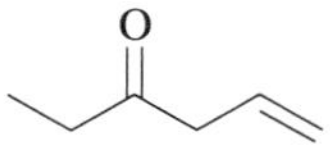 (hex-5-en-3-one) | Sn(OTf)$_2$, NEPIP | 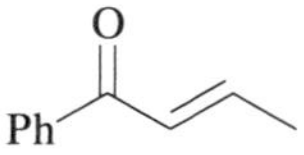 (Ph-CO-CH=CH-CH$_3$) | 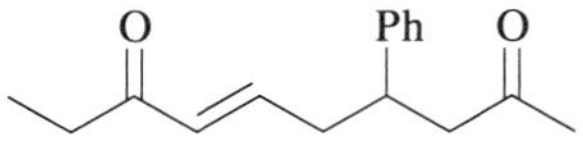 −45°, (60) | 41 |

TABLE II. MICHAEL REACTION WITH ACHIRAL SUBSTRATES; NO OR SIMPLE DIASTEREOSELECTION (*Continued*)

| Starting Material | Enolization Reagent | Acceptor | Product(s) and Yield(s) (%), *syn/anti* | Refs. |
|---|---|---|---|---|
| | Sn(OTf)$_2$, NEPIP | 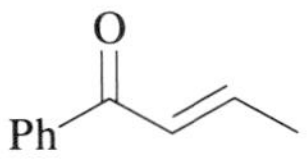 | 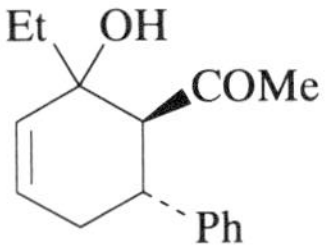 −45 to 0°, (52) | 41 |
| | Sn(OTf)$_2$, NEPIP | 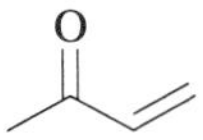 | 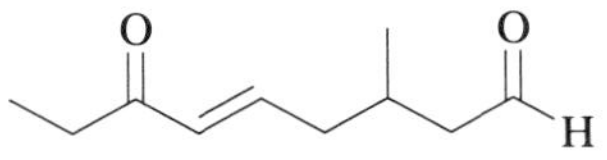 −45° (38) | 41 |
| 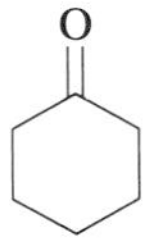 | Sn(OTf)$_2$, NEPIP | 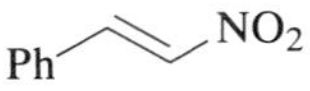 | 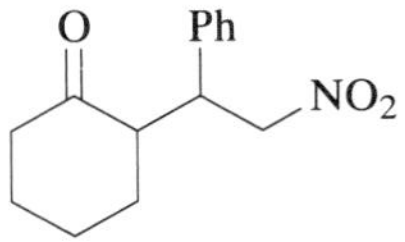 (64) <7/93 | 154 |
| 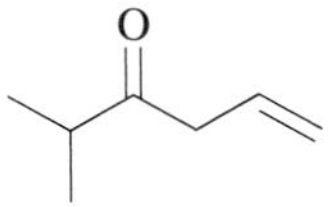 | Sn(OTf)$_2$, NEPIP | 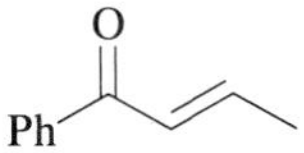 | 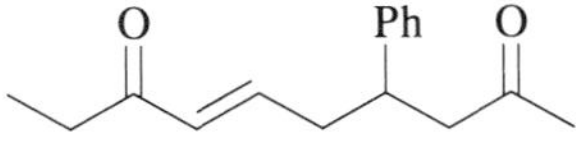 −45°, (72) | 41 |
| | Sn(OTf)$_2$, NEPIP | 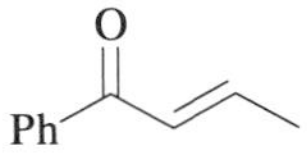 | 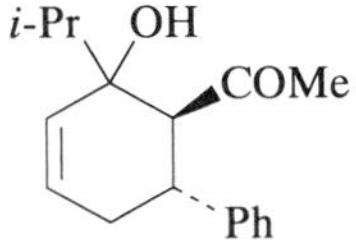 −45 to 0°, (66) | 41 |
| | Sn(OTf)$_2$, NEPIP | 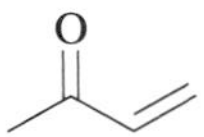 | 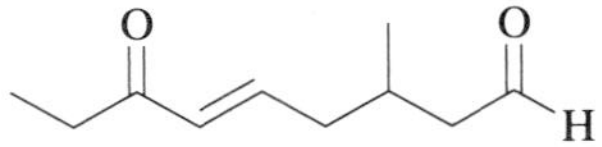 −45° (55) | 41 |

C$_7$

TABLE II. MICHAEL REACTION WITH ACHIRAL SUBSTRATES; NO OR SIMPLE DIASTEREOSELECTION (*Continued*)

| Starting Material | Enolization Reagent | Acceptor | Product(s) and Yield(s) (%), syn/anti | Refs. |
|---|---|---|---|---|
| C$_8$ 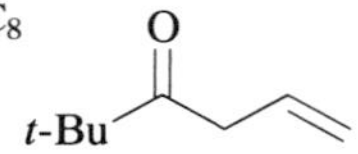 | Sn(OTf)$_2$, NEPIP | 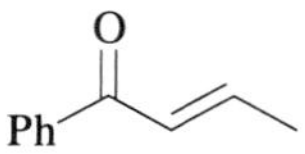 | 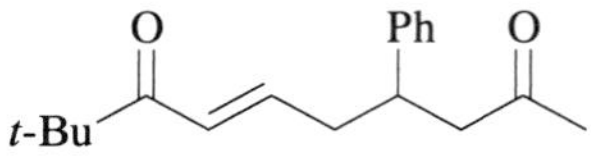 -45°, (83) | 41 |
| | Sn(OTf)$_2$, NEPIP | 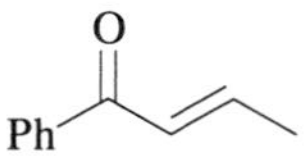 | 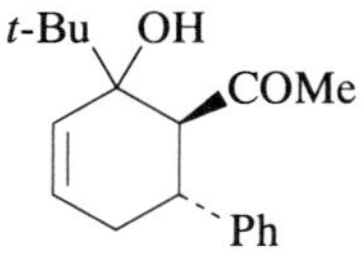 -45 to 0°, (73) | 41 |
| | Sn(OTf)$_2$, NEPIP | 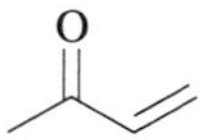 | 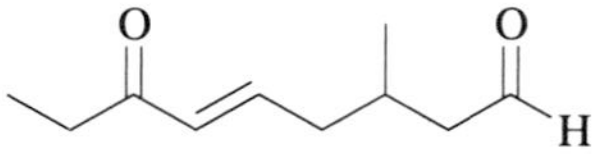 -45° (75) | 41 |
| | Sn(OTf)$_2$, NEPIP | 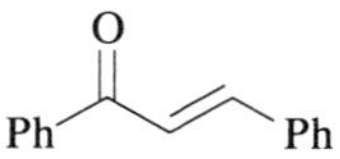 | 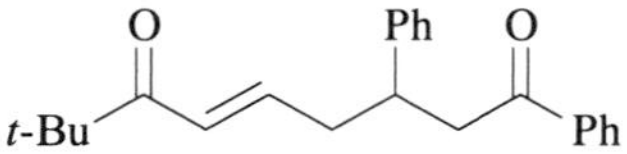 -45°, (81) | 41 |
| | Sn(OTf)$_2$, NEPIP | 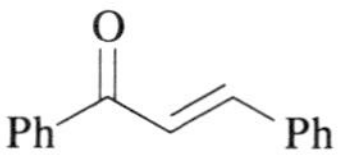 | 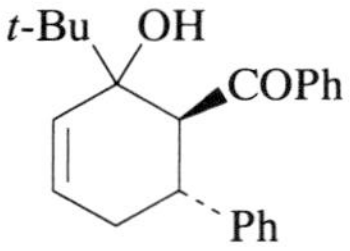 -45 to 0°, (70) | 41 |

72

| Starting Material | Enolization Reagent | Acceptor | Product(s) and Yield(s) (%), *syn/anti* | Refs. |
|---|---|---|---|---|
| C$_9$ | | | | |
| (propiophenone, Ph–C(=O)–) | Sn(OTf)$_2$, NEPIP | 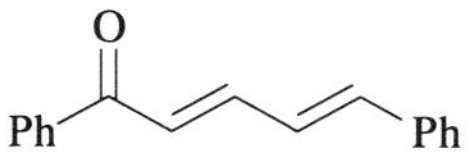 (t-Bu, Ph dienone) | 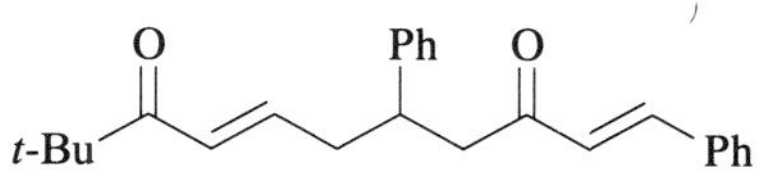 –45°, (71) | 41 |
| | Sn(OTf)$_2$, NEPIP | 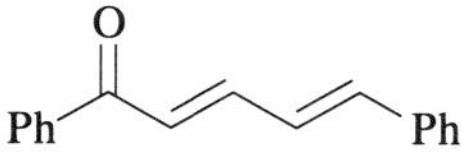 (Ph, Ph dienone) | 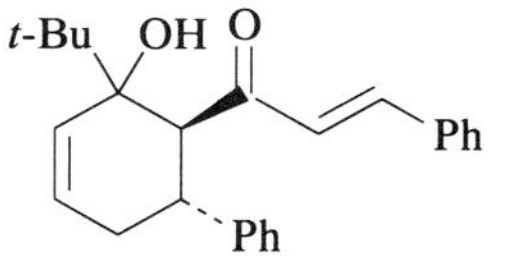 –45 to 0°, (72) | 41 |
| 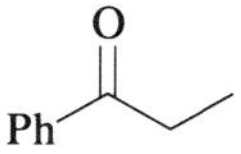 | Sn(OTf)$_2$, NEPIP | 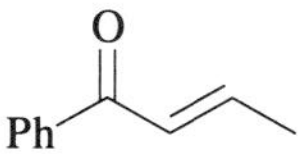 (Ph enone), TMSCl | 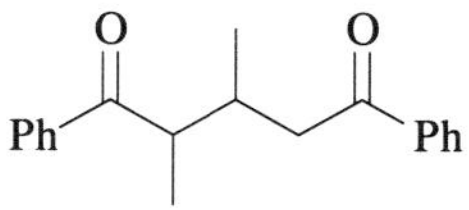 (54) | 111, 153 |
| | Sn(OTf)$_2$, NEPIP | 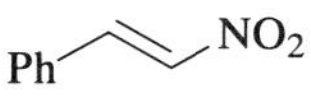 (Ph–NO$_2$) | 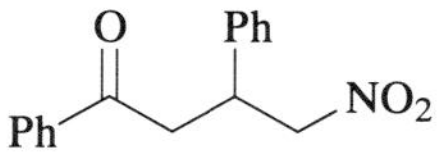 (67) 24/76 | 154 |
| C$_{10}$ | | | | |
| (4-t-Bu-cyclohexanone) | Sn(OTf)$_2$, NEPIP | (Ph–NO$_2$) | (76) 38/62 | 154 |

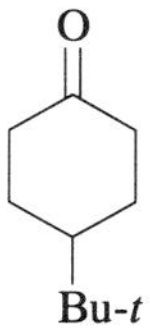

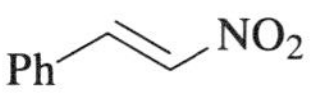

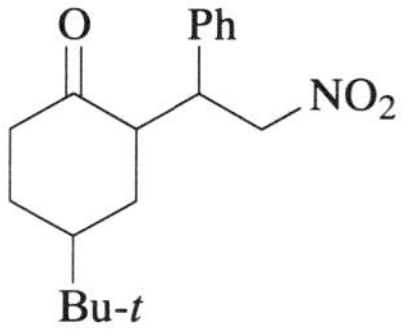

# TABLE III. ENANTIOSELECTIVE ALDOL REACTION

| Starting Material | Acceptor | Chiral Ligand | Product(s) and Yield(s) (%), *syn/anti* | Refs. |
|---|---|---|---|---|
| C₄ (MeO–C(=S)–Et) | PhCHO | (7) piperidine–pyrrolidine ligand, Me | 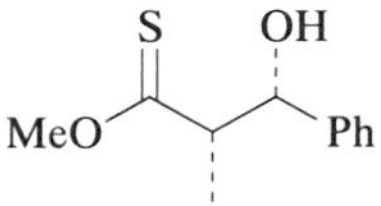 (73) 78/22, 90% ee | 78 |
| C₅ (3-pentanone, O=C) | PhCHO | 7 | R = Ph, (54) 75/25, 45% ee | 36 |
| | PhCHO | 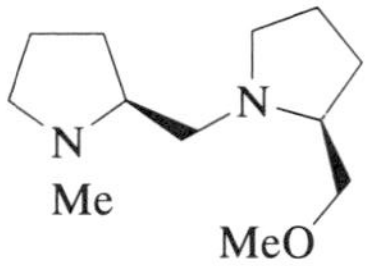 (11) MeO-substituted pyrrolidine ligand, Me | R = Ph, (60) 75/25, 55% ee | 36 |
| | p-MeOC₆H₄CHO | 7 | R = p-MeOC₆H₄, (69) 89/11, 50% ee | 36 |
| (thiazolidinethione acetyl) | PhCHO | 7 | R = Ph, (79), 65% ee | 36, 165 |
| | i-PrCHO | 7 | R = i-Pr, (63), > 90% ee | 36, 165 |
| | Ph(CH₂)₂CHO | 7 | R = Ph(CH₂)₂, (76), 90% ee | 36, 165 |

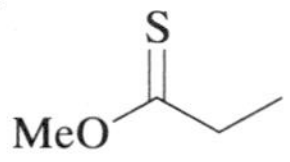

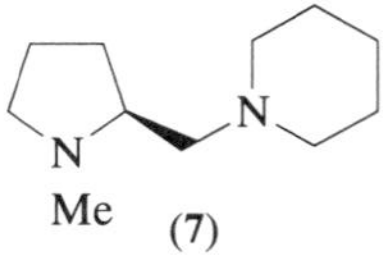

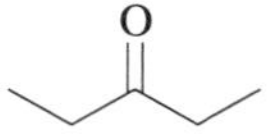

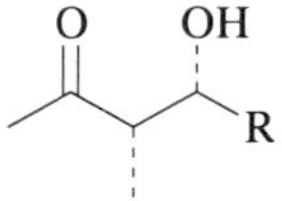

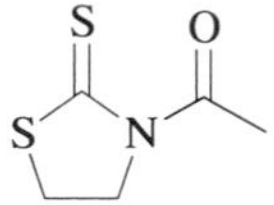

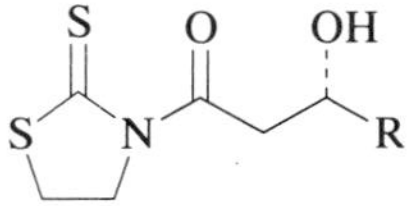

74

TABLE III. ENANTIOSELECTIVE ALDOL REACTION (*Continued*)

| Starting Material | Acceptor | Chiral Ligand | Product(s) and Yield(s) (%), *syn/anti* | Refs. |
|---|---|---|---|---|
| $C_6H_{11}CHO$ | | 7 | R = $C_6H_{11}$, (81), 88% ee | 36, 165 |
| EtCHO | | 7 | R = Et, (70), 90% ee | 36, 165 |
| $n$-$C_5H_{11}CHO$ | | 7 | R = $n$-$C_5H_{11}$, (65), 90% ee | 36, 165 |

| Starting Material | Acceptor | Chiral Ligand | 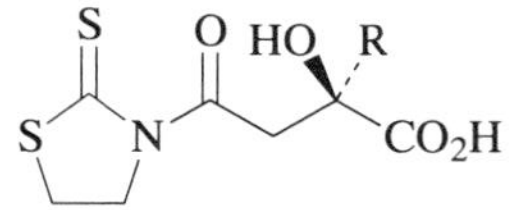 Product(s) and Yield(s) (%), *syn/anti* | Refs. |
|---|---|---|---|---|
| | 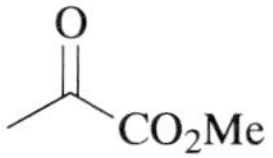 | 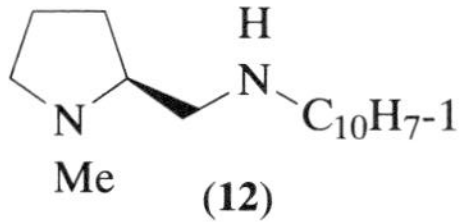 **12** | R = Me, (74), 85% ee | 166 |
| | 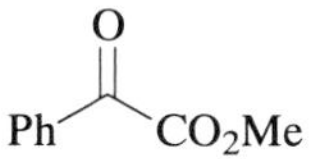 Ph—CO$_2$Me | **12** | R = Ph, (78), >95% ee | 166 |
| | 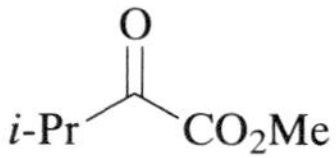 $i$-Pr—CO$_2$Me | **12** | R = $i$-Pr, (75), >95% ee | 166 |
| | 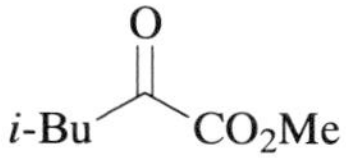 $i$-Bu—CO$_2$Me | **12** | R = $i$-Bu, (65), >95% ee | 166 |
| | 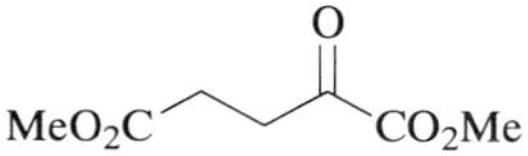 MeO$_2$C—CO$_2$Me | **12** | R = MeO$_2$C(CH$_2$)$_2$, (80), >95% ee | 166 |

Chiral Ligand **(12)**

75

TABLE III. ENANTIOSELECTIVE ALDOL REACTION (*Continued*)

| Starting Material | Acceptor | Chiral Ligand | Product(s) and Yield(s) (%), *syn/anti* | Refs. |
|---|---|---|---|---|
| 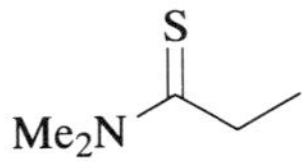 (S, Me₂N) | PhCHO | 7 | 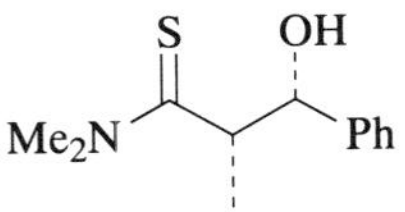 (93), 92/8, 85% ee | 78 |
| (O, t-Bu) | PhCHO | 7 | R = Ph, (22), >95/5, 53% ee | 36 |
| | PhCHO | 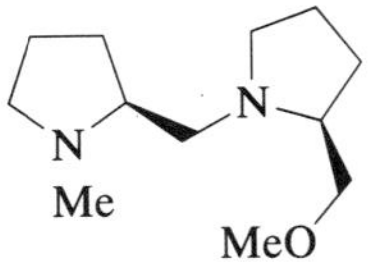 (11) | R = Ph, (29), >95/5, 80% ee | 36 |
| | Ph(CH₂)₂CHO | 7 | R = Ph(CH₂)₂, (35), >95/5, 53% ee | 36 |
| | Ph(CH₂)₂CHO | 11 | R = Ph(CH₂)₂, (40), >95/5, 70% ee | 36 |
| | p-ClC₆H₄CHO | 11 | R = p-ClC₆H₄, (30), >95/5, 80% ee | 36 |
| | p-MeOC₆H₄CHO | 11 | R = p-MeOC₆H₄, (24), >95/5, 77% ee | 36 |
| 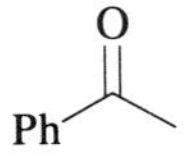 C₈ (O, Ph) | PhCHO | 7 | 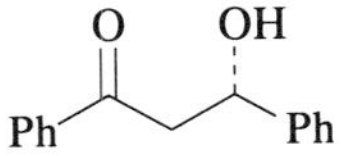 (35), 75% ee | 36, 164 |

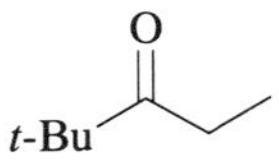

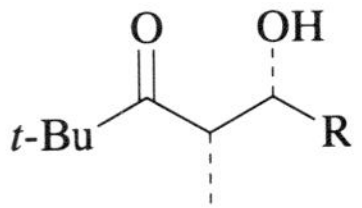

TABLE III. ENANTIOSELECTIVE ALDOL REACTION (*Continued*)

| Starting Material | Acceptor | Chiral Ligand | Product(s) and Yield(s) (%), *syn/anti* | Refs. |
|---|---|---|---|---|
| $C_9$ 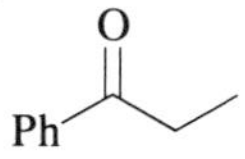 <br><br> Ph–C(=O)–CH$_2$CH$_3$ | | | 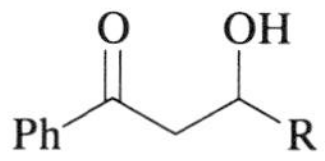 | |
| | PhCHO | 7 | R = Ph, (74), 85/15, 80% ee | 36, 164 |
| | p-MeC$_6$H$_4$CHO | 7 | R = p-MeC$_6$H$_4$, (72), 89/11, 80% ee | 36, 164 |
| | p-ClC$_6$H$_4$CHO | 7 | R = p-ClC$_6$H$_4$, (72), 85/15, 85% ee | 36, 164 |
| | p-MeOC$_6$H$_4$CHO | 7 | R = p-MeOC$_6$H$_4$, (78), 89/11, 80% ee | 36, 164 |
| | i-PrCHO | 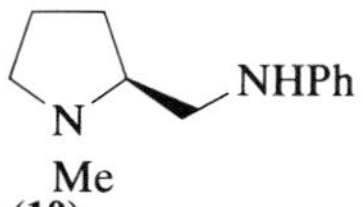 **NHPh** <br> **(10)** | R = i-Pr, (69), >95/5, 75% ee | 36, 164 |
| | t-BuCHO | 10 | R = t-Bu, (57), 100/0, 90% ee | 36, 164 |
| | C$_6$H$_{11}$CHO | 10 | R = C$_6$H$_{11}$, (24), >95/5, 77% ee | 36, 164 |
| $C_{10}$ 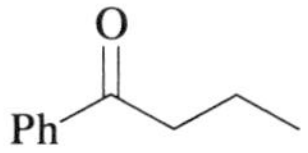 | PhCHO | 7 | 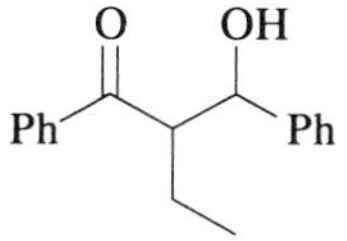 (72), 83/17, 75% ee | 36, 164 |

77

## TABLE IV. ENANTIOSELECTIVE MICHAEL REACTION[a]

| Starting Material | Acceptor | Chiral Ligand | Product(s) and Yield(s) (%), syn/anti | Refs. |
|---|---|---|---|---|
| C₃ — MeS–C(=S)–Me 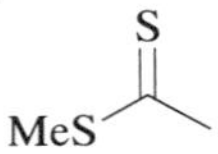 | 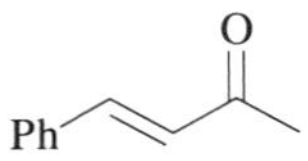 Ph–CH=CH–C(=O)–Me | 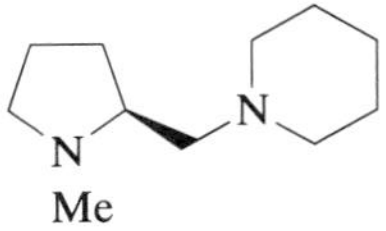 (7) | 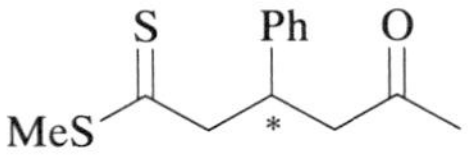 MeS–C(=S)–CH₂–CH(Ph)*–CH₂–C(=O)–Me (73), 26% ee | 167 |
| | 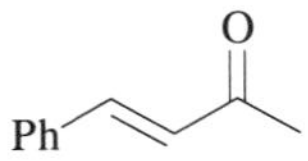 Ph–CH=CH–C(=O)–Me | 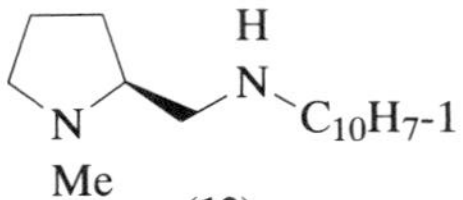 (12) | " (82), 70% ee | 167 |
| | 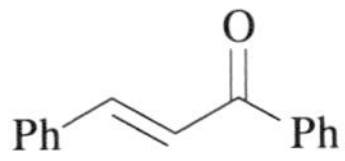 Ph–CH=CH–C(=O)–Ph | 12 | 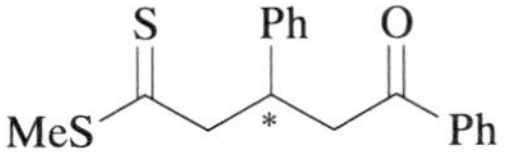 MeS–C(=S)–CH₂–CH(Ph)*–CH₂–C(=O)–Ph (75), 40% ee | 167 |
| | 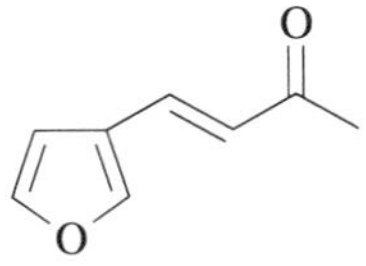 (furyl)–CH=CH–C(=O)–Me | 12 | 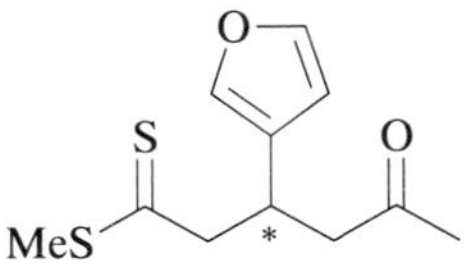 MeS–C(=S)–CH₂–CH(furyl)*–CH₂–C(=O)–Me (79), 60% ee | 167 |
| | 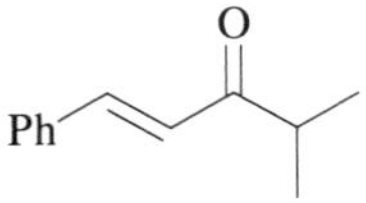 Ph–CH=CH–C(=O)–CH(Me)₂ | 12 | 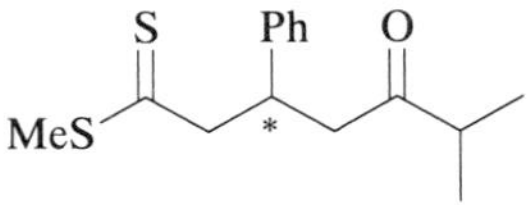 MeS–C(=S)–CH₂–CH(Ph)*–CH₂–C(=O)–CH(Me)₂ (79), 60% ee | 167 |
| | 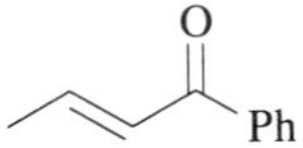 Me–CH=CH–C(=O)–Ph | 12 | 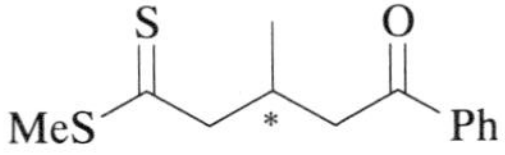 MeS–C(=S)–CH₂–CH(Me)*–CH₂–C(=O)–Ph (62), 15% ee | 167 |
| | 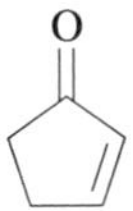 2-cyclopentenone | 12 | 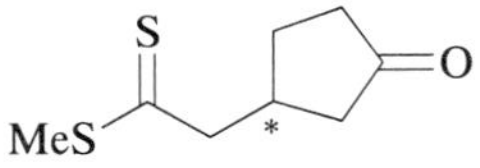 MeS–C(=S)–CH₂–(3-cyclopentanon-1-yl)* (44), 30% ee | 167 |

Chiral ligands:

(7) = 1-[(1-methylpyrrolidin-2-yl)methyl]piperidine

(12) = N-(1-naphthyl)-1-methylpyrrolidine-2-methanamine

TABLE IV. ENANTIOSELECTIVE MICHAEL REACTION (Continued)

| Starting Material | Acceptor | Chiral Ligand | Product(s) and Yield(s) (%), syn/anti | Refs. |
|---|---|---|---|---|
| C₆ | | | | |
| | | **7** | (70), >95/5, 80% ee (syn) | 167 |
| | | **12** | " (72), >95/5, 93% ee (syn) | 167 |
| STMS | | Sn(OTf)₂ (10 mol%) + **12** | (80), 70% ee | 78,168 |
| | | Sn(OTf)₂ (10 mol%) + **12** | (82), 60% ee | 78,168 |
| | | Sn(OTf)₂ (10 mol%) + **12** | (79), 40% ee | 78,168 |

<sup>a</sup> The absolute configurations at * were not determined.

79

TABLE V. **E**NANTIOSELECTIVE **S**ULFENYLATION[a]

| Starting Material | Acceptor[b] | Product(s) and Yield(s) (%) | Refs. |
|---|---|---|---|
| C$_6$ | PhS—S—C$_{10}$H$_7$-1 (with O, O) | (72), 50% ee | 169 |
| | PhS—S—C$_{10}$H$_7$-1 (with O, O) | (93), 81% ee | 169 |
| C$_7$ t-Bu | PhS—S—C$_{10}$H$_7$-1 (with O, O) | t-Bu (52), 70% ee | 169 |
| C$_9$ Ph | PhS—S—C$_{10}$H$_7$-1 (with O, O) | Ph (78), 85% ee | 169 |
|  | PhSCl | " (63), 54% ee | 169 |
|  | | " (27), 60% ee | 169 |
|  | PhS—S—Ph (with O, O) | " (58), 75% ee | 169 |
| C$_{10}$ Ph | PhS—S—Ph (with O, O) | Ph (80), 75% ee | 169 |
| C$_{12}$ Ph | PhS—S—Ph (with O, O) | (91), 82% ee | 169 |

[a] The absolute configurations at * were not determined.
[b] All reactions were carried out with Sn(OTf)$_2$, NEPIP, and chiral diamine **7**.

TABLE VI. DIASTEREOSELECTIVE ALDOL REACTION

| Starting Material | Enolization Reagent | Acceptor | Product(s) and Yield(s) (%), diastereomer ratio | Refs. |
|---|---|---|---|---|
| 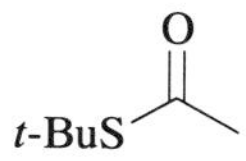 C₆ | LDA, SnCl₂ | 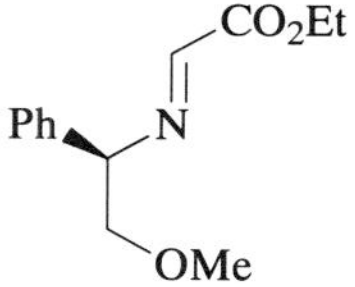 | 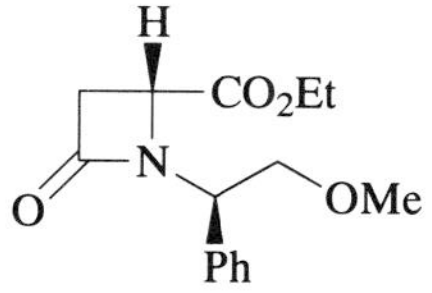 (52), 92/8 | 174 |
| 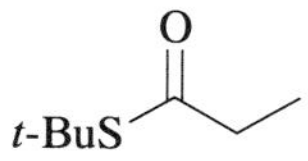 C₇ | LDA, SnCl₂ | 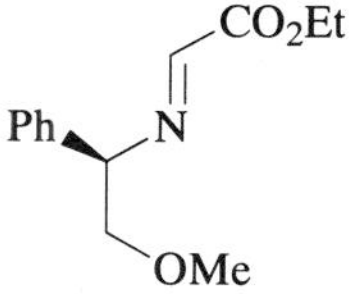 | 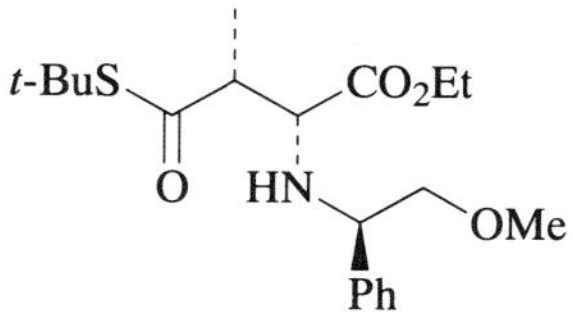 (85), 75/25 | 174 |
|  | LDA, Sn(OTf)₂ | " | " (81), 67/33 | 174 |
|  | LDA, SnCl₂ | 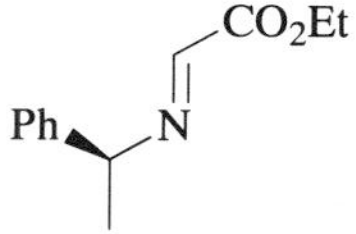 | 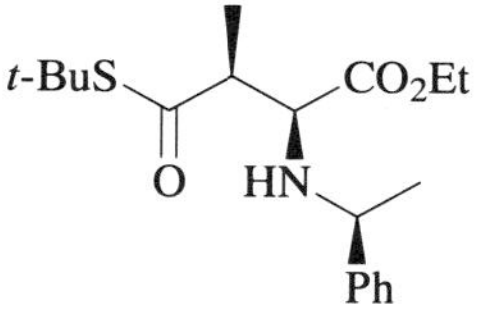 (78), 91/9 | 174 |
| 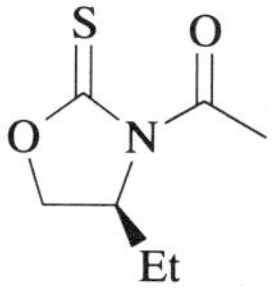 | Sn(OTf)₂, NEPIP | i-PrCHO | 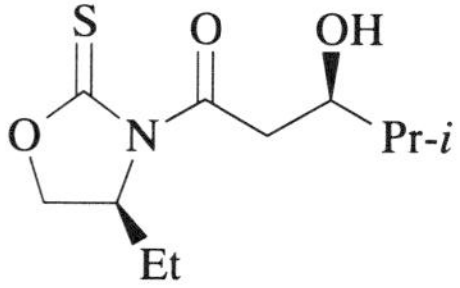 (60), 91/9 | 38 |

81

TABLE VI. DIASTEREOSELECTIVE ALDOL REACTION (*Continued*)

| Starting Material | Enolization Reagent | Acceptor | Product(s) and Yield(s) (%), diastereomer ratio | Refs. |
|---|---|---|---|---|
| C₈ *t*-BuS–C(O)–CH₂CH₂CH₃ | Sn(OTf)₂, NEPIP | Ph–CH=CH–CHO | 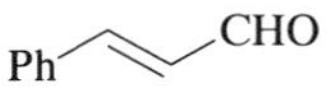 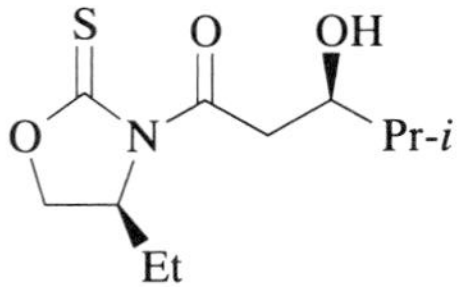 (thiazolidinethione-N-acyl product with OH, Pr-*i*) (77), 93/7 | 38 |
| | Sn(OTf)₂, NEPIP | CH₂=C(CH₃)–CHO | 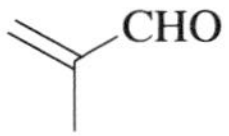 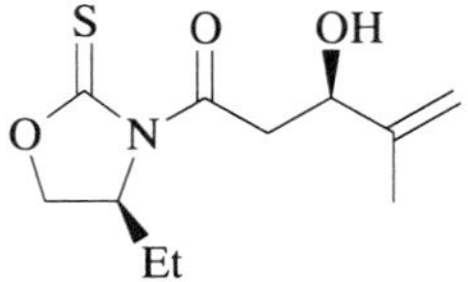 (thiazolidinethione-N-acyl product with OH) (74), 93/7 | 38 |
| | Sn(OTf)₂, NEPIP | (CH₃)₂C=CH–CHO | 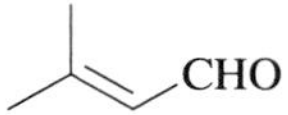 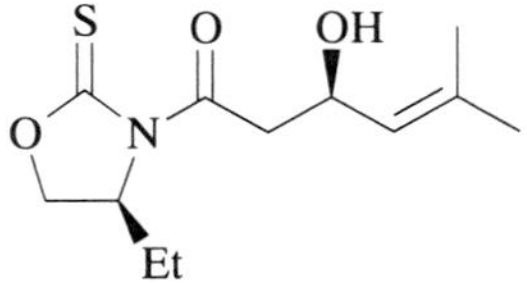 (thiazolidinethione-N-acyl product with OH) (74), 89/11 | 38 |
| 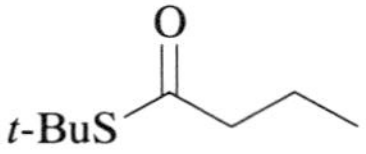 | 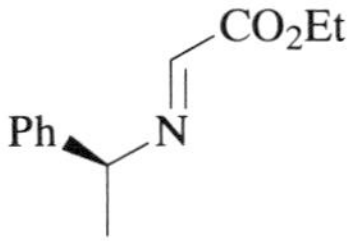 LDA, SnCl₂ | 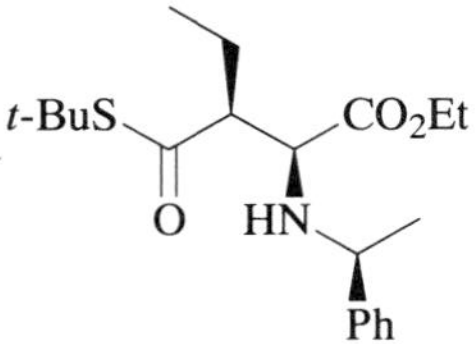 EtO₂C–CH=N–CH(CH₃)Ph | (*t*-BuS–C(O)–CH(Et)–CH(CO₂Et)–NH–CH(CH₃)Ph) (78), 95/5 | 174 |

82

TABLE VI. DIASTEREOSELECTIVE ALDOL REACTION (*Continued*)

| Starting Material | Enolization Reagent | Acceptor | Product(s) and Yield(s) (%), diastereomer ratio | Refs. |
|---|---|---|---|---|
| 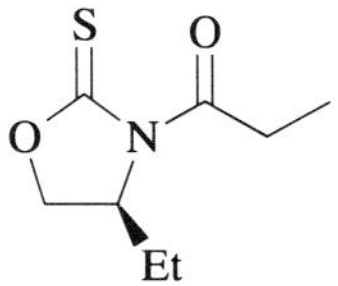 | Sn(OTf)₂, NEPIP | PhCHO | 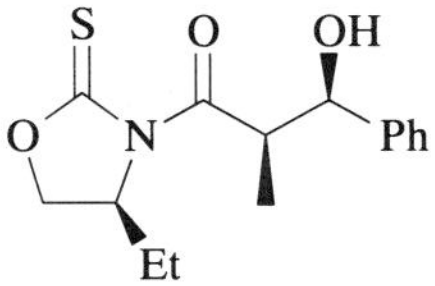 (74), 86/14 | 38 |
| 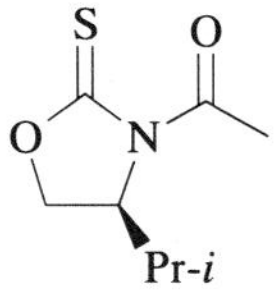 | Sn(OTf)₂, NEPIP | Ph⌒CHO | 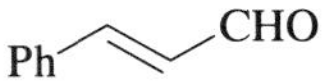 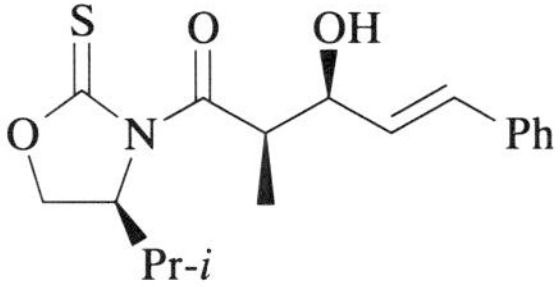 (81), 97/3 | 38 |
| | Sn(OTf)₂, NEPIP | ⌒CHO | 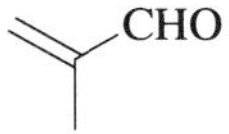 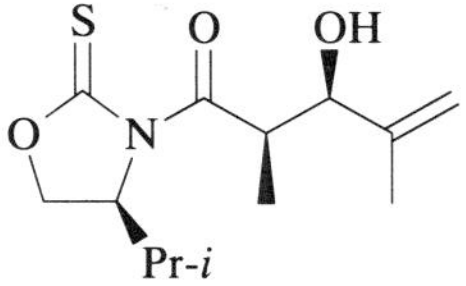 (70), 97/3 | 38 |
| | Sn(OTf)₂, NEPIP | | R¹ = R² = H (86), >99/1 | 177 |
| | | | R¹ = H, R² = Me (85), 99/1 | 177 |
| | | | R¹ = H, R² = Et (90), 99/1 | 177 |

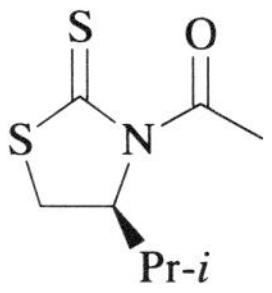

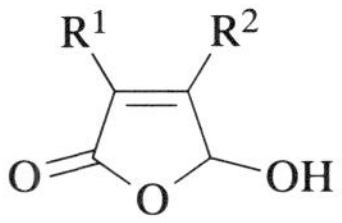

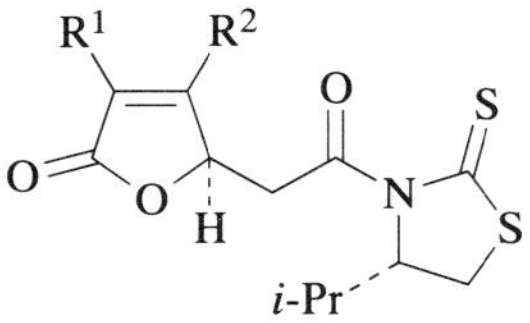

83

TABLE VI. DIASTEREOSELECTIVE ALDOL REACTION (*Continued*)

| Starting Material | Enolization Reagent | Acceptor | Product(s) and Yield(s) (%), diastereomer ratio | Refs. |
|---|---|---|---|---|
| | | | $R^1 = H, R^2 = n\text{-}Pr$ (85), >99/1 | 177 |
| | | | $R^1 = H, R^2 = i\text{-}Pr$ (93), 99/1 | 177 |
| | | | $R^1 = H, R^2 = Ph$ (81), 99/1 | 177 |
| | | | $R^1 = R^2 =$ 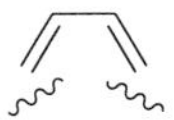 (87), 99/1 | 177 |
| | | | $R^1 = R^2 = (CH_2)_4$ (90), 99/1 | 177 |
| | Sn(OTf)$_2$, NEPIP | $n\text{-}C_5H_{11}CHO$ | R = $n$-C$_5$H$_{11}$, (84), **I/II/III/IV** = 50/25/6/19 | 39 |
| | LDA, SnCl$_2$ | $n\text{-}C_5H_{11}CHO$ | R = $n$-C$_5$H$_{11}$, (84), **I/II/III/IV** = 73/12/6/9 | 39 |
| | Sn(OTf)$_2$, NEPIP | $i\text{-}BuCHO$ | R = $i$-Bu, (68), **I/II/III/IV** = 61/21/9/9 | 39 |
| | LDA, SnCl$_2$ | $i\text{-}BuCHO$ | R = $i$-Bu, (73), **I/II/III/IV** = 78/11/4/7 | 39 |
| | Sn(OTf)$_2$, NEPIP | $i\text{-}PrCHO$ | R = $i$-Pr, (64), **I/II/III/IV** = 50/25/10/15 | 39 |
| | LDA, SnCl$_2$ | $i\text{-}PrCHO$ | R = $i$-Pr, (77), **I/II/III/IV** = 74/18/4/4 | 39 |

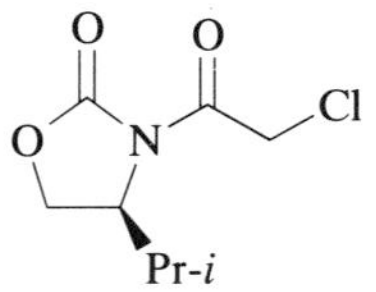

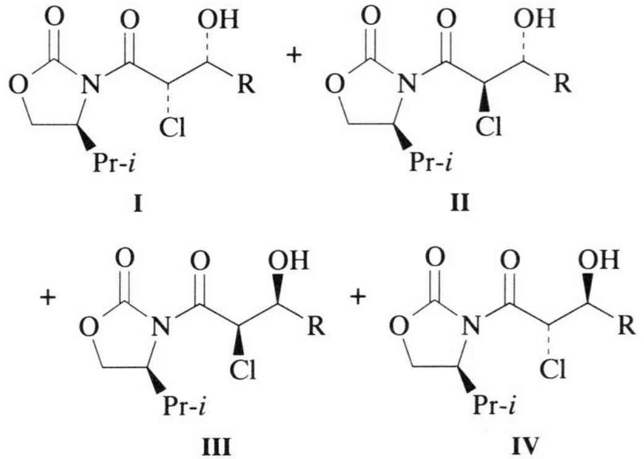

TABLE VI. DIASTEREOSELECTIVE ALDOL REACTION (*Continued*)

| Starting Material | Enolization Reagent | Acceptor | Product(s) and Yield(s) (%), diastereomer ratio | Refs. |
|---|---|---|---|---|
| (oxazolidinone with Pr-*i*, bromoacetyl) | LDA, SnCl₂ | *i*-PrCHO | **I** + **II** + **III** + **IV**; R = *i*-Pr, (65), **I/II/III/IV** = 80/14/3/3 | 39 |
| (oxazolidinone with CO₂Et, propionyl) | Sn(OTf)₂, *i*-Pr₂NEt | PhCHO | (71), 86/14 | 178 |
| C₉ (*t*-BuS isovaleryl) | LDA, SnCl₂ | PhCH(CH₃)N=CHCO₂Et | (79), 95/5 | 174 |

85

TABLE VI. DIASTEREOSELECTIVE ALDOL REACTION (*Continued*)

| Starting Material | Enolization Reagent | Acceptor | Product(s) and Yield(s) (%), diastereomer ratio | Refs. |
|---|---|---|---|---|
| C<sub>10</sub> | SnCl<sub>2</sub>, LAH | PhCHO | R = Ph, (69), 97/3 | 180 |
| | | | R = (76), >99.5/0.5 | 180, 181 |
| C<sub>12</sub> | Sn(OTf)<sub>2</sub>, NEPIP | MeCHO | R = Me, (62), 76/24 | 38 |
| | Sn(OTf)<sub>2</sub>, NEPIP | *i*-PrCHO | R = *i*-Pr, (68), 89/11 | 38 |
| | Sn(OTf)<sub>2</sub>, NEPIP | *i*-BuCHO | R = *i*-Bu, (64), 82/18 | 38 |

TABLE VI. DIASTEREOSELECTIVE ALDOL REACTION (*Continued*)

| Starting Material 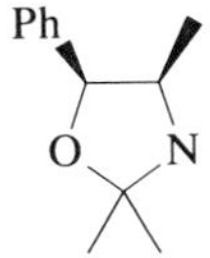 | Enolization Reagent | Acceptor | Product(s) and Yield(s) (%), diastereomer ratio 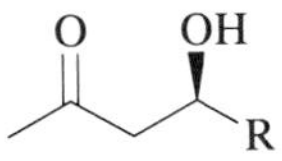 | Refs. |
|---|---|---|---|---|
| | LDA, SnCl₂ | Ph(CH₂)₂CHO | R = Ph(CH₂)₂, (59), 58% ee | 46 |
| | " | *n*-PrCHO | R = *n*-Pr, (60), 58% ee | 46 |
| | " | Et₂CHCHO | R = Et₂CH, (60) , 69% ee | 46 |
| | " | C₆H₁₁CHO | R = C₆H₁₁, (65), 73% ee | 46 |
| | " | *t*-BuCHO | R = *t*-Bu, (65), 86% ee | 46 |
| 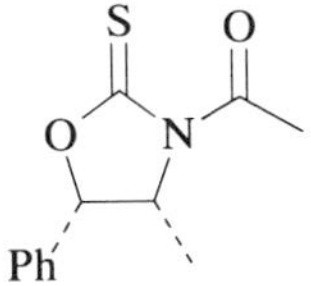 | Sn(OTf)₂, NEPIP | *i*-PrCHO | 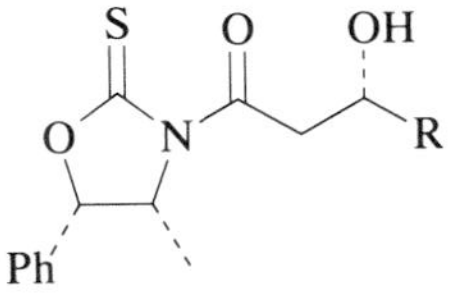 (65), 83.7/16.3 | 38 |
| C₁₃ | Sn(OTf)₂, NEPIP | MeCHO | R = Me, (75), 91/9 | 179 |
| | " | *i*-PrCHO | R = *i*-Pr, (92), 99/1 | 179 |
| | " | PhCHO | R = Ph, (91), 99/1 | 179 |

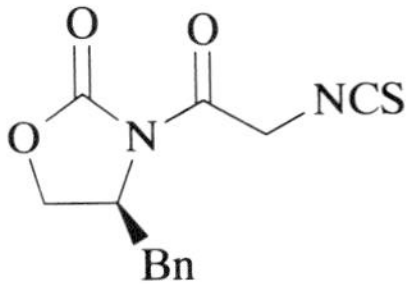

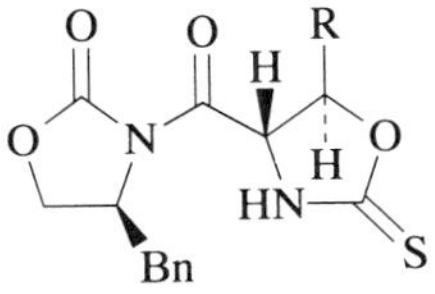

TABLE VI. DIASTEREOSELECTIVE ALDOL REACTION (*Continued*)

| Starting Material | Enolization Reagent | Acceptor | Product(s) and Yield(s) (%), diastereomer ratio | Refs. |
|---|---|---|---|---|
| C14 | Sn(OTf)$_2$, NEPIP | 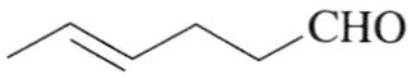 | R  = (81), 93/7 | 179 |
| | Sn(OTf)$_2$, NEPIP |  | R 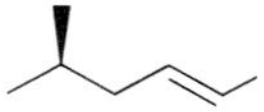 = (71), 97/3 | 179 |
| | Sn(OTf)$_2$, NEPIP | 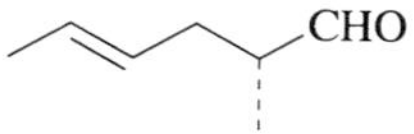 | R 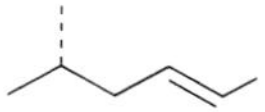 = (73), 94/6 | 179 |
| 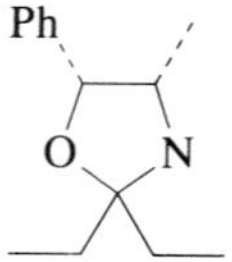 | | | 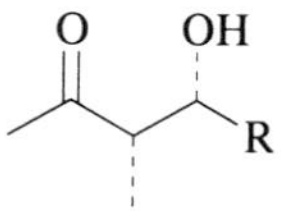 | |
| | LDA, SnCl$_2$ | Ph(CH$_2$)$_2$CHO | R = Ph(CH$_2$)$_2$, (77), 87/13, 92% ee (*syn*) | 182 |
| | " | EtCHO | R = Et, (61), 88/12, 95% ee (*syn*) | 182 |
| | " | C$_6$H$_{11}$CHO | R = C$_6$H$_{11}$, (75), 90/10, 92% ee (*syn*) | 182 |
| | " | t-BuCHO | R = t-Bu, (56), 86/14, >95% ee (*syn*) | 182 |
| C15 | | | | |
| | LDA, SnCl$_2$ | Ph(CH$_2$)$_2$CHO | R = Ph(CH$_2$)$_2$, (64), 85% ee | 46 |
| | " | C$_6$H$_{11}$CHO | R = C$_6$H$_{11}$, (54), 84% ee | 46 |
| | " | t-BuCHO | R = t-Bu, (56), >95% ee | 46 |

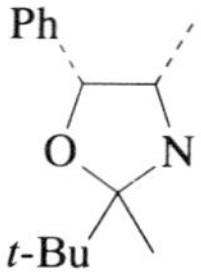

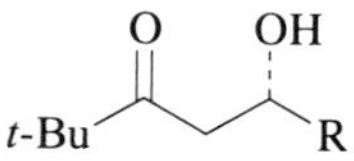

88

TABLE VI. DIASTEREOSELECTIVE ALDOL REACTION (*Continued*)

| Starting Material | Enolization Reagent | Acceptor | Product(s) and Yield(s) (%), diastereomer ratio | Refs. |
|---|---|---|---|---|
| C16 | | | | |
| | LDA, SnCl₂ | PhCHO | + (66), 93/7 | 195 |
| | LDA, SnBr₂ | " | (33), 92/8 | 195 |
| | LDA, Sn(OTf)₂ | " | (38), 66/34 | 195 |
| | Sn(OTf)₂, Et₃N | EtCHO | R = Et, (71), 79/21 | 194 |
| | " | i-PrCHO | R = i-Pr, (83), 95/5 | 194 |
| | " | PhCHO | R = Ph, (85), 89/11 | 194 |
| | " | 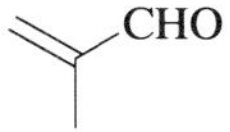 CHO | R  = , (77), 95/5 | 194 |

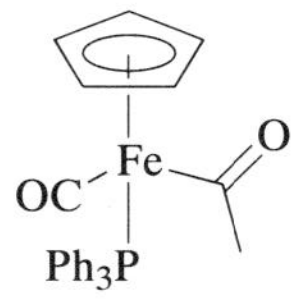

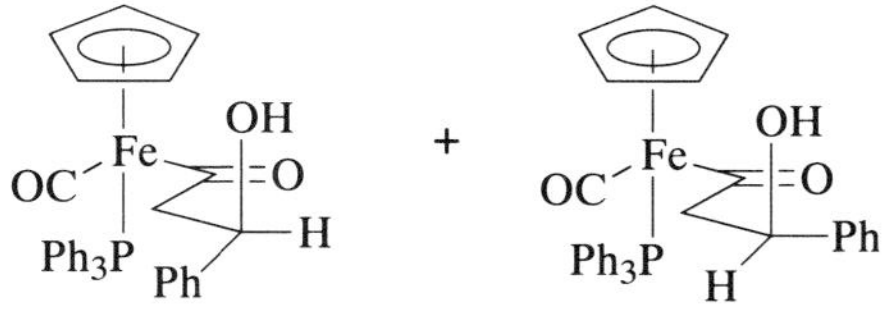

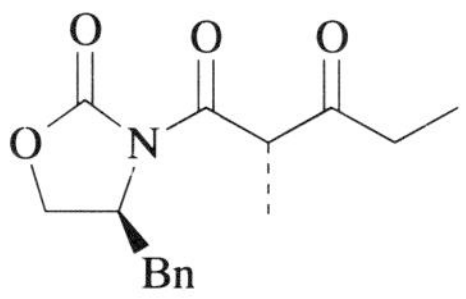

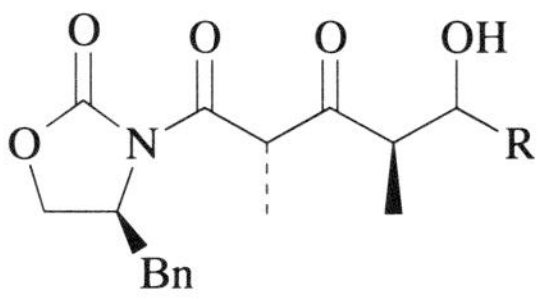

TABLE VI. DIASTEREOSELECTIVE ALDOL REACTION (Continued)

| Starting Material | Enolization Reagent | Acceptor | Product(s) and Yield(s) (%), diastereomer ratio | Refs. |
|---|---|---|---|---|

$C_{17}$

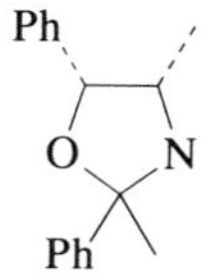

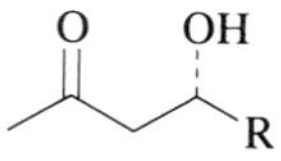

| | LDA, SnCl$_2$ | Ph(CH$_2$)$_2$CHO | R = Ph(CH$_2$)$_2$, (68), 70% ee | 46 |
| | " | n-PrCHO | R = n-Pr, (69), 76% ee | 46 |
| | " | C$_6$H$_{11}$CHO | R = C$_6$H$_{11}$, (64), 77% ee | 46 |
| | " | t-BuCHO | R = t-Bu, (66), 93% ee | 46 |

$C_{19}$

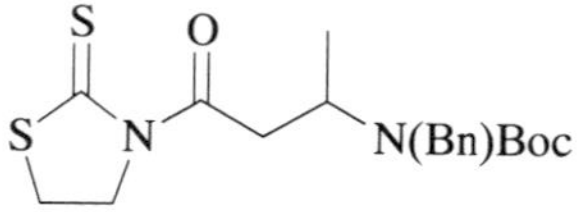

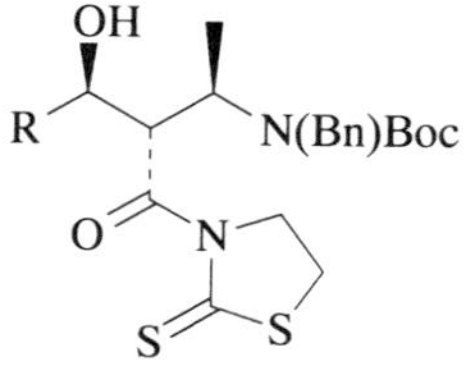

| | Sn(OTf)$_2$, NEPIP | Ph(CH$_2$)$_2$CHO | R = Ph(CH$_2$)$_2$, (91), >98/2 | 175 |
| | " | PhCHO | R = Ph, (99), >98/2 | 175 |
| | " | n-C$_5$H$_{11}$CHO | R = n-C$_5$H$_{11}$, (91), >98/2 | 175 |
| | " | n-PrCHO | R = n-Pr, (98), >98/2 | 175 |
| | " | MeCHO | R = Me, (98), >98/2 | 175 |

TABLE VI. DIASTEREOSELECTIVE ALDOL REACTION (*Continued*)

| Starting Material | Enolization Reagent | Acceptor | Product(s) and Yield(s) (%), diastereomer ratio | Refs. |
|---|---|---|---|---|

C$_{33}$

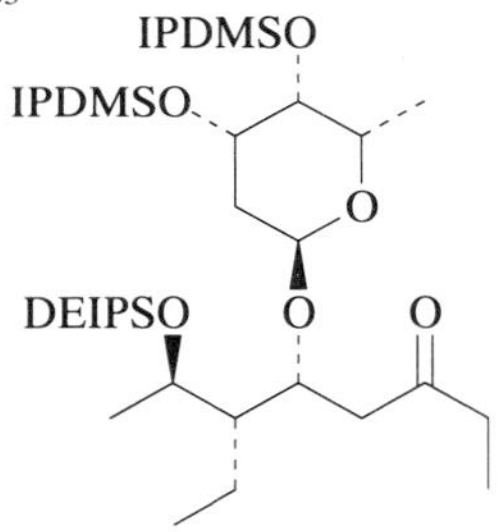

Sn(OTf)$_2$, NEPIP

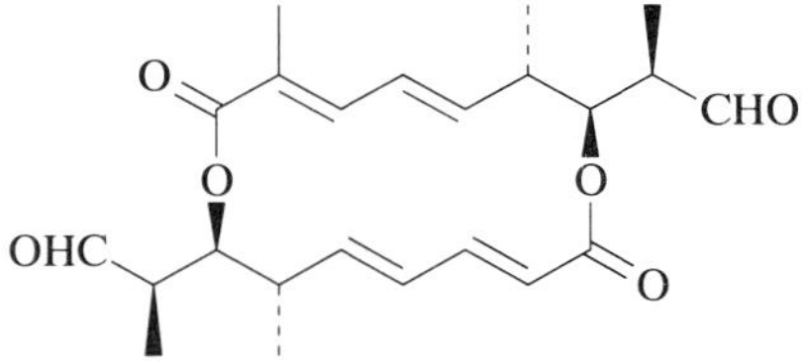

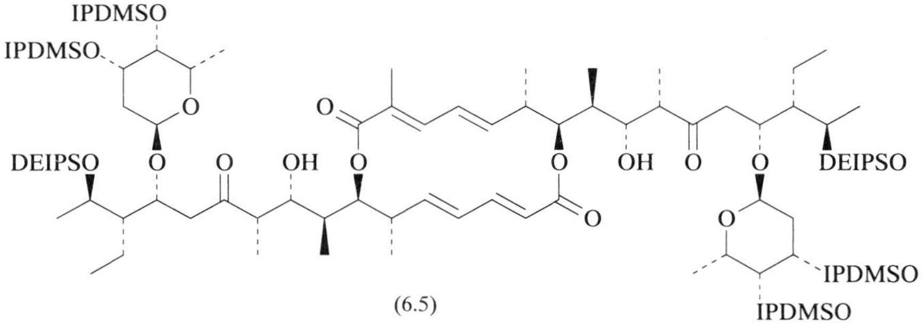

220

(6.5)

TABLE VI. DIASTEREOSELECTIVE ALDOL REACTION (*Continued*)

| Starting Material | Enolization Reagent | Acceptor | Product(s) and Yield(s) (%), diastereomer ratio | Refs. |
|---|---|---|---|---|

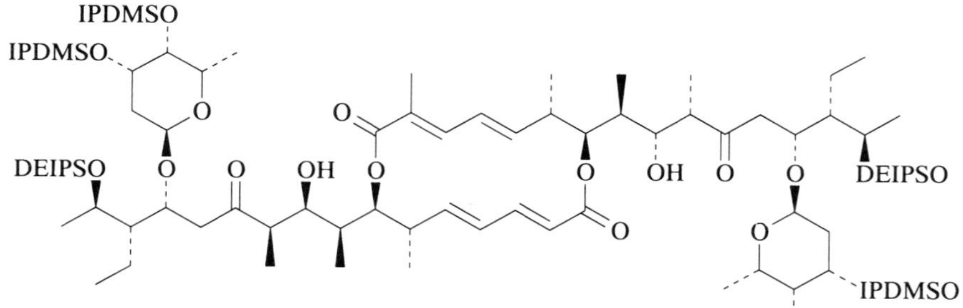

(21)

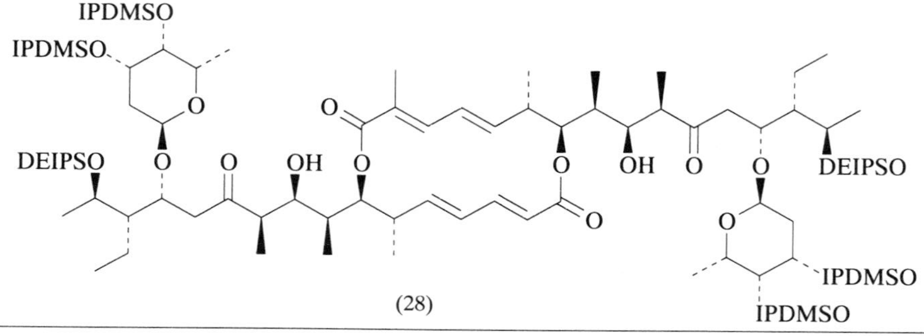

(28)

TABLE VII. DIASTEREOSELECTIVE ALDOL ALKYLATION

| Starting Material | Enolization Reagent | Acceptor | Product(s) and Yield(s) (%), Diastereomer Ratio | Refs. |
|---|---|---|---|---|
| C$_7$ | 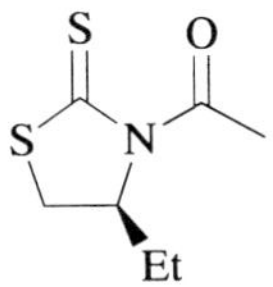 Sn(OTf)$_2$, NEPIP | | 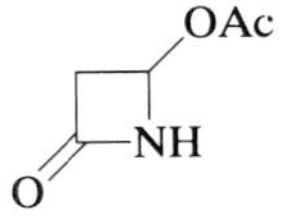 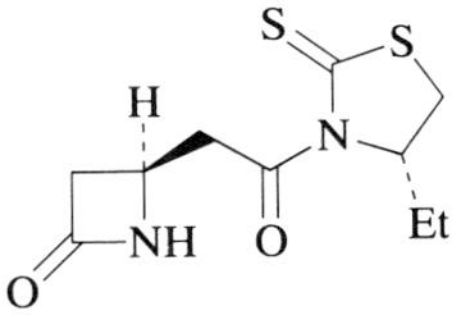 (82), 95/5 | 183 |
| C$_8$ | 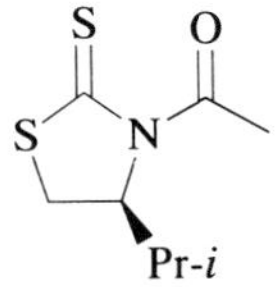 Sn(OTf)$_2$, NEPIP | | 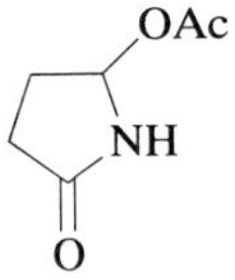 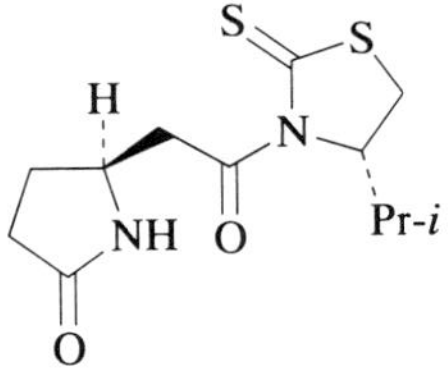 (67), >97/3 | 189 |
| 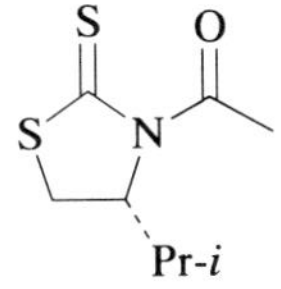 | Sn(OTf)$_2$, NEPIP | | 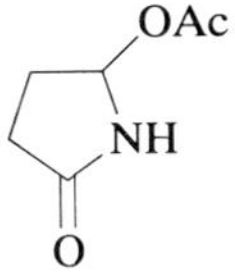 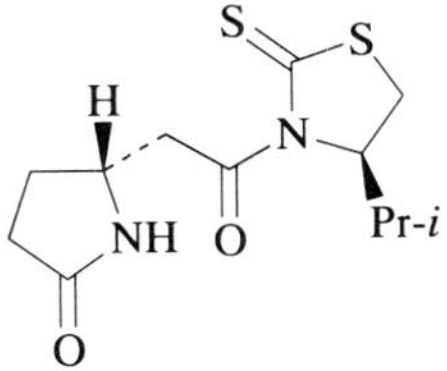 (66), >97/3 | 189 |
| | Sn(OTf)$_2$, NEPIP | | 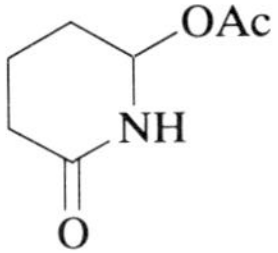 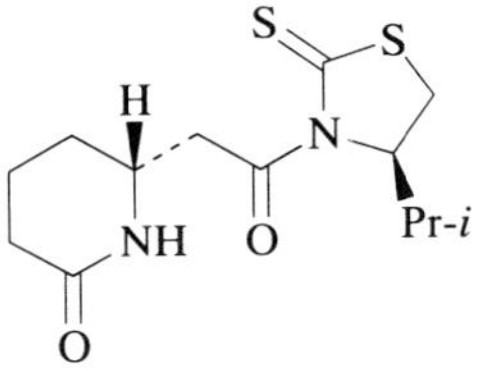 (63), >98/2 | 189 |

TABLE VII. DIASTEREOSELECTIVE ALDOL ALKYLATION (*Continued*)

| Starting Material | Enolization Reagent | Acceptor | Product(s) and Yield(s) (%), Diastereomer Ratio | Refs. |
|---|---|---|---|---|
| 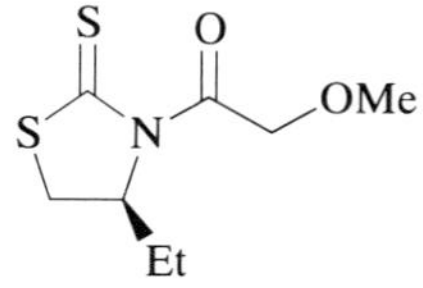 | Sn(OTf)$_2$, NEPIP | 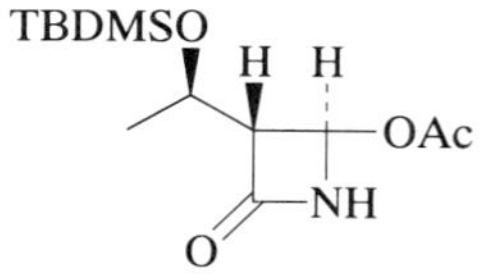 | 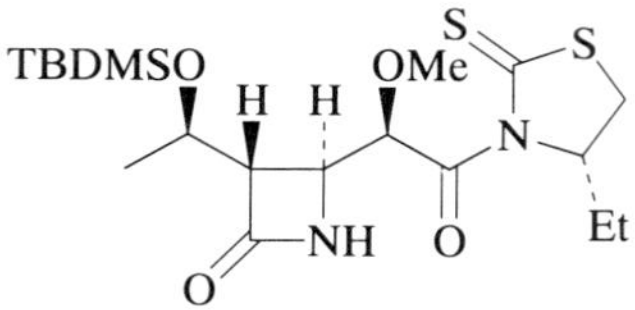 (96), 98/2 | 186 |
| 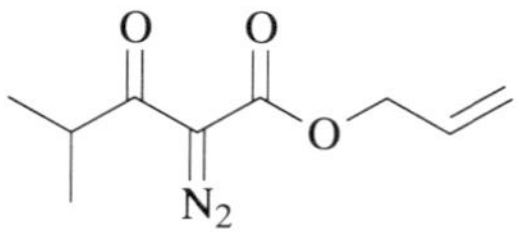 | Sn, I$_2$ (cat.), AgBF$_4$ (5 mol%) | 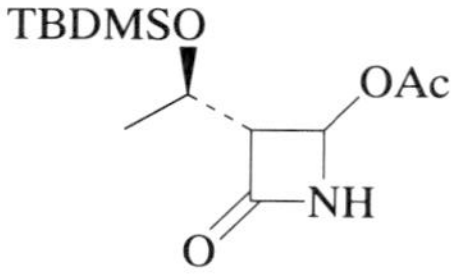 | 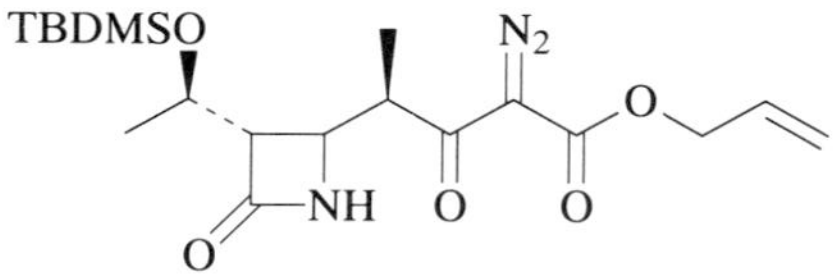 (75-80), 75/25 | 190 |
| C$_9$ | Sn(OTf)$_2$, *i*-Pr$_2$NEt, ZnBr$_2$ | | +    59/41 | 184 |

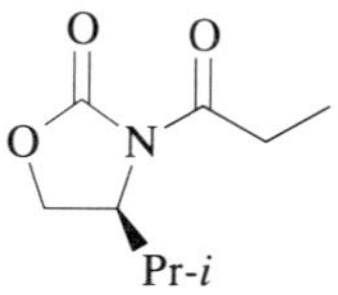

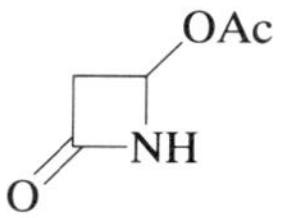

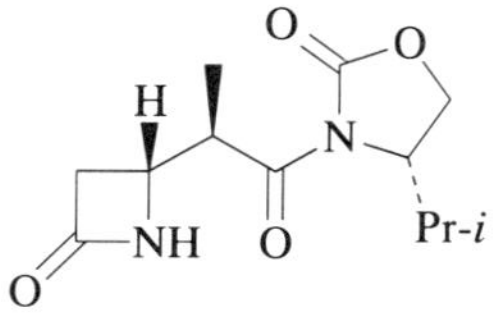

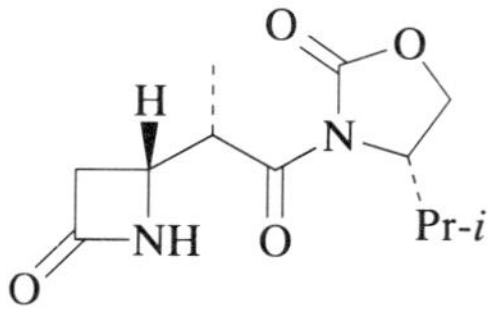

TABLE VII. DIASTEREOSELECTIVE ALDOL ALKYLATION (*Continued*)

| Starting Material | Enolization Reagent | Acceptor | Product(s) and Yield(s) (%), Diastereomer Ratio | Refs. |
|---|---|---|---|---|
| | Sn(OTf)₂, *i*-Pr₂NEt, ZnBr₂ | | (80), 92/8 | 184 |
| | Sn(OTf)₂, NEPIP | | (55), 97/3 | 187 |
| | Sn(OTf)₂, NEPIP | | (55), 97/3 | 187 |
| C₁₀ | Sn(OTf)₂, NEPIP | | (64), >99/1 | 188, 189 |

95

TABLE VII. DIASTEREOSELECTIVE ALDOL ALKYLATION (*Continued*)

| Starting Material | Enolization Reagent | Acceptor | Product(s) and Yield(s) (%), Diastereomer Ratio | Refs. |
|---|---|---|---|---|
| C$_{11}$ | | | | |
| | Sn(OTf)$_2$, NEPIP | | 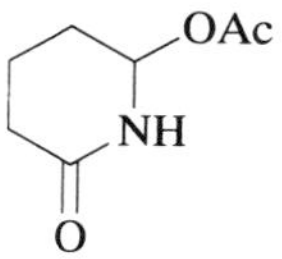 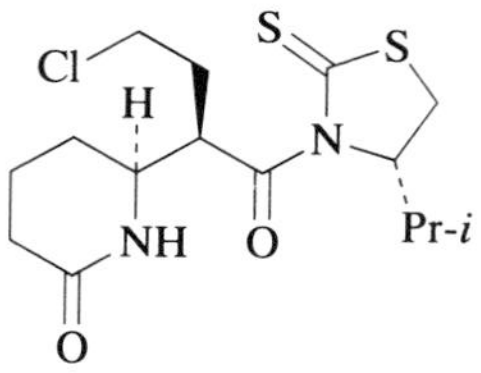 (57), >98/2 | 188, 189 |
| 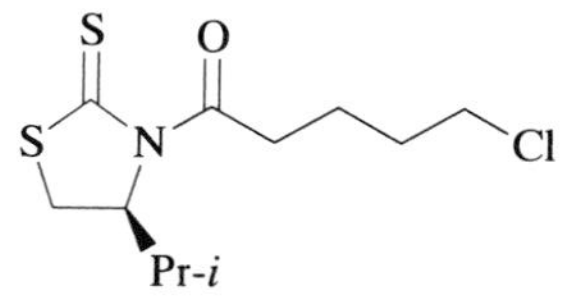 | Sn(OTf)$_2$, NEPIP | | 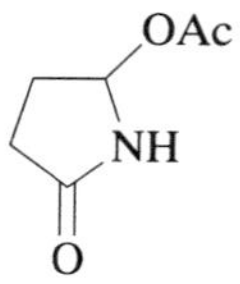 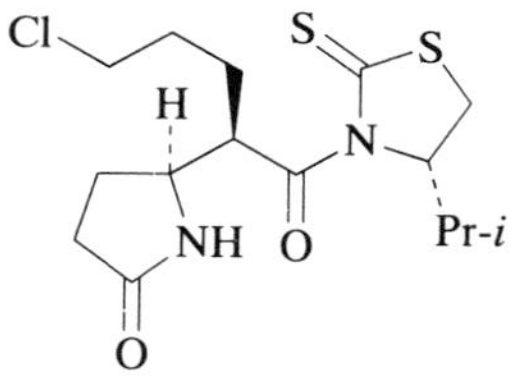 (72), >99/1 | 188, 189 |
| | Sn(OTf)$_2$, NEPIP | | 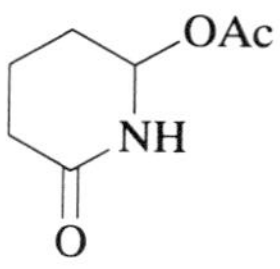 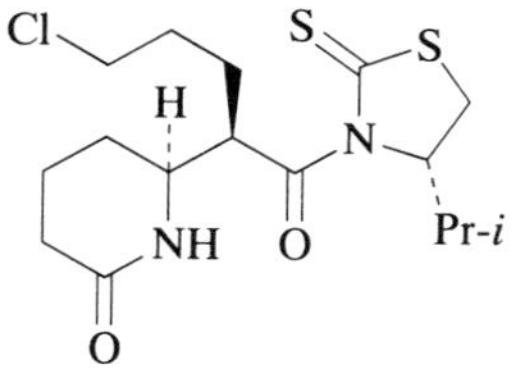 (73), >96/4 | 188, 189 |
| 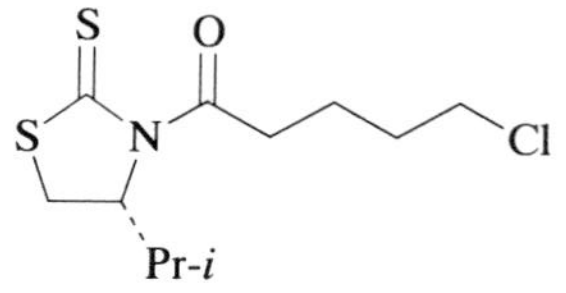 | Sn(OTf)$_2$, NEPIP | | 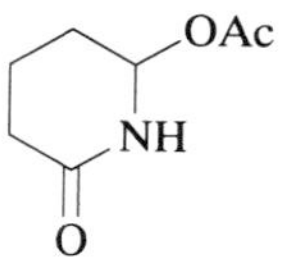 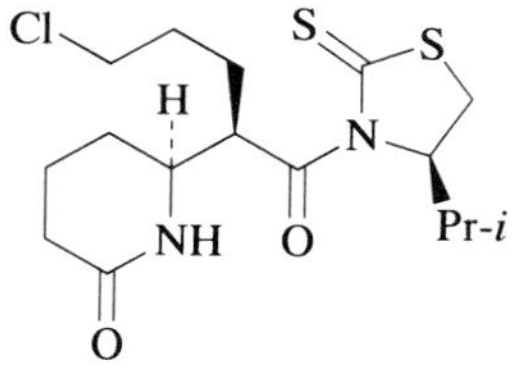 (73), >96/4 | 189 |

96

TABLE VII. DIASTEREOSELECTIVE ALDOL ALKYLATION (*Continued*)

| Starting Material | Enolization Reagent | Acceptor | Product(s) and Yield(s) (%), Diastereomer Ratio | Refs. |
|---|---|---|---|---|
| C₁₄ | | | | |
| 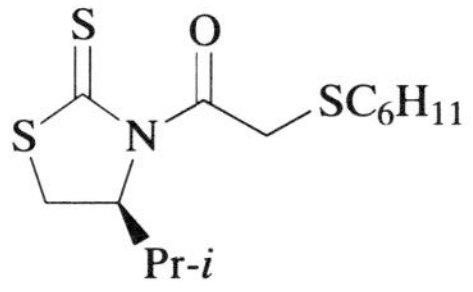 | Sn(OTf)₂, NEPIP | | 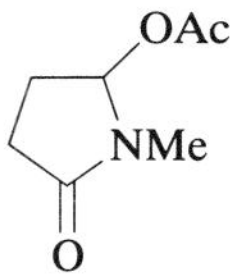 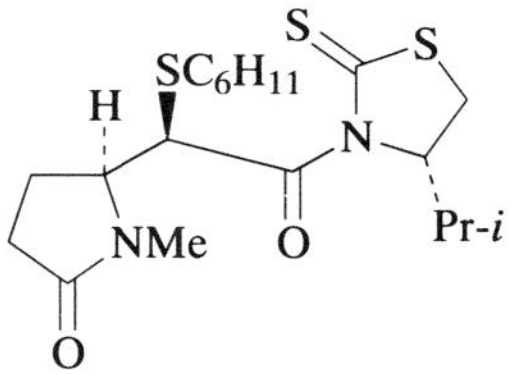 (46), 83/17 | 185 |
| 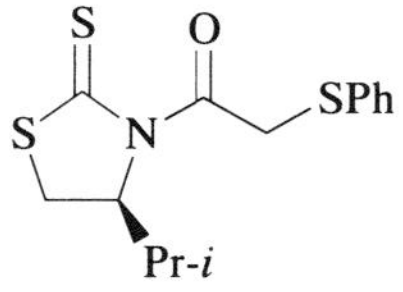 | Sn(OTf)₂, NEPIP | | 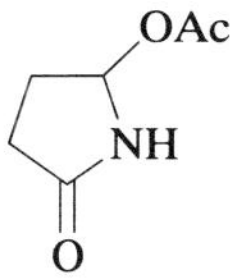 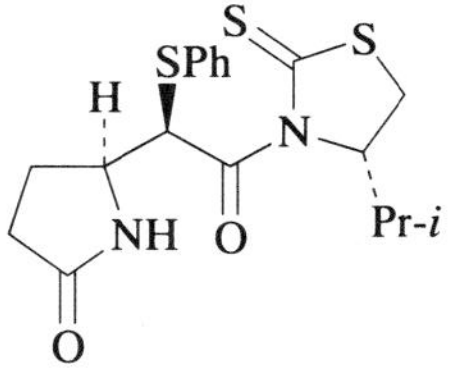 (92), >95/5 | 189 |
| | Sn(OTf)₂, NEPIP | | 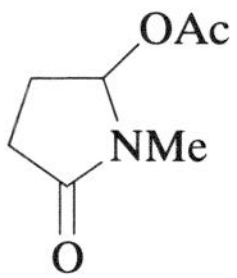 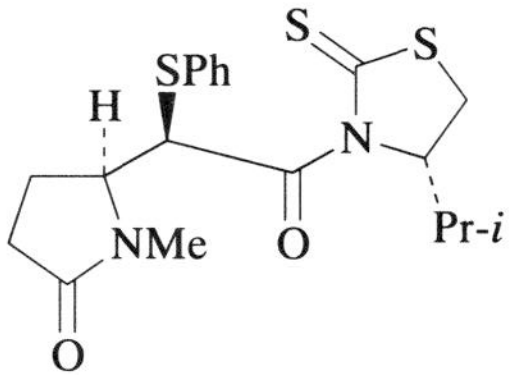 (57), 91/9 | 185 |
| C₁₅ | | | | |
| 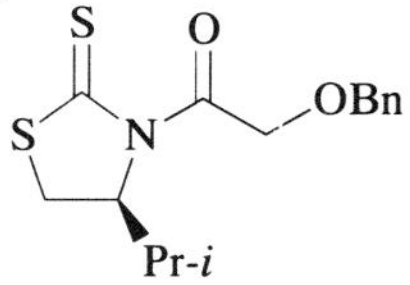 | Sn(OTf)₂, NEPIP | | 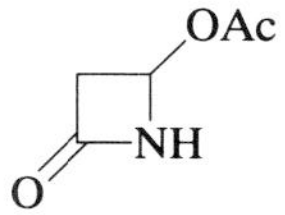 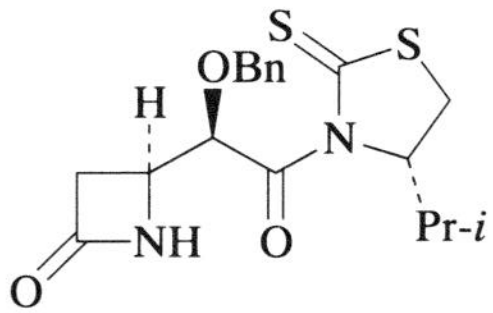 (79), 97/3 | 187 |

97

TABLE VII. DIASTEREOSELECTIVE ALDOL ALKYLATION (*Continued*)

| Starting Material | Enolization Reagent | Acceptor | Product(s) and Yield(s) (%), Diastereomer Ratio | Refs. |
|---|---|---|---|---|
| | Sn(OTf)$_2$, NEPIP | | (84), 97/3 | 187 |
| | Sn(OTf)$_2$, NEPIP | | (57), 84/16 | 185 |
| | Sn(OTf)$_2$, NEPIP | | (78), >98/2 | 189 |
| C$_{16}$ | Sn(OTf)$_2$, NEPIP | | (52), 99/1 | 187 |

98

## REFERENCES

[1] R. C. Poller, *Organic Compounds of Group IV Metals,* in *Comprehensive Organic Chemistry,* D. N. Jones, Ed., Pergamon Press, Oxford, 1979, Vol. 3, p. 1073.

[2] A. G. Davies and P. J. Smith, *Tin,* in *Comprehensive Organometallic Chemistry, Vol. 2,* G. Wilkinson, Ed., Pergamon Press, Oxford, 1982, p. 519.

[3] M. Pereyre, J-P. Quintard, and A. Rahm, *Tin in Organic Synthesis,* Butterworths, London, 1987.

[4] T. Mukaiyama, *Pure Appl. Chem., 58,* 505 (1986).

[5] T. Mukaiyama, *Org. React., 28,* 203 (1982).

[6] C. H. Heathcock in *Asymmetric Synthesis,* J. D. Morrison, Ed., Academic Press, New York, (1984), Vol. 3, p. 111.

[7] D. A. Evans, J. V. Nelson, T. R. Taber, and T. R. Topics, *Top. Stereochem., 13,* 1 (1982).

[8] S. Masamune, W. Choy, J. S. Petersen, and L. R. Sita, *Angew. Chem., Int. Ed. Engl., 24,* 1 (1985).

[9] T. Mukaiyama, *Pure Appl. Chem., 55,* 1749 (1983).

[10] T. Mukaiyama, *Isr. J. Chem., 22,* 162 (1984).

[11] T. Mukaiyama and M. Murakami, *Croat. Chem. Acta, 59,* 221 (1986).

[12] C. H. Heathcock in *Comprehensive Organic Synthesis,* B. M. Trost, Ed., Pergamon Press, Oxford, 1991, Vol. 2, p. 133.

[13] C. H. Heathcock in *Comprehensive Organic Synthesis,* B. M. Trost, Ed., Pergamon Press, Oxford,1991, Vol. 2, p. 181.

[14] B. M. Kim, S. F. Williams, and S. Masamune in *Comprehensive Organic Synthesis,* B. M. Trost, Ed., Pergamon Press, Oxford, 1991, Vol. 2, p. 239.

[15] M. W. Rathke and P. Weipert in *Comprehensive Organic Synthesis,* B. M. Trost, Ed., Pergamon Press, Oxford, 1991, Vol. 2, p. 277.

[16] I. Paterson in *Comprehensive Organic Synthesis,* B. M. Trost, Ed., Pergamon Press, Oxford, 1991, Vol. 2, pp. 301.

[17] Y. Yamamoto, H. Yatagai, and K. Maruyama, *J. Chem. Soc., Chem. Commun.,* **1981,** 162.

[18] S. Shenvi and J. K. Stille, *Tetrahedron Lett., 23,* 627 (1982).

[19] T. Mukaiyama, T. Harada, and S. Shoda, *Chem. Lett.,* **1980,** 1507.

[20] T. Harada and T. Mukaiyama, *Chem. Lett.,* **1981,** 1109.

[21] T. Mukaiyama and T. Harada, *Chem. Lett.,* **1981,** 621.

[22] S. Shoda and T. Mukaiyama, *Chem. Lett.,* **1981,** 723.

[23] T. Mukaiyama, M. Yamaguchi, and J. Kato, *Chem. Lett.,* **1981,** 1505.

[24] T. Mukaiyama and T. Harada, *Chem. Lett.,* **1981,** 1527.

[25] T. Harada and T. Mukaiyama, *Chem. Lett.,* **1982,** 161.

[26] T. Mukaiyama, R. W. Stevens, N. Iwasawa, *Chem. Lett.,* **1982,** 353.

[27] M. T. Reetz, *Angew. Chem., Int. Ed. Engl., 23,* 556 (1984).

[28] C. Gennari in *Comprehensive Organic Synthesis,* B. M. Trost, Ed., Pergamon Press, Oxford, 1991, Vol. 2, pp. 629.

[29] M. Guette, J. Capillon, and J. P. Guette, *Tetrahedron, 29,* 3659 (1973).

[30] D. Seebach, H. O. Kalinowsky, B. Bastani, G. Crass, H. Daum, H. Dörr, N. P. DuPreez, V. Ehrig, W. Langer, C. Nüssler, H.-A. Oei, and M. Schmidt, *Helv. Chim. Acta,* 60, 301 (1977).

[31] S. Brandänge, S. Josephson, L. Mörch, S. Vallen, *Acta Chem. Scand. Ser. B, 35,* 273 (1981).

[32] M. Asami, H. Ohno, S. Kobayashi, and T. Mukaiyama, *Bull. Chem. Soc. Jpn., 51,* 1869 (1978).

[33] T. Mukaiyama, *Tetrahedron, 37,* 4111 (1981).

[34] T. Mukaiyama, *Chem. Scr., 25,* 13 (1985).

[35] T. Mukaiyama and M. Asami, *Top. Curr. Chem., 127,* 133 (1985).

[36] T. Mukaiyama, N. Iwasawa, R. W. Stevens, and T. Haga, *Tetrahedron, 40,* 1381 (1984).

[37] T. Mukaiyama and N. Iwasawa, *Chem. Lett.,* **1982,** 1903.

[38] Y. Nagao, S. Yamada, T. Kumagai, M. Ochiai, and E. Fujita, *J. Chem. Soc., Chem. Commun.,* **1985,** 1418.

[39] A. Abdel-Magid, I. Lantos, and L. N. Pridgen, *Tetrahedron Lett., 25,* 3273 (1984).

[40] D. A. Evans and A. E. Weber, *J. Am. Chem. Soc.*, **108**, 6757 (1986).

[41] R. W. Stevens and T. Mukaiyama, *Chem. Lett.*, **1985**, 851.

[42] T. Harada and T. Mukaiyama, *Chem. Lett.*, **1982**, 467.

[43] T. Mukaiyama, N. Yamasaki, R. W. Stevens, and M. Murakami, *Chem. Lett.*, **1986**, 213.

[44] T. Mukaiyama, H. Suzuki, and T. Yamada, *Chem. Lett.*, **1986**, 915.

[45] L. S. Liebeskind, M. E. Welker, and R. W. Fengl, *J. Am. Chem. Soc.*, **108**, 6328 (1986).

[46] K. Narasaka, T. Miwa, H. Hayashi, and M. Ohta, *Chem. Lett.*, **1984**, 139.

[47] T. Yura, N. Iwasawa, and T. Mukaiyama, *Chem. Lett.*, **1986**, 187.

[48] T. Mukaiyama, J. Kato, and M. Yamaguchi, *Chem. Lett.*, **1982**, 1291.

[49] J. M. McIntosh and G. M. Masse, *J. Org. Chem.*, **40**, 1294 (1975).

[50] H. C. Brown, M. M. Rogic, and M. W. Rathke, *J. Am. Chem. Soc.*, **90**, 6218 (1968).

[51] E. M. Schultz and S. Mickey, *Organic Synthesis, Collected Volume III*, Wiley, New York, 1955, p. 343.

[52] R. W. Stevens, N. Iwasawa, and T. Mukaiyama, *Chem. Lett.*, **1982**, 1459.

[53] T. Mukaiyama and T. Inoue, *Chem. Lett.*, **1976**, 559.

[54] T. Inoue and T. Mukaiyama, *Bull. Chem. Soc. Jpn.*, **53**, 174 (1980).

[55] D. A. Evans, J. Bartroli, T. L. Sih, *J. Am. Chem. Soc.*, **103**, 2127 (1981).

[56] S. Masamune, W. Choy, F. A. J. Kerdesky, B. Imperiali, *J. Am. Chem. Soc.*, **103**, 1566 (1981).

[57] T. Imamoto, T. Kusumoto, and M. Yokoyama, *Tetrahedron Lett.*, **24**, 5233 (1983).

[57a] K. Nagasawa, H. Kanbara, K. Matsushita, and K. Ito, *Tetrahedron Lett.*, **26**, 6477 (1985).

[58] R. W. Stevens and T. Mukaiyama, *Chem. Lett.*, **1983**, 595.

[59] T. Mukaiyama, T. Haga, and N. Iwasawa, *Chem. Lett.*, **1982**, 1601.

[60] E. öhler, H. -S. Kang, E. Zbiral, *Chem. Ber.*, **121**, 299 (1988).

[61] G. Berti, *Top. in Stereochem.*, **7**, 93 (1973).

[62] M. S. Newman and B. J. Magerlein, *Org. React.*, **5**, 413 (1949).

[63] M. Ballester, *Chem. Rev.*, **55**, 283 (1955).

[64] E. Weitz and A. Scheffer, *Chem. Ber.*, **54**, 2327 (1921).

[65] C. H. Heathcock, M. C. Pirrung, C. T. Buse, J. P. Hagen, S. D. Young, and J. E. Sohn, *J. Am. Chem. Soc.*, **101**, 7077 (1979).

[66] T. Mukaiyama, T. Yura, and N. Iwasawa, *Chem. Lett.*, **1985**, 809.

[67] Masamune, S.; McCarthy, P. A. in *Macrolide Antibiotics. Chemistry, Biology, and Practice*, Omura, S., Ed., Academic Press, New York, **1984**.

[68] *Trends in Synthetic Carbohydrate Chemistry*, D. Horton, L. D. Hawkins, and G. J. McGarvey, Eds., ACS Symposium Series 386, American Chemical Society, Washington, DC, 1989.

[69] T. Mukaiyama, *Angew. Chem., Int. Ed. Engl.*, **18**, 707 (1979).

[70] T. Izawa and T. Mukaiyama, *Bull. Chem. Soc. Jpn.*, **52**, 555 (1979).

[71] E. Fujita, *Pure. Appl. Chem.*, **53**, 1141 (1981).

[72] T. Mukaiyama and N. Iwasawa, *Chem. Lett.*, **1982**, 1903.

[73] T. Mukaiyama and N. Iwasawa, *Chem. Lett.*, **1984**, 753.

[74] N. Iwasawa, H. Huang, and T. Mukaiyama, *Chem. Lett.*, **1985**, 1045.

[75] T. Mukaiyama, R. Tsuzuki, and J. Kato, *Chem. Lett.*, **1983**, 1825.

[76] J. Kato and T. Mukaiyama, *Chem. Lett.*, **1983**, 1727.

[77] T. Yura, N. Iwasawa, and T. Mukaiyama, *Chem. Lett.*, **1986**, 187.

[78] N. Iwasawa, T. Yura, and T. Mukaiyama, *Tetrahedron*, **45**, 1197 (1989).

[79] D. A. Armitage and A. W. Sinden, *J. Organomet. Chem.*, **90**, 285 (1975).

[80] A. G. Davies and P. J. Smith in *Comprehensive Organometallic Chemistry*, G. Wilkinson, Ed., F. G. A. Stone, and E. W. Abel, Pergamon Press Ltd., Oxford, Vol. 2, pp. 604–608, 1982.

[81] *The Chemistry of Carbon-Nitrogen Double Bond*, S. Patai, Ed., Wiley Interscience, London, 1970.

[82] R. W. Layer, *Chem. Ber.*, **96**, 489 (1963).

[83] T. Iimori and M. Shibasaki, *Tetrahedron Lett.*, **26**, 1523 (1985).

[84] Y. Yamamoto, W. Ito, and K. Maruyama, *J. Chem. Soc., Chem. Commun.*, **1985**, 1131.

[85] T. Chiba, M. Nagatsuma, and T. Nakai, *Chem. Lett.*, **1984**, 1927.

[86] J. -C. Fiaud and H. B. Kagan, *Tetrahedron Lett.*, **1971**, 1019.

[87] N. Yamasaki, M. Murakami, and T. Mukaiyama, *Chem. Lett.,* **1986**, 1013.

[88] M. Ohshima, M. Murakami, and T. Mukaiyama, *Chem. Lett.,* **1985**, 1871.

[89] *The Chemistry of the Quinonoid Compounds,* Part I, S. Patai, Ed., John Wiley and Sons, London (1974).

[90] *The Chemistry of the Quinonoid Compounds,* Part II, S. Patai, Ed., John Wiley and Sons, London (1974).

[91] G. A. Kraus and B. Roth, *J. Org. Chem.,* **43**, 4923 (1978).

[92] L. Strzelecki and B. Marric, *Bull. Soc. Chim. Fr.,* **1969**, 4413.

[93] G. A. Kraus and B. Roth, *Tetrahedron Lett.,* **1977**, 3129.

[94] A. Fischer and G. N. Henderson, *Tetrahedron Lett.,* **1980**, 701.

[95] L. C. Lasne, J. L. Ripoll, and A. Thuillier, *Chem. Ind.,* **20**, 830 (1980).

[96] A. Fischer and G. N. Henderson, *Tetrahedron Lett.,* **24**, 131 (1983).

[97] M. F. Hawthorne and M. Reintjes, *J. Am. Chem. Soc.,* **86**, 951 (1964).

[98] M. F. Hawthorne and M. Reintjes, *J. Am. Chem. Soc.,* **87**, 4585 (1965).

[99] B. M. Mikhailov and G. S. Ter-Sarkisyan, *Izv. Akad. Nauk USSR, Ser. Khim.,* **1966**, 380; *Chem. Abstr.,* **64**, 15907h (1966).

[100] B. M. Mikhailov, G. S. Ter-Sarkisyan, and N. A. Nikolaeva, *Izv. Akad. Nauk USSR, Ser. Khim.,* **1968**, 541; CA, 69, 67448v (1968).

[101] G. W. Kabalka, *J. Organomet. Chem.,* **33**, C25 (1971).

[102] J. Majnusz and R. W. Lenz, *Eur. Polym. J.,* **21**, 565 (1985).

[103] L. S. Hegedus, E. L. Waterman, and J. Catlin, *J. Am. Chem. Soc.,* **94**, 7155 (1972).

[104] L. S. Hegedus, B. R. Evans, D. E. Korte, E. L. Waterman, and K. Sjöberg, *J. Am. Chem. Soc.,* **98**, 3901 (1976).

[105] K. Maruyama and Y. Naruta, *J. Org. Chem.,* **43**, 3796 (1978).

[106] Y. Naruta, *J. Am. Chem. Soc.,* **102**, 3774 (1980).

[107] Y. Naruta, *J. Org. Chem.,* **45**, 4097 (1980).

[108] Y. Naruta, N. Nagai, Y. Arita, and K. Maruyama, *Chem. Lett.,* **1983**, 1683.

[109] K. Mori, M. Sakakibara, and M. Waku, *Tetrahedron Lett.,* **25**, 1085 (1984).

[110] T. Mukaiyama, R. S. J. Clark, and N. Iwasawa, *Chem. Lett.,* **1987**, 479.

[111] T. Mukaiyama, N. Iwasawa, T. Yura, and R. S. J. Clark, *Tetrahedron,* **43**, 5003 (1987).

[112] E. D. Bergmann, D. Ginsburg, and R. Pappo, *Org. React.,* **10**, 179 (1959).

[113] V. J. Lee in *Comprehensive Organic Synthesis,* B. M. Trost, Ed., Pergamon Press, Oxford, 1991, Vol. 4, p. 69.

[114] V. J. Lee in *Comprehensive Organic Synthesis,* B. M. Trost, Ed., Pergamon Press, Oxford, 1991, Vol. 4, p. 139.

[115] K. Narasaka, K. Soai, and T, Mukaiyama, *Chem. Lett.,* **1974**, 1223.

[116] K. Narasaka, K. Soai, Y. Aikawa, and T. Mukaiyama, *Bull. Chem. Soc. Jpn.,* **49**, 779 (1976).

[117] T. Yanami, M. Miyashita, and A. Yoshikoshi, *J. Org. Chem.,* **45**, 607 (1980).

[118] D. Seebach and M. A. Brook, *Helv. Chim. Acta,* **68**, 319 (1985).

[119] C. H. Heathcock, M. H. Norman, and D. E. Uehling, *J. Am. Chem. Soc.,* **107**, 2797 (1985).

[120] S. Kobayashi, M. Murakami, and T. Mukaiyama, *Chem. Lett.,* **1985**, 1535.

[121] T. Mukaiyama, M. Tamura, and S. Kobayashi, *Chem. Lett.,* **1986**, 1017.

[122] T. Mukaiyama, M. Tamura, and S. Kobayashi, *Chem. Lett.,* **1986**, 1817.

[123] T. Mukaiyama, M. Tamura, and S. Kobayashi, *Chem. Lett.,* **1987**, 743.

[124] T. Mukaiyama, S. Kobayashi, M. Tamura, and Y. Sagawa, *Chem. Lett.,* **1987**, 491.

[125] N. Iwasawa and T. Mukaiyama, *Chem. Lett.,* **1987**, 463.

[126] M. Kawai, M. Onaka, and Y. Izumi, *J. Chem. Soc., Chem. Commun.,* **1987**, 1203.

[127] N. Minowa and T. Mukaiyama, *Chem. Lett.,* **1987**, 1719.

[128] H. Ohki, M. Wada, and K. Akiba, *Tetrahedron Lett.,* **29**, 4719 (1988).

[129] S. Kobayashi, M. Tamura, and T. Mukaiyama, *Chem. Lett.,* **1988**, 91.

[130] T. Mukaiyama and R. Hara, *Chem. Lett.,* **1989**, 1171.

[131] T. Sato, Y. Wakahara, J. Otera, and H. Nozaki, *Tetrahedron Lett.,* **31**, 1581 (1990).

[132] M. Yamaguchi, M. Tsukamoto, and I. Hirao, *Chem. Lett.,* **1984**, 375.

[133] M. Yamaguchi, M. Tsukamoto, S. Tanaka, and I. Hirao, *Tetrahedron Lett.,* **25**, 5661 (1984).

[134] M. Yamaguchi, K. Hasebe, S. Tanaka, and T. Minami, *Tetrahedron Lett.*, **27**, 959 (1986).

[135] D. A. Oare and C. H. Heathcock, *Tetrahedron Lett.*, **27**, 6169 (1986).

[136] C. H. Heathcock, M. A. Henderson, D. A. Oare, and M. A. Sanner, *J. Org. Chem.*, **50**, 3019 (1985).

[137] C. H. Heathcock and D. A. Oare, *J. Org. Chem.*, **50**, 3022 (1985).

[138] D. A. Oare and C. H. Heathcock, *Top. Stereochem.* **19**, 227 (1989).

[139] D. A. Hunt, *Org. Prep. Proced. Int.*, **21**, 705 (1989).

[140] M. J. Chapdelaine and M. Hulce, *Org. React.*, **38**, 225 (1990).

[141] J. Mulzer, G. Hartz, U. Kühl, and G. Brüntrup, *Tetrahedron Lett.*, **1978**, 2949.

[142] J. Metzner, A. Chucholowski, O. Lammer, I. Jibril, and G. Huttner, *J. Chem. Soc., Chem. Commun.*, **1983**, 869.

[143] M. Zuger, T. Weller, and D. Seebach, *Helv. Chim. Acta*, **63**, 2005 (1980).

[144] F. E. Ziegler and K. -J. Hwang, *J. Org. Chem.*, **48**, 3349 (1983).

[145] W. Oppolzer, R. Pitteloud, G. Bernardinelli, and K. Baettig, *Tetrahedron Lett.*, **24**, 4975 (1983).

[146] J. Bertrand, L. Gorrichon, and P. Maroni, *Tetrahedron*, **40**, 4127 (1984).

[147] E. J. Corey and R. T. Peterson, *Tetrahedron Lett.*, **26**, 5025 (1985).

[148] G. H. Posner and S. -B. Lu, *J. Am. Chem. Soc.*, **107**, 1424 (1985).

[149] G. H. Posner, S. B. Lu, and E. Asirvatham, *Tetrahedron Lett.*, **27**, 659 (1986).

[150] G. H. Posner and E. Asirvatham, *Tetrahedron Lett.*, **27**, 663 (1986).

[151] K. Kpegba, P. Metzner, and R. Rakotonirina, *Tetrahedron Lett.*, **27**, 1505 (1986).

[152] M. E. Krafft, R. M. Kennedy, and R. A. Holton, *Tetrahedron Lett.*, **27**, 2087 (1986).

[153] T. Yura, N. Iwasawa, and T. Mukaiyama, *Chem. Lett.*, **1987**, 791.

[154] R. W. Stevens and T. Mukaiyama, *Chem. Lett.*, **1985**, 855.

[155] T. Mukaiyama and A. Ishida, *Chem. Lett.*, **1975**, 319.

[156] T. Mukaiyama and A. Ishida, *Chem. Lett.*, **1975**, 1201.

[157] T. Mukaiyama and A. Ishida, *Chem. Lett.*, **1977**, 467.

[158] T. Mukaiyama and A. Ishida, *Bull. Chem. Soc. Jpn.*, **51**, 2077 (1978).

[159] I. Fleming, J. Goldhill, and I. Paterson, *Tetrahedron Lett.*, **1979**, 3205.

[160] I. Fleming, J. Goldhill, and I. Paterson, *Tetrahedron Lett.*, **1979**, 3209.

[161] I. Fleming and T. V. Lee, *Tetrahedron Lett.*, **1981**, 705.

[162] P. Albaugh-Robertson and J. A. Katzenellenbougen, *Tetrahedron Lett.*, **1982**, 723.

[163] J. D. Donaldson in *Progress in Inorganic Chemistry*, F. A. Cotton, Ed., Wiley, New York, 1967, Vol. 8, p. 287.

[164] N. Iwasawa and T. Mukaiyama, *Chem. Lett.*, **1982**, 1441.

[165] N. Iwasawa and T. Mukaiyama, *Chem. Lett.*, **1983**, 297.

[166] R. W. Stevens and T. Mukaiyama, *Chem. Lett.*, **1983**, 1799.

[167] T. Yura, N. Iwasawa, and T. Mukaiyama, *Chem. Lett.*, **1988**, 1021.

[168] T. Yura, N. Iwasawa, K. Narasaka, and T. Mukaiyama, *Chem. Lett.*, **1988**, 1025.

[169] T. Yura, N. Iwasawa, R. S. J. Clark, and T. Mukaiyama, *Chem. Lett.*, **1986**, 1809.

[170] D. J. Cram and D. R. Wilson, *J. Am. Chem. Soc.*, **85**, 1245 (1963).

[171] M. Cherest, H. Felkin, and N. Prudent, *Tetrahedron Lett.*, **1968**, 2199.

[172] M. T. Anh, *Top. Curr. Chem.* **88**, 145 (1980)

[173] T. Mukaiyama, K. Suzuki, and T. Yamada, *Chem. Lett.*, **1982**, 929.

[174] T. Yamada, H. Suzuki, and T. Mukaiyama, *Chem. Lett.*, **1987**, 293.

[175] N. Iwasawa and T. Mukaiyama, *Chem. Lett.*, **1986**, 637.

[176] Y. Nagao, Y. Hagiwara, T. Kumagai, M. Ochiai, T. Inoue, K. Hashimoto, and E. Fujita, *J. Org. Chem.*, **51**, 2391 (1986).

[177] Y. Nagao, W.-M. Dai, M. Ochiai, and M. Shiro, *J. Org. Chem.*, **54**, 5211 (1989).

[178] C. Hsiao, L. Liu, and M. J. Miller, *J. Org. Chem.*, **52**, 2201 (1987).

[179] D. A. Evans and A. E. Weber, *J. Am. Chem. Soc.*, **108**, 6757 (1986).

[180] A. S. Kende, K. Kawamura, and M. J. Orwat, *Tetrahedron Lett.*, **30**, 5821 (1989).

[181] A. S. Kende, K. Kawamura, and R. J. DeVita, *J. Am. Chem. Soc.*, **112**, 4070 (1990).

[182] K. Narasaka and T. Miwa, *Chem. Lett.*, **1985**, 1217.

[183] Y. Nagao, T. Kumagai, S. Tamai, T. Abe, Y. Kuramoto, T. Taga, S. Aoyagi, Y. Nagase, M. Ochiai, Y. Inoue, and E. Fujita, *J. Am. Chem. Soc.*, **108**, 4673 (1986).

[184] L. M. Fuentes, I. Shinkai, and T. N. Salzmann, *J. Am. Chem. Soc.*, **108**, 4675 (1986).

[185] Y. Nagao, W.-M. Dai, and M. Ochiai, *Tetrahedron Lett.*, **29**, 6133 (1988).

[186] Y. Nagao, T. Abe, H. Shimizu, T. Kumagai, and Y. Inoue, *J. Chem. Soc., Chem. Commun.*, **1989**, 821.

[187] Y. Nagao, T. Kumagai, T. Abe, M. Ochiai, T. Taga, K. Machida, and Y. Inoue, *J. Chem. Soc., Chem. Commun.*, **1987**, 602.

[188] Y. Nagao, W.-M. Dai, M. Ochiai, S. Tsukagoshi, and E. Fujita, *J. Am. Chem. Soc.*, **110**, 289 (1988).

[189] Y. Nagao, W.-M. Dai, M. Ochiai, S. Tsukagoshi, and E. Fujita, *J. Org. Chem.*, **55**, 1148 (1990).

[190] R. Deziel and M. Endo, *Tetrahedron Lett.*, **29**, 61 (1988).

[191] H. Huang, N. Iwasawa, and T. Mukaiyama, *Chem. Lett.*, **1984**, 1465.

[192] S. S. Labadie and J. K. Stille, *Tetrahedron*, **40**, 2329 (1984).

[193] E. Nakamura and I. Kuwajima, *Tetrahedron Lett.*, **24**, 3347 (1983).

[194] D. A. Evans, J. S. Clark, R. Matternich, V. J. Novack, and G. S. Sheppard, *J. Am. Chem. Soc.*, **112**, 866 (1990).

[195] L. S. Liebeskind, M. E. Welker, and R. W. Fengl, *J. Am. Chem. Soc.*, **108**, 6328 (1986).

[196] S. Kobayashi, T. Mukaiyama, *Chem. Lett.*, **1989**, 297.

[197] T. Mukaiyama, H. Uchiro, S. Kobayashi, *Chem. Lett.*, **1989**, 1001.

[198] S. Kobayashi, T. Sano, T. Mukaiyama, *Chem. Lett.*, **1989**, 1319.

[199] T. Mukaiyama, H. Uchiro, S. Kobayashi, *Chem. Lett.*, **1989**, 1757.

[200] S. Kobayashi, Y. Fujishita, T. Mukaiyama, *Chem. Lett.*, **1989**, 2069.

[201] T. Mukaiyama, S. Kobayashi, H. Uchiro, I. Shiina, *Chem. Lett.*, **1990**, 129.

[202] T. Mukaiyama, S. Kobayashi, *J. Organomet. Chem.*, **382**, 39 (1990).

[203] T. Mukaiyama, H. Uchiro, I. Shiina, S. Kobayashi, *Chem. Lett.*, **1990**, 1019.

[204] T. Mukaiyama, H. Uchiro, S. Kobayashi, *Chem. Lett.*, **1990**, 1147.

[205] T. Mukaiyama, S. Kobayashi, T. Sano, *Tetrahedron*, **46**, 4653 (1990).

[206] S. Kobayashi, Y. Fujishita, T. Mukaiyama, *Chem. Lett.*, **1990**, 1455.

[207] T. Mukaiyama, I. Shiina, S. Kobayashi, *Chem. Lett.*, **1990**, 2201.

[208] S. Kobayashi, Y. Tsuchiya, T. Mukaiyama, *Chem. Lett.*, **1991**, 541.

[209] S. Kobayashi, A. Ohtsubo, T. Mukaiyama, *Chem. Lett.*, **1991**, 831.

[210] S. Kobayashi, H. Uchiro, Y. Fujishita, I. Shiina, T. Mukaiyama, *J. Am. Chem. Soc.*, **113**, 4247 (1991).

[211] T. Mukaiyama, M. Furuya, A. Ohtsubo, S. Kobayashi, *Chem. Lett.*, **1991**, 989.

[212] T. Mukaiyama, H. Asanuma, I. Hachiya, S. Kobayashi, *Chem. Lett.*, **1991**, 1209.

[213] S. Kobayashi, T. Harada, J. S. Han, *Chem. Express*, **6**, 563 (1991).

[214] S. Kobayashi, M. Furuya, A. Ohtsubo, T. Mukaiyama, *Tetrahedron Asym.*, **7**, 635 (1991).

[215] T. Mukaiyama, I. Shiina, S. Kobayashi, *Chem. Lett.*, **1991**, 1901.

[216] S. Kobayashi, I. Shiina, J. Izumi, T. Mukaiyama, *Chem. Lett.*, **1992**, 373.

[217] K. Sisido, Y. Takeda, and Z. Kinugasa, *J. Am. Chem. Soc.*, **83**, 538 (1961).

[218] B. J. Batchelor, J. N. R. Ruddick, J. R. Sams, and F. Aubke, *Inorg. Chem.*, **16**, 1414 (1977).

[219] J. A. Dale, D. L. Dull, and H. S. Mosher, *J. Org. Chem.*, **34**, 2543 (1969).

[220] K. Toshima, K. Tatsuta, and M. Kinoshita, *Bull. Chem. Soc. Jpn.*, **61**, 2369 (1988).

# CHAPTER 2

# THE [2,3]-WITTIG REARRANGEMENT

Takeshi Nakai and Koichi Mikami

*Tokyo Institute of Technology*
*Meguro, Tokyo 152, Japan*

## CONTENTS

|  | PAGE |
|---|---|
| Acknowledgments . | 106 |
| Introduction . | 106 |
| Mechanism . | 107 |
|   [2,3] Shift vs. [1,2] Shift . | 107 |
|   [2,3] Shift vs. [1,4] Shift . | 108 |
|   Transition State Conformation. | 109 |
| Scope and Limitations . | 109 |
|   Acyclic Substrates . | 109 |
|   Cyclic Substrates . | 112 |
|   Competing Reactions . | 115 |
| Stereochemical Control . | 116 |
|   Olefinic Stereoselection . | 117 |
|   Diastereoselection . | 119 |
|     Transition-State Model . | 119 |
|     Highly Selective Variants . | 121 |
|     Cyclic Variants . | 122 |
|   Asymmetric Synthesis . | 123 |
|   Asymmetric Transition Type . | 124 |
|   Asymmetric Induction Type . | 128 |
|   Chiral Base-Induced Type . | 132 |
|   Configurationally Defined Carbanion Type . | 133 |
| Synthetic Applications . | 134 |
|   Acyclic Stereocontrol . | 134 |
|   Steroid Side-Chain Synthesis . | 136 |
|   Medium and Large Ring Natural Product Synthesis . | 139 |
|   Sigmatropic Sequences . | 141 |
| Experimental Procedures . | 145 |
|   *threo*-4-Methyl-5-hexen-1-yn-3-ol [Rearrangement of (*E*)-Crotyl Propargyl Ether]. | 145 |
|   *erythro*-4-Methyl-5-hexen-1-yn-3-ol [Rearrangement of | |
|   (*Z*)-Crotyl γ-(Trimethylsilyl)propargyl Ether] . | 145 |
|   (−)-(3*R*,4*S*,5*E*)-2,4,7-Trimethyl-1,5-octadien-3-ol [Rearrangement of a Bis(allylic) Ether] | 145 |

*Organic Reactions, Vol. 46*, Edited by Leo A. Paquette et al.
ISBN 0-471-08619-3    © 1994 Organic Reactions, Inc. Published by John Wiley & Sons, Inc.

(1S,3aS,7aS)-7,7a-Dihydro-4-isobutyl-7a-methyl-1-[(2-(trimethylsilylethoxy)methoxy]-
3a(6H)-indanmethanol [Stille Rearrangement via Tin–Lithium Exchange]    .    .    .    146
1-Undecen-4-ol [Reductive Lithiation-Mediated [2,3]-Wittig Rearrangement] .    .    .    146
(S,S)-anti-2-Isopropyl-9-cyclodecyn-1-ol ([2,3]-Wittig Ring Contraction of a
13-Membered Propargylic Allylic Ether through Use of a Chiral Base) .    .    .    .    147
(2S,3R)-2-Hydroxy-3-methylpentanoic acid [Rearrangement via the Transmetalation
of a Lithium Enolate to a Zirconium Enolate]    .    .    .    .    .    .    .    .    148
8-Phenylmenthyl 2-Hydroxy-3-methyl-4-pentenoate [Rearrangement via Transmetalation
of an Enol Silyl Ether with Titanium Tetrachloride]    .    .    .    .    .    .    .    148
Methyl 2-Hydroxy-3-methyl-4-pentenoate [Silyl Triflate-Catalyzed-Rearrangement]    .    149
TABULAR SURVEY    .    .    .    .    .    .    .    .    .    .    .    .    .    .    149
  Table I. Rearrangements of Benzylic Ethers .    .    .    .    .    .    .    .    .    151
  Table II. Rearrangements of Alkyl Ethers    .    .    .    .    .    .    .    .    .    156
  Table III. Rearrangement of Allylic Ethers    .    .    .    .    .    .    .    .    .    168
  Table IV. Rearrangements of Propargylic Ethers .    .    .    .    .    .    .    .    174
  Table V. Enolate Rearrangements of α-Allyloxy Carbonyl Compounds    .    .    .    .    182
  Table VI. [2,3]-Wittig Ring Contractions    .    .    .    .    .    .    .    .    .    200
REFERENCES    .    .    .    .    .    .    .    .    .    .    .    .    .    .    .    206

## ACKNOWLEDGMENTS

We wish to thank Professor Shinji Murai of Osaka University and Professor Andrew S. Kende of The University of Rochester for assistance in the literature searches. In addition, we would like to acknowledge the significant contribution of Dr. Robert Joyce in the preparation of this chapter.

## INTRODUCTION

The [2,3]-sigmatropic rearrangement, generalized by Eq. 1, constitutes a versatile type of bond reorganization which encompasses a number of variations in

$$\text{(Eq. 1)}$$

terms of both the atom pair (X,Y) and the type of electron pair on Y (anions, nonbonding electron pairs, or ylides). The Sommelet–Hauser rearrangement is representative.[1]

This chapter focuses on the special class of [2,3]-sigmatropic rearrangement that involves an oxycarbanion (X = oxygen, Y = carbanion) as the migrating terminus (Eq. 2). This type of rearrangement is now termed the [2,3]-Wittig (sig-

$$\text{(Eq. 2)}$$

matropic) rearrangement. The reaction name clearly originates from the fact that this rearrangement formally represents a [2,3]-sigmatropic version of the classic Wittig rearrangement,[2,3] a well-known 1,2-alkyl shift of oxycarbanions (Eq. 3).

(Eq. 3)

The [2,3]-Wittig rearrangement has a rather recent history. Perhaps the first observation of the [2,3]-Wittig shift is the rearrangement of the allyl fluorenyl ether **1** (Eq. 4),[4-7] which was made in 1960 in the context of mechanistic studies on the Wittig rearrangement.

(Eq. 4)

The period of the 1960s to the early 1970s witnessed slow progress with a focus on mechanistic studies mainly of allyl benzyl ether systems.[8-10] The synthetic power of this carbanion rearrangement as a general method was recognized when Still (1978)[11] and Nakai (1981)[12] established the highly stereoselective variants of the genuine [2,3]-Wittig rearrangement. In recent years the [2,3]-Wittig rearrangement has enjoyed widespread application in many facets of organic synthesis. Various aspects of the reaction have been reviewed.[13-17]

This chapter deals with the mechanism, scope and limitation, stereochemistry, and synthetic applications of the [2,3]-Wittig rearrangement with emphasis on the stereochemical aspects and the synthetic utility. Other hetero [2,3]-Wittig rearrangements such as thio-[2,3]-Wittig variants are not covered.

## MECHANISM

### [2,3] Shift vs. [1,2] Shift

The competition and mechanistic distinction of the [1,2] and [2,3] shift in the carbanion rearrangement concerned was the subject of early mechanistic studies.[8-10] As exemplified in Eq. 4, the [1,2] shift often competes with the [2,3] shift to an extent that depends markedly on the substrate structure and the reaction temperature. For instance, the rearrangement of benzyl ether **2** affords a mixture of the [1,2] and [2,3] products, with the ratio varying with the temperature.[18]

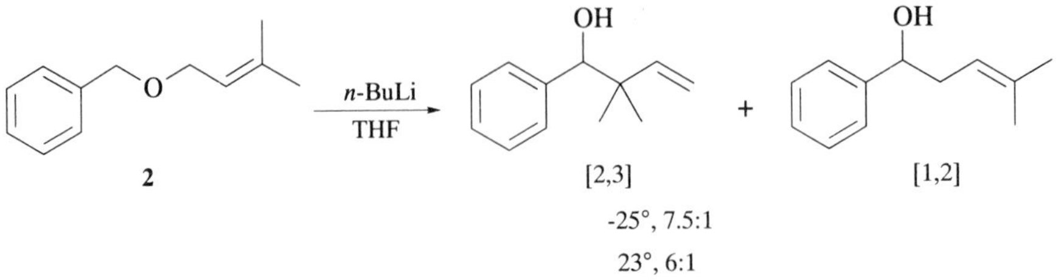

$$2 \xrightarrow[\text{THF}]{n\text{-BuLi}} \quad [2,3] \quad + \quad [1,2]$$

-25°, 7.5:1

23°, 6:1

However, it is now generally recognized that concurrence of the [1,2] shift can be minimized and often suppressed completely when the reaction is carried out at the proper temperature.

Moreover, it is now widely accepted that the [1,2]-Wittig rearrangement proceeds via a radical dissociation–recombination mechanism,[8–10] whereas the [2,3]-Wittig rearrangement is a concerted, thermally allowed sigmatropic reaction following the Woodward–Hoffman rules[19] or Fukui's frontier orbital theory.[20] Thus the [2,3]-Wittig rearrangement is a concerted reaction that proceeds through a six-electron, five-membered transition state in a suprafacial fashion along the allylic array as depicted in A or B. Theoretically, it is thus evident that

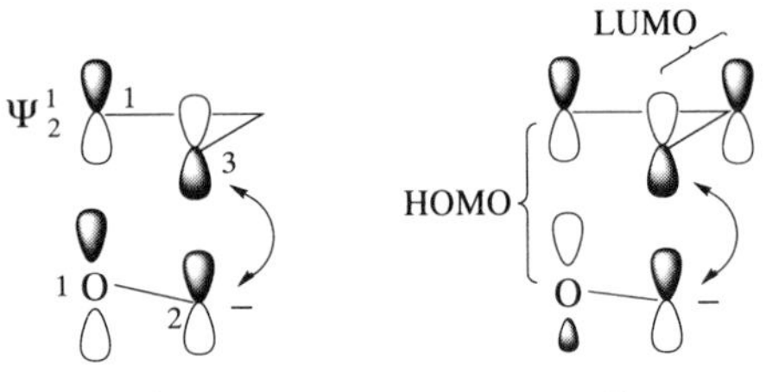

A                                    B

the smaller the energy gap between the HOMO (carbanion) and the LUMO (allyl), the more readily the rearrangement occurs. This means roughly that the less stable the carbanion involved, the faster the rearrangement.

## [2,3] Shift vs. [1,4] Shift

An additional problem of periselectivity arises in the carbanion rearrangement of the bis(allyl) ether system. In this case, both the [1s,4s] and [2s,3s] shifts are allowed by orbital symmetry (Eq. 5), together with the nonconcerted [1,2] and

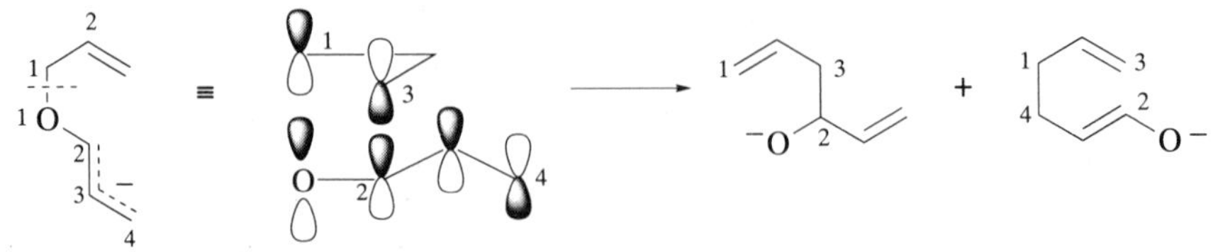

(Eq. 5)

[3,4] shifts.[8-10] In fact, the rearrangement of bis(prenyl) ether affords 8% of the [1,4] product, along with 67% of the [2,3], 14% of the [1,2], and 10% of the [3,4] products.[18] The exact mechanism of the [1,4] shift (concerted vs. dissociation–recombination) is still controversial.[8-10,21] The problem of periselectivity for the [2,3] vs. [1,4] shifts becomes more serious with certain cyclic ether substrates.

## Transition State Conformation

To understand the stereochemistry of the [2,3]-Wittig rearrangement, the conformation of the five-membered transition state needs to be determined. That is an extremely difficult task.[22] Even given the reasonable postulate that the [2,3] shift proceeds via a "folded envelope" conformation, there are still the three options **A–C**.

A                          B                          C

While conformers **A** and **B** are often used to explain the stereochemistry of [2,3]-sigmatropic processes,[8-10,18,23] conformer **C** has been proposed by the authors of this review as the preferable conformation for the transition state of the [2,3]-Wittig process.[24] While molecular mechanics (MM2) calculations have led to conflicting results,[25,26] a recent ab initio molecular orbital calculation has shown that the transition state conformation quite similar to conformer **A** is located by the 3-21G and 6-31+G basis set levels.[27] In the literature, however, both conformers **A** and **C** have often been used. It appears that the basic conformers **A** and **C** work equally well to explain the stereochemical outcome in most [2,3]-Wittig processes. Thus the basic conformer of type **C**, more familiar to the authors of this chapter, is used as a working hypothesis throughout this review *only* for the sake of consistency.

### SCOPE AND LIMITATIONS

Generally speaking, the [2,3]-Wittig rearrangement can be achieved with any $\alpha$-(allyloxy)carbanions with different substituents (G) on the migrating terminus. Thus its synthetic utility is determined principally by the availability of methods for generating the oxycarbanions at temperatures low enough to minimize the undesirable [1,2] shift. Nonetheless, the [2,3]-Wittig rearrangement has limitations in the range of applicable substrates, especially for some cyclic substrates.

## Acyclic Substrates

The [2,3]-Wittig rearrangement of acyclic ethers can usually be achieved in a highly periselective manner as long as carbanion generation and rearrangement can be carried out at temperatures ranging from $-60$ to $-85°$. The most popular method for carbanion generation is direct lithiation (deprotonation) with butyllithium or lithium diisopropylamide (LDA). A standard procedure is illustrated in Eq. 6.[28,29] The deprotonation method is widely applicable to various types of substrates, including bis(allyl) ethers, allyl benzyl ethers, allyl propargyl ethers,

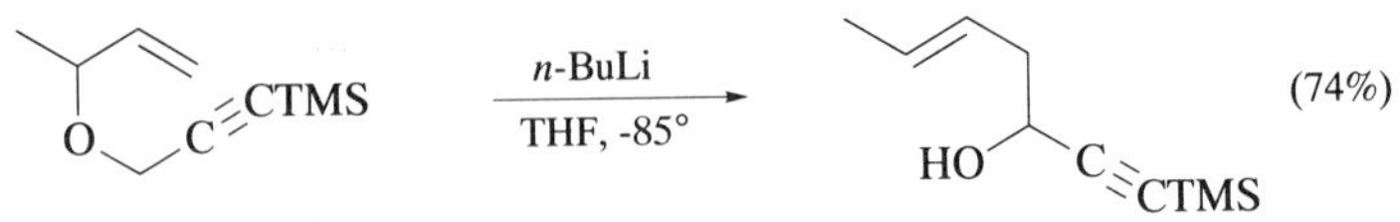

(74%)   (Eq. 6)

and $\alpha$-allyloxy carbonyl compounds. Of course, the applicability of this method is restricted to those compounds possessing relatively acidic $\alpha$ hydrogens. This limitation can be overcome by transmetalation methods such as the tin–lithium exchange reaction (Eq. 7)[11] or reductive lithiations of $O,S$-acetals (Eq. 8).[30,31] The

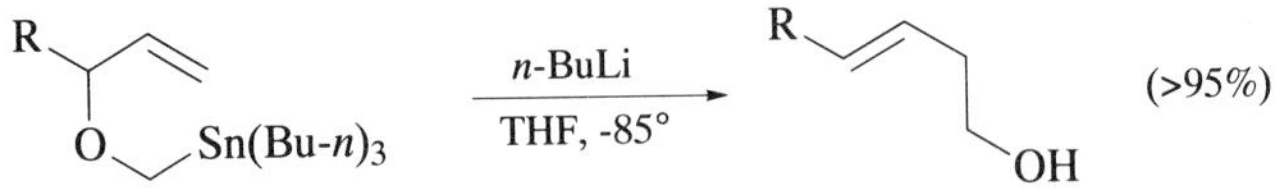

(>95%)   (Eq. 7)

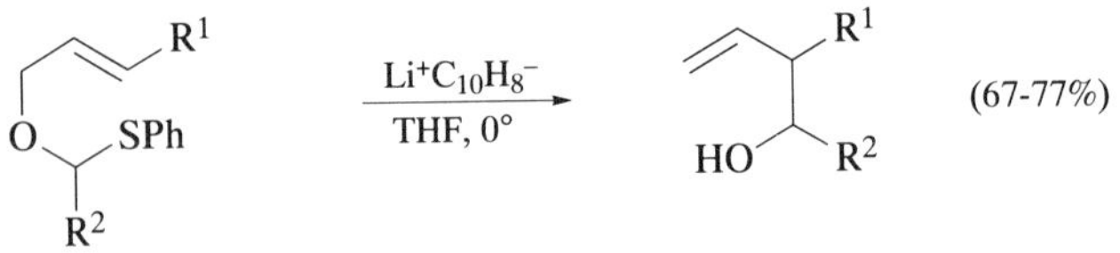

(67-77%)   (Eq. 8)

use of the tin–lithium exchange procedure by Still and Mitra was significant in the history of the [2,3]-Wittig rearrangement, since it uncovered the great synthetic potential of this type of rearrangement.

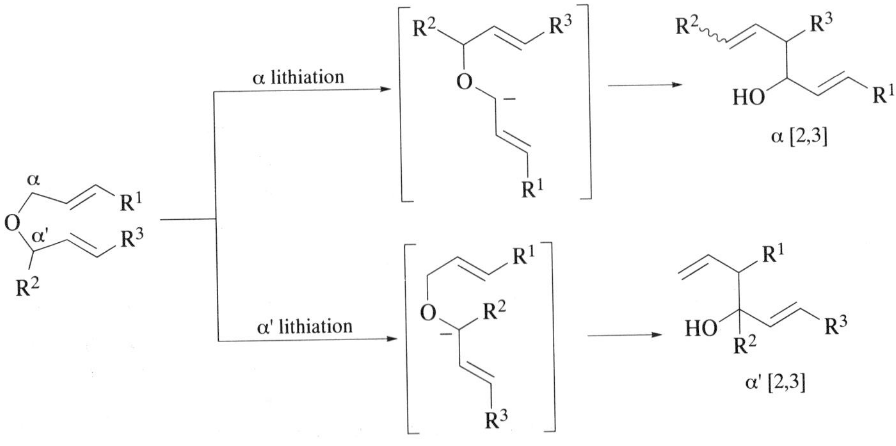

Particularly notable is the [2,3]-Wittig rearrangement of unsymmetrical bis(allylic) ethers, another important variant which considerably enhanced the synthetic potential of the [2,3]-Wittig rearrangement.[12] In this reaction, an additional regiochemical problem arises from the possibilities for $\alpha$ vs. $\alpha'$ lithiation. A general regioselection rule has now been established from systematic studies of the rearrangement of a variety of bis(allylic) ethers under standard conditions (butyllithium, tetrahydrofuran, −85°).[12] In the rearrangements of unsymmetrical substrates with different substitution patterns at the $\alpha$ and $\gamma$ positions of the two allylic moieties, lithiation takes place exclusively on the less-substituted allylic moiety, thus leading to the exclusive formation of the $\alpha$-[2,3] Wittig product as the single regioisomer. Thus **3** provides the single regioisomer **4**, while **5** affords a 1:2 mixture of the $\alpha$- and $\alpha'$-[2,3] products. The $\beta$-alkyl substitution has little effect. Thus **6** provides a 3:4 mixture of the regioisomers, whereas **7** affords only the $\alpha$-[2,3] product.

3

4   (65%)

5

6, R = H
7, R = Me

However, no problem with regioselectivity arises when a carbanion-stabilizing group such as methylthio or trimethylsilyl is at the $\gamma$ position of one allyl moiety; both **8**[32] and **9**[33] provide the $\alpha$-[2,3] product exclusively, independent of the

8

9

substitution pattern. Such positional ambiguities are not present in the rearrangements of allyl propargyl ethers since lithiation occurs exclusively on the propargylic moiety.[12]

Replacement of the allyl migrating group by a propargylic group constitutes another general class of [2,3]-Wittig variants that afford allenic alcohols (Eq. 9).[34-37] While the yields are only modest because of the strained allenic

G = CN, C≡CMe

(50-64%)

(Eq. 9)

transition state, Still's variant (G = SnR$_3$) and the carboxylic acid variant (G = CO$_2$H) have been reported to give allenic products in respectable yields.[38,39]

## Cyclic Substrates

The [2,3]-Wittig rearrangement is also applicable to a wide variety of cyclic substrates. These variants are classified according to Ziegler's convention originally
proposed for the cyclic Claisen variants.[40] In this convention the carbons to which
the tether bridging the pericyclic array is attached are expressed in the form {m,n}.
The twelve types shown in Fig. 1 are conceptually possible. The subscripts *endo*

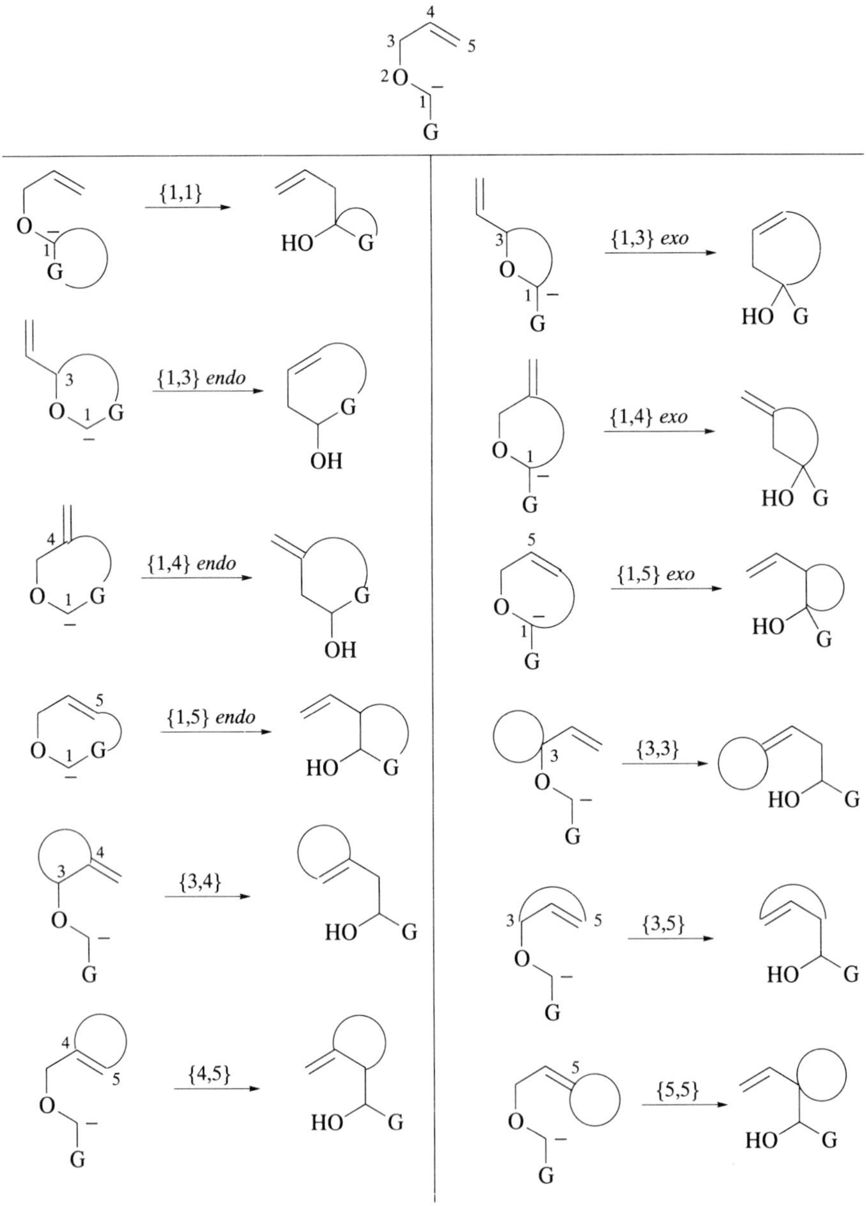

**FIG. 1**

and *exo* designate the presence of substituent G as a part of the tether in the ring, and as a substituent on the ring, respectively.

Each of these cyclic variants provides a different synthetic consequence, namely, ring enlargement for {1,3} rearrangement, ring contraction for {1,4} and {1,5} rearrangements, introduction of a new side chain onto the ring for {1,1}, {3,4}, {3,5}, and {4,5} rearrangements, and creation of a quaternary center on the ring for {5,5} rearrangement. Despite their great potential, however, only a limited number of cyclic rearrangements have been exploited thus far.

The {1,5}-[2,3]-Wittig rearrangement is relatively well studied in the context of synthesis of medium and large ring natural products. The rearrangements of the 13-membered bis(allyl) ether **10** [25,41] and the 17-membered allyl propargyl ether **11** [42,43] are the first reported prototypes of this group. The stereochemical features of this type of ring contraction and its application in natural product synthesis are described later. It should be noted here that the rearrangement of cyclic bis(allyl) ethers no longer obeys the regioselection rule mentioned above for acyclic substrates. In fact, **10** provides a 1:3 mixture of the $\alpha$- and $\alpha'$-[2,3] products [41] and **12** affords only the $\alpha'$-[2,3] product via exclusive lithiation on the more substituted moiety. [44]

There are several examples of {3,4}-[2,3]-Wittig rearrangements, particularly in steroid side chain synthesis. The earliest examples are shown in Eq. 10. [45-47] The stereochemical features and synthetic applications of this methodology are the subject of a subsequent section.

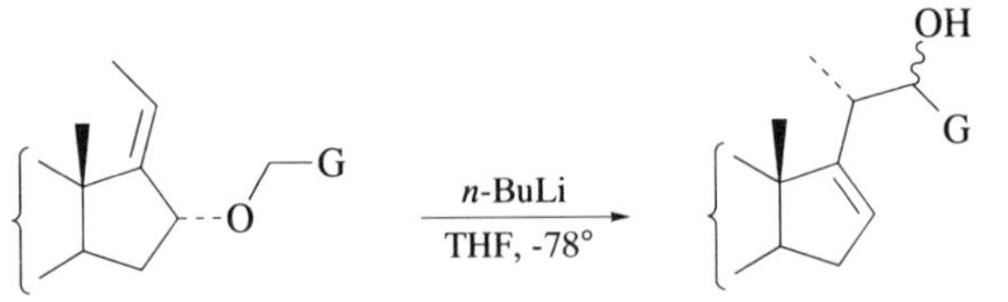

(Eq. 10)

G = Sn(Bu-$n$)$_3$, C≡CX (X = H, TMS)                    (75 to >95%)

The {3,5} rearrangement was also relatively well studied. However, this type of cyclic variant often suffers the serious drawback that the undesired [1,2] shift competes seriously with it (Eq. 11).[48] Of special interest are the allyl cyclohexenyl

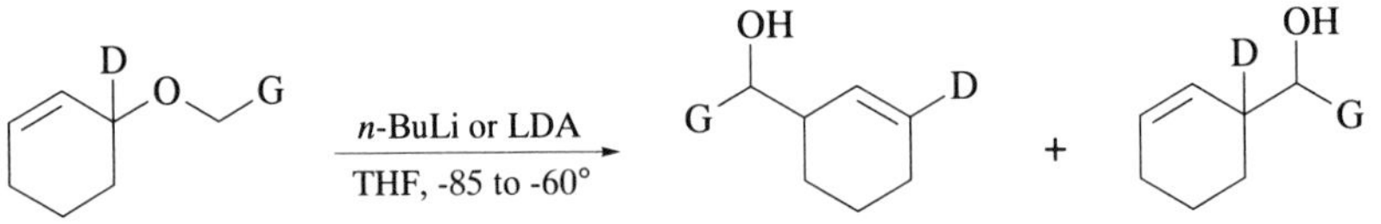

$$\xrightarrow[\text{THF, -85 to -60°}]{n\text{-BuLi or LDA}}$$

[2,3]            [1,2]

G = C≡CH, 38:62
G = Ph, 37:63
G = CO$_2$H, 67:33                    (Eq. 11)

ethers **13**[49] where the [1,4] product is also formed, with the ratio dependent on the substitution pattern on the ring; a conformational effect is suggested to explain the variation in periselectivity. Interestingly, however, the [2,3] product is obtained mainly with the {3,5} prototype of the cyclohexenyl (Eq. 12)[50,51] and dihydrofuryl substrates (Eq. 13).[52-54]

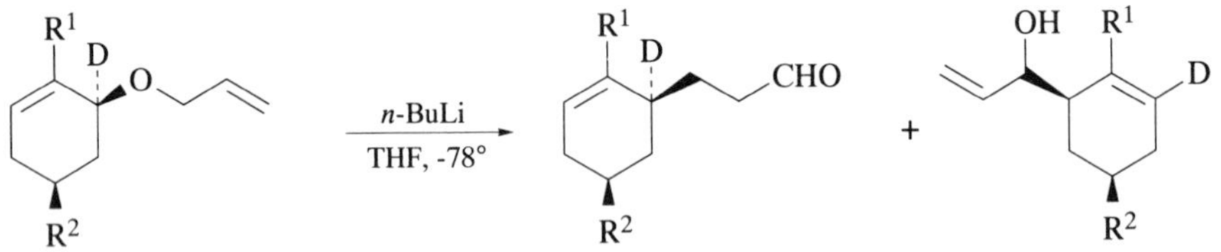

$$\xrightarrow[\text{THF, -78°}]{n\text{-BuLi}}$$

**13**                    [1,4]                    [2,3]

| R$^1$ | R$^2$ | [1,4] | : | [2,3] | : | [1,2] |
|-------|-------|-------|---|-------|---|-------|
| H | H | 12 | : | 24 | : | 64 |
| H | Me | 18 | : | 54 | : | 28 |
| Me | C(Me)=CH$_2$ | 48 | : | 6 | : | 46 |

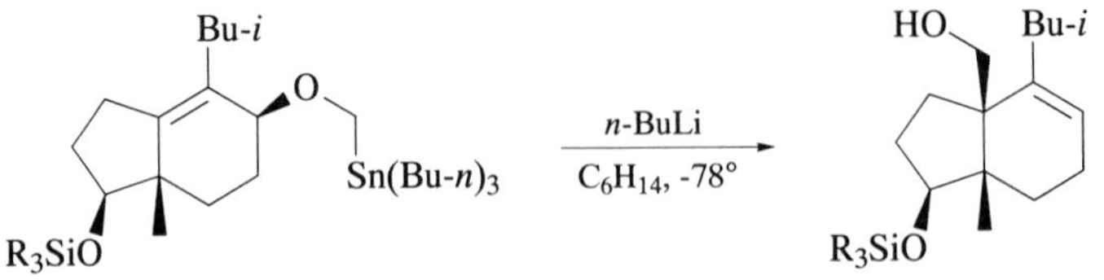

$$\xrightarrow[\text{C}_6\text{H}_{14}, \text{-78°}]{n\text{-BuLi}}$$

(34%)

(Eq. 12)

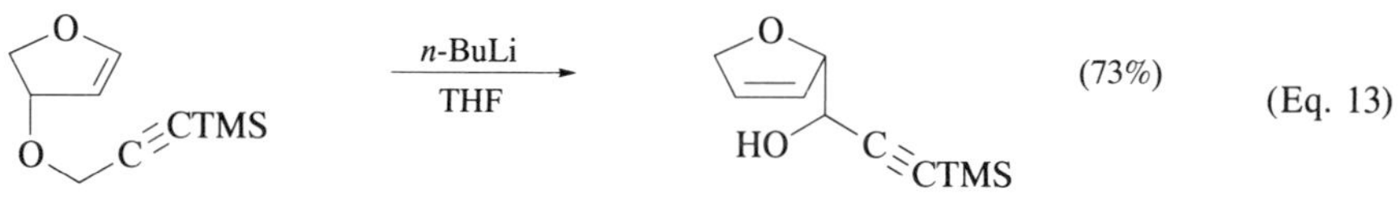

(73%)   (Eq. 13)

Examples of other types of {m,n} rearrangement are quite rare. An example of the {1,1} type is given in Eq. 4. A single example was reported for the {1,3}*exo*(six-membered ring with G = COMe),[55] the {3,3} (tetrahydrofuran ring with G = CO$_2$H),[56] the {4,5} (cyclohexane ring with G = C≡CMe),[57] and the {5,5} (cyclohexane ring with G = CO$_2$H).[58] No examples were reported of the {1,3}*endo*, {1,4}*exo*, and {1,5}*exo* rearrangements.

## Competing Reactions

The enolate rearrangement of α-allyloxy carbonyl systems **14** deserves special comment since two competing modes of sigmatropic processes, the [2,3]-Wittig rearrangement and the [3,3]-Claisen rearrangement, are conceivable.

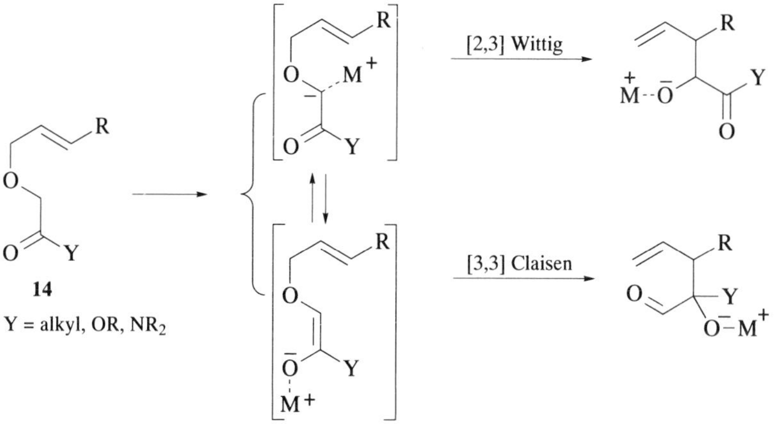

Conflicting observations have been reported for the allyloxy ketone system. Treatment of ketone **15** with potassium *tert*-butoxide[55] and of ketone **16a** with LDA in a mixture of hexamethylphosphoramide (HMPA) and tetrahydrofuran[59] affords the [2,3]-Wittig product exclusively, whereas treatment of ketone **16b** with potassium hydride (or sodium hydride) produces the [3,3]-Claisen product predominantly.[60]

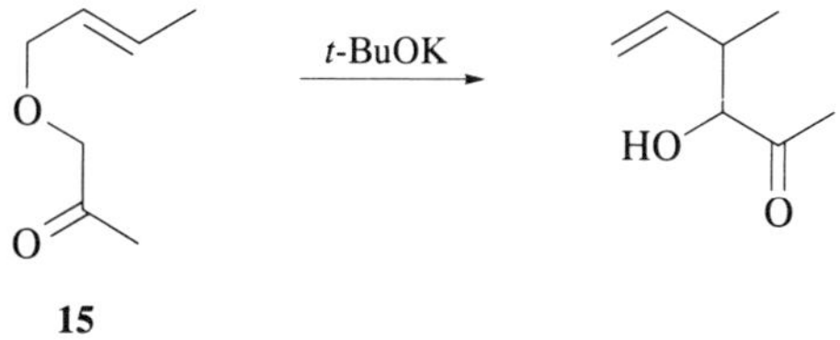

**16a**, R = H
**16b**, R = Me

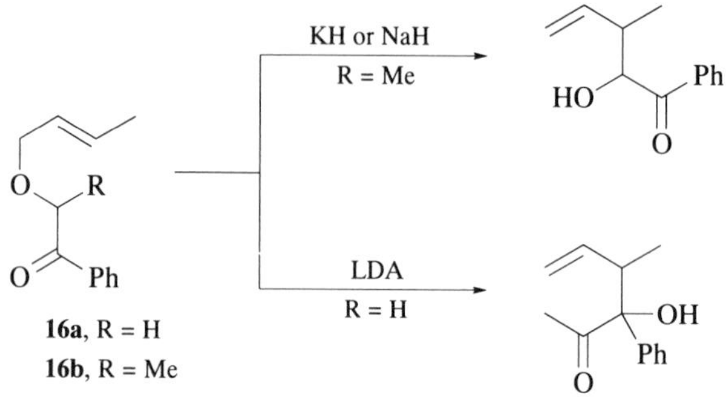

Interesting rearrangements were also reported for the acid **17,** the ester **18,** the amide **19,** and the silylated derivatives **20** and **21.** When the lithium enolate is generated with LDA in tetrahydrofuran, both acid **17**[58] and amide **19**[61] undergo the [2,3]-Wittig rearrangement at ca. −80°, whereas ester **18** does not.[62] In contrast, the lithium enolate of **18** generated in HMPA–tetrahydrofuran undergoes the [2,3]-Wittig shift at −70°.[62] Heating of **20** at 80° affords the [3,3]-Claisen product quantitatively,[62–64] whereas transmetalation of **20** with tin(IV) chloride or titanium(IV) chloride at −50° leads to the exclusive formation of the [2,3]-Wittig product.[65] On treatment with tetrabutylammonium fluoride at −70°, **21** affords the [2,3]-Wittig product quantitatively, while **20** does not rearrange under the same conditions.[62]

**21**

**17**, Y = OH
**18**, Y = OMe
**19**, Y = $N(CH_2)_4$

**20**

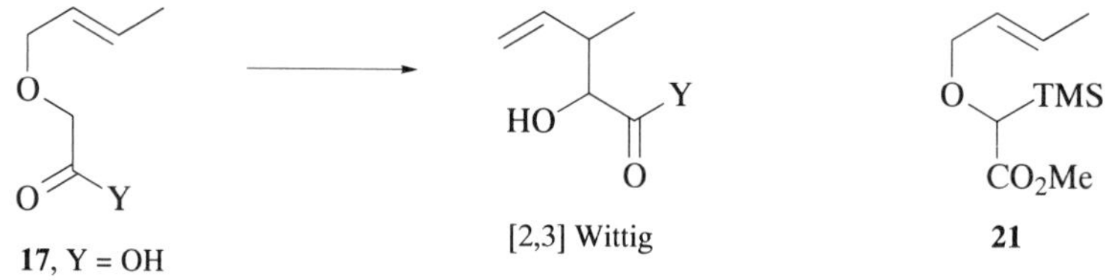

## STEREOCHEMICAL CONTROL

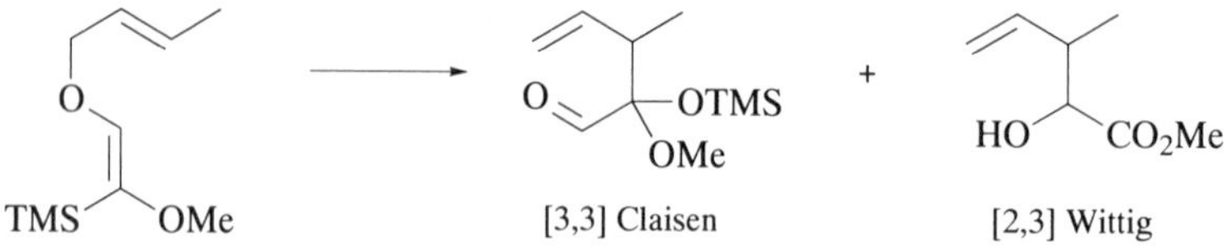

The [2,3]-Wittig rearrangement usually proceeds through a highly ordered cyclic transition state to create a new C-C double bond and a new C-C single bond. Consequently, this type of rearrangement should allow stereochemical control, including stereoselective generation of an olefinic bond, with both internal and relative asymmetric induction, and chirality transfer along the pericyclic array. While studies on olefinic stereoselection began simultaneously with the initial mechanistic studies, stereocontrol over the newly created chiral centers is the subject of recent investigations, mainly because the [2,3]-sigmatropic rearrangement was believed to show only modest levels of asymmetric induction[66] except for a single example.[18] Over the past decade, however, remarkable progress has been made in the development of stereoregulated [2,3]-Wittig variants, hence the [2,3]-Wittig technology is now increasingly utilized to great advantage in stereocontrolled organic synthesis. This section deals with the stereochemical principles that govern the [2,3]-Wittig rearrangement.

### Olefinic Stereoselection

The rearrangement of ethers derived from secondary allylic alcohols affords the $E$ and $Z$ products. Examination of the transition-state conformations suggests that the R group should prefer the *exo* orientation, thus leading to preferential formation of the $E$ isomer (Eq. 14). The $E$ preference is amply confirmed and widely recognized as a general attribute of the [2,3]-Wittig rearrangement.

(Eq. 14)

| R | G | |
|---|---|---|
| Me | C(R')=CH$_2$  (R' = H, Me)[12] | 98-100% $E$ |
| Me | C≡CR'  (R' = H, TMS)[67] | 93-98% $E$ |
| Me | Ph[18] | 100% $E$ |
| Me | CO$_2$H[12] | >75% $E$ |
| Me | CO$_2$Me[59] | 75% $E$ |

Two notable exceptions to this $E$ selectivity were reported. One is the rearrangement of the tin-substituted ethers **22** which affords the $Z$ product predominantly.[11] This unusual $Z$ selectivity was applied to the synthesis of the Cecropia juvenile hormone.[68] The other exception is the zirconium enolate rearrangement of allyloxy ester **23** which gives the $Z$ product exclusively.[69] It is also notable that the introduction of a bulky group such as trimethylsilyl at the $\beta$ position of the allyl group results in decreased $E$ selectivity because of the destabilization of the generally preferred *exo* transition state by the additional interaction of $\beta$-Me$_3$Si with

**22**

| $R^1$ | $R^2$ | |
|---|---|---|
| $n$-$C_7H_{15}$ | H | 60% *Z* |
| $n$-Bu | Me | 100% *Z* |

100% *Z*

**23** R = Me, $i$-Pr, $n$-Bu

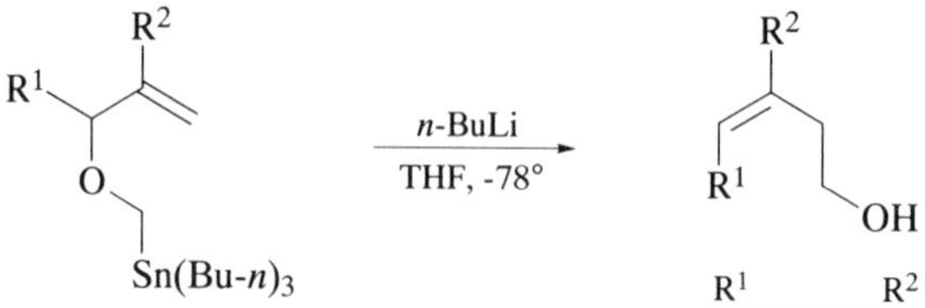

R.[70] The rearrangement of **24,** which proceeds with 80% *E* selectivity, is particularly interesting, leading eventually to the *Z* olefin in a leukotriene synthesis.[71]

**24** R = $n$-$C_5H_{11}$

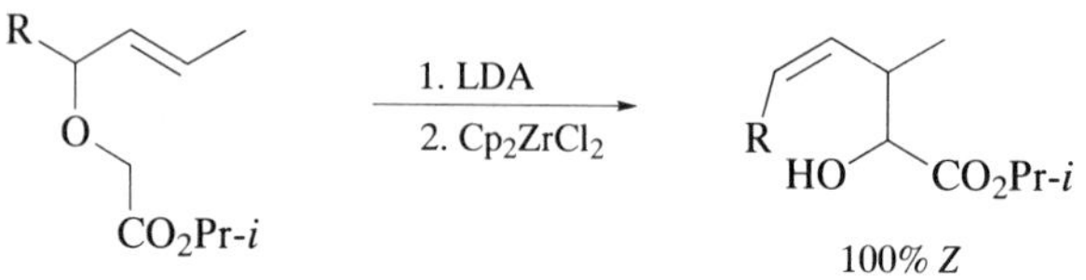

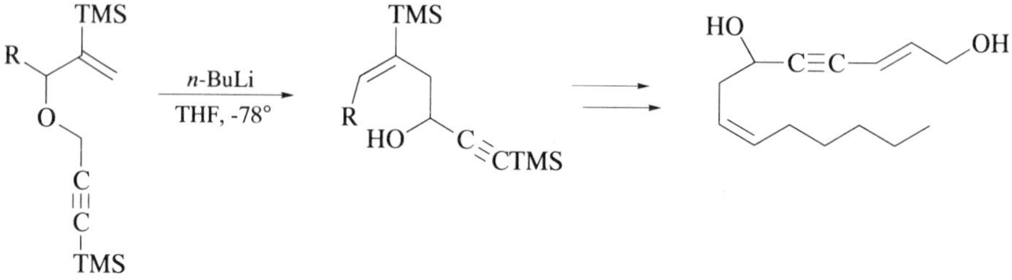

The rearrangement of ethers derived from tertiary allylic alcohols (Eq. 15) would not be expected to show high *E* selectivity, because of the small difference

(Eq. 15)

$R^1 > R^2$

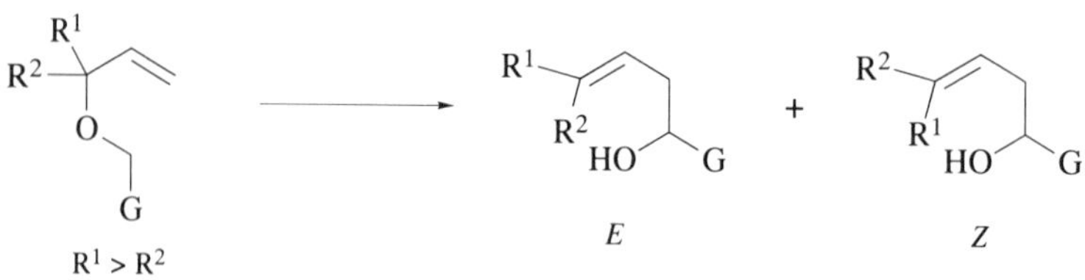

in energy between the two transition states. However, exclusive formation of the *E* trisubstituted olefin was reported in the rearrangement of **25** (Eq. 16),[72] where the methoxymethoxy (MOMO) group plays a key role in defining the transition state geometry. Of special interest is the {3,4} rearrangement of **26,** which affords varying amounts of the *E* and *Z* products, depending on the reaction conditions (Eq. 17).[73]

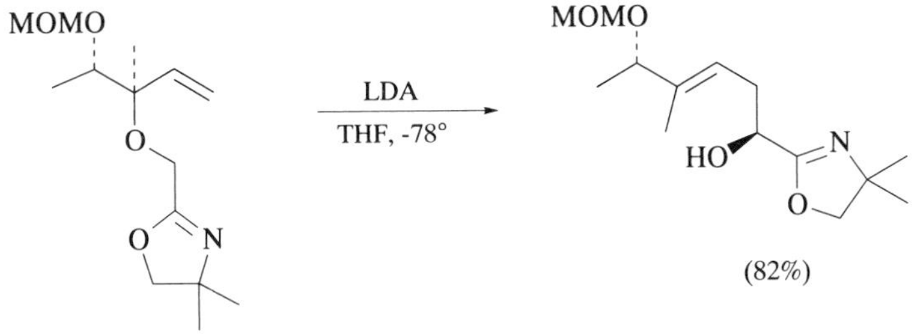

(Eq. 16)

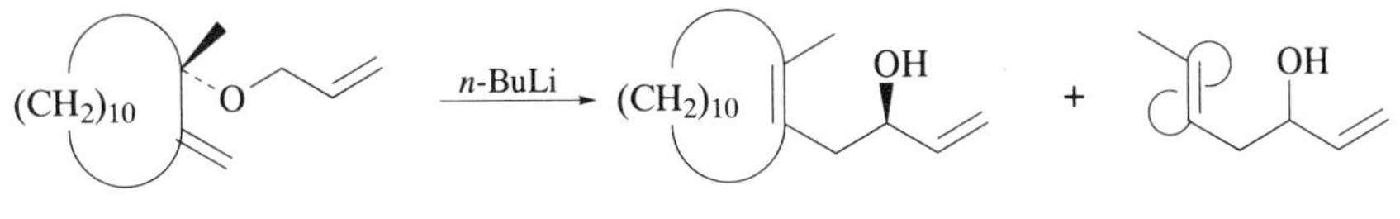

(82%)

**25**

THF 1:1
HMPA-THF 50:1

(Eq. 17)

**26**

## Diastereoselection

**Transition-State Model.** Of the stereoselections achievable with the [2,3]-Wittig rearrangement, diastereoselection (internal 1,2 asymmetric induction) with respect to the newly created vicinal chiral centers is the most important from the standpoint of stereocontrol. The rearrangement of γ-substituted allyl ethers provides the two racemic diastereomers through the two pairs of transition states (Fig. 2). Thus the $E$ substrate can proceed through transition states $T_1$ and $T_2$ which lead to the *threo* and *erythro* products, respectively. Similarly, the $Z$ substrate affords the *threo* and *erythro* products via $T_3$ and $T_4$, respectively. The two transition states ($T_1$ vs. $T_2$ or $T_3$ vs. $T_4$) are unequal in energy, and the *threo/erythro* ratio reflects the transition state geometry.

A detailed study of the rearrangements of a variety of the geometric pairs of crotyl ethers with different G groups revealed significant trends of diastereoselection. These results led to the proposal of the transition state model depicted below, which provides a logical basis for explaining and predicting the diastereoselection of a wide range of [2,3]-Wittig rearrangements.[24] The essentials are as follows: (1) The sense of diastereoselection is dictated primarily by the olefin geometry of the substrate, and the degree is determined critically by the nature of the G group on the migrating terminus. (2) As a general rule, an $E$ substrate exhibits *threo* selection, whereas a $Z$ substrate shows *erythro* selection. The general selection is rationalized in terms of the pseudo-1,3-diaxial interaction of G with $H_\beta$ in $T_2$ and $T_3$. For instance, the $Z$ to *erythro* selection occurs because $T_3$ is sterically less favorable than $T_4$. (3) The *erythro* selectivity observed for the $Z$ series is in the order: G = Ph (100%) > C(CH$_3$)=CH$_2$ > CH=CH$_2$ > C≡CH (90%), whereas the *threo* selectivity observed for the $E$ series is in the opposite order: G = C≡CH (99%) > CH=CH$_2$ > C(CH$_3$)=CH$_2$ (70%) > Ph (53%). While this order of *erythro* selection is consistent with the expected order of the 1,3 re-

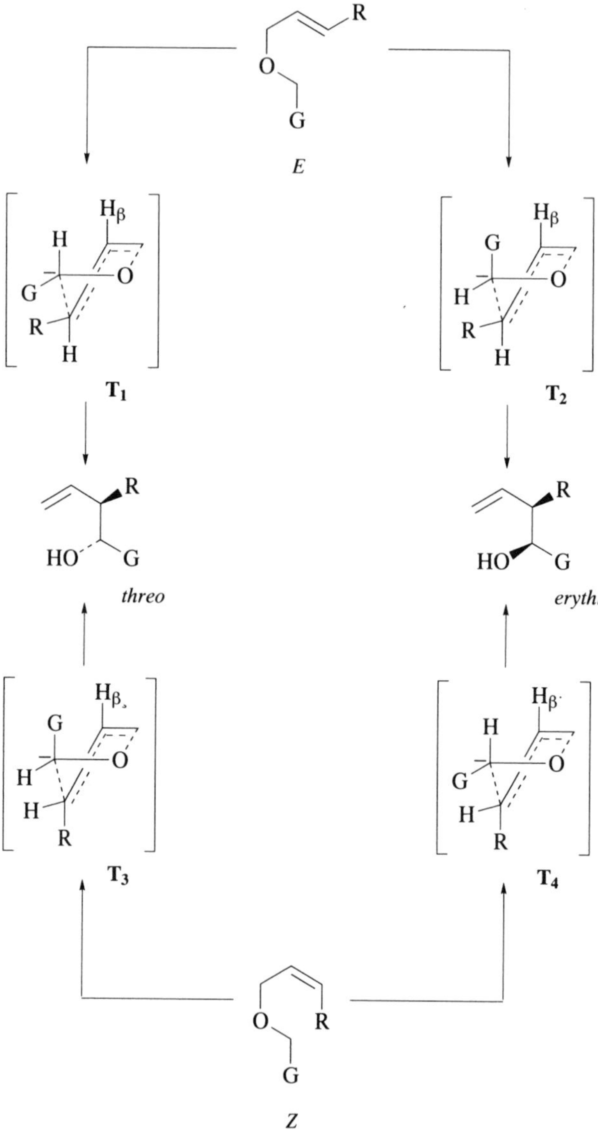

**FIG. 2**

pulsion concerned, the order of *threo* selection is best explained by assuming an additional gauche interaction of G with R in the preferred $T_1$, which operates to diminish the *threo* selectivity. (4) A notable exception to this general selection rule is the enolate [2,3]-Wittig family including G = $CO_2H$, $CO_2R$, and $CONR_2$,[58,61–65] where a relatively high degree of E to *erythro* selection is generally observed, while the Z substrates show a low level of either *threo* or *erythro* selection. The unusual E to *erythro* selection reflects the fact that the gauche interaction in $T_1$ prevails overwhelmingly over the 1,3 repulsion in $T_2$. This transition state model is further strengthened by many other examples and also by theoretical calculations,[26] thus providing a logical basis for developing highly diastereoselective variants described below. It should be noted, however, that a different transition state model has recently been proposed.[27]

**Highly Selective Variants.** Among the variants described so far, only two (G = Ph for Z to *erythro* and G = C≡CH for E to *threo*) provide a synthetically useful level (>98%) of diastereoselection. However, other variants have been developed that exhibit >98% for either *threo* or *erythro* selection. Equations 18 and 19 summarize the highly E to *threo* and Z to *erythro* selective variants, respectively.[28-29] It should be noted that the G group leading to an enhanced E to *threo*

$$G = C≡CH \quad 99\% \; threo$$
$$G = C≡CMe \quad 99\% \; threo$$

(Eq. 18)

$$G = Ph \quad >99\% \; erythro$$
$$G = C≡CMe \quad 100\% \; erythro$$
$$G = C≡CTMS \quad 100\% \; erythro$$

(Eq. 19)

selectivity does not provide an increased Z to *erythro* selectivity and vice versa, except for the pair of G = C≡CMe.

Of special interest is the case of G = C≡CSiMe$_3$, in which the Z substrate exhibits an extremely high *erythro* selectivity that surprisingly exceeds the geometric purity of the substrate used, whereas the E counterpart also shows *erythro* selection, although the degree is moderate (73%).[28,29] From the synthetic point of view, the propargylic variants thus developed possess particular advantages; they can provide an extremely high level of either diastereoselection by the proper choice of crotyl geometry and ethynyl group, and the rearrangement product possesses unique multifunctionality which readily allows a variety of further transformations, as demonstrated in the formal total synthesis of (+)-oudemansin.[28,29]

On the other hand, two highly E to *erythro* selective variants were also developed which afford $\alpha$-hydroxy-$\beta$-alkyl carboxylic acid derivatives of synthetic value (Eq. 20).[61,74] In these cases, the Z counterparts show a much lower *erythro* selectivity. Of further interest is that the rearrangement of (E)-**27a** shows a high *erythro* selectivity, whereas (E)-**27b** provides a high *threo* selectivity.[75,76] Another notable exception is the rearrangement of an E/Z mixture of **28** where both the E and Z substrates show extremely high *threo* selectivity.[77]

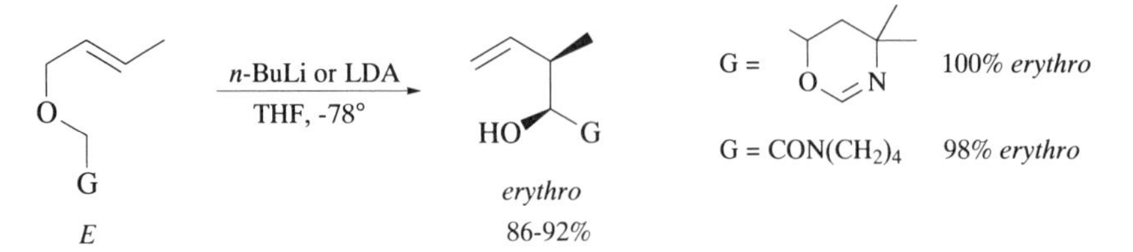

$$G = \text{(oxazoline structure)} \qquad 100\% \; erythro$$

$$G = CON(CH_2)_4 \qquad 98\% \; erythro$$

erythro
86-92%

(Eq. 20)

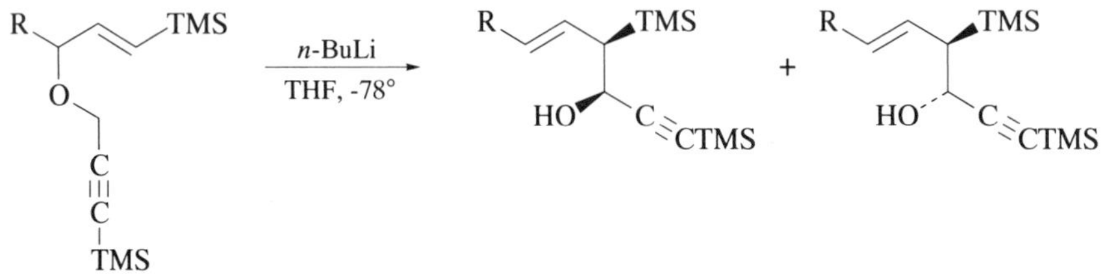

(E)-**27a**, R = H                                                97:3        (85-97%)
(E)-**27b**, R = Me                                              3:97

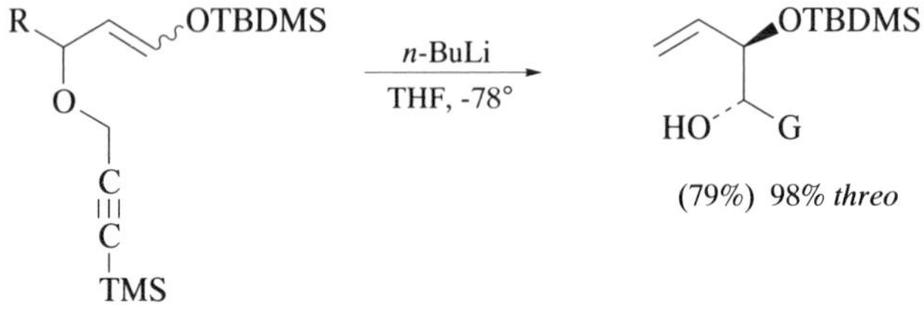

(79%) 98% threo

**28** E/Z = 35:65

**Cyclic Variants.** The diastereoselection in cyclic variants is different from
that in the acyclic counterparts discussed so far, since the proximity of the reacting
centers enforced by bridging is expected to profoundly influence the transition
state geometry. The diastereoselection in ring contraction via the {1,5} rearrangement is relatively well studied. For instance, the rearrangement of the 13-
membered bis(allyl) ether **29** provides an entirely different selection from that of
the acyclic counterpart.[41] Both the E, E and Z, Z substrates show *threo* (*trans*) selectivity to afford (E)-*trans*-**30** and (Z)-*trans*-**31** as single products, respectively,

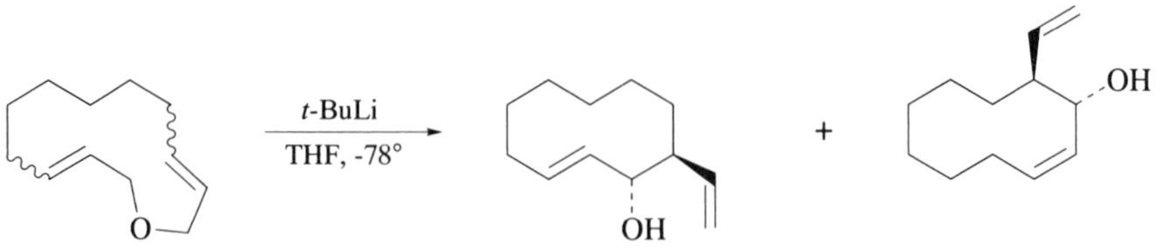

**29**                              **30**                              **31**

(94-96%)

while the *E*, *Z* substrate produces a complex isomeric mixture. In contrast, rearrangement of the 13-membered allyl propargyl system **32** follows the general selection rule.[78,79] The *E* substrate provides the *threo* (*trans*) product **33** exclusively, whereas the *Z* substrate affords the *erythro* (*cis*) product **34** exclusively. Reasonable mechanistic grounds were advanced for these diastereoselections. For instance, the high *E* to *threo* (*trans*) selection reflects the fact that conformer **C** is sterically less congested, since conformer **D** suffers from the steric interaction indicated.[25,42,43] Interestingly, diastereoselection in the rearrangement of the 17-membered ether **11** depends markedly upon the solvent used; hexane–tetrahydrofuran affords the *threo* (*trans*) product predominantly, whereas HMPA–tetrahydrofuran produces the *erythro* (*cis*) isomer as the major product.[42,43] These observations suggest that ring size of the substrate and/or the nature of the base used are additional factors in determining diastereoselectivity in the [2,3]-Wittig ring contractions.

**32** **33** **34**

(>90%)

**C** ⟶ *threo* (*trans*) **D** ⟶ *erythro* (*cis*)

**11**

10% THF-$C_6H_{14}$, -20° 4.5:1

4% HMPA-THF, -78° 1:6

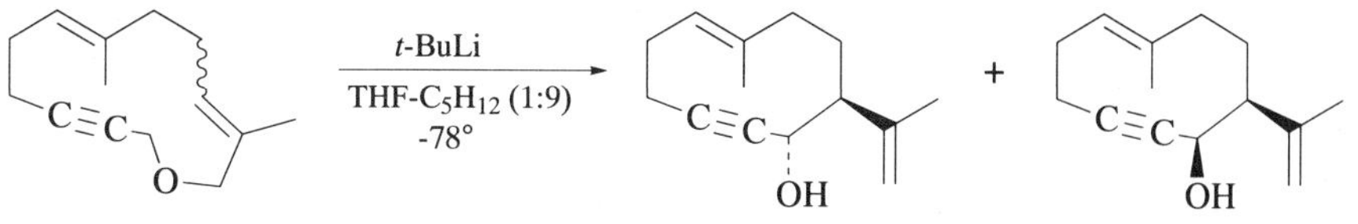

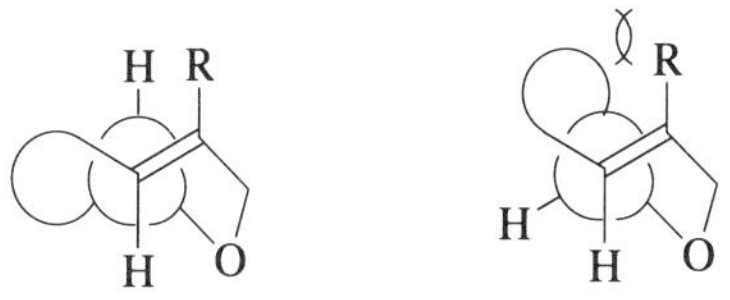

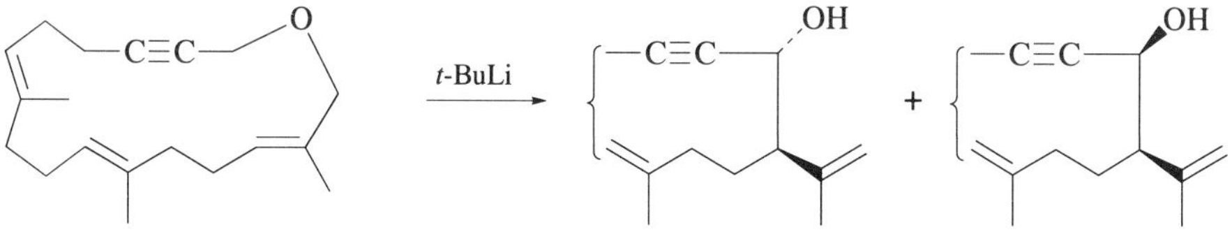

## Asymmetric Synthesis

In principle, the asymmetric versions of the [2,3]-Wittig rearrangement leading to optically active products can be classified into four types in terms of the key stereoselection involved. The first is the so-called "asymmetric transition type" that employs a chiral substrate derived from an *enantiomerically enriched* allylic alcohol and hence involves asymmetric transmission (chirality transfer) along the allylic array. The second is the "asymmetric induction type," which employs a substrate having a *chiral* substituent somewhere on the pericyclic array; hence the problem of diastereofacial selection (relative asymmetric induction) arises between the preexisting stereocenter and the newly created chiral center(s). The third is the "chiral base-induced type," which is an enantioselective reaction of an achiral substrate with a chiral *nonracemic* base. The fourth is the "configurationally defined carbanion type," which involves a diastereo- or enantiomerically defined lithium-bearing terminus, wherein the steric course of inversion vs. retention at the migrating terminus is the key issue.

**Asymmetric Transmission Type.**   The most synthetically valuable feature of the [2,3]-Wittig rearrangement is its ability to transmit the chirality at C-1 of the allylic moiety into the newly created chiral centers at C-3 and C-4 as generalized in Eq. 21. This type of asymmetric rearrangement destroys the original chirality

(Eq. 21)

while simultaneously creating new ones, and hence there is conservation of optical purity. The asymmetric transmission technology is widely applicable to both acyclic and cyclic substrates, thus providing an efficient tool for concurrent control of absolute and relative stereochemistry with considerable stereopredictability.

Guided by the transition state model advanced above, one can readily predict both the olefinic geometry and the absolute and relative stereochemistry of the product from the three variables in the substrates: the absolute configuration, the double bond geometry, and the nature of the G group. In an asymmetric version of the highly (Z to *erythro*)-selective variants shown in Eq. 19, for instance, (S)-(Z)-**35** would rearrange predominantly to (E)-(3S, 4S)-**36** through the transition

(Z)-(S)-**35**

(E)-(3S,4S)-**36**

state with *exo*-$R^1$ and *equatorial*-G. The key to success is the proper choice of the combination of olefin geometry and G group.

The first success of such asymmetric transmission was in the rearrangement of (S)-(Z)-**35a,** which attains complete chirality transfer along with practically

100% of both *E* and *erythro* selectivity, as predicted, to eventually afford insect pheromone **37** of the same optical purity as that of the substrate.[80] Essentially

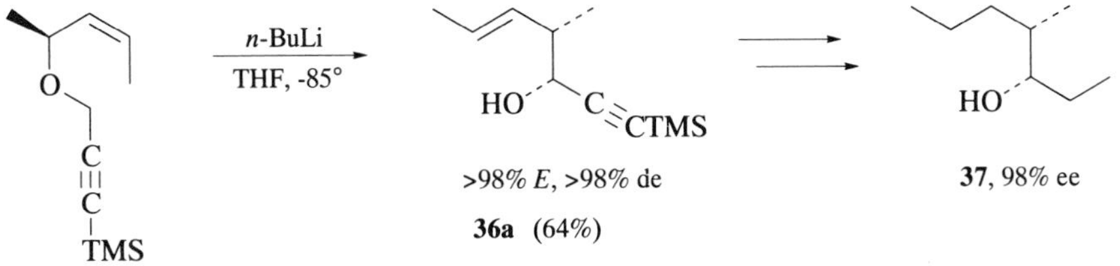

**35a**, 98% ee

>98% *E*, >98% de

**36a**  (64%)

**37**, 98% ee

complete chirality transfer (>95%) has also been achieved in other (*Z* to *erythro*)-selective variants with different G groups such as $CH=CH_2$[81] and $C_6H_5$[81,82] (Eqs. 22 and 23).

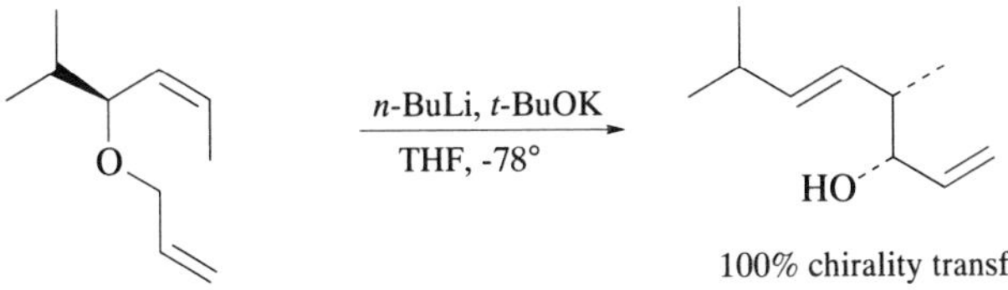

(Eq. 22)

100% chirality transfer

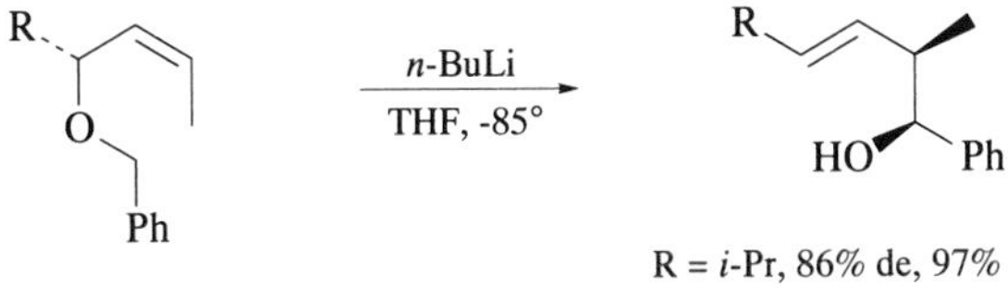

(Eq. 23)

R = *i*-Pr, 86% de, 97% chirality transfer

R = *i*-Bu, ~100% de, 94% chirality transfer

This asymmetric transmission technology is also applicable to the rather unusual (*E* to *erythro*)-selective variants. The Zr–enolate rearrangement of **38** gives (*Z*)-**39** with >96% chirality transfer,[69] whereas the Ti–enolate rearrangement of **40** produces (*E*)-**41** with complete chirality transfer.[83] Effective asymmetric transmission is also feasible in rearrangements of tertiary allylic ethers, if they

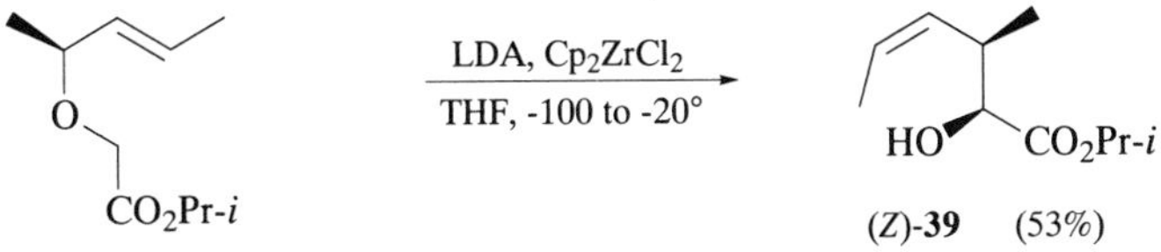

**38**

LDA, Cp$_2$ZrCl$_2$

THF, -100 to -20°

(*Z*)-**39**   (53%)

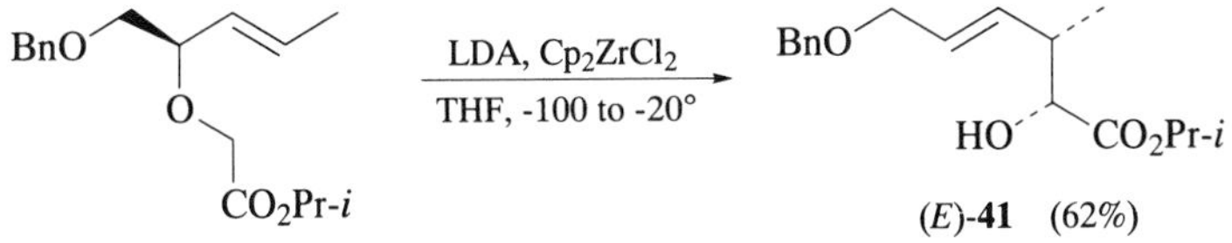

**40**

LDA, Cp$_2$ZrCl$_2$

THF, -100 to -20°

(*E*)-**41**   (62%)

are well suited for the exclusive formation of a single olefinic geometry. The rearrangements of $(E)$- and $(Z)$-**42** afford exclusively $(E)$-*erythro*- and $(E)$-*threo*-**43**,

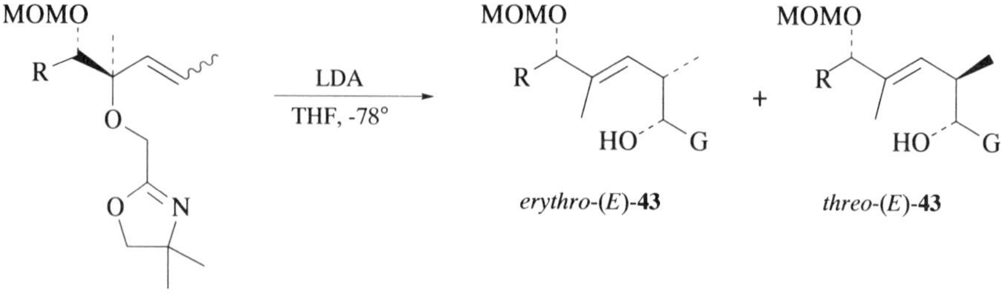

$(E)$- or $(Z)$-**42**
R = Me, BnOCH$_2$

erythro-$(E)$-**43**          threo-$(E)$-**43**

respectively, whereas the epimers at the tertiary stereogenic center produce a stereoisomeric mixture.[72,84]

The asymmetric transmissions via cyclic [2,3]-Wittig rearrangements are also known, although they are relatively small in number. Equation 24 shows an ex-

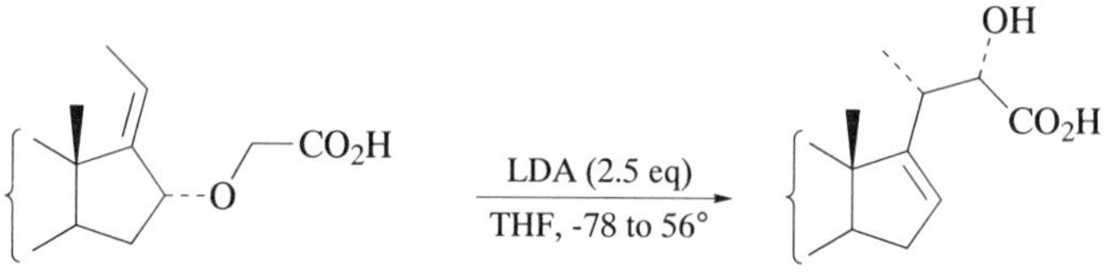

(Eq. 24)

ample of {3,4} rearrangement in the steroid side chain synthesis.[85,86] More applications of this type of rearrangement to steroid side chain synthesis will be described later. Equation 25 illustrates an example of {3,5} rearrangement.[54] An-

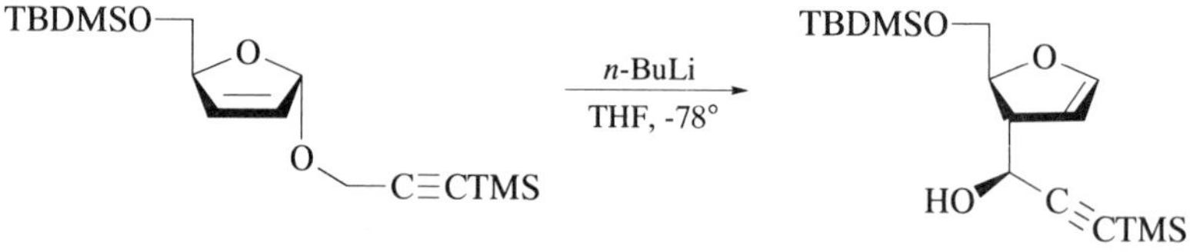

(Eq. 25)

other interesting example is the dianion {3,3} rearrangement on the carbohydrate template (Eq. 26), where $(E)$- and $(Z)$-**44** afford *threo*- and *erythro*-**45**, respec-

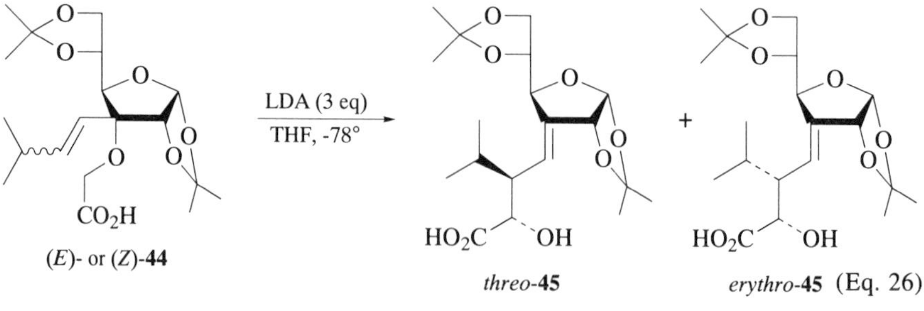

$(E)$- or $(Z)$-**44**

threo-**45**          erythro-**45**  (Eq. 26)

tively.[56] It is noted that the senses of diastereoselection in Eqs. 24 and 26 are opposite to those observed with the acyclic counterparts.

While the asymmetric transmission processes discussed so far involve chirality transfer from C-1 to both C-3 and C-4, the C-1 chirality can, of course, be transferred to either C-3 or C-4 when one employs an enantiomerically enriched substrate where G = H and $R^2 \neq$ H or G $\neq$ H and $R^2 =$ H, respectively (cf. Eq. 21).

Such 1,3-chirality transfer has been achieved via the Wittig–Still variant within acyclic and cyclic frameworks. The simplest example is the rearrangement of (Z)-(S)-**46** that affords (E)-(R)-**47** exclusively, while (E)-(S)-**46** provides a 1:1 mixture of (E)-(S)- and (Z)-(R)-**47**. This 1,3-chirality transfer technology has been widely utilized in cyclic frameworks to effect a suprafacial hydroxymethylation.[50,51,87-90]

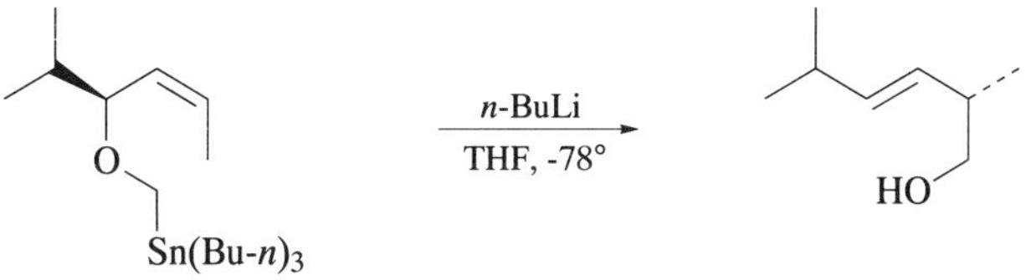

(Z)-(S)-**46**  (E)-(S)-**47**

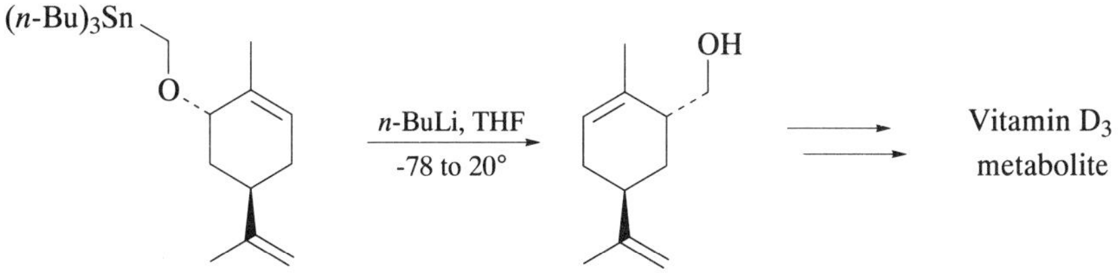

(77%)

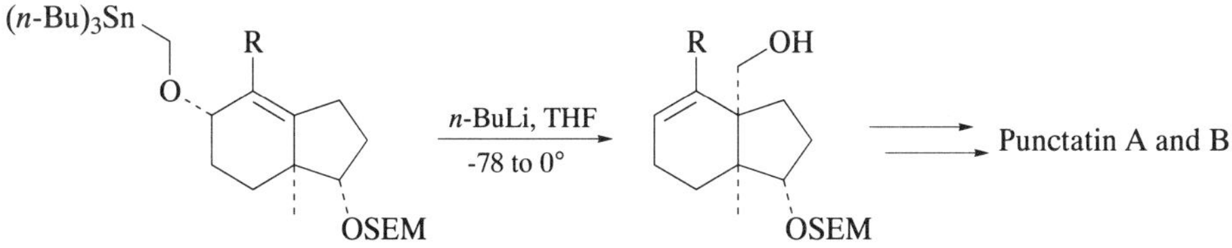

R = H, i-Pr  (34-45%)

1,3-Chirality transfer from a tertiary stereocenter is also known. The rearrangements of (E)- and (Z)-**48** afford (E)-(S)- and (E)-(R)-**49**, respectively. Again, the epimers at the tertiary stereocenter produce an E/Z mixture.[91-93]

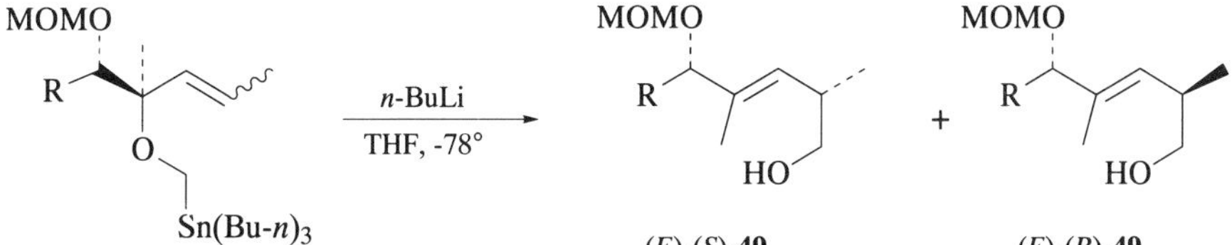

(E)- or (Z)-**48**
R = CH(Me)Pr-n

(E)-(S)-**49**  +  (E)-(R)-**49**

Chirality transfer from C-1 to C-4 has been reported in different variants. Again, the minimum requirement for complete 1,4-chirality transfer is the exclusive formation of only one double-bond geometry. Equation 27 shows acyclic ex-

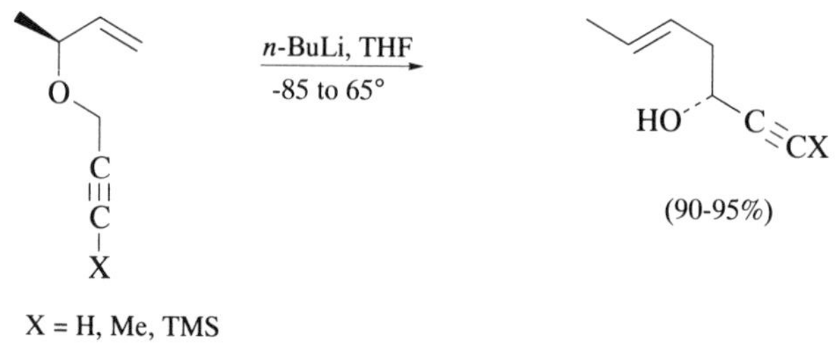

$$\text{(Eq. 27)}$$

X = H, Me, TMS

amples which attain 81–98% of chirality transfer together with 93–98% $E$ selectivity.[67] An example of complete 1,4-chirality transfer from a tertiary stereocenter is already shown in Eq. 16. An example of the {3,4}-rearrangement is given in Eq. 28.[94]

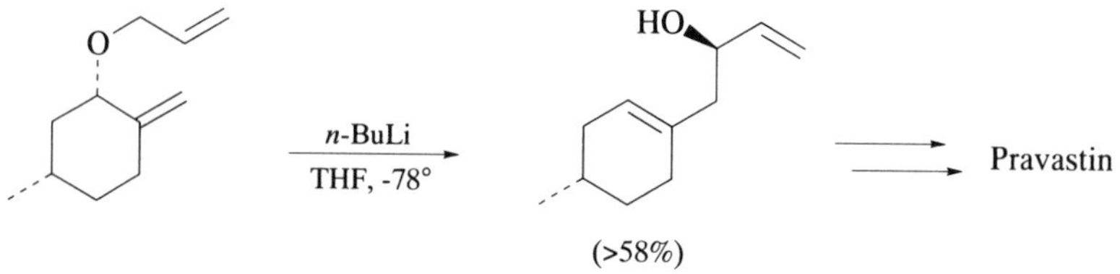

$$\text{(Eq. 28)}$$

**Asymmetric Induction Type.** This type of rearrangement can be divided into the three types **A–C** in terms of the location of a chiral substituent ($G_c^*$ or $R_c^*$) in the substrate.

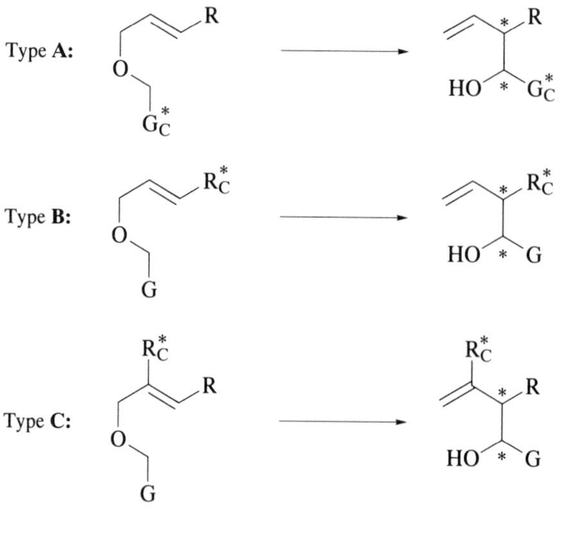

The rearrangement of type **A** has been well studied using various chiral enolates as the migrating terminus, despite the potential competition posed by the
[3,3]-Claisen process. The first example reported, **50a**, employed the chiral 2-ox-
azoline ring as the $G_c^*$ to provide a reasonably high level of both diastereofacial
selection (% de) and *erythro* selection (Eq. 29).[95] However, a similar reaction of

*erythro:threo* = 90:10                2R, 78% ee

**50a** R = Me                                                      (Eq. 29)

**50b** (R = H) shows only 38% de (2R). Interestingly, the dianion rearrangement
of **51** with potassium hydride shows the opposite sense of $\pi$-facial selection to
give (2S)-**52** in 84% de, whereas the KH-induced process in the presence of 18-
crown-6 exhibits a further reversal in $\pi$-facial selection to afford (2R)-**52** in 96%
de.[96] It should be noted that similar KH-induced rearrangements of the (E)-crotyl
analog of **51** provided an extremely low level of *erythro* selection.

(2R)-**52**                (2S)-**52**

|                           |      |
|---------------------------|------|
| KH, THF, 15-20°           | 8:92 |
| KH, 18-crown-6, -20°      | 98:2 |

**51**

Other examples have been reported which involve a chiral amide enolate[61,97] and
a chiral ester enolate.[98] In view of the tremendous recent progress in chiral enolate
chemistry, this type of asymmetric enolate [2,3]-Wittig strategy provides a powerful tool for asymmetric synthesis of a variety of $\alpha$-hydroxy-$\beta$-alkylcarboxylic
acid derivatives, an important class of intermediates for natural product synthesis.

LDA, THF, -78°        (2S), 95% *erythro*, 52% de

$Y_C^* =$ [structure: N-methylpyrrolidine with CH₂OMEM]    LDA, Cp₂ZrCl₂,    (2R), 98% erythro, 60% de
                                                           THF, -78°

$Y_C^* =$ MOMO [pyrrolidine structure] OMOM    LDA, Cp₂ZrCl₂,    (2S), 97% erythro, 96% de
                                               THF, -70°

$Y_C^* = -O$ [bicyclic structure with Ph]    LDA, HMPA-THF,    (2S), 90% erythro, 97% de
                                             -70°

A novel example employs an enantiomerically pure (*o*-methoxybenzene) chromium complex as the $G_c^*$ to afford exclusively the *erythro* product (97%) in 99% ee after oxidative demetalization (Eq. 30).[99-101] Also interesting is the ring

[reaction scheme]    $\xrightarrow[\text{TMEDA-THF, -78°}]{n\text{-BuLi}}$    $\xrightarrow{h\nu, O_2}$    [product structure with OMc, HO]

(67%) 99% ee

(Eq. 30)

contractive rearrangement of **53**, in which a high asymmetric induction by a remote stereocenter is observed.[102,103]

[structure 53 with OTBDMS]    $\xrightarrow[\substack{\text{THF, C}_5\text{H}_{12}, \\ -78°}]{n\text{-BuLi}}$    [product with OTBDMS, HO]    +    other isomers

**53**                                            (90%) 89:11

The rearrangement of type **B** is also amply precedented. The successful examples are shown in Eqs. 31[104-108] and 32,[109] where high asymmetric induction by the chiral γ substituent is observed together with high *erythro* or *threo* selection, depending on the olefin geometry. Thus this type of asymmetric rearrangement provides an efficient way to achieve stereocontrol over three contiguous chiral centers.

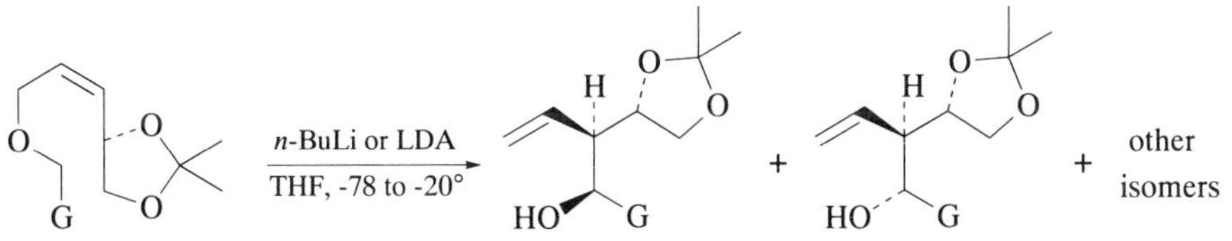

(Z)-**54**

(Z)-**54a**  G = C≡CTMS[104, 105]    >99 : 0 : 0

(E)-**54a**[104]    12 : 81 : 7

(Z)-**54b**  G = CO$_2$Bu-$t$[106]    >99 : 0 : 0

(Z)-**54c**  G = Sn(Bu-$n$)$_3$[107]    >99 : 0 : 0    G = H    (Eq. 31)

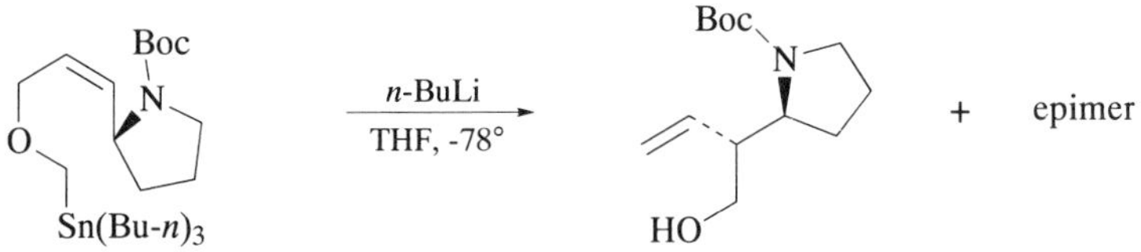

(88%) >99:1     (Eq. 32)

This methodology is also applicable to cyclic variants, as shown in Eqs. 25 and 26. Equations 33[110] and 34[58] illustrate the applicability to {4,5} and {5,5} rearrangements.

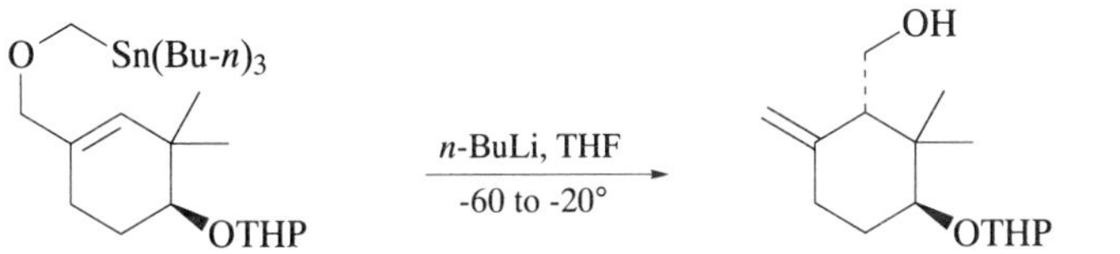

(98%) *trans:cis* = 93:7     (Eq. 33)

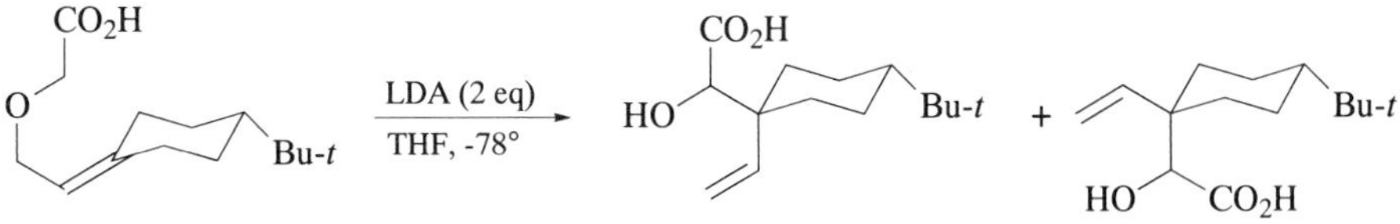

(63%) 100:0

(Eq. 34)

Rearrangements of type **C** are also known but with few examples. High levels of 1,3 asymmetric induction have been reported in the rearrangements of **55** and **56**, but their senses are opposite each other.[111] An extremely high *anti* asymmetric induction has also been observed in the {4,5} rearrangement of **57** to establish a quaternary stereocenter.[112,113]

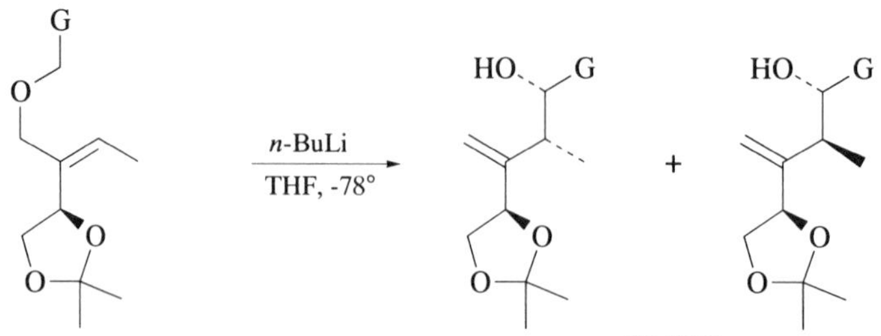

(70-89%)

(E)-55, G = Sn(Bu-n)₃          3:97,  G = H

(Z)-55                         95:5

(E)-56, G = C≡CTMS            88:6

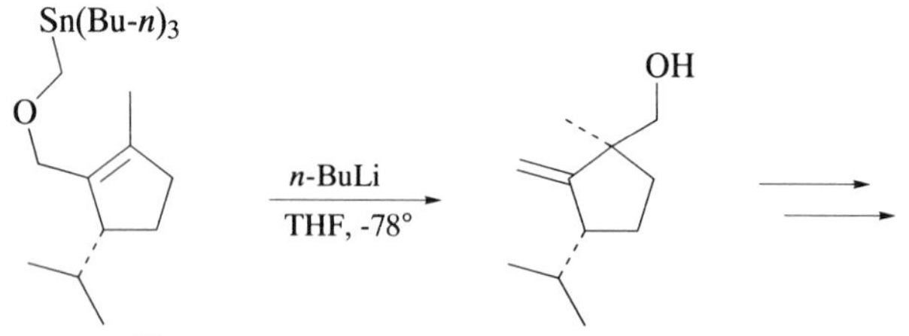

**57**

(>75%)                         refigeranic acid

Of special interest is the rearrangement of **58**, which involves asymmetric induction of both types **B** and **C** to give the single stereoisomer **59**, a potentially useful intermediate for prostanoid synthesis.[114]

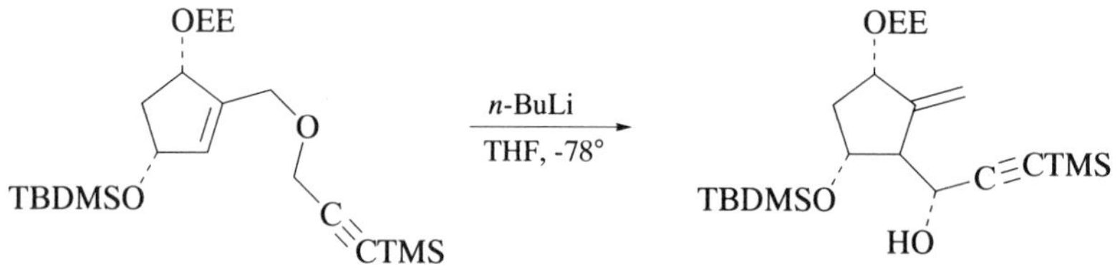

**58**                         **59**   (89%)

prostanoid

**Chiral Base-Induced Type.**  In principle, the [2,3]-Wittig rearrangement with a chiral *nonracemic* base could produce an optically active product since the chiral terminus involved could be *nonracemic*. Thus this type of asymmetric process is enantioselective, and hence is synthetically more valuable than the diastereoselective reactions described above. However, this is more difficult to practice. The first success was achieved in the {1,5} rearrangement of **60** with lithium amide (+)-**61**, which afforded (+)-alcohol **62** in 70% ee.[115,116] The product was converted to aristolactone (+)-**63**. However, a similar rearrangement of the closely related 9- and 17-membered substrates with (+)-**61** provided only 25 and 29% ee, respectively.[115-117] Accordingly, the level of enantioselection appears to depend critically upon the chiral environment provided by the cyclic framework. In fact, no appreciable levels of ee have been observed in the rearrangements of acyclic substrates with (+)-**61** (Eq. 35).[115,116] However, the rearrangement of **64** with (+)-**61** has been shown to provide a higher optical yield.[118,119]

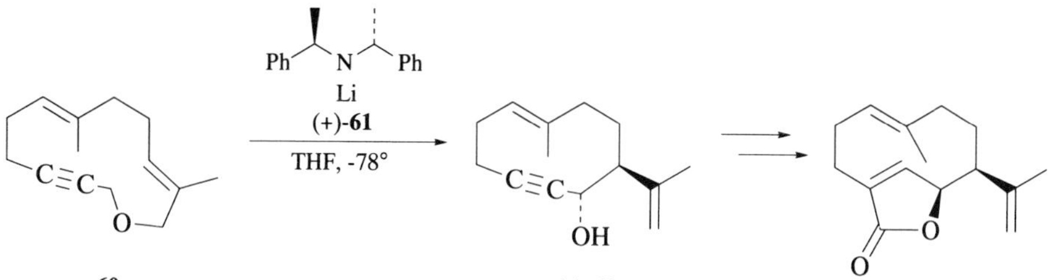

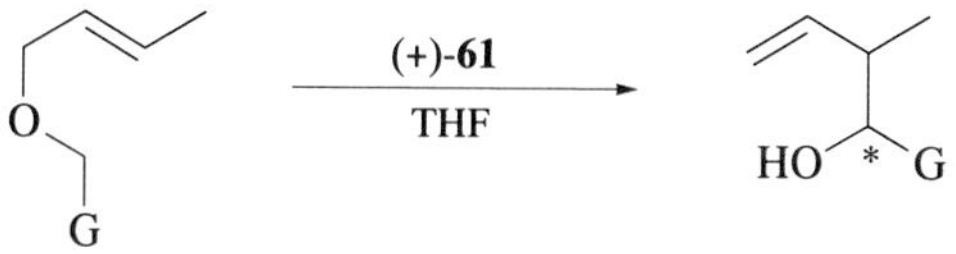

$$G = CO_2H, C\equiv CH, CON(CH_2)_4 \qquad <5\% \text{ ee}$$

(Eq. 35)

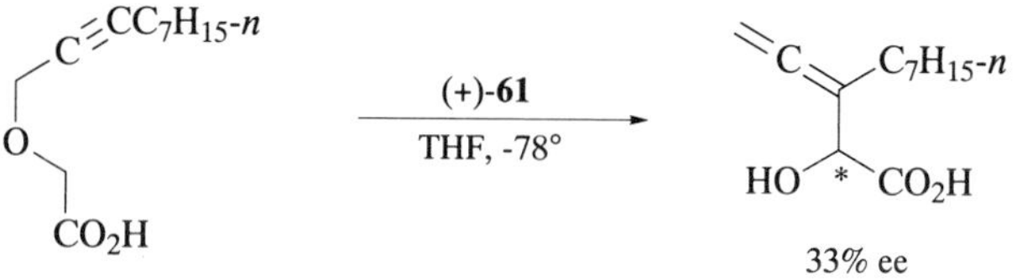

33% ee

**64**

**Configurationally Defined Carbanion Type.** Since most asymmetric [2,3]-Wittig variants involve a chiral carbanion terminus, the fundamental problem encountered, which concerns the steric course, is inversion vs. retention at the lithium-bearing terminus (Eq. 36). That has been the subject of controversy.

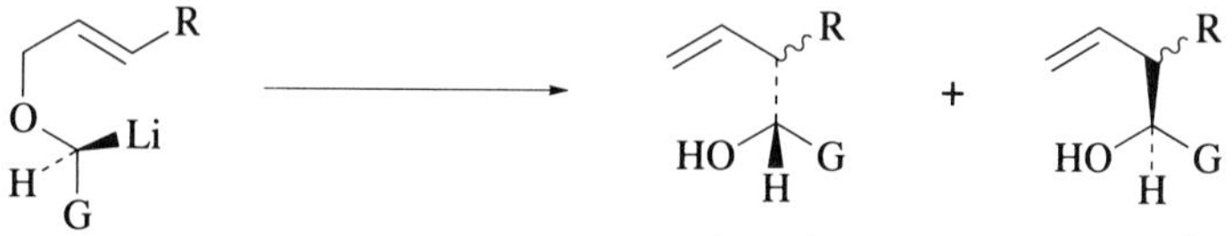

inversion          retention

(Eq. 36)

Conflicting conclusions have been drawn from the theoretical work based on ab initio molecular orbital calculations.[25,27]

The first experimental evidence in support of the inversion course was reported in the rearrangement of the diastereomerically defined tetrahydropyran **65** which afforded the inverted [2,3]-Wittig product **66** along with the inverted [1,2]-Wittig product **67**.[120] More recently, rearrangement of the enantiomerically de-

fined stannane (*R*)-**68** has been shown to proceed with complete inversion of configuration together with high *threo* selectivity in the synthesis of insect pheromone (+)-**69**.[121] Thus in general the [2,3]-Wittig rearrangement is likely to proceed with inversion of configuration at the lithium-bearing terminus.

## SYNTHETIC APPLICATIONS

The [2,3]-Wittig rearrangement has found many applications in stereocontrolled total syntheses of a vast array of natural products. In this section, a brief representative sampling of some uses of [2,3]-Wittig technology is described within selected contexts such as acyclic stereocontrol, steroid side chain synthesis, synthesis of medium- and large-ring natural products, and sigmatropic sequences.

### Acyclic Stereocontrol

In the preceding sections some asymmetric [2,3]-Wittig variants were shown to create vicinal chiral centers with high asymmetric induction. Since the resulting stereodefined products possess a unique multifunctionality well suited for further stereocontrolled transformations, asymmetric [2,3]-Wittig technology provides a versatile method for acyclic stereocontrol over three or more chiral centers.

A typical example is the synthesis of (+)-blastmycinone (**73**).[122] This synthesis features the sequential combination of the rearrangement of (*S*)-**70** with zinc borohydride reduction[123] of hydroxy ketone **72** derived from **71**. This silylpropargyl ether protocol has also been utilized in the asymmetric synthesis of talanomycin A (**74**),[124] serricornin (**75**),[125,126] and the fragment **76** of amphotericin B.[127]

(S)-**70** **71** (96%) **72** **73**

**74** **75** **76**

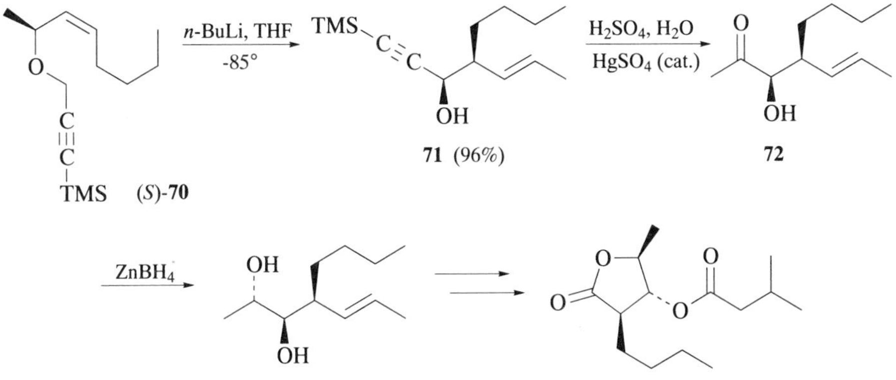

The asymmetric rearrangement of methallyl ether system **77a** has been used in the synthesis of the Prelog–Djerassi lactonic aldehyde **78**[128] and its acid from **77b**,[129] where the third chiral center is introduced by stereoselective hydrobora-

**77a**, R = *i*-Pr
**77b**, R = *i*-Bu

**78**

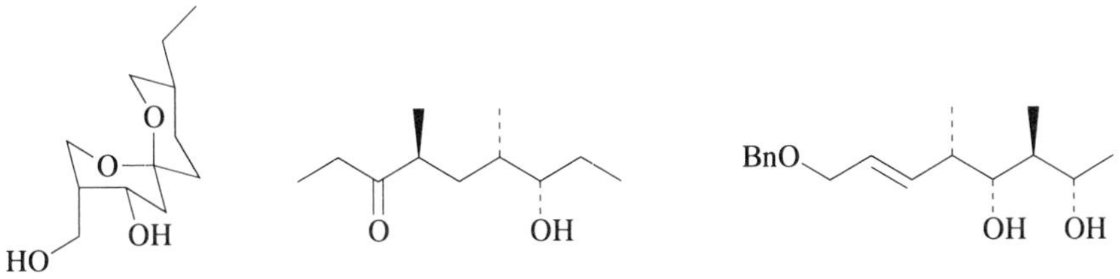

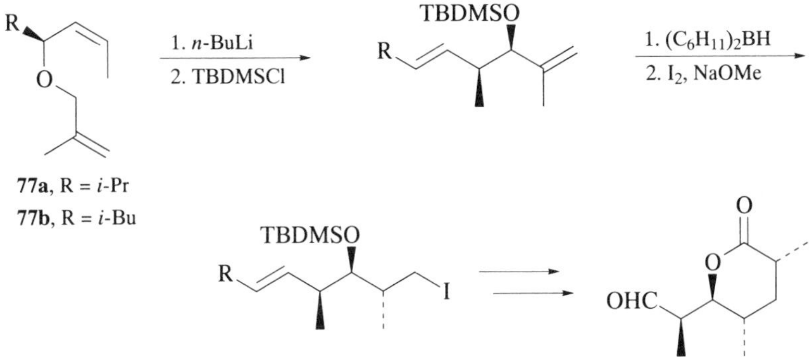

tion. The [2,3]-Wittig product **41** from **40** described earlier has been elaborated to (+)-Ireland's acid **79** having four contiguous chiral centers, a key precursor of tirandamycin A (**80**).[130]

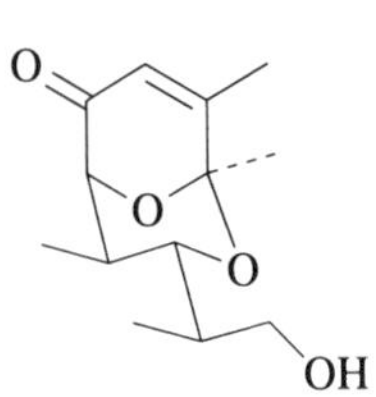

**79**

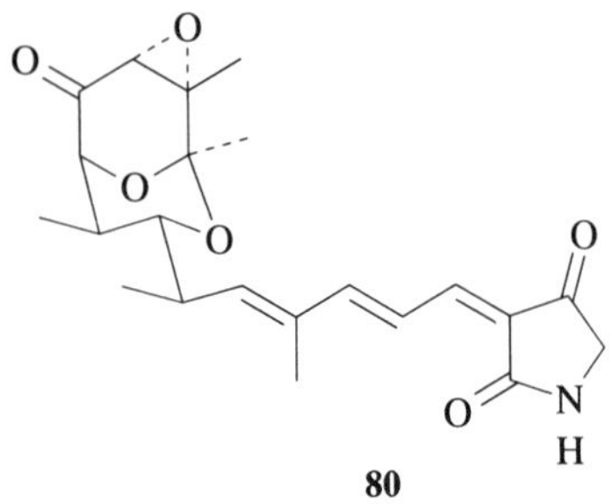

**80**

The asymmetric rearrangement of tertiary allylic ethers has also found application in natural product synthesis. For example, the previously described rearrangement of (Z)-**48** provides a quick entry to (+)-2-epi-invictolide (**81**).[91–93] The asymmetric synthesis of subunit **82** of zincophorin has been achieved by rearrangement of a tertiary allylic ether of type (E)-**42** described earlier.[84] This type

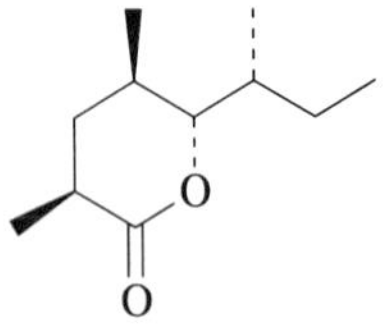

**81**

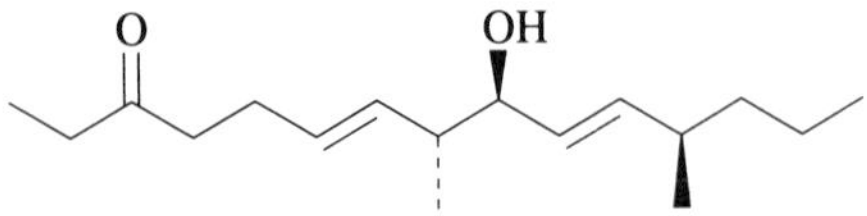

**82**

of tertiary [2,3]-Wittig protocol has also been utilized in the synthesis of key intermediate **83** to 18-deoxynargenicin A$_1$.[131] These examples demonstrate the

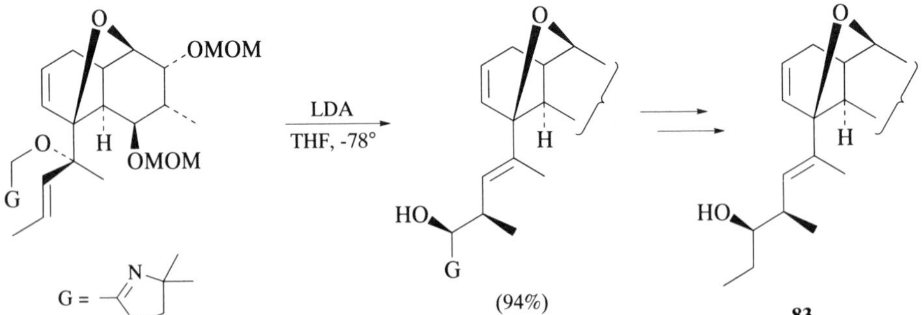

LDA
THF, -78°

G =

(94%)

**83**

great potential of asymmetric [2,3]-Wittig technology as a general synthetic strategy for complex natural products with multiple chiral centers.

## Steroid Side Chain Synthesis

Steroid side chain synthesis has attracted considerable synthetic effort over the decade,[132,133] and within this context the concept of asymmetric transmission via [2,3]-Wittig rearrangement has found many applications. One general and simple application is to employ [2,3]-Wittig technology for effecting chirality transfer within side chain frameworks. The propargyl ether protocol has been used in the synthesis of petrosterol (**84**), a marine steroid.[134] Still's variant has been utilized

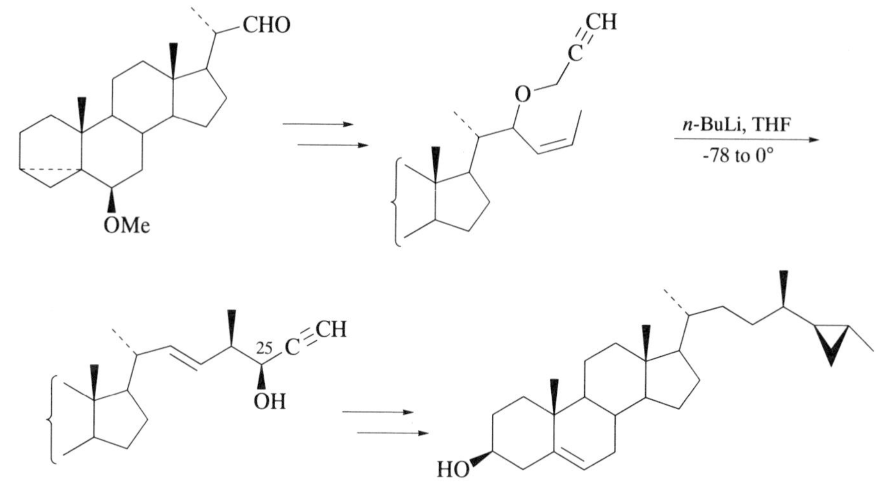

(50%)   >95% *S* at C-25

**84**

in the synthesis of (24*R*)-25,28-dihydroxy side chain (**85**)[135,136] and (25*R*)-26-hydroxycholesterol (**86**), a potent inhibitor of cholesterol biosynthesis.[137] Notable

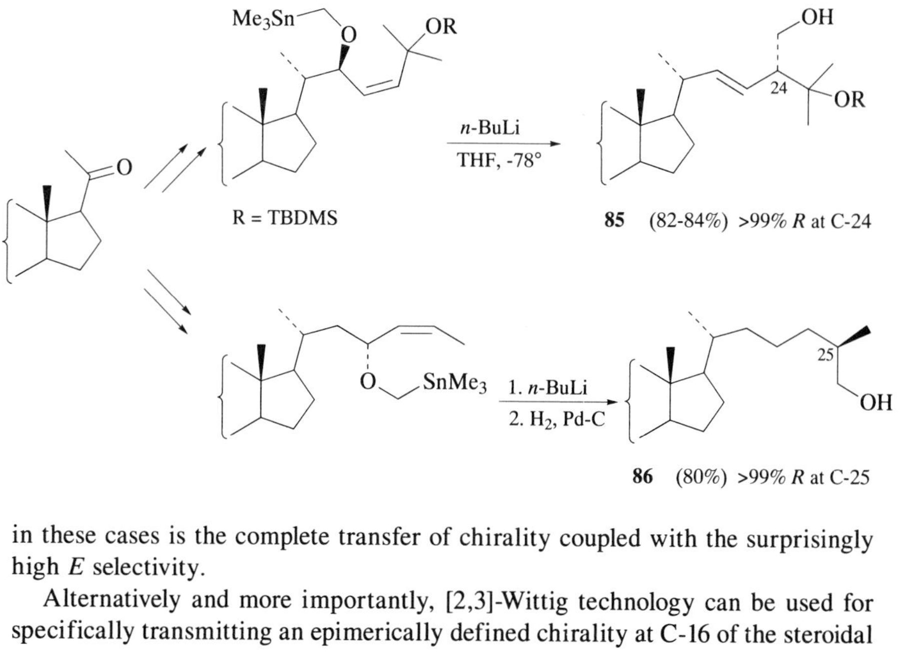

R = TBDMS

**85**   (82-84%) >99% *R* at C-24

**86**   (80%) >99% *R* at C-25

in these cases is the complete transfer of chirality coupled with the surprisingly high *E* selectivity.

Alternatively and more importantly, [2,3]-Wittig technology can be used for specifically transmitting an epimerically defined chirality at C-16 of the steroidal nucleus to the new chiral centers at C-20 and C-22 of the side chain as generalized by Eq. 37. As already illustrated in Eqs. 10 and 24, this approach allows for the

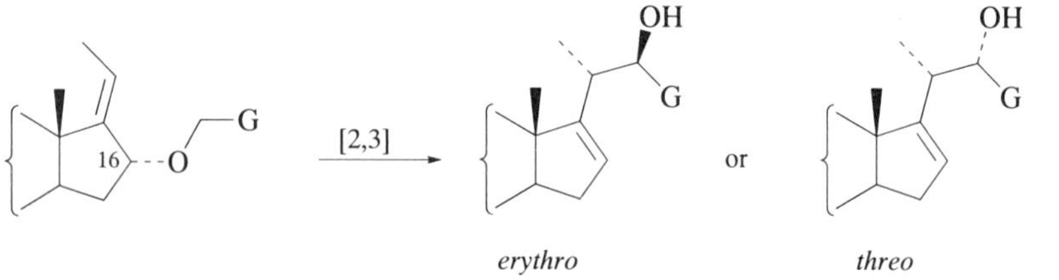

erythro          threo

(Eq. 37)

concurrent control of absolute and relative configuration at C-20 and C-22 through the proper combination of the *exo*-olefin geometry, the configuration ($\alpha$ and $\beta$) at C-16, and the G group. The most remarkable example is the stereocontrolled synthesis of either (22S)- or (22R)-hydroxy-23-acetylenic side chain from the single precursor **88a**, easily derived from the commercially available epoxypregnenolone **87**.[47] The dianion rearrangement of **88a** affords (20S, 22S)-**89** as a single stereoisomer, whereas introduction of the silyl group induces the reversal of diastereoselection to give (20S, 22R)-**90** as a single stereoisomer. Interestingly, the $\beta$-face rearrangement of (Z)-16$\beta$-**88b** also affords **90** as a single stereoisomer.[138] These Wittig products can undoubtedly serve as key intermediates for the synthesis of many important sidechain modified steroids. For instance, **89** and **90** can be converted to the insect hormone ecdysone[139] and the plant growth regulator brassinolide,[140] respectively.

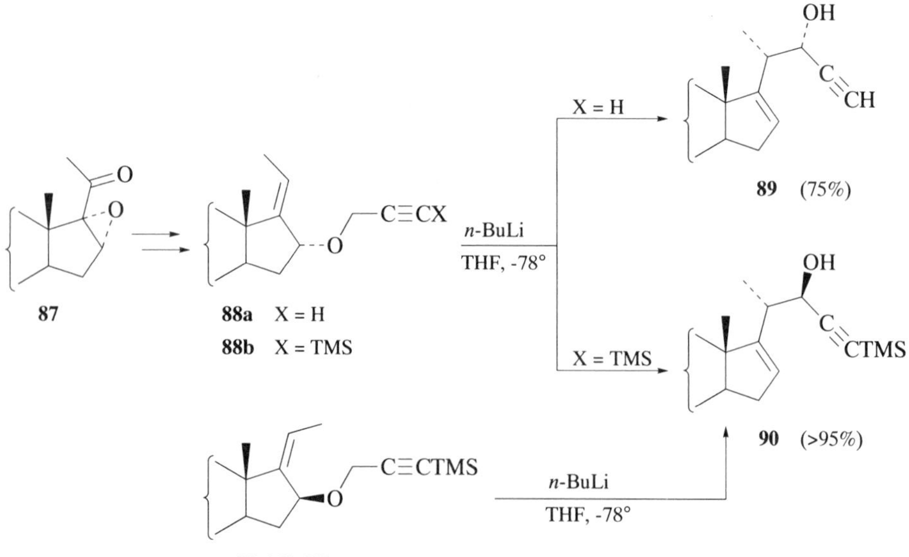

## Medium and Large Ring Natural Product Synthesis

As mentioned in the preceding sections, various types of cyclic [2,3]-Wittig variants have found many applications in total synthesis of a variety of complex natural products. In this section, the synthetic utility of the {1,5} rearrangement ("[2,3]-Wittig ring contraction technology") is described within the context of total synthesis of medium and large ring natural products.

In the germacranolide area, (±)-costunolide (93) was synthesized via the rearrangement of 13-membered bis(allyl) ether 91, which gave a mixture of 92 and its regioisomer(s) in a 75:25 ratio.[25,41] A similar rearrangement of 94 provides a quick synthetic entry to (±)-hagenolide (95).[25,141] The previously described rearrangements of the 13-membered (32) and 17-membered propargylic ethers (11) have been used as key steps in the synthesis of germacranolide (±)-aristolactone (96)[78,79] and cembranoids, (±)-epimukulol (97), and (±)-desoxyasperdiol (98),[42,43] respectively. Later, an enantioselective synthesis of (+)-96 was accomplished via the chiral base-induced rearrangement of 60 as previously described.[116]

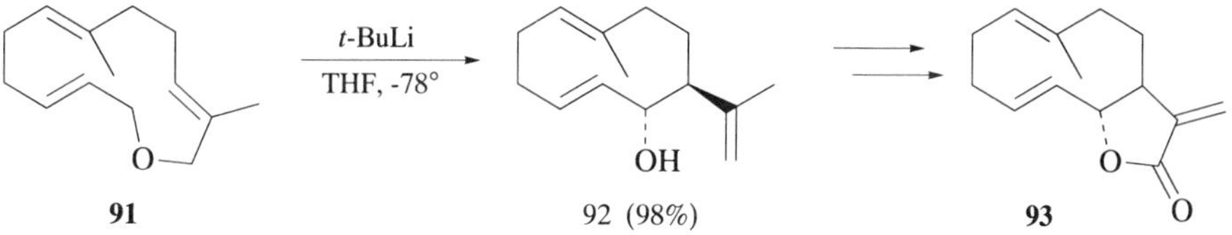

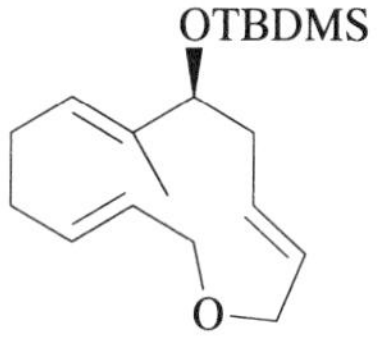

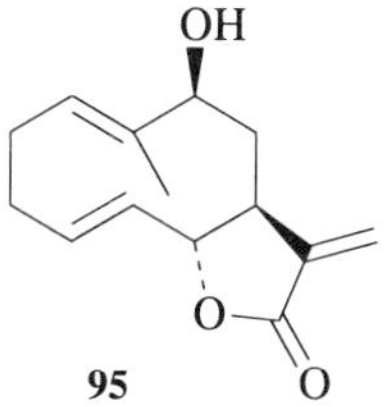

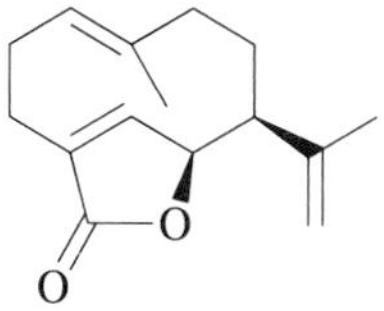

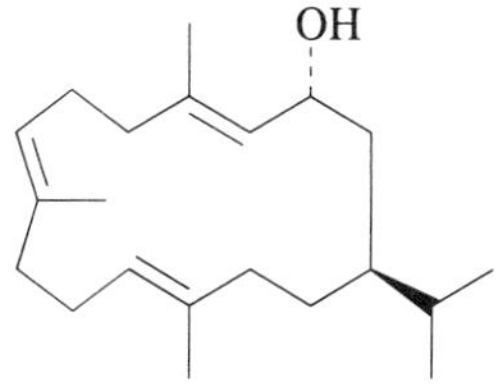

A similar [2,3]-Wittig ring contraction has also been used in the synthesis of a key precursor of kallolide A (99), an antiinflammatory diterpene,[142] and of (+)-α-2,7,11-cembratriene-4,5-diol (α-CBT, 100) and its β-isomer, tumor inhibitory constituents of tobacco.[143,144]

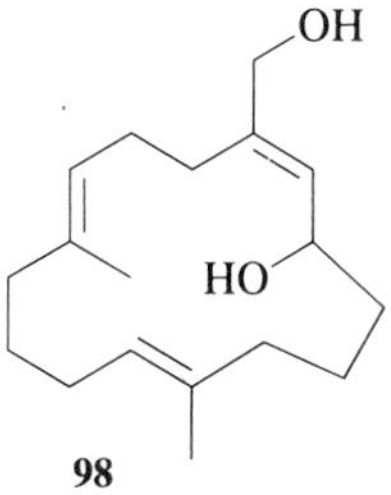

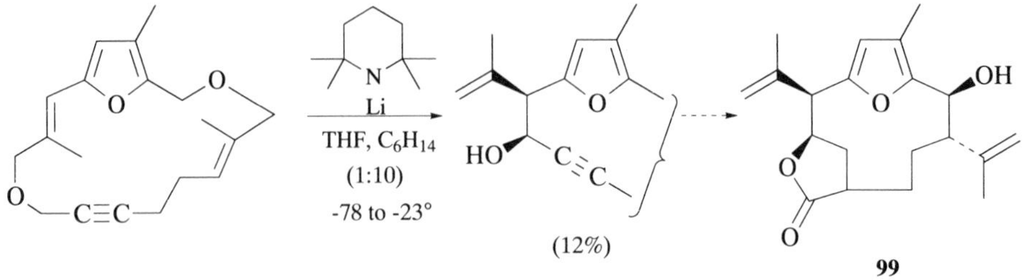

(12%)

**99**

1. *n*-BuLi, TMEDA,
   THF, C$_5$H$_{12}$

2. Swern oxidation

(70%)  *syn/anti* = 9:1

Na, NH$_3$

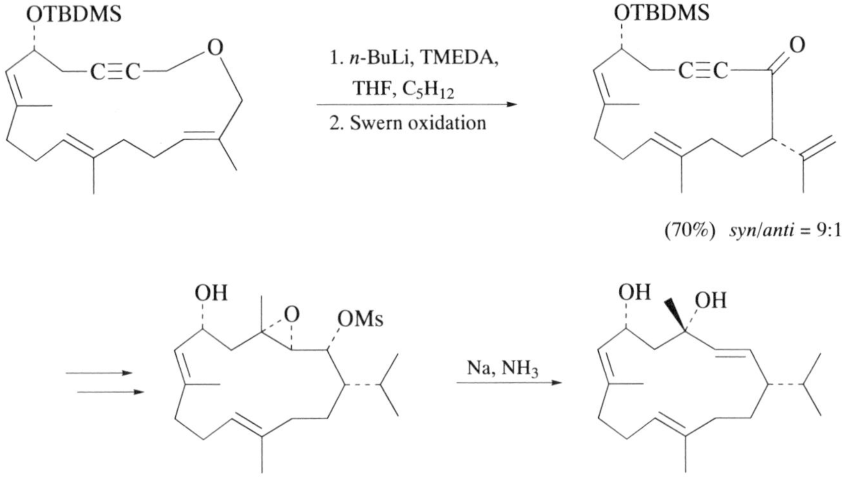

**100**

The utility of the [2,3]-Wittig rearrangement of 12-membered cyclic ether **101** has been demonstrated in the synthesis of the chromophore analogs **102** and **103** of antitumor antibiotic neocarzinostatin and esperamicin–calichemicin analog **104**.[145,146] Thus [2,3]-Wittig ring contraction strategy provides a general and

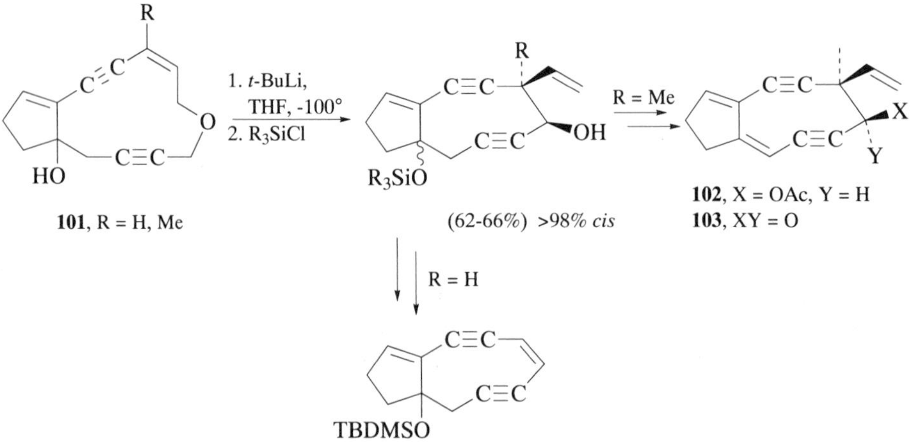

1. *t*-BuLi,
   THF, -100°

2. R$_3$SiCl

R = Me

**101**, R = H, Me

(62-66%) >98% *cis*

**102**, X = OAc, Y = H
**103**, XY = O

R = H

**104**

efficient approach to stereocontrolled construction of medium and large ring carbocycles.

## Sigmatropic Sequences

In the preceding sections the [2,3]-Wittig rearrangement of unsymmetrical bis(allyl) ethers (or allyl propargyl ether variants followed by semihydrogenation) has been shown to afford 1,5-dien-3-ols in regio- and stereoselective fashions. Since the [2,3]-Wittig products and their derivatives are well qualified as substrates for different sigmatropic rearrangements, the particular [2,3]-Wittig variants may constitute various types of sigmatropic sequences.[147,148] Equations 38–41 illustrate the four sequences developed thus far, which provide unique and facile synthetic methods for various classes of unsaturated carbonyl compounds possessing interesting molecular frameworks.[149,150] Of particular value is that the net effect of these sequences allows two or three allylic moieties initially linked by an ether bond(s) to be recombined into a new C–C bond(s) in a regio- and stereocontrolled manner.

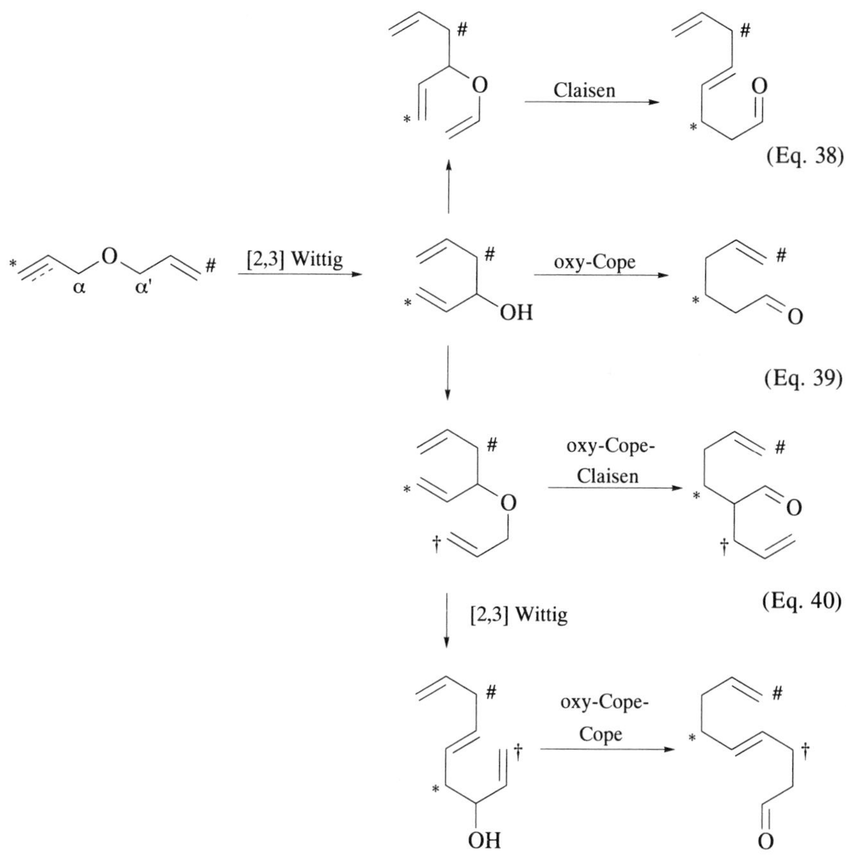

(Eq. 38)

(Eq. 39)

(Eq. 40)

(Eq. 41)

As illustrated by Eq. 38, [2,3]-Wittig products can be used as substrates for the Claisen rearrangement to afford a variety of the functionalized 1,4-dienes.[151] For instance, the enol ether Claisen process[152] of **105a** affords $(E,E)$-dienal **106a** and the Johnson–Claisen modification[153] of **105b** gives $(E,E)$-dienoate **107b**, whereas the Ireland–Claisen modification[154] of the acetate of **105b** produces $(E,E)$-dienoic acid **108b**. Thus, the [2,3]-Wittig–Claisen sequence permits ready access to a wide variety of functionalized $(E,E)$-dienes which are often found in natural products as well as synthetic intermediates thereof. Its synthetic potential has been illustrated in the synthesis of butenolide **109**, a well-known precursor of antibiotic $(\pm)$-cerulenin (**110**).[151]

**105a**   $R^1$ = Me, $R^2$ = H
**105b**   $R^1$ = $n$-$C_5H_{11}$, $R^2$ = H

**106**   Y = H
**107**   Y = OEt
**108**   Y = OH

**109**

**110**

As depicted by Eq. 39, the sequential [2,3]-Wittig-oxy-Cope rearrangement provides a versatile route to $\delta,\varepsilon$-unsaturated carbonyl compounds. Of particular significance is that the easy availability of stereodefined 1,3-dien-3-ols via diastereoselective [2,3]-Wittig variants makes it possible to analyze the hitherto unsettled stereochemistry of the acyclic oxy-Cope process.[155] For instance, the $E/Z$ selection in the oxy-Cope process of [2,3]-Wittig product of type **111** is not dependent on the substrate stereochemistry, but on the rearrangement procedure (Eq. 42).[149,150] The anionic oxy-Cope procedure[156,157] provides a modest $E$ selec-

**111**                               $E/Z$   (Eq. 42)

tivity (67–72%), whereas thermolysis in decane (ca. 170°) affords an increased $E$ selectivity (92–95%). The utility of the [2,3]-Wittig-thermolytic oxy-Cope sequence has been illustrated in the synthesis of insect pheromone (±)-brevicomine (**112**) and marine natural product oxocrinal (**113**)[158] and of functionalized vinyl silanes (Eq. 43).[159–161]

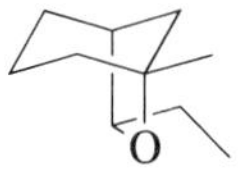

**112** 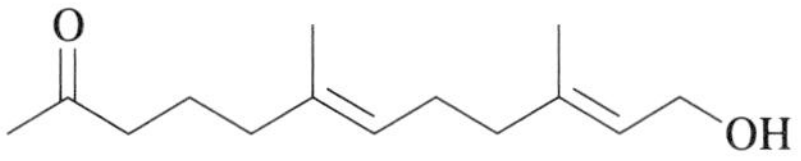                                    **113**

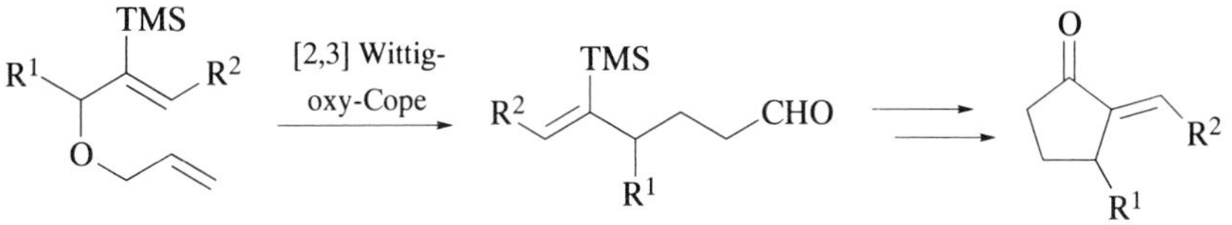

(Eq. 43)

In contrast, however, the stereochemistry of the anionic oxy-Cope rearrangement of type **114** has been reported to depend critically on the substrate stereochemistry; e.g., (*Z, erythro*)- and (*Z, threo*)-**114** exhibit 98%*E* and 80%*Z*

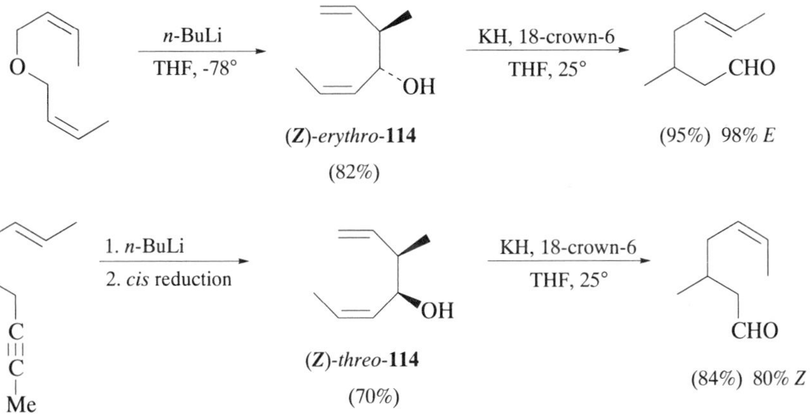

(*Z*)-*erythro*-**114**            (95%) 98% *E*

(82%)

(*Z*)-*threo*-**114**            (84%) 80% *Z*

(70%)

selectivity, respectively, while the *E* counterparts provide a moderate *E* selectivity.[162] More interestingly, the anionic oxy-Cope process of (*erythro*)-**116** derived from (*S*)-**115** proceeds with high *E* and *threo* selectivity together with nearly complete asymmetric transmission in the formal total synthesis of insect pheromone (+)-farnal (**117**).[163] Thus acyclic oxy-Cope technology, when properly designed in terms of the substrate stereochemistry, provides a new efficient method for acyclic stereocontrol.

The triple sigmatropic sequence depicted by Eq. 44 can be achieved by thermolysis of allylic ether **118**, prepared via etherification of the [2,3]-Wittig product, to afford dienal **119** as the major product.[149,150] In certain cases cyclic alcohol **120** is also formed as a byproduct which arises from an intramolecular ene reaction of **119**.

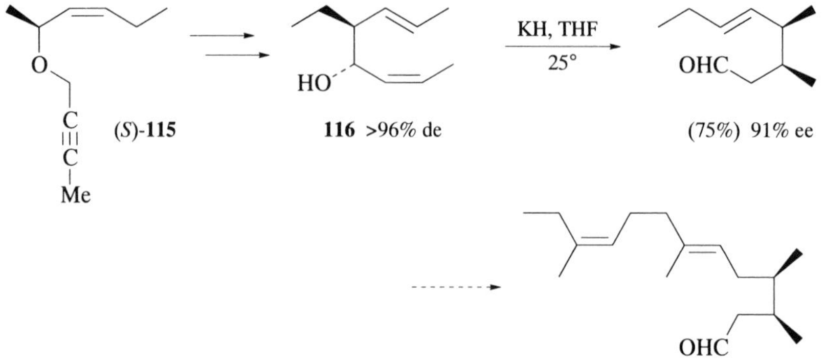

(*S*)-**115** **116** >96% de (75%) 91% ee

**117**

**118**

(Eq. 44)

**119** (41-46%) **120**

Furthermore, the quadruple sequence depicted by Eq. 45 can be effected by another [2,3]-Wittig process of ether **118** followed by thermolysis of the resulting trienol **121** to give trienal **122**.

**118**
**118a**, R = $(CH_2)_2CH{=}CMe_2$

**121**

(Eq. 45)

**122** (44%)

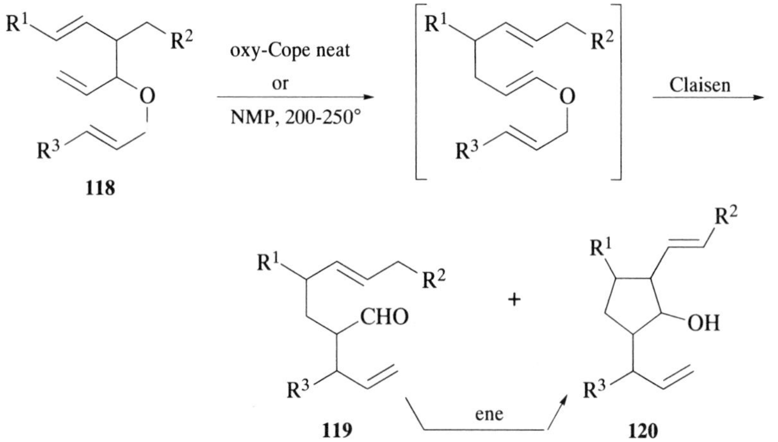

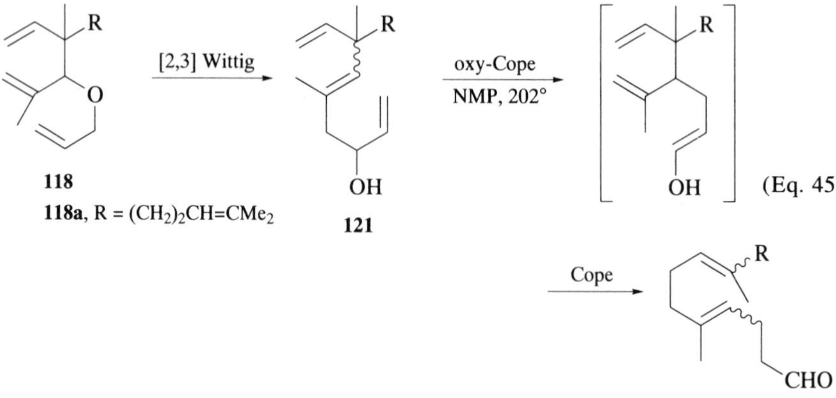

## EXPERIMENTAL PROCEDURES

***threo*-4-Methyl-5-hexen-1-yn-3-ol [Rearrangement of (*E*)-Crotyl Propargyl Ether].**[29]  A solution of *n*-butyllithium in hexane (160 mL, 208 mmol) was added dropwise to a cold (−85°) solution of (*E*)-crotyl propargyl ether (8.8 g, 80 mmol) in tetrahydrofuran (160 mL) (in an ethanol/liquid nitrogen/dry ice bath). The resulting mixture was stirred at that temperature for 8 hours, allowed to warm to 0° (5 hours), and quenched with saturated aqueous ammonium chloride solution (50 mL). Usual workup followed by distillation afforded a predominantly *threo* mixture of 4-methyl-5-hexen-1-yn-3-ols (6.32 g, 72%) as an oil: bp 52–53°/10 mm Hg; IR (neat) 3500, 3300, 1645, 1000, 920 cm$^{-1}$; $^1$H NMR δ 1.10 (d, $J$ = 7.2 Hz, 3H), 2.10–2.63 (m, 1H), 2.32 (d, $J$ = 2.3 Hz, 1H), 4.13 (dd, $J$ = 5.4 and 2.3 Hz, 1H), 4.97–5.28 (m, 2H), 5.82 (ddd, $J$ = 17.3, 10.1, and 7.5 Hz, 1H); GC (130°) $t_R$ = 25 and 28 minutes (7:93). On addition of Eu(fod)$_3$ (8 mg) to a solution of the alcohol in carbon tetrachloride (7.6 mg in 0.3 mL), the doublet at δ 1.10 was changed into the two doublets at δ 1.88 and 2.13 (relative intensity = 1:13); the $^1$H NMR ratio was identical with the GC ratio.

***erythro*-4-Methyl-5-hexen-1-yn-3-ol [Rearrangement of (Z)-Crotyl γ-(Trimethylsilyl)propargyl Ether].**[29]  A solution of (Z)-crotyl γ-(trimethylsilyl) propargyl ether (9.12 g, 50 mmol) in tetrahydrofuran (50 mL) was treated with *n*-butyllithium (50 mL, 60 mmol) in the predescribed manner. To the resulting mixture was added a mixture of cesium fluoride (0.25 g, 1.67 mmol), water (2.7 mL), and methanol (4.65 mL), and the mixture was heated at 50° for 15 minutes. Usual workup followed by distillation gave a predominantly *erythro* mixture of 4-methyl-5-hexen-1-yn-3-ol (5.5 g, 75%): GC (130°) $t_R$ = 25 and 28 minutes (98:2).

**(−)-(3*R*,4*S*,5*E*)-2,4,7-Trimethyl-1,5-octadien-3-ol [Rearrangement of a Bis(allylic) Ether].**[128]  Potassium *tert*-butoxide (6.17 g, 55 mmol) was dissolved in 100 mL of tetrahydrofuran. The solution was cooled to −78° and 8.51 g (50.6 mmol) of (Z)-1-(2-propyl)-2-butenyl 2-methylpropenyl ether was added. A solution of *n*-butyllithium (40 mL, 1.55 M, 62 mmol) was slowly added. The mixture was warmed to 0° over 4 hours. The reaction was quenched with water and the product was extracted with ether. The organic phase was dried over magnesium sulfate and concentrated. Kugelrohr distillation (90°/25 mm Hg) gave 7.41 g (87%) of the product. Analysis by HPLC (methanol:ethyl acetate:hexane = 1:5:100, Partisil M9 column) indicated a 97:3 mixture of *erythro* and *threo* products (the *erythro* isomer was eluted second). The ratio was confirmed by $^1$H NMR (200 MHz): *erythro* δ 3.87 (d, $J$ = 5.9 Hz); *threo* δ 3.66 (d, $J$ = 8.3 Hz). A diastereomerically pure material was obtained by flash chromatography (0.5 methanol, 5 ethyl acetate, 100 hexane): $^1$H NMR (200 MHz) δ 0.97 (d, $J$ = 6.6 Hz, 6H), 0.99 (d, $J$ = 6.8 Hz, 3H), 1.64 (br, OH), 1.70 (s, 3H), 2.20–2.40 (m, 2H), 3.87 (d, $J$ = 5.9 Hz, 1H), 4.87 (m, 1H), 4.93 (m, 1H), 5.30 (dd, $J$ = 15.6 and 6.8 Hz, 1H), 5.46 (dd, $J$ = 15.6 and 6.2 Hz, 1H); $^{13}$C NMR

(50.1 MHz) $\delta$ 14.6, 18.5, 22.6, 31.1, 39.7, 79.0, 111.7, 129.3, 138.1, 145.4; $[\alpha]_D^{24}$ + 2.56° (c 3.36, THF). $^1$H NMR analysis with the aid of Eu(hfc)$_3$ indicated a 95:5 mixture of enantiomers. Exact mass calcd for $C_{11}H_{20}O$: m/e 168.1514. Found: m/e 168.1517.

**(1S,3aS,7aS)-7,7a-Dihydro-4-isobutyl-7a-methyl-1-[(2-trimethylsilylethoxy) methoxy]-3a(6H)-indanmethanol [Stille Rearrangement via Tin–Lithium Exchange.**[51] A solution of (1S,5S,7aS)-5,6,7,7a-tetrahydro-7a-methyl-1-[(2-trimethylsilylethoxy)methoxy]-5-indanol (9.75 g, 27.5 mmol) in dry tetrahydrofuran (50 mL) was added to a suspension of potassium hydride (2.0 g, 1.8 eq) in the same solvent (50 mL), and the mixture was stirred at room temperature for 2 hours. At this point, iodomethyl(tri-n-butyl)stannane (12.5 g, 1.06 eq) was added, and stirring was continued for 1.5 hours. A small amount of water was carefully introduced and then 50 mL of saturated ammonium chloride solution was added. The product was extracted into petroleum ether (3 times), and the combined organic layers were dried and concentrated. The residue was chromatographed on 40 g of silica gel (elution with 3% ethyl acetate in petroleum ether) to furnish 17.22 g (95.1%) of the (tri-n-butylstannyl)methyl ether as a yellowish oil.

A cold (−78°) magnetically stirred solution of the above material (17.22 g, 26.2 mmol) in dry hexane (250 mL, distilled from calcium hydride) was blanketed with argon and treated dropwise with n-butyllithium (17.8 mL of 1.55 N in hexane) during 5 minutes. After 2 hours, another 9 mL of the n-butyllithium solution was introduced, and the reaction mixture was allowed to warm to room temperature during 6 hours and stand overnight. After careful addition of water, the product was extracted into dichloromethane (three times), and the combined extracts were dried and evaporated. The crude product (18.43 g) was purified by HPLC on silica gel (eluted with 10% ethyl acetate in petroleum ether) to give 3.45 g (34%) of (1S,3aS,7aS)-7,7a-dihydro-4-isobutyl-7a-methyl-1-[(2-trimethylsilylethoxy)methoxy]-3a(6H)-indanmethanol and 2.82 g (27.8%) of the [1,2] rearrangement product: colorless oil, $[\alpha]_D^{22}$ 59.3° (c 4.0, benzene): IR (neat) 3500, 2942, 1244, 1050 cm$^{-1}$; $^1$H NMR (300 MHz, CDCl$_3$) $\delta$ 5.69 (t, J = 3.8 Hz, 1H), 4.68 (d, J = 6.9 Hz, 1H), 4.65 (d, J = 6.9 Hz, 1H), 3.90 (t, J = 7.0 Hz, 1H), 3.68–3.52 (m, 3H) 3.44 (d, J = 11.3 Hz, 1H), 2.10–2.07 (m, 2H), 1.96 (m, 1H), 1.82–1.79 (m, 3H), 1.68–1.43 (m, 6H), 0.99 (s, 3H), 0.95–0.87 (m, 8H), 0.01 (s, 9H); MS, m/z (M$^+$-C$_4$H$_8$) calcd 310.1964, obsd 310.2000. Its p-nitrobenzoate was a pale yellow oil.

**1-Undecen-4-ol [Reductive Lithiation-Mediated [2,3]-Wittig Rearrangement].**[30] Lithium naphthalenide in tetrahydrofuran was prepared by allowing equimolar amounts of lithium metal and naphthalene to stir overnight in a volume of tetrahydrofuran sufficient to afford a final concentration of 0.5 M. 1-(2-Propenyloxy)octyl phenyl sulfide (0.47 mmol) in 3 mL of tetrahydrofuran was cooled to 0° and 3 mL of the 0.5 M solution of lithium naphthalenide in tetrahydrofuran was introduced. After 20 minutes another 1-mL portion of lithium naphthalenide was added. After 1 hour the mixture was poured into water and extracted with ether. Flash chromatography on silica gel furnished 1-undecen-4-ol (67%).

**(S,S)-anti-2-Isopropyl-9-cyclodecyn-1-ol ([2,3]-Wittig Ring Contraction of a 13-Membered Propargylic Allylic Ether through Use of a Chiral Base).**[116] The chiral amide base was prepared from 910 mg (7.5 mmol) of (S,S)-bis(1-phenylethyl)amine in 10 mL of tetrahydrofuran at 0° under a nitrogen atmosphere by dropwise addition of 2.7 mL (6.8 mmol) of 2.5 M n-butyllithium in hexane. After 30 minutes at 0°, the solution of the amide was slowly added via cannula to a stirred, cooled (−41°) solution of 438 mg (2.15 mmol) of the cyclic ether in 10 mL of tetrahydrofuran. The resulting mixture and the bath were allowed to warm slowly for 20 minutes to 30°, and the mixture was diluted with water. The separated aqueous layer was extracted with ether. The combined ether layers were washed with 5% hydrochloric acid, water, and brine, and dried over anhydrous magnesium sulfate. Filtration and removal of solvent gave an oil that was purified by column chromatography on silica gel (15% ethyl acetate-hexanes) to give 303 mg (70%) of (−)-(S,S)-anti-2-isopropenyl-9-cyclodecyn-1-ol: IR (film) 3450, 3050, 2900, 2895, 2250, 2200, 1640, 1440, 1025, 880 cm$^{-1}$; $^1$H NMR (300 MHz, CDCl$_3$) δ 1.4–2.4 (m, CH$_2$s), 1.68 (s, vinyl CH$_3$s), 4.05 (m, carbinyl H), 4.80 (s, vinyl Hs, 2H), 5.08 (s, vinyl H); [α]$_D$ −27.1° (c, 5.4, CHCl$_3$; ee = 57%). A material of 60–80% ee was obtained in 70–85% yield from rearrangements conducted under comparable conditions to those described above.

**(1R,2R)-(5E)-2-Isopropenyl-5-methyl-5-cyclodecen-9-ynyl (S)-O-Methylman-delate. (Determination of the Absolute Configuration of (+)-(1R,2R)-anti-2-iso-propenyl-9-cyclodecyn-1-ol)**[116] To a solution of 330 mg (1.6 mmol) of (+)-(1R,2R)-anti-2-isopropenyl-9-cyclodecyn-1-ol [α]$_D$ +33.4° (c 4.2, CHCl$_3$; ee = 70%)], 0.5 g (3 mmol) of (S)-O-methylmandelic acid, and 320 mg of dicyclohexylcarbodiimide in 7 mL of dichloromethane was added a catalytic amount of 4-dimethylamino-pyridine. After being stirred overnight at room temperature under nitrogen, the mixture was concentrated under reduced pressure to afford a yellow solid. This solid was washed with ether and filtered, and the organic layer was concentrated to afford a white viscous oil that was chromatographed on silica gel (2% ethyl acetate-hexanes), yielding 322 mg (71%) of the (S)-O-methyl-mandelic ester as a colorless viscous oil (>90% pure by $^1$H NMR analysis): $^1$H NMR (300 MHz, CDCl$_3$) δ 2.46 (dt, J = 11 and 4 Hz, H-2).

**(1S,2S)-(5E)-2-Isopropenyl-5-methyl-5-cyclodecen-9-ynyl (S)-O-Methylman-delate. (Determination of the Absolute Configuration of (−)-(1S,2S)-anti-2-iso-propenyl-9-cyclodecyn-1-ol)**[116] The procedure described above was followed with 326 mg (1.60 mmol) of (−)-(1S,2S)-anti-2-isopropenyl-9-cyclodecyn-1-ol [[α]$_D$ −36.5° (c 4.94, CHCl$_3$; ee = 77%)], 0.5 g (3 mmol) of (S)-O-methylman-delic acid, 0.4 g (2.0 mmol) of dicyclohexylcarbodiimide, and a catalytic amount of 4-dimethylaminopyridine in 12 mL of dichloromethane. The mixture was stirred overnight, and the product was isolated by removal of solvent and chromatography on silica gel (2% ethyl acetate/hexanes) to give 360 mg (73%) of (S)-O-methylmandelate (>95% pure by $^1$H NMR analysis) as a colorless viscous oil: $^1$H NMR (300 MHz, CDCl$_3$) $\delta$ 2.58 (dt, $J$ = 11 and 4 Hz, H-2).

**(2S,3R)-2-Hydroxy-3-methylpentanoic Acid [Rearrangement via the Transmetalation of a Lithium Enolate to a Zirconium Enolate].[97]** A solution of n-butyllithium (1.6 mol/L, 258 $\mu$L, 1.2 eq) was added slowly to a tetrahydrofuran solution (330 $\mu$L) of (2S,5S,2′E)-N-2′-butenyloxyacetyl-2,5-bis(methoxymethyl)pyrrolidine (114.0 mg) at $-100°$ under a nitrogen atmosphere. After stirring at $-100°$ for 2 hours, a solution of dicyclopentadienylzirconium dichloride (120 mg, 1.2 eq) in tetrahydrofuran (1.5 mL) was added slowly to the solution and the mixture was stirred at the same temperature for 3 hours and then at $-70°$ for 3 hours. Saturated aqueous potassium fluoride (150 $\mu$L) was added and the mixture was allowed to warm to room temperature and then passed through a short column of silica gel. Concentration and chromatrography on silica gel gave N-2′-hydroxy-3′-methyl-4′-pentenoyl-2,5-bis(methoxymethyl) pyrrolidine (47.4 mg, 42%). The product (32 mg) was hydrogenated (on palladium in ethanol) and hydrolyzed (1 mol/L, HCl, 100°) to afford a quantitative yield of (2S, 3R)-2-hydroxy-3-methylpentanoic acid (11 mg); $[\alpha]_D^{25}$ +11.2° ($c$ = 0.36, H$_2$O).

**8-Phenylmenthyl 2-Hydroxy-3-methyl-4-pentenoate [Rearrangement via the Transmetalation of an Enol Silyl Ether with Titanium Tetrachloride].[65]** To a solution of diisopropylamine (0.21 mL, 1.5 mmol) in tetrahydrofuran (3 mL) was added a 1.55 N solution of n-butyllithium in hexane (0.9 mL, 1.3 mmol) at 0°, and the mixture was stirred for 30 minutes. To this solution was added dropwise a solution of 8-phenylmenthyl (E)-3-oxa-5-heptenoate (94% E, 344 mg, 1.0 mmol) in tetrahydrofuran (3 mL) at $-70°$, and the solution was stirred for 5 minutes at that temperature, followed by rapid addition of chlorotrimethylsilane (0.19 mL, 1.5 mmol). The reaction mixture was warmed to room temperature, diluted with hexane, filtered through Celite, and evaporated. The residue was dissolved in hexane and filtered again. The filtrate was evaporated to afford the enol silyl ether of 8-phenylmenthyl (E)-3-oxa-5-heptenoate.

To a solution of the enol silyl ether (417 mg, 1.0 mmol) in dichloromethane (2 mL) was added titanium tetrachloride (0.11 mL, 1.0 mmol) at $-50°$. The mixture was stirred at that temperature for ca. 10 hours, then allowed to warm to 0°, poured into water, and extracted with dichloromethane. The organic layer was washed with an aqueous sodium bicarbonate solution, dried over magnesium sulfate, and evaporated. Subsequent silica gel chromatography gave 8-phenylmenthyl 2-hydroxy-3-methyl-4-pentenoate (276 mg, 80%): GC (XE-60, 30 m, 180°) $t_R$ 24.3 (2S,3S), 24.6 (2S,3R), 29.9 (2R,3R), 30.9 (2R,3S) minutes; HPLC (Finepak SIL-5, hexane:ethyl acetate = 30:1) 18.3 (2S,3S), 13.8 (2S,3R), 13.6 (2R) minutes; TLC (hexane:ether = 3:1) $R_f$ 0.28 (2S), 0.34 (2R); $^1$H NMR (CDCl$_3$) for 2S $\delta$ 3.03 (m, 1H), 4.6–5.0 (m, 3H), 5.60 (ddd, $J$ = 7.2, 9.0, and 18.0 Hz, 1H),

7.20 (m, 5H); for 2R δ 3.68 (m, 1H), 4.7–5.1 (m, 3H), 5.76 (ddd, J = 7.2, 9.6, and 16.5 Hz, 1H), 7.32 (m, 5H).

**Methyl 2-Hydroxy-3-methyl-4-pentenoate [Silyl Triflate-Catalyzed Rearrangement].**[164] To a solution of methyl (2-butenyloxy)acetate (93% E, 0.144 g, 1.0 mmol) and triethylamine (0.15 mL, 1.1 mmol) in dichloromethane (5 mL) was added trimethylsilyl triflate (0.23 mL, 1.2 mmol) at 0°, and the mixture was stirred overnight at room temperature. The mixture was then diluted with water. The separated aqueous layer was extracted with ethyl acetate. The combined organic layers were washed with 5% hydrochloric acid, water, and brine, and dried over anhydrous magnesium sulfate. Filtration and removal of solvent gave an oil which was chromatographed on silica gel (ethyl acetate-hexane) to give 0.12 g (83%) of methyl 2-hydroxy-3-methyl-4-pentenoate: GC (PEG 20M): $t_R$ 16 and 14 (erythro:threo = 92:8) minutes; [1]H NMR (CDCl₃) for erythro δ 1.00 (d, J = 7.2 Hz, 3H), 2.43–2.83 (m, 2H), 3.77 (s, 3H), 4.03–4.25 (m, 2H), 5.06 (d, J = 10.5 Hz, 1H), 5.09 (d, J = 16.8 Hz, 1H), 5.87 (ddd, J = 7.0, 10.5, and 16.8 Hz, 1H); for threo δ 1.14 (d, J = 7.2 Hz, 3H).

## TABULAR SURVEY

The [2,3]-Wittig rearrangements of benzyl ethers, alkyl ethers, allyl ethers, propargyl ethers, and α-(allyloxy)carbonyl compounds are grouped in Tables I–V. Table VI contains examples of [2,3]-Wittig contractions. Within each table entries are listed by increasing numbers of carbon atoms, using the *Chemical Abstracts* convention. Yields, in parentheses, are based on isolated products. A dash (—) indicates that no yield was reported. Numbers not in parentheses are product or diastereomeric ratios, and values in brackets [ ] are corresponding ratios corrected to 100% stereochemical purity of the products.

The literature has been reviewed to the end of 1991, and includes some papers that appeared up to mid-1992.

The following abbreviations are used in the tables:

| | |
|---|---|
| 18-c-6 | 18-crown-6 |
| Bn | benzyl |
| Boc | *tert*-butoxycarbonyl |
| BOM | benzyloxymethyl |
| Cp | cyclopentadienyl |
| DB-18-c-6 | dibenzo-18-crown-6 |
| DC-18-c-6 | dicyclohexyl-18-crown-6 |
| ds | diastereoselectivity |
| EE | ethoxyethyl |
| HMPA | hexamethylphosphoric triamide |
| KHDS | potassium hexamethyldisilazane |
| LBPEAR | lithiobis[(R)-1-phenylethyl]amide |

| LBPEA*S* | lithiobis[(*S*)-1-phenylethyl]amide |
| LDA | lithium diisopropylamide |
| LDCA | lithium dicyclohexylamide |
| LHDS | lithium hexamethyldisilazane |
| LTBB | lithium 4,4'-di-*tert*-butylbiphenylide |
| LTMP | lithium 2,2,6,6-tetramethylpiperidide |
| MEM | (2-methoxyethoxy)methyl |
| MOM | methoxymethyl |
| SEM | 2-(trimethylsilyl)ethoxymethyl |
| TBAB | tetrabutylammonium bromide |
| TBDMS | *tert*-butyldimethylsilyl |
| TBDPS | *tert*-butyldiphenylsilyl |
| TBPB | tetrabutylphosphonium bromide |
| Tf | trifluoromethanesulfonyl (trifyl) |
| THF | tetrahydrofuran |
| THP | tetrahydropyranyl |
| TMEDA | $N,N,N',N'$-tetramethylethylenediamine |
| TMS | trimethylsilyl |
| Ts | *p*-toluenesulfonyl |

Table I. REARRANGEMENTS OF BENZYL ETHERS

| Ether | Conditions | Product(s) and Yield(s) (%) | Refs. |
|---|---|---|---|
| C$_{10}$ 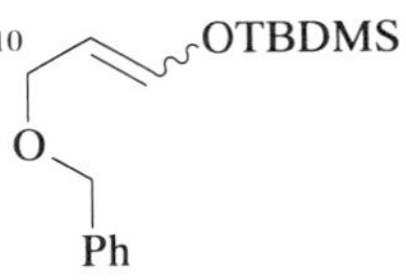 (structure: OTBDMS allylic benzyl ether) | | 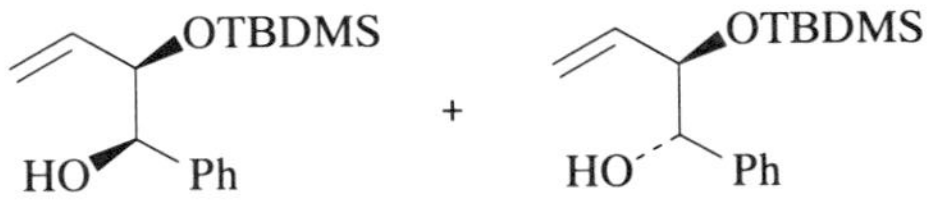 I (structure OTBDMS, Ph, HO) + II (structure OTBDMS, Ph, HO) | |
| $E:Z = 29:71$ | $n$-BuLi, THF-C$_6$H$_{14}$, -78 to -65° | (73), **I:II** = 18:82 | 77 |
| $E:Z = 7:93$ | " | (81), **I:II** = 23:77 | 77 |
| C$_{11}$ 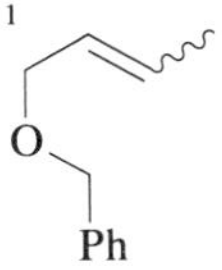 (structure: benzyl ether) | | 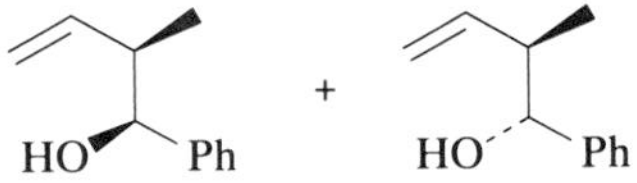 I (structure) + II (structure) | |
| $E$ | $n$-BuLi, TMEDA-Et$_2$O-C$_6$H$_{14}$, -80 to -25° | **I:II** = 1:1 | 18 |
| $Z$ | " | I only | 18 |
| $E$ | MeLi, THF-Et$_2$O, rt | **I:II** = 2:1 | 9 |
| $E/Z = 93:7$ | $n$-BuLi, THF-C$_6$H$_{14}$ (1:1), -85° | **I:II** = 63:37 [61:39] | 24 |
| $E/Z = 5:95$ | " | **I:II** = 93:7 [95:5] | 24 |
| $E/Z = 93:7$ | $n$-BuLi, THF-C$_6$H$_{14}$ (1:4), -85° | **I:II** = 68:32 | 24 |
| $E/Z = 93:7$ | TMEDA-Et$_2$O-THF (1:3:7) | **I:II** = 56:44 | 24 |
| $E/Z = 93:7$ | TMEDA-Et$_2$O-THF (1:1:1.2) | **I:II** = 49:51 [45:55] | 24 |
| $E/Z = 5:95$ | " | **I:II** = 95:5 [98:2] | 24 |
| 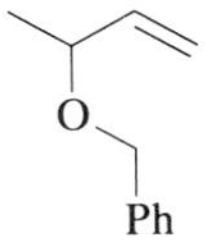 C$_{11}$ (structure: allylic benzyl ether, Ph, O) | $n$-BuLi, TMEDA-Et$_2$O-C$_6$H$_{14}$, -80 to -25° | 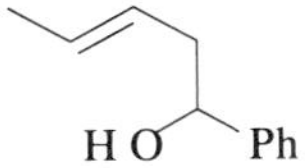 *E* only (structure: HO, Ph) | 18 |

151

Table I. REARRANGEMENTS OF BENZYL ETHERS (*Continued*)

| Ether | Conditions | Product(s) and Yield(s) (%) | Refs. |
|---|---|---|---|

$C_{12}$

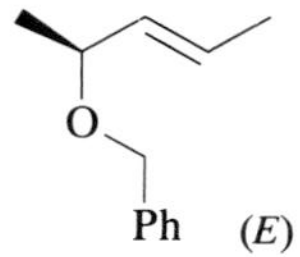

(E) 72% ee

n-BuLi; THF, 0°

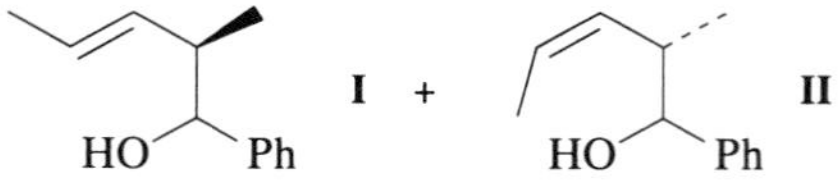

I + II

I:II = 83:17, 72% ee

10

| $R^1$ | $R^2$ | Conditions | 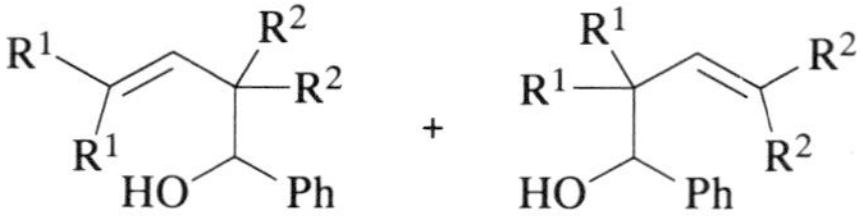 [2,3]:[1,2] | Refs. |
|---|---|---|---|---|
| H | Me | n-BuLi, THF, 25° | 95:5 | 165 |
| H | Me | n-BuLi, THF, -10° | 98:2 | 165 |
| H | Me | n-BuLi, TMEDA-Et$_2$O-C$_6$H$_{14}$, 25° | 6:1 | 18 |
| H | Me | n-BuLi, TMEDA-Et$_2$O-C$_6$H$_{14}$, -20° | 7.5:1 | 18 |
| H | Me | n-BuLi, TMEDA-Et$_2$O-C$_6$H$_{14}$, -80° | 8:1 | 18 |
| Me | H | n-BuLi, THF, 25° | 77:23 | 165 |
| Me | H | n-BuLi, THF, -20° | 83:17 | 165 |

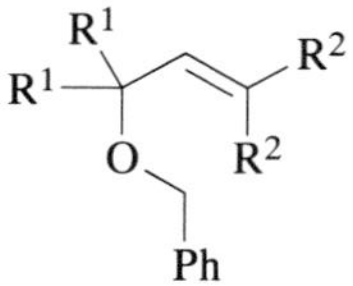

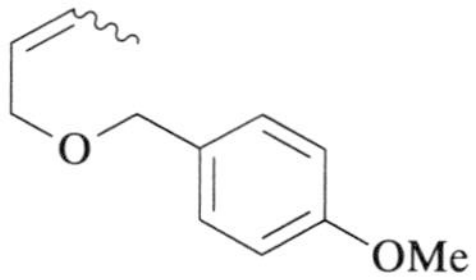

[2,3]

[1,2]

I + II

OMe

| | | | |
|---|---|---|---|
| Z | | | |
| E | | | |

n-BuLi; THF-C$_6$H$_{14}$, -85°

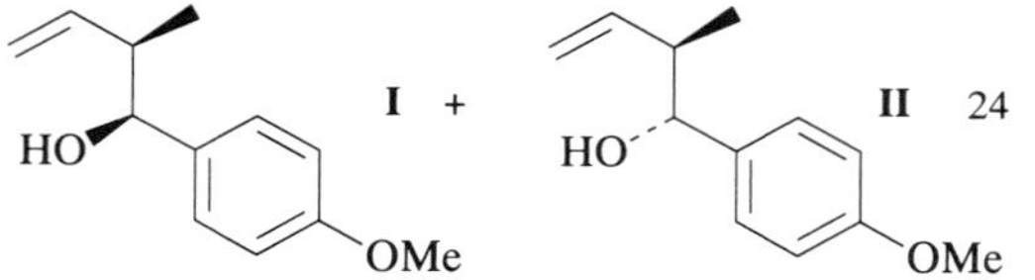

(79-98) I:II = 100:0

(70-98) I:II = 40:60

24

Table I. REARRANGEMENTS OF BENZYL ETHERS (*Continued*)

| Ether | Conditions | Product(s) and Yield(s) (%) | Refs. |
|---|---|---|---|

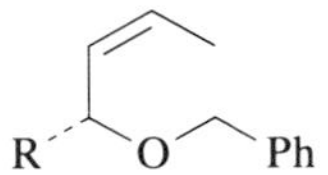

$C_{12}$   R = Me (Z) 95% ee
$C_{14}$   R = i-Pr (Z) 91% ee
    R = i-Pr (E) 91% ee
$C_{15}$   R = i-Bu (Z) 68% ee

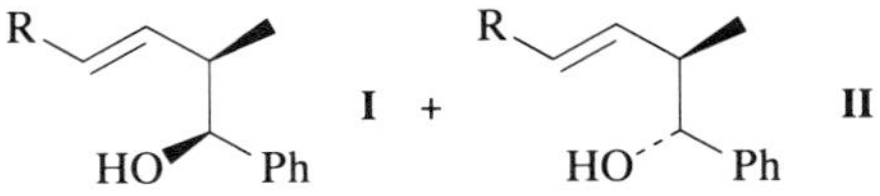

Conditions: *n*-BuLi, THF, –85°

Products **I** + **II**:

(—), **I:II** = 80:20, 92% ee   81
(89), **I:II** = 93:7, 91% ee   82
(95), **I:II** = 20:62, 91% ee   135,136
(—), **I:II** = 96:4, 64% ee   81

---

$C_{14}$

| $R^1$ | $R^2$ |
|---|---|
| Me | H |
| H | Me |
| Me | H |
| H | Me |

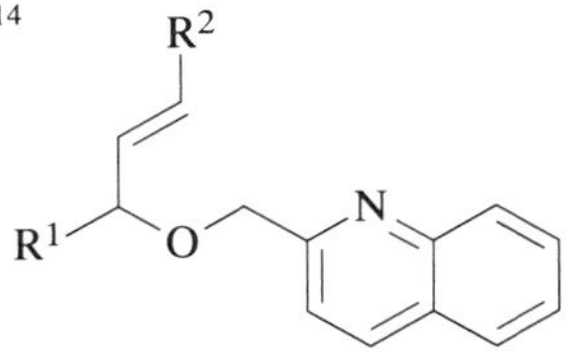

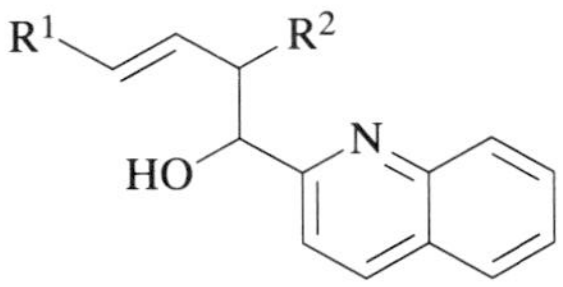

Conditions:
NaNH₂, NH₃, –33°, 2 h
NaNH₂, NH₃, –33°, 2 h
*n*-BuLi, THF, –25°, 2 h
*n*-BuLi, THF, –25°, 2 h

Product:

(74)   6
(78)   6
(90)   6
(quant)   6

---

$C_{14}$   L = CO, E:Z = 96:4
    L = CO, E:Z = 12:88
$C_{31}$   L = PPh₃

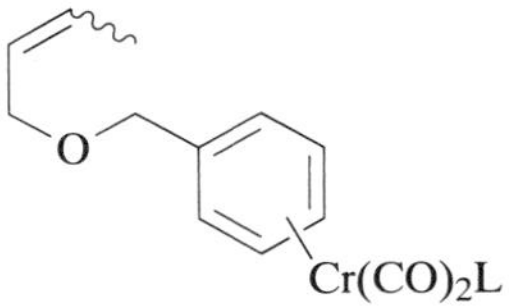

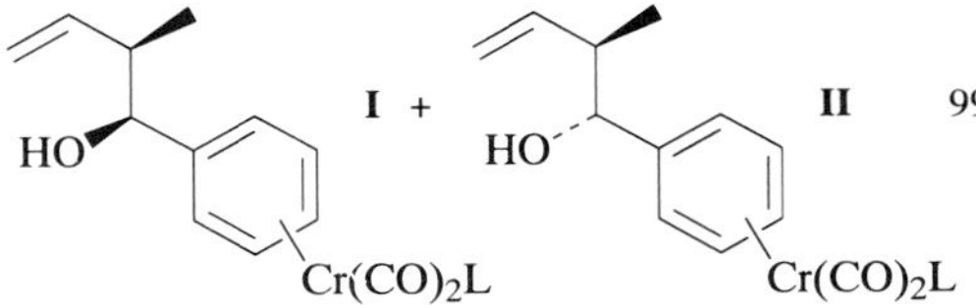

Conditions: LDA, THF, –78°, 7 h

Products **I** + **II**:

(69), **I:II** = 95:5
(40), **I:II** = 48:52
(95), **I:II** = 88:12

99, 100

---

153

Table I. REARRANGEMENTS OF BENZYL ETHERS (*Continued*)

| Ether 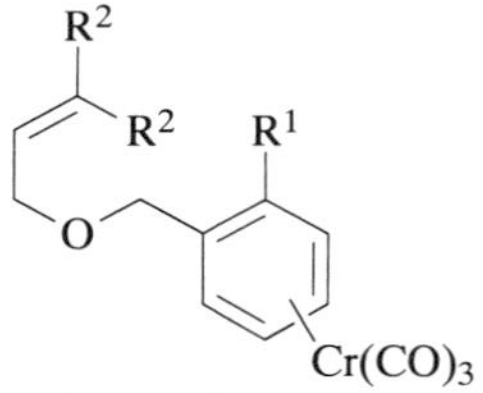 | Conditions | Product(s) and Yield(s) (%) 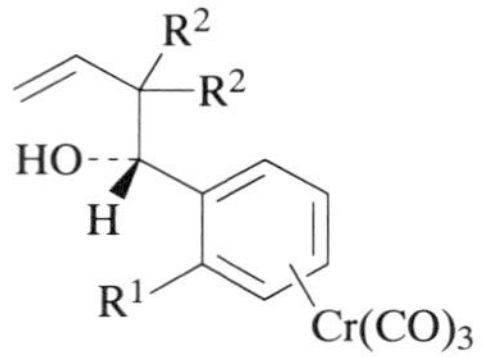 | Refs. |
|---|---|---|---|

C_14 (structure with $R^2$, $R^1$, $Cr(CO)_3$, benzyl ether)

| $R^1$ | $R^2$ | | |
|---|---|---|---|
| OMe | H | *n*-BuLi, THF, 0°, 20 min | (16) | 101 |
| OMe | H | *n*-BuLi, THF, -40°, 2 h | (30) | 101 |
| OMe | H | *n*-BuLi, THF, -50°, 4 h | (48) | 101 |
| C_16 Me | Me | *n*-BuLi, THF, -60°, 4 h | (88) | 101 |
| Me | Me | *n*-BuLi, THF, -20°, 4 h | (90) | 101 |

Product: $R^2$, $R^2$, $R^1$, HO, H, $Cr(CO)_3$

C_15 (structure, R, $Cr(CO)_3$)

R = Me, *E:Z* = 80:20
R = OMe, >99.3% *E*

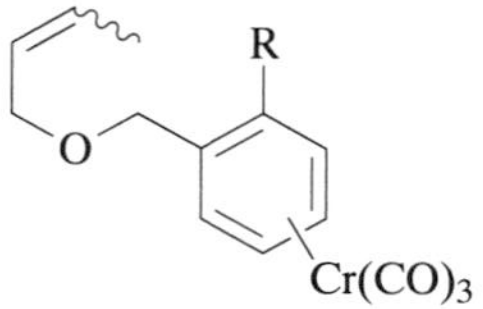

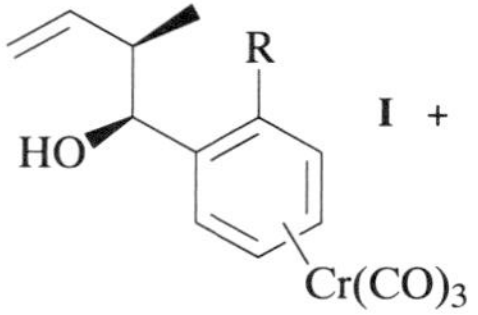

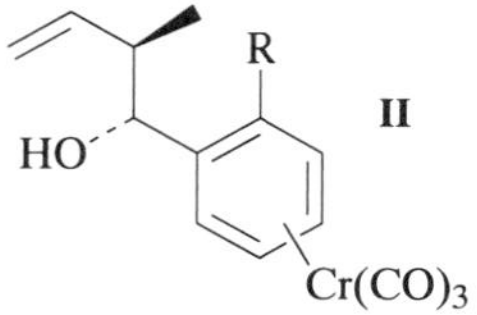

*n*-BuLi, THF, -60°, 6 h
*n*-BuLi, THF-TMEDA, -78°

Product: I + II
I: HO, R, $Cr(CO)_3$
II: R, HO, $Cr(CO)_3$

(90), **I:II** = 80:20    101
(67), **I:II** = 97:3    99, 100

(+)-*R*, >99.3% *E* (OMe, $Cr(CO)_3$)

*n*-BuLi, THF-TMEDA, -78°

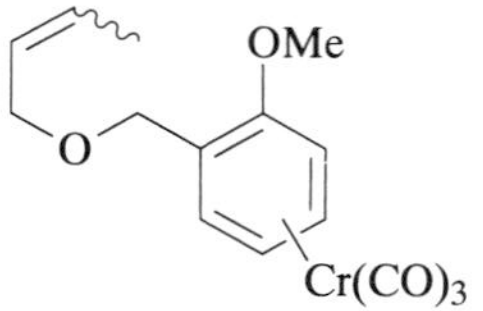

Product: HO, OMe, $Cr(CO)_3$

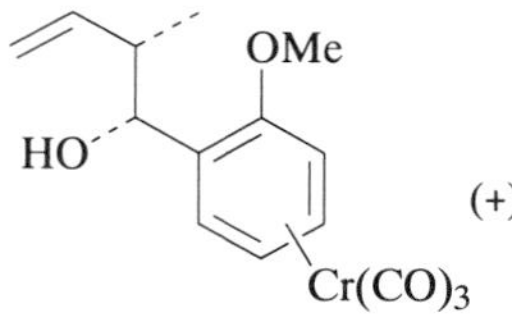

(+)-(1*R*,2*S*), >99.5% ee    99, 100

154

Table I. REARRANGEMENTS OF BENZYL ETHERS (*Continued*)

| Ether | Conditions | Product(s) and Yield(s) (%) | Refs. |
|---|---|---|---|

$C_{17}$

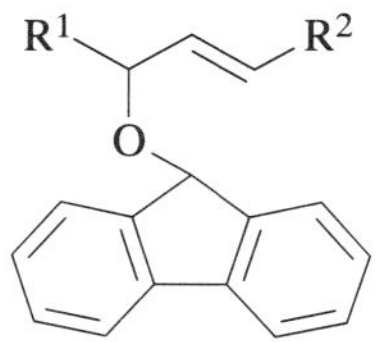

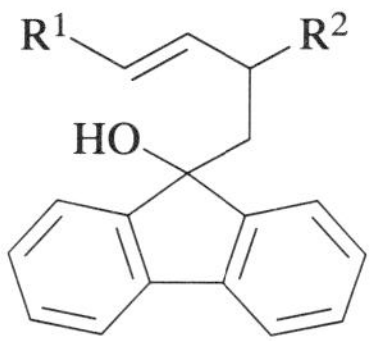

| $R^1$ | $R^2$ | | | |
|---|---|---|---|---|
| Me | H | MeLi, THF, -50°, 3 h | (95) | 5 |
| H | Me | MeLi, THF, -50°, 3 h | (94) | 5 |
| Me | H | $NaNH_2$, $NH_3$, -33°, 2 h | (70) | 6 |
| H | Me | $NaNH_2$, $NH_3$, -33°, 2 h | (85) | 6 |
| H | Me | KOH (20%), TBAB, $C_6H_6$, rt, 6 h | (93) | 7 |
| H | Me | KOH (20%), TBPB, $C_6H_6$, rt, 6 h | (87) | 7 |
| H | Me | KOH (20%), 18-c-6, $C_6H_6$, rt, 6 h | (89) | 7 |
| H | Me | KOH (20%), DC-18-c-6, $C_6H_6$, rt, 6 h | (90) | 7 |
| H | Me | KOH (20%), DB-18-c-6, $C_6H_6$, rt, 6 h | (55) | 7 |

$C_{18}$

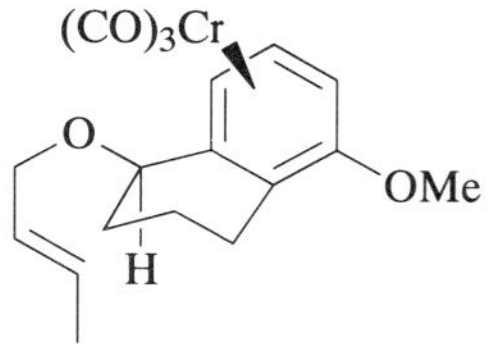

*n*-BuLi

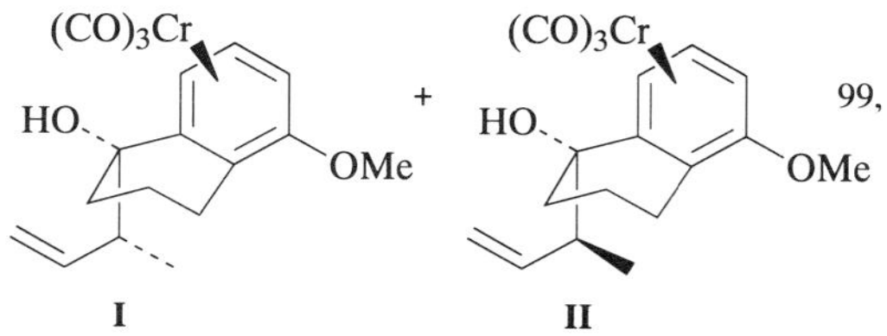

(60), **I:II** = 88:12    99, 100

Table II. REARRANGEMENTS OF ALKYL ETHERS

| Ether | Conditions | Product(s) and Yield(s) (%) | Refs. |
|---|---|---|---|
| 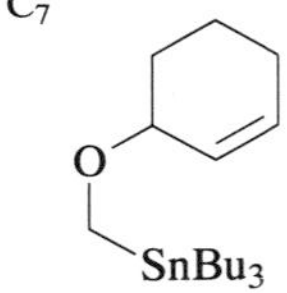 C₇ | *n*-BuLi, THF, -78° | 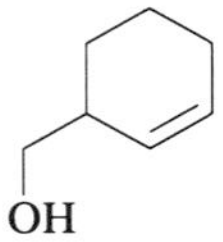 (93) | 11 |
| C₈ 92% ee | *n*-BuLi, THF, -78°, 40 min; 0°, 40 min | 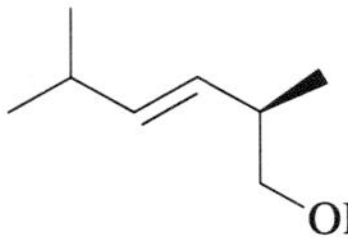 (72) 92% ee | 136 |
| 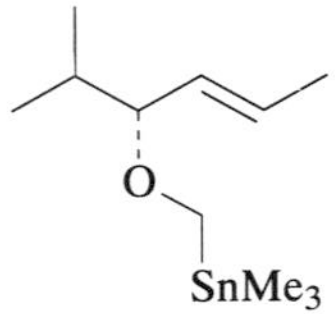 92% ee | *n*-BuLi, THF, -78°, 40 min; 0°, 40 min | 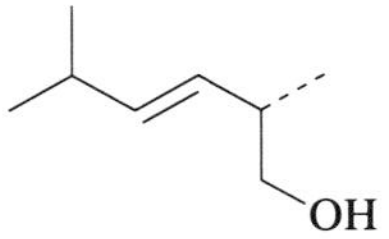 92% ee (72) 53:47 + 92% ee | 136 |
| | *n*-BuLi, THF, -78°, 1 h | (78) 93:7 (67) 45:55 (71) 89:11 (81) 96:4 | 107 |

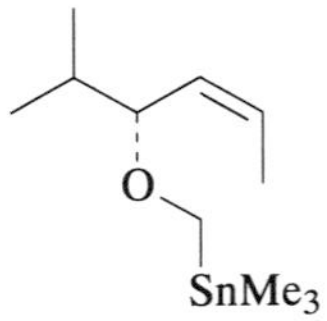

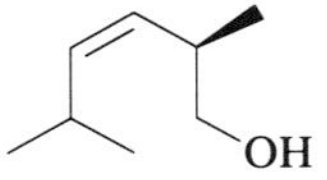

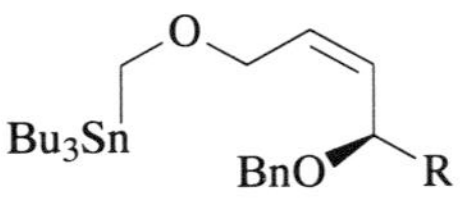

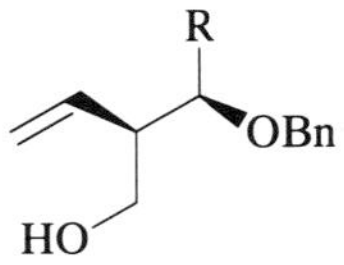

R
*i*-Pr
*t*-Bu
Ph
Bn

C₈
C₉
C₁₁
C₁₂

156

Table II. REARRANGEMENTS OF ALKYL ETHERS (*Continued*)

| Ether | Conditions | Product(s) and Yield(s) (%) | Refs. |
|---|---|---|---|
| $C_8$ | $n$-BuLi, THF, -78°, 1 h | <br>R<br>MOM (66) 98:2<br>MEM (89) 98:2<br>Bn (78) 93:7<br>$i$-Pr (81) 93:7 | 107 |
| $C_8$ | 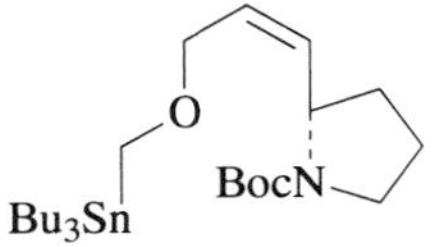 $n$-BuLi, THF, HMPA, -70°, 1.5-6 h | 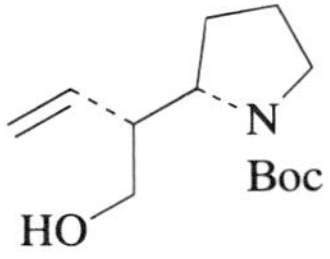 (88)  ds >99.8:0.2 | 109 |
| 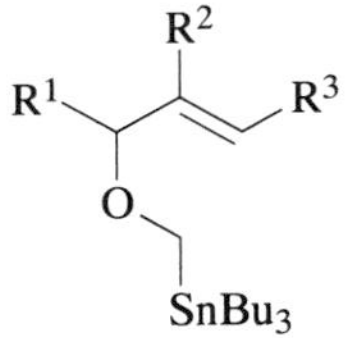 | $n$-BuLi, THF, -78° | 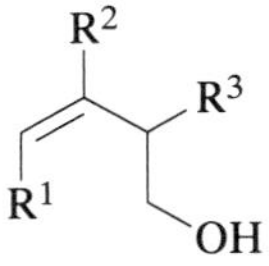 | 11 |

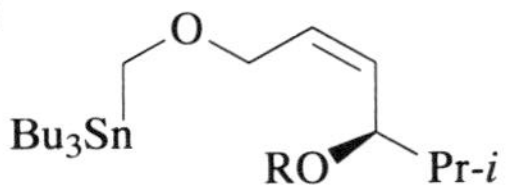

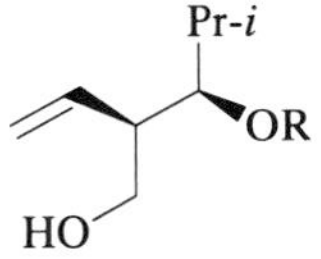

|  | $R^1$ | $R^2$ | $R^3$ | | |
|---|---|---|---|---|---|
| $C_9$ | $n$-Bu | Me | H | (quant) | 96-97% Z |
| $C_{11}$ | $n$-C$_7$H$_{15}$ | H | H | (>95) | 60% Z |
| $C_{12}$ | $n$-C$_7$H$_{15}$ | H | Me $E$ | (96) | 100% $E$ |
|  | $n$-C$_7$H$_{15}$ | H | Me $Z$ | (91) | 65% Z |
| $C_{14}$ | 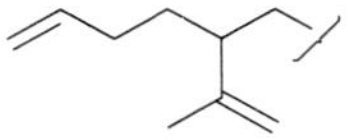 | Me | H | (83) | 95-96% Z |

157

Table II. REARRANGEMENTS OF ALKYL ETHERS (*Continued*)

| Ether | Conditions | Product(s) and Yield(s) (%) | Refs. |
|---|---|---|---|
| 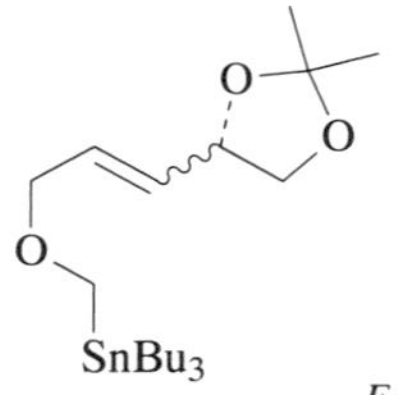 $E$<br>$Z$ | *n*-BuLi, THF, -78° | 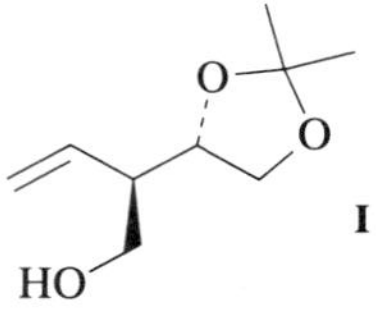 **I** 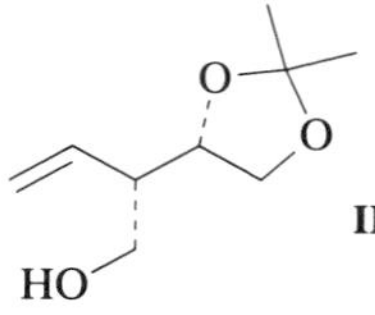 + **II**<br>**I** (79) + **II** (6)<br>**I:II** = >99:1 | 107-109 |
| 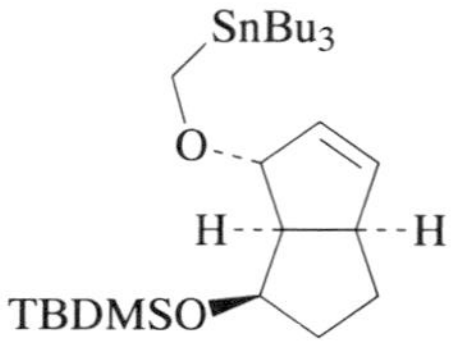 | *n*-BuLi, C$_6$H$_{14}$, -78°, 2 h | 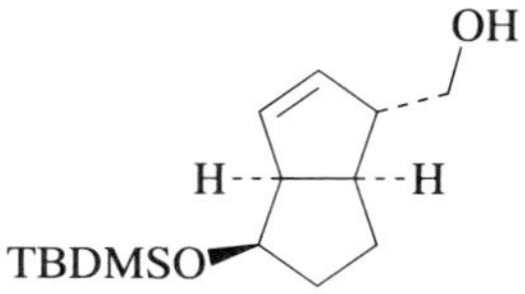 (quant) | 87 |
| C$_{10}$ | *n*-BuLi, THF,<br><-60 to -20°, 1 h | (—)  ds 92.7:7.3 | 110 |
| 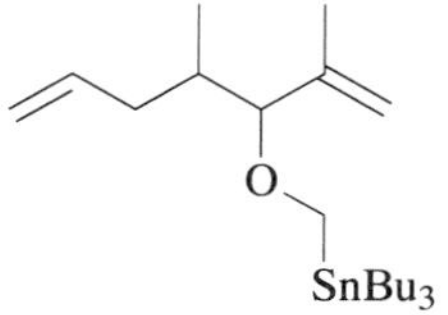 | *n*-BuLi, -78° | 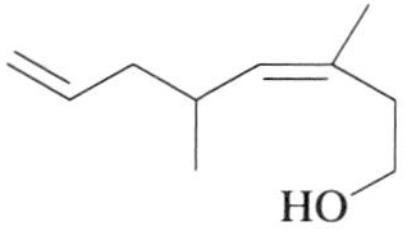 (>75) + *E*-diene (tr) | 166 |

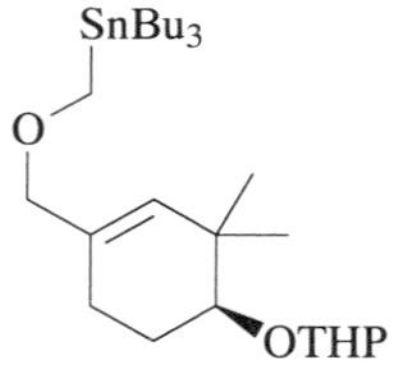

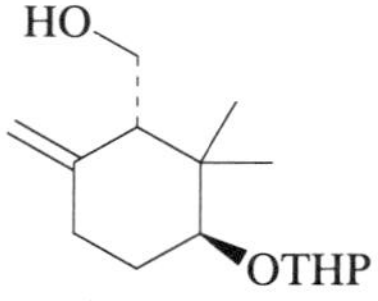

158

Table II. REARRANGEMENTS OF ALKYL ETHERS (*Continued*)

| Ether | Conditions | Product(s) and Yield(s) (%) | Refs. |
|---|---|---|---|
| 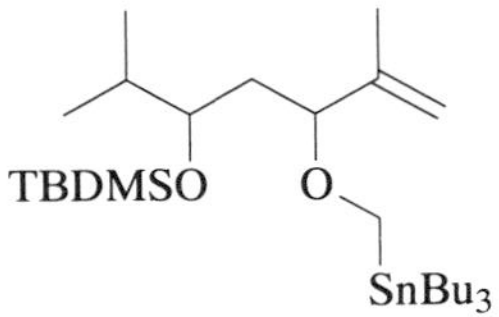 | *n*-BuLi | 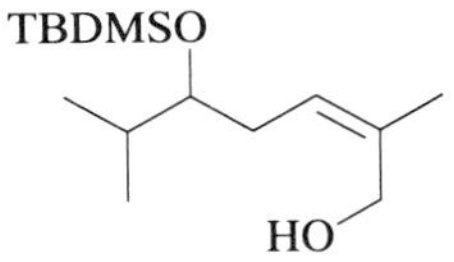 (>65) | 30, 31 |
| 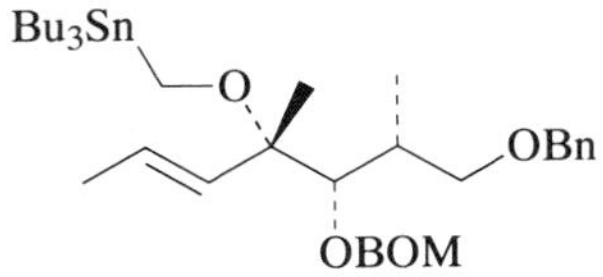 | *n*-BuLi, THF, -78° | 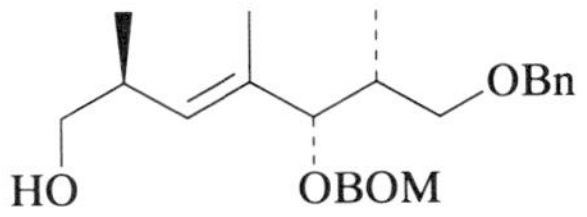 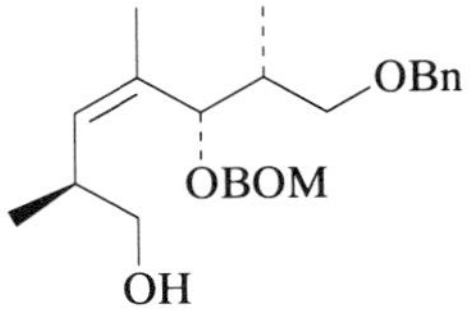 (65) 71:29 | 91–93 |
| 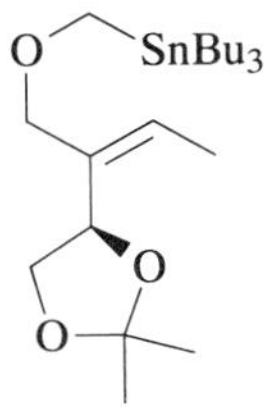 | *n*-BuLi, THF, -78°, 2 h | 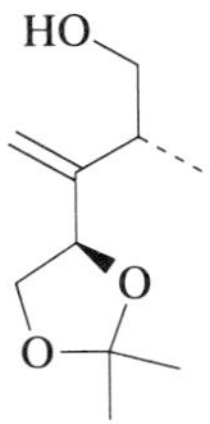 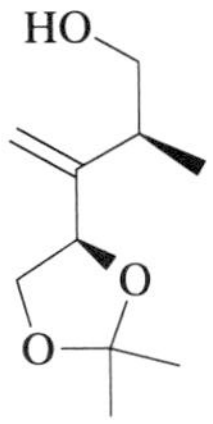 (89) 95:5<br>(70) 3:97 | 111 |

159

Table II. REARRANGEMENTS OF ALKYL ETHERS (*Continued*)

| Ether | Conditions | Product(s) and Yield(s) (%) | Refs. |
|---|---|---|---|
| C₁₁ | | | |
| 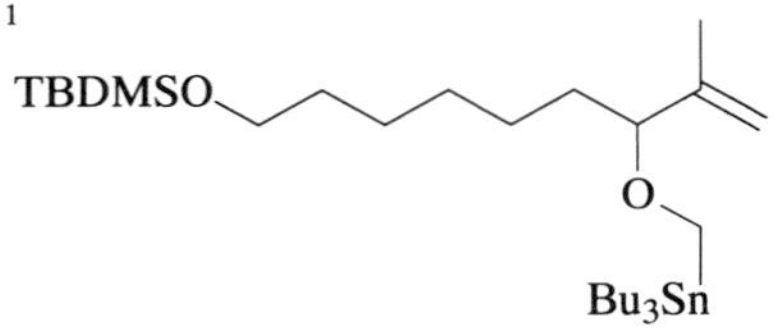 TBDMSO (structure with Bu₃Sn) | *n*-BuLi, -78° | 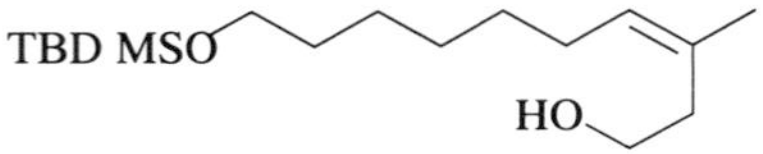 TBD MSO (structure, HO) (—) | 167 |
| 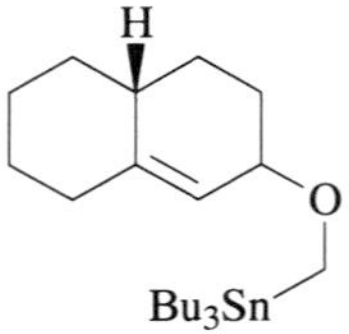 (bicyclic structure with H, O—Bu₃Sn) | *n*-BuLi; THF, -78° | 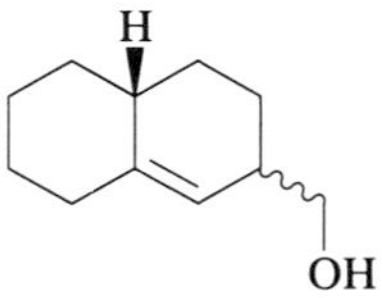 (bicyclic structure with H, OH) (—) | 11 |
| C₁₁ C₁₄ (bicyclic structure with R, SEMO, O—SnBu₃) | *n*-BuLi, C₆H₁₄, -78 to 0° | 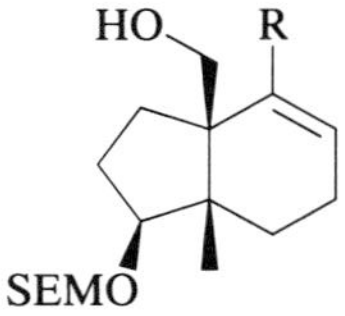 (bicyclic structure with R, HO, SEMO) R = H (45) R = *i*-Pr (34) | 50, 51 |
| 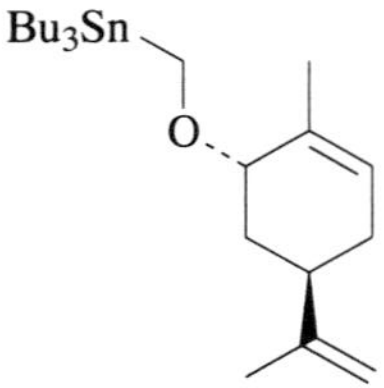 C₁₁ (cyclohexene structure with Bu₃Sn, O) | *n*-BuLi; THF, -78° to rt, 20 min | 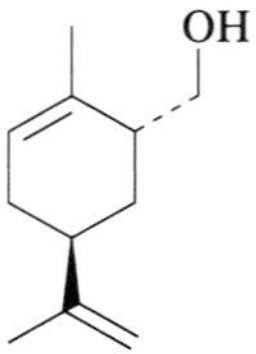 (cyclohexene structure with OH) (77) | 46 |

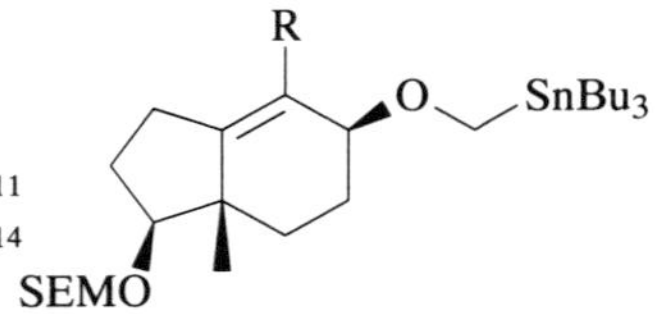

Table II. REARRANGEMENTS OF ALKYL ETHERS (*Continued*)

| Ether | Conditions | Product(s) and Yield(s) (%) | Refs. |
|---|---|---|---|
| 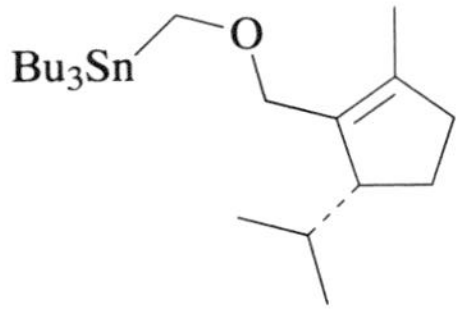 | *n*-BuLi, THF, -78°, 2 h | 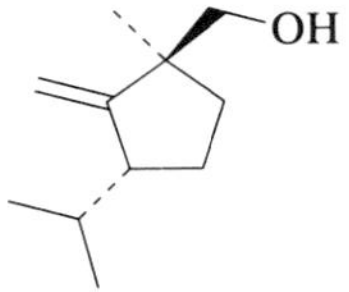 (>75) | 112, 113 |
| C$_{12}$ 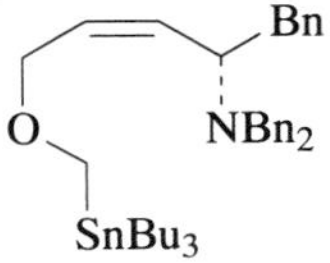 | *n*-BuLi, THF, -78°, 1 h | 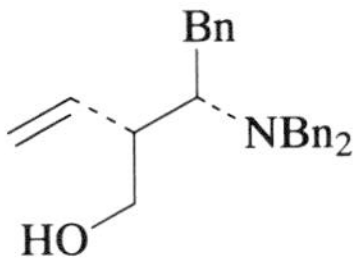 (45)  ds >95:5 | 109 |
| 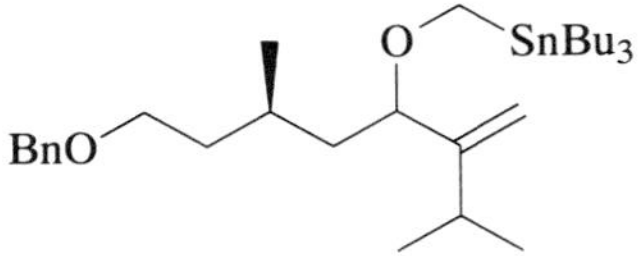 | *n*-BuLi, -65 to -70°, 45 min | 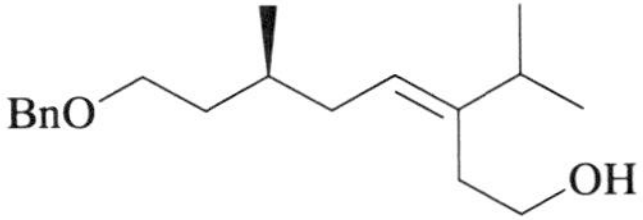 (>74) | 168 |
| 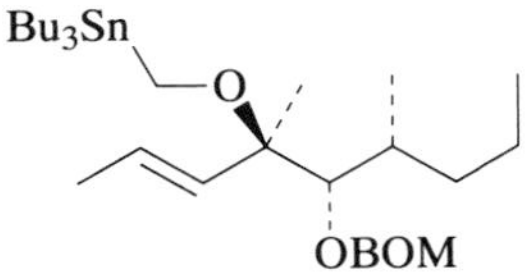 | *n*-BuLi, THF, -78° | 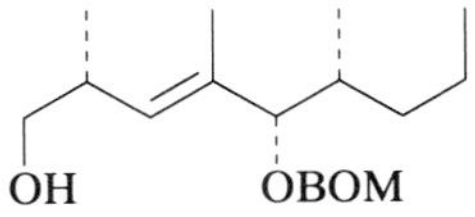 (57)  >100:1 | 91-93 |
| 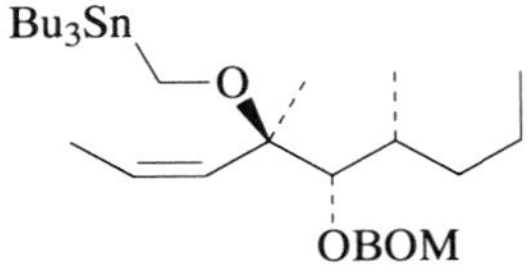 | *n*-BuLi, THF, -78° | 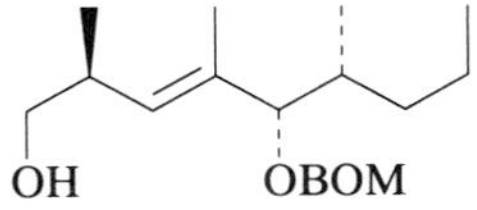 (64)  >100:1 | 91-93 |

161

Table II.  REARRANGEMENTS OF ALKYL ETHERS (*Continued*)

| Ether | Conditions | Product(s) and Yield(s) (%) | Refs. |
|---|---|---|---|
| 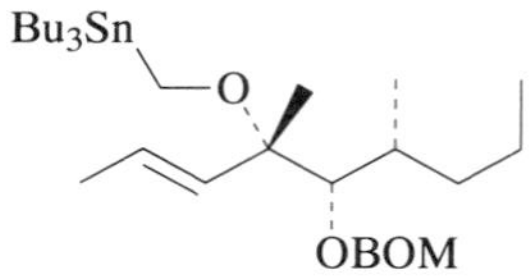 Bu₃Sn— structure with OBOM | *n*-BuLi, THF, -78° | 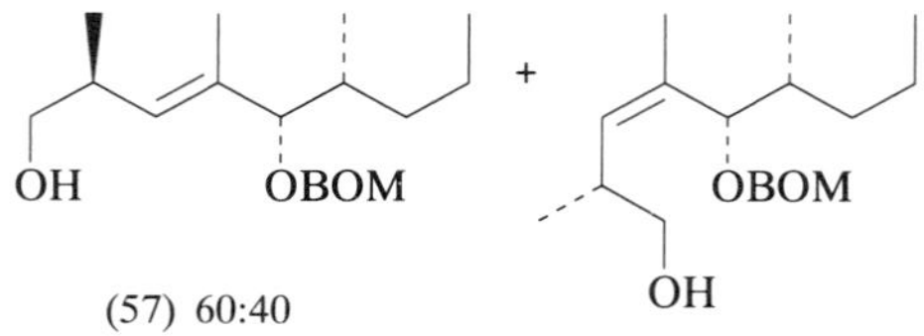 OBOM / OH structures (57) 60:40 | 91-93 |
| 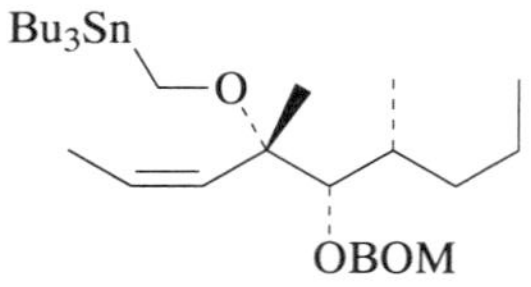 Bu₃Sn— structure with OBOM | *n*-BuLi, THF, -78° | 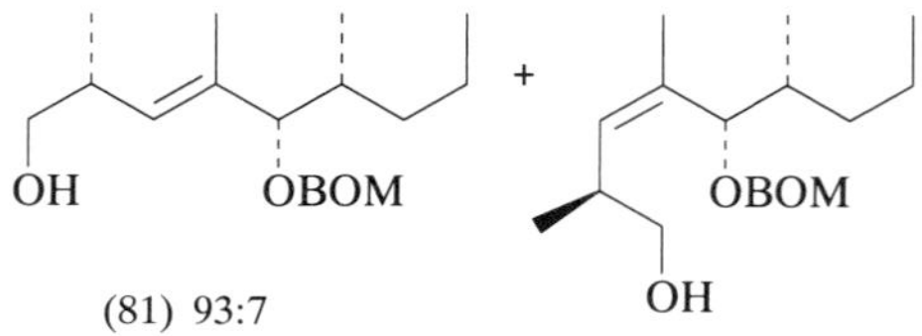 OBOM / OH structures (81) 93:7 | 91-93 |
| 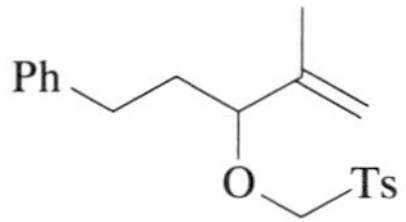 C₁₃ Ph— structure with O—Ts | RLi, HMPA, THF, -78° | 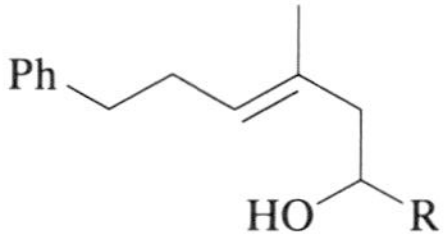 Ph— HO—R  R = Bu (42); R 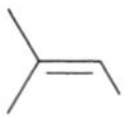 = (59); R 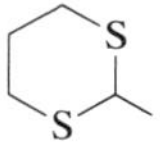 = (41) | 169 |
|  | 3 *n*-BuLi, Et₂O, HMPA/THF (3%), -78°, 1 h; to -40°, 1 h | " R = Bu (63) | 169 |

162

Table II. REARRANGEMENTS OF ALKYL ETHERS (*Continued*)

| Ether | Conditions | Product(s) and Yield(s) (%) | Refs. |
|---|---|---|---|

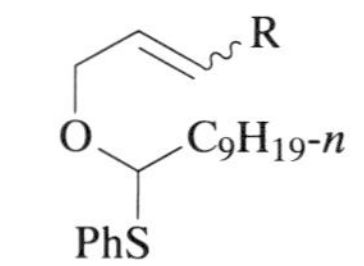

Row 1:

Conditions: Li•C$_{10}$H$_8$, THF, 0°, 1 h

Products:

| R | |
|---|---|
| H | (67) |
| Me, *E* | (71) 12:1 |
| Me, *Z* | (77) 1:2.5 |

Refs.: 30, 31

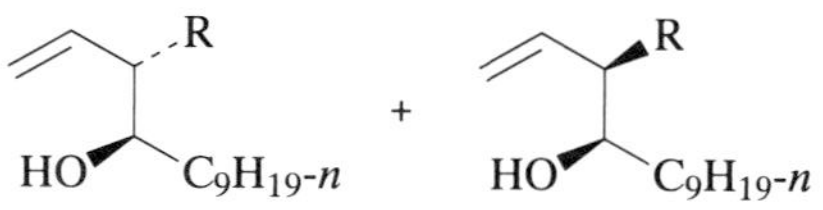

Row 2 (C$_{14}$):

Conditions: *n*-BuLi, –78°

Products:

(>75)
96.5% *Z*

Refs.: 170

Row 3 (C$_{14}$):

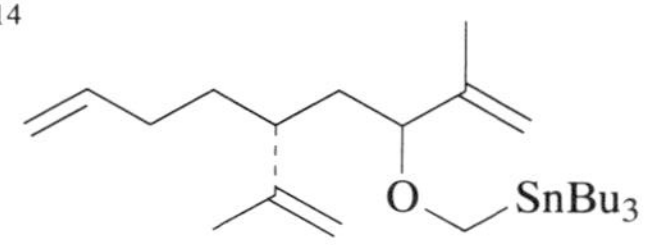

Conditions: *n*-BuLi, THF, –78° to rt, 0.5–12 h

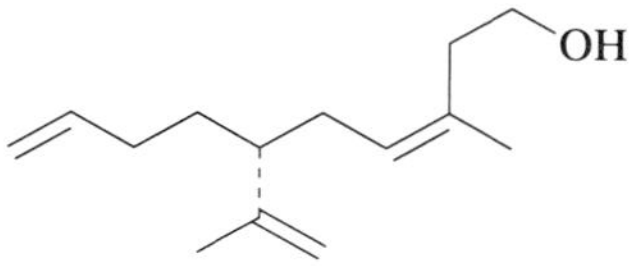

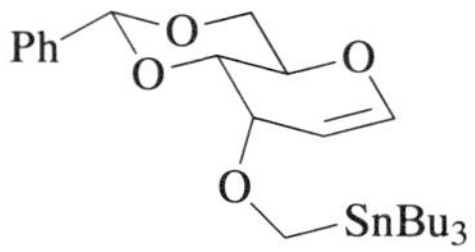

Products:

(—)
25:75

Refs.: 52, 53

Row 4 (C$_{15}$):

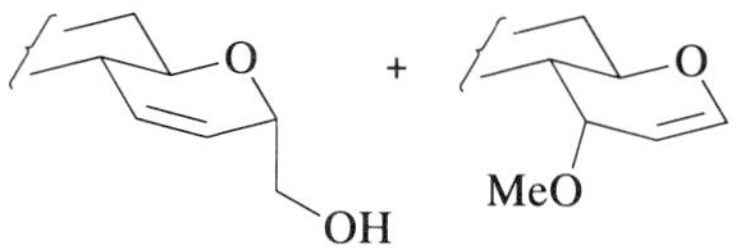

Conditions: 2 BuLi, THF, –5 to 5°, 2 h

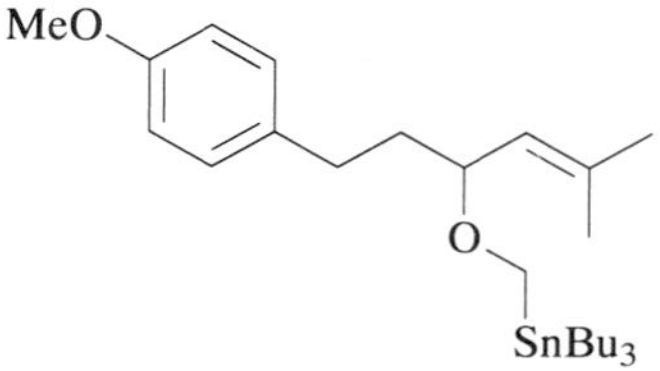

Products:

(61)
*trans* only

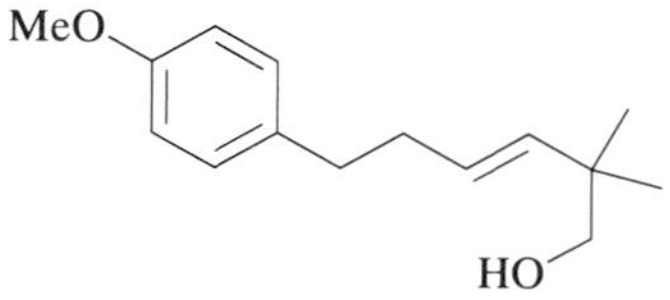

Refs.: 11

163

Table II. REARRANGEMENTS OF ALKYL ETHERS (*Continued*)

| Ether | Conditions | Product(s) and Yield(s) (%) | Refs. |
|---|---|---|---|
| 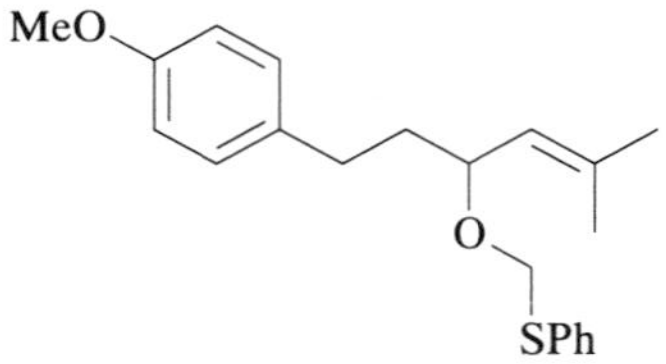 | Li•C$_{10}$H$_8$, THF, -5 to 5°, 2 h | 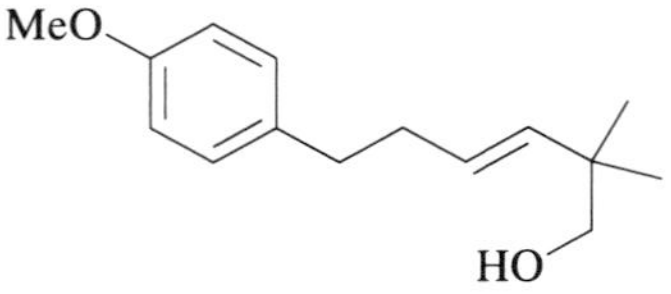 | 11 |
| 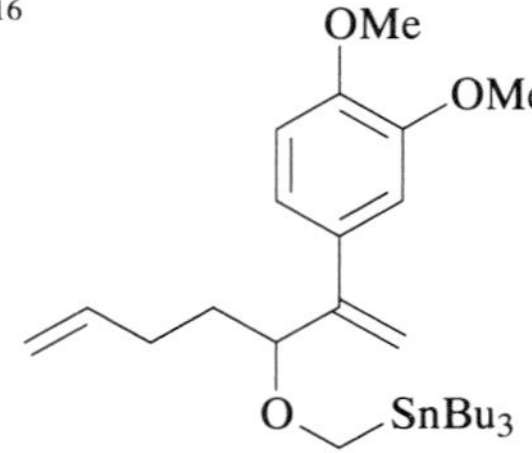 | *n*-BuLi, THF, -78°, 30 min; to 0°, 10 min | 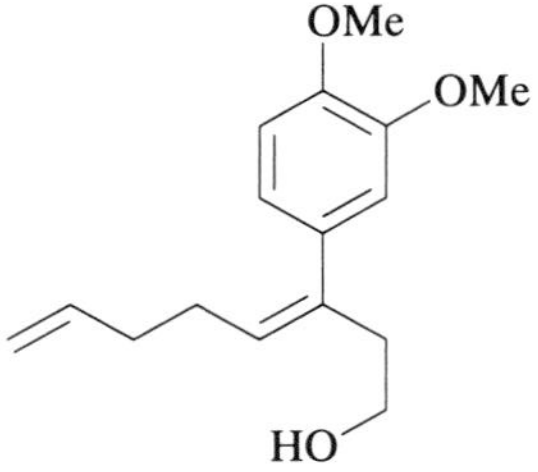 (94)<br> | 171 |
| 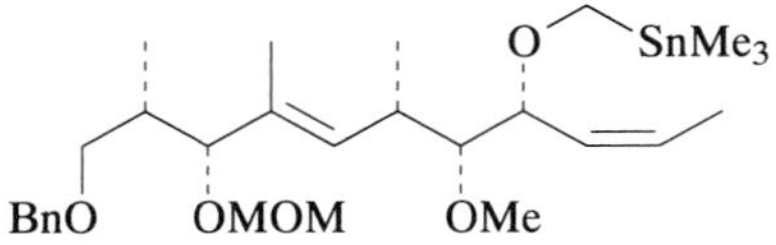 | MeLi, THF, -78° | 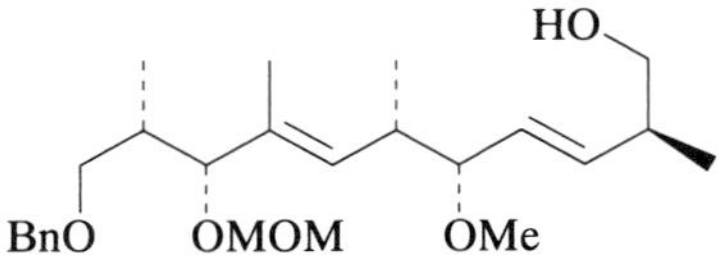 (78)<br> | 91-93 |
| C$_{17}$ 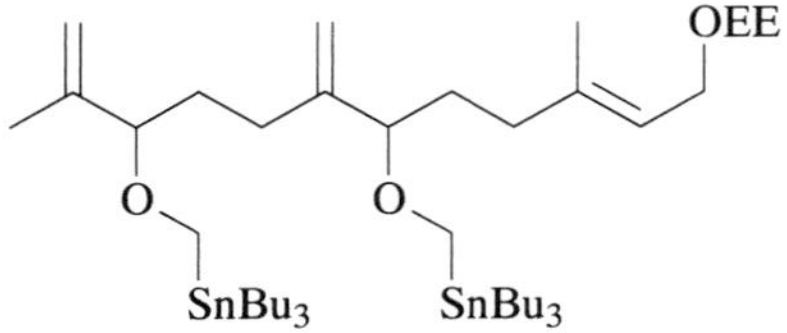<br> | *n*-BuLi, THF, -78 to -20° | 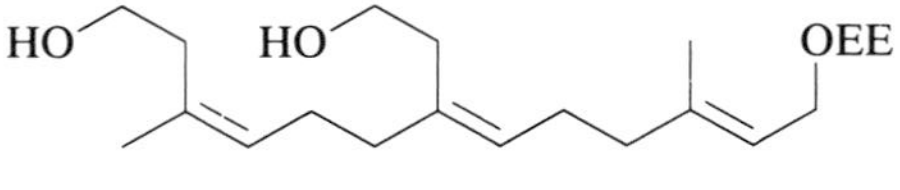 (93)<br> | 68 |

C$_{16}$

164

Table II. REARRANGEMENTS OF ALKYL ETHERS (*Continued*)

| Ether | Conditions | Product(s) and Yield(s) (%) | Refs. |
|---|---|---|---|
| 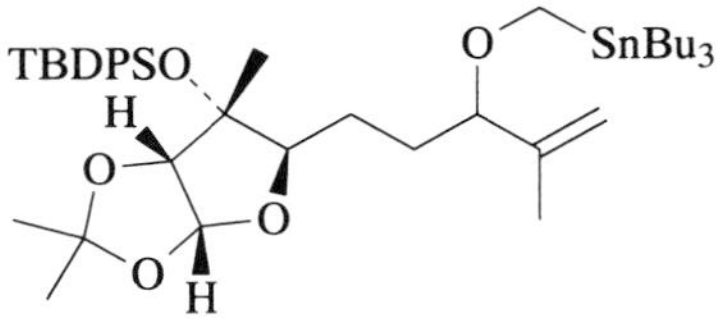 | *n*-BuLi, THF | 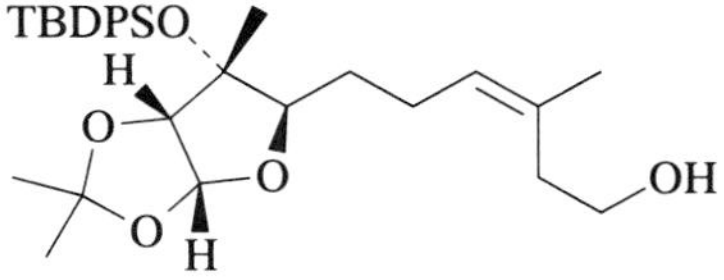 (>53) | 30, 31 |
| $C_{18}$ 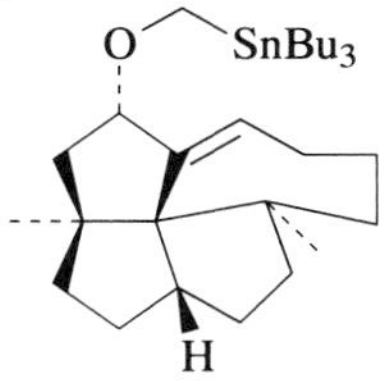 | *n*-BuLi, $C_6H_{14}$, 0° | 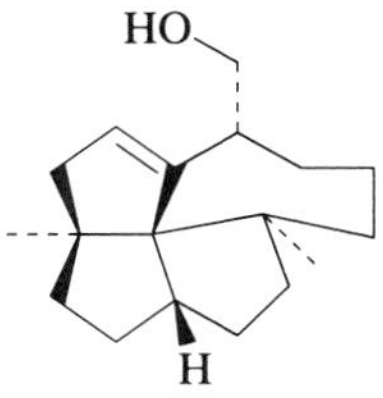 (>57) | 89 |
| 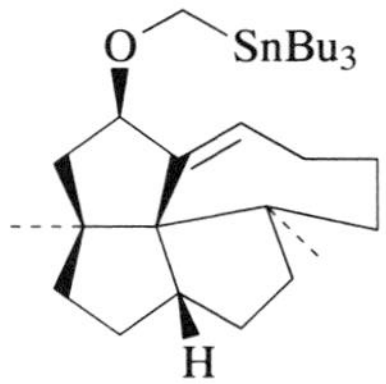 | *n*-BuLi, $C_6H_{14}$, 0° | 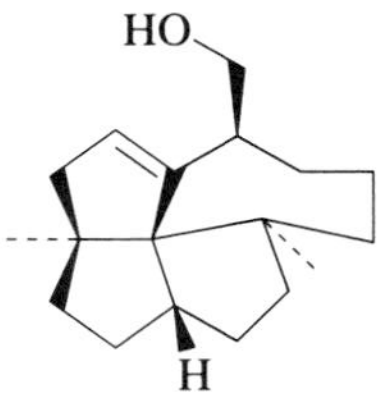 (>60) | 89 |

165

Table II. REARRANGEMENTS OF ALKYL ETHERS (*Continued*)

| Ether | Conditions | Product(s) and Yield(s) (%) | Refs. |
|---|---|---|---|
| 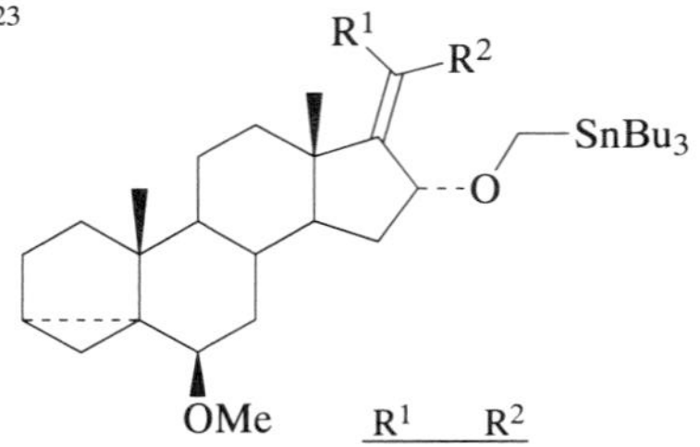 $C_{23}$ | $n$-BuLi, THF, $-78°$, 15 min | 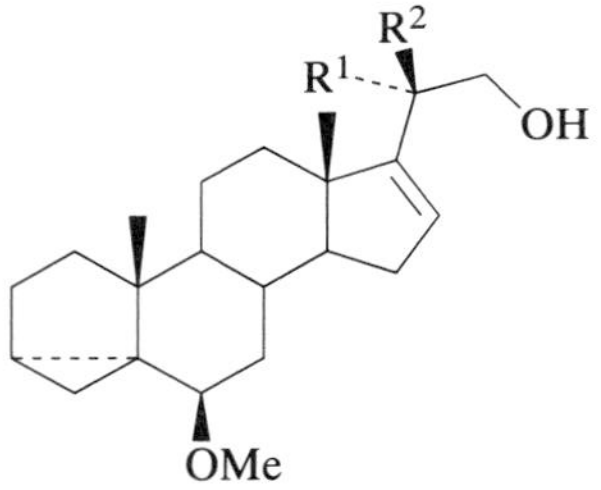 (70)<br>(83) | 45 |
| | | | |
| 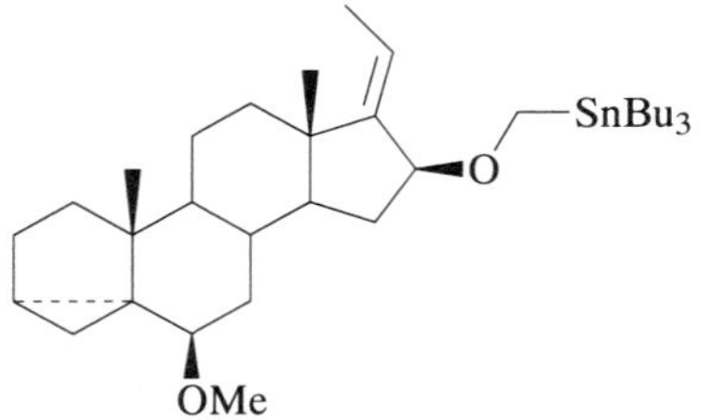 $C_{27}$ | $t$-BuLi, THF, reflux | 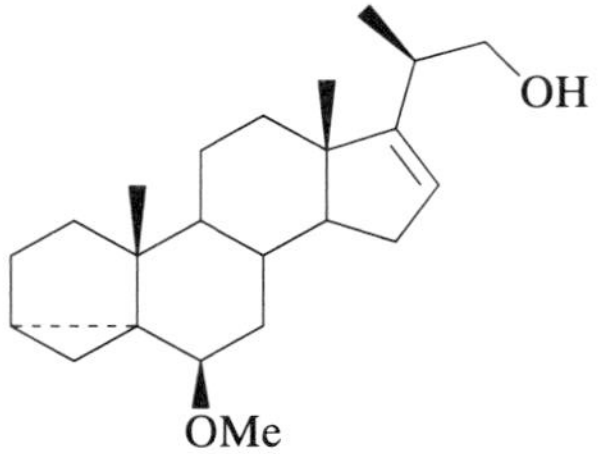 (89) | 46 |
| 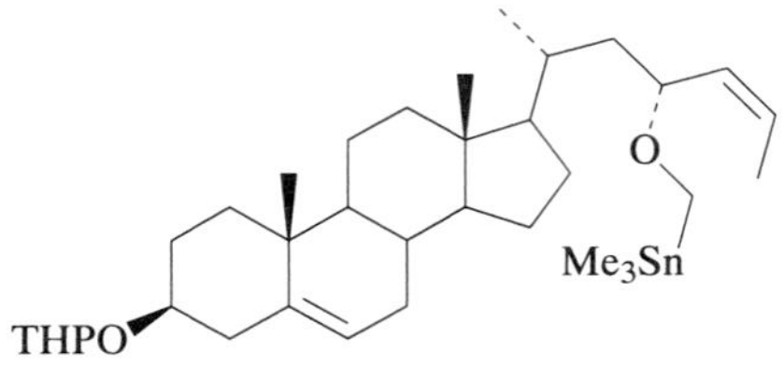 | $n$-BuLi, THF, $-78$ to $0°$ | 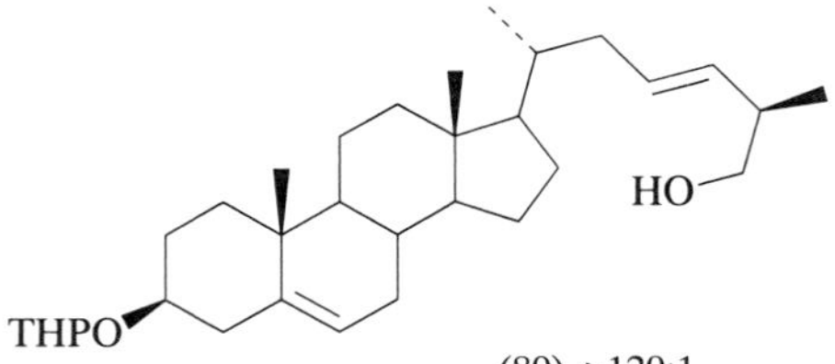 (80) >120:1 | 137 |

$\dfrac{R^1 \quad R^2}{Me \quad H}$  
$H \quad Me$

166

Table II. REARRANGEMENTS OF ALKYL ETHERS (*Continued*)

| Ether | Conditions | Product(s) and Yield(s) (%) | Refs. |
|---|---|---|---|
| 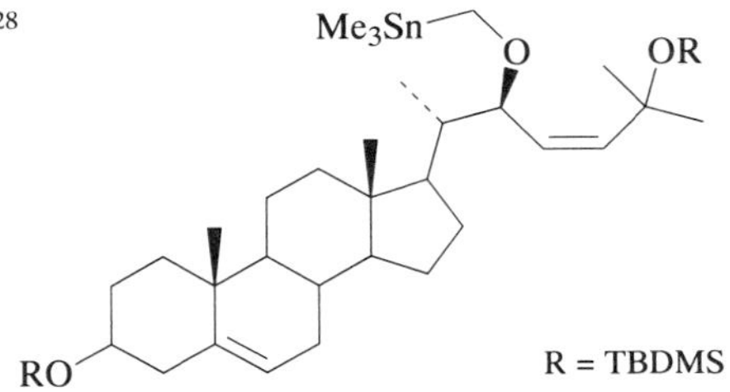 | *n*-BuLi, THF, -78°, 40 min; 0°, 40 min<br><br>R = TBDMS | 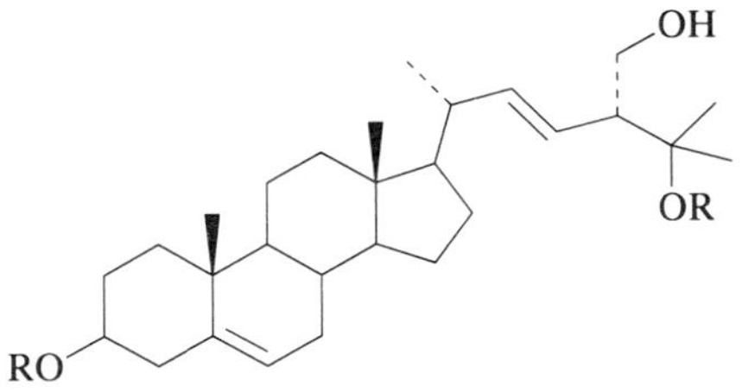 | 135 |
| 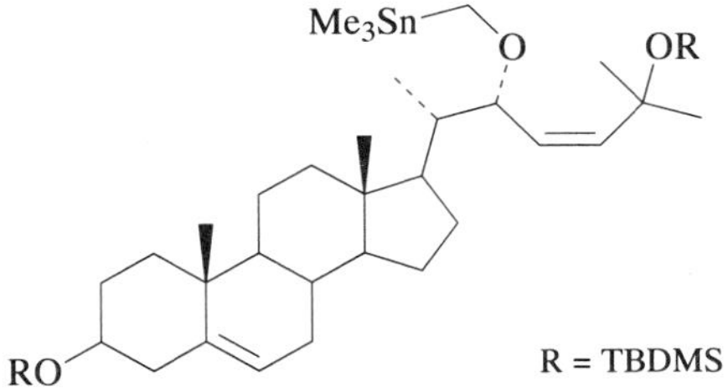 | *n*-BuLi, THF, -78°, 40 min; 0°, 40 min<br><br>R = TBDMS | 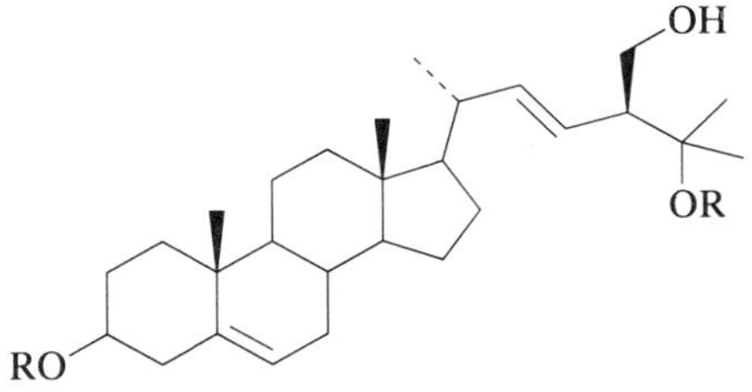 | 135 |
| 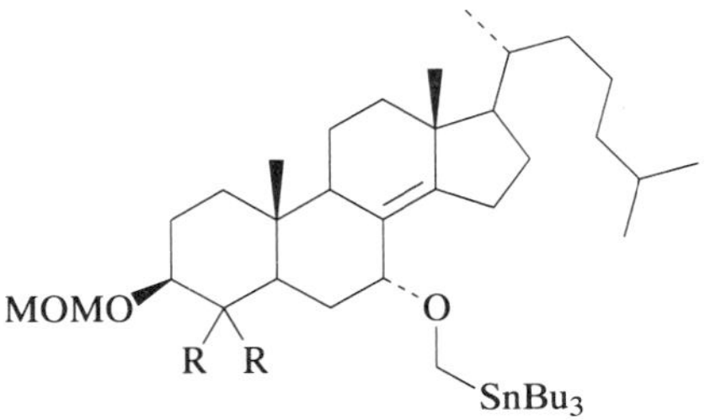 | *n*-BuLi, THF, -78 to rt | 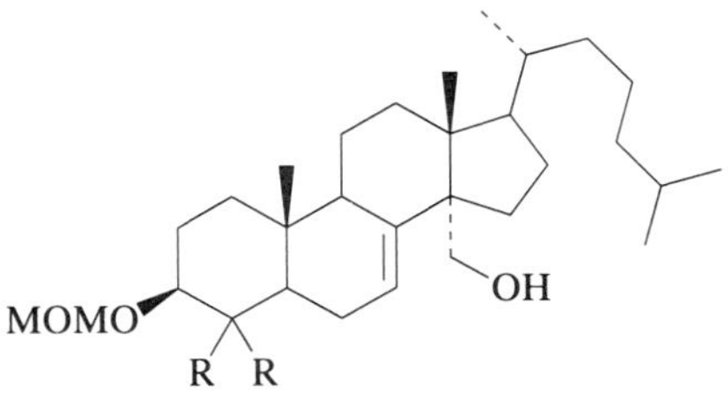 <br><br>R = H (79)<br>R = Me (83) | 90 |

C$_{28}$

C$_{28}$
C$_{29}$

167

Table III. REARRANGEMENTS OF ALLYL ETHERS

| Ether | Conditions | Product(s) and Yield(s) (%) | Refs. |
|---|---|---|---|
| C$_6$   C$_7$   OTBDMS structure, 70% Z | n-BuLi, THF-C$_6$H$_{14}$ (1:1), -78° | R = H   (53) 95:5 <br> R = Me   (74) 91:9 | 77 |
| C$_7$   C$_8$ <br> R = H, 93% E <br> R = H, 95% Z <br> R = Me, 93% E <br> R = Me, 83% Z | n-BuLi, THF-C$_6$H$_{14}$, -85°, 6-8 h | (81) 79:21 [84:16] <br> (88) 12:88 [8:92] <br> (70) 67:33 [72:28] <br> (71) 16:84 [5:95] | 12 |
| C$_7$ 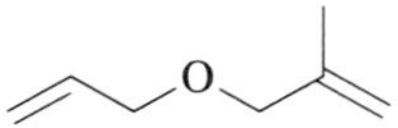 | n-BuLi, THF-C$_6$H$_{14}$, -85°, 6-8 h | 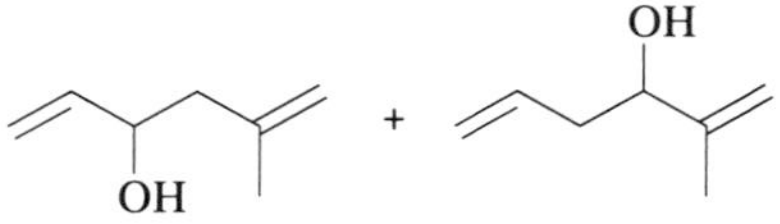 (77) 4:3 | 12 |
| 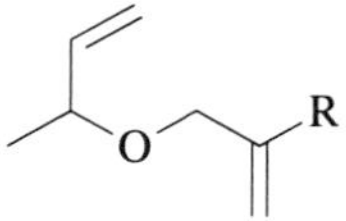 C$_7$   R = H <br> C$_8$   R = Me | n-BuLi, THF-C$_6$H$_{14}$, -85°, 6-8 h | 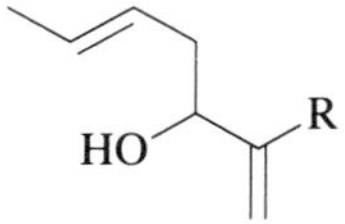 (79) E <br> (60) E | 12 |

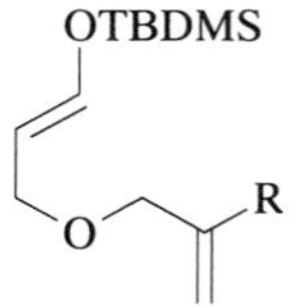

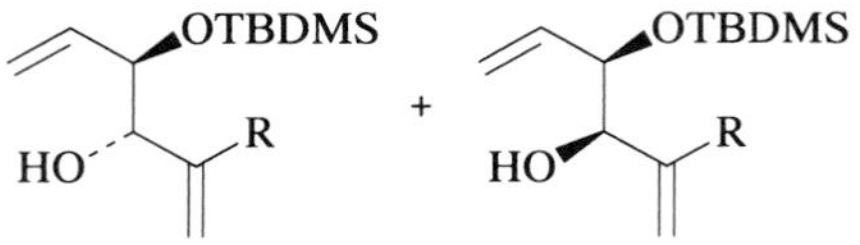

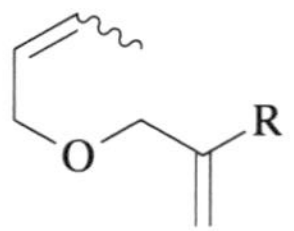

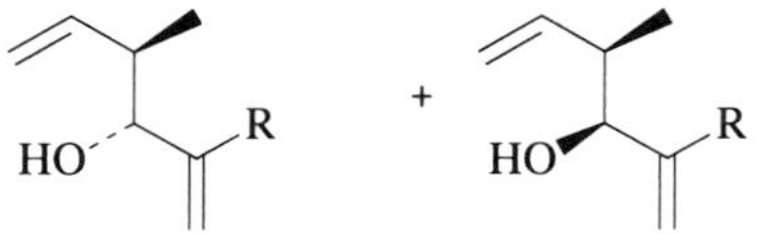

Table III. REARRANGEMENTS OF ALLYL ETHERS (*Continued*)

| Ether | Conditions | Product(s) and Yield(s) (%) | Refs. |
|---|---|---|---|
| $C_8$ 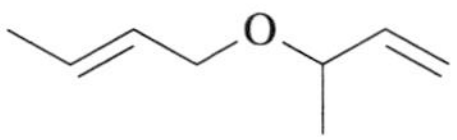 | n-BuLi, THF-C$_6$H$_{14}$, -85°, 6-8 h | 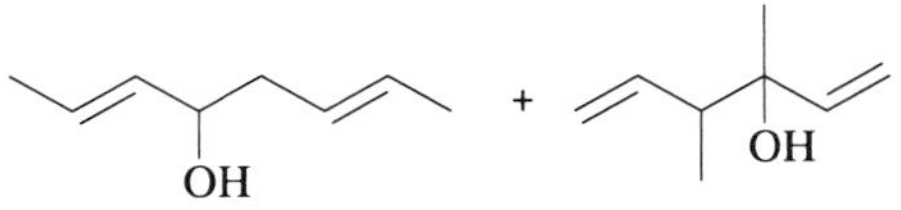  (82) 1:2 | 12 |
| 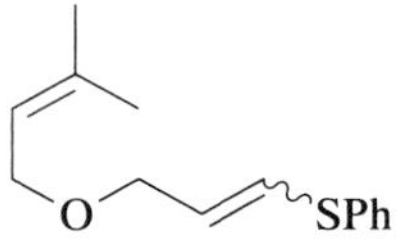 | n-BuLi, THF, -40 to -50°, 2 h | 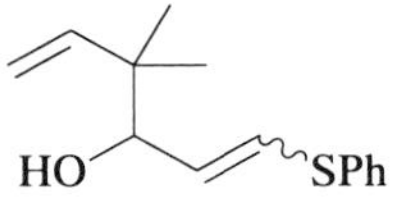  (—) | 32 |
| $C_9$   R = H<br>$C_{10}$   R = Me | n-BuLi, THF-C$_6$H$_{14}$, -78° |  (93)<br>(50) | 33 |
| $C_9$ <br>R$^1$ = Me, R$^2$ = H<br>R$^1$ = H, R$^2$ = Me | n-BuLi, THF-C$_6$H$_{14}$, -85°, 6-8 h |  (65)<br>(66) | 12 |

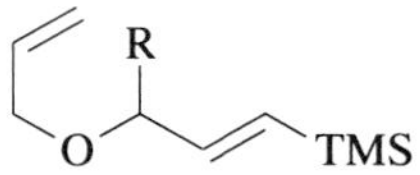

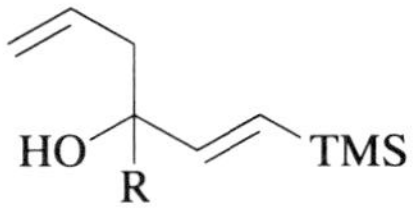

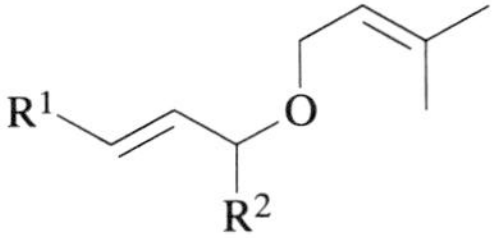

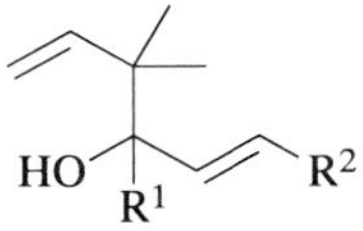

169

Table III. REARRANGEMENTS OF ALLYL ETHERS (*Continued*)

| Ether | Conditions | Product(s) and Yield(s) (%) | Refs. |
|---|---|---|---|

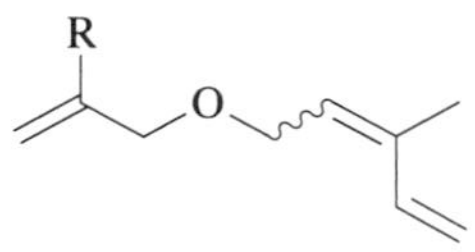

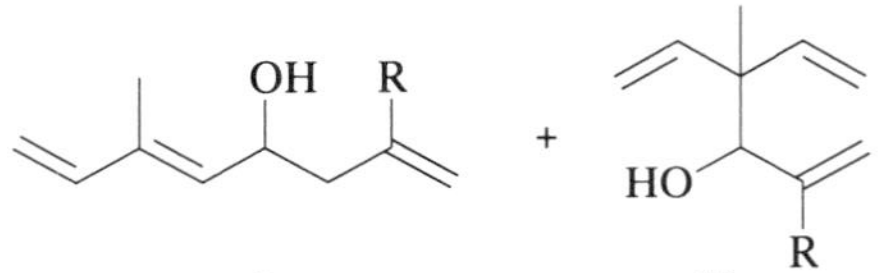

C$_9$   R = H     *n*-BuLi, THF, -90°, 2 h     **I** (27) + **II** (33)     172
R = H     KNH$_2$, NH$_3$, -33°     **I** (68)
C$_{10}$   R = Me     *n*-BuLi, THF, -90°     **I** (36)
R = Me     KNH$_2$, NH$_3$, -33°     **I** (50)

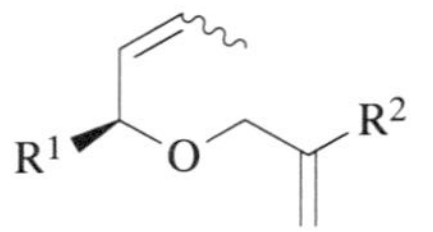

*n*-BuLi, THF-C$_6$H$_{14}$ (1:1), -78°

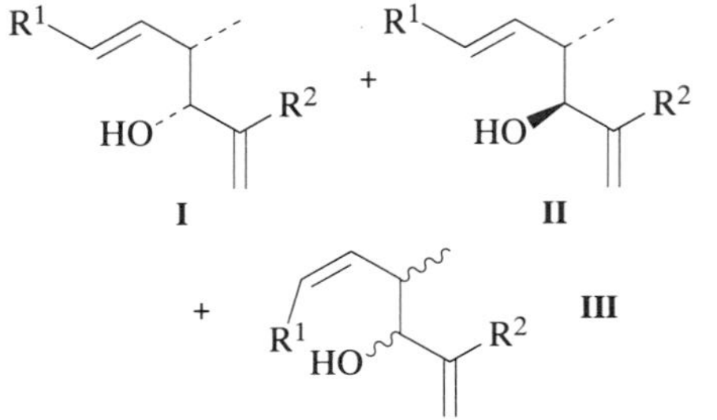

| | R$^1$ | R$^2$ | |
|---|---|---|---|
| C$_{10}$ | *i*-Pr | H *Z* | 91% ee |
| | *i*-Pr | H *E* | |
| C$_{11}$ | *i*-Pr | Me *Z* | 91% ee |
| | *i*-Pr | Me *E* | |
| C$_{12}$ | *i*-Bu | Me *Z* | |

(84) **I:II:III** = 92:8:0 } 91% ee
(75) **I:II:III** = 40:60:0 }
(93) **I:II:III** = 93:7:0 } 91% ee
(89) **I:II:III** = 40:50:10 }
(95)

128, 81

170

Table III. REARRANGEMENTS OF ALLYL ETHERS (*Continued*)

| Ether | Conditions | Product(s) and Yield(s) (%) | Refs. |
|---|---|---|---|

$C_{10}$

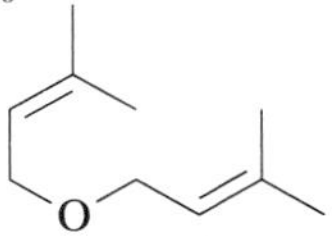

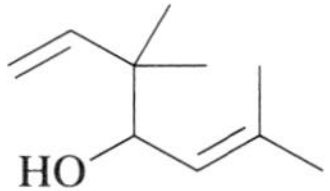

| | $n$-BuLi, THF, 25° | **I + II** (—) 1:1 | 165 |
| | $n$-BuLi, THF, -15° | **I** (—) | 165 |
| | $n$-BuLi, THF, -25° | **I** (67) + **II** (14) | 18 |

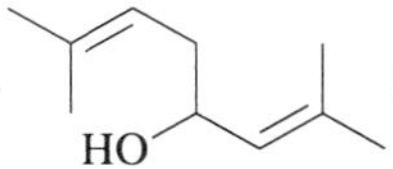

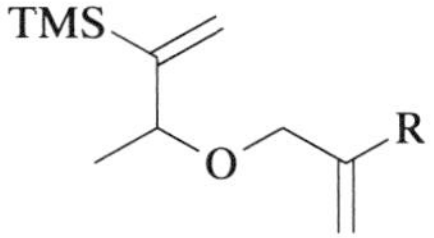

R = H
R = Me

| $C_{10}$ | LDCA, THF, -30°, 6 h | (62) 64% Z | 70 |
| $C_{11}$ | | (89) 59% Z | |

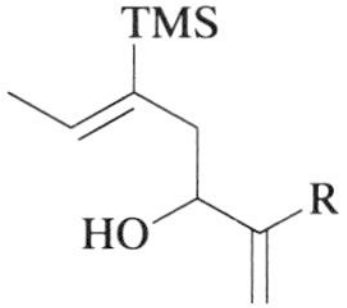

R = H 92% Z
R = Me 92% Z

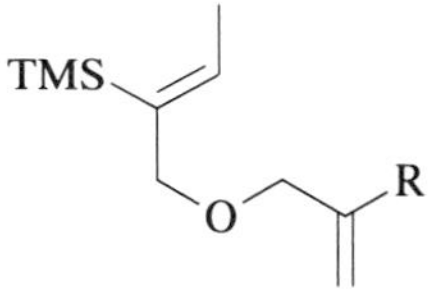

| $C_{10}$ | LDCA, THF, -30°, 6 h | (80) 82:18 | 70 |
| $C_{11}$ | | (73) 91:9 | |

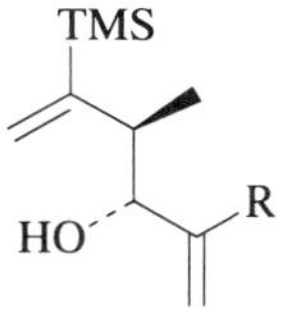

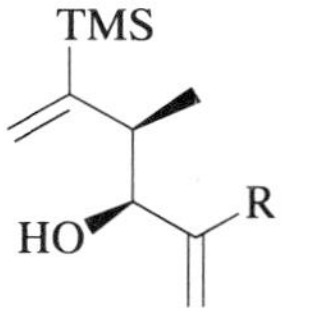

171

Table III. REARRANGEMENTS OF ALLYL ETHERS (*Continued*)

| Ether | Conditions | Product(s) and Yield(s) (%) | Refs. |
|---|---|---|---|

$C_{10}$

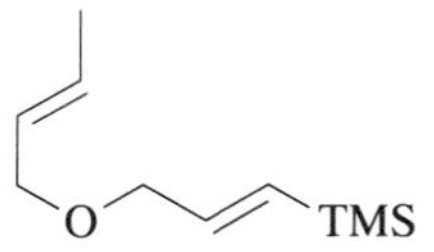

95% *E*
98% *Z*

*n*-BuLi, THF-C$_6$H$_{14}$, -78°

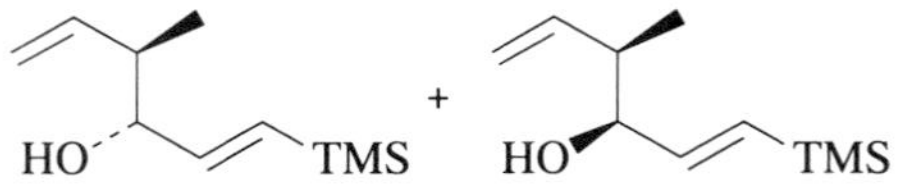

(58)  51:49
(72)  7:93

33

$C_{11}$

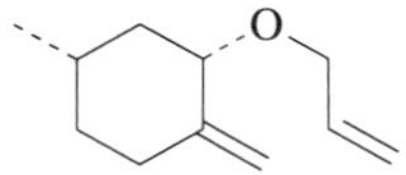

*n*-BuLi; THF, -78°

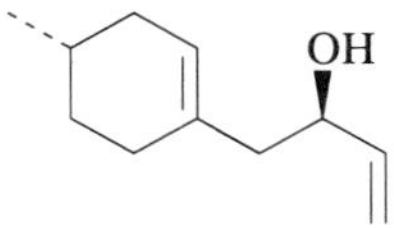

(>58)

94

$C_{13}$

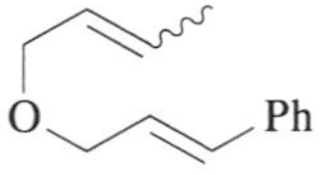

*Z*
*E*

*n*-BuLi, THF-C$_6$H$_{14}$, -78°

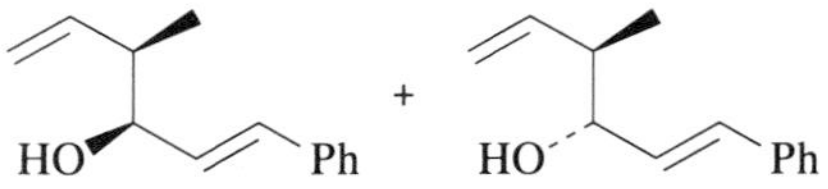

(70-98)  72:28
(70-98)  24:76

28, 29

$C_{15}$

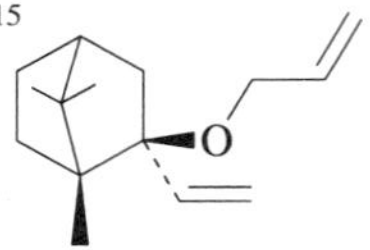

*t*-BuLi, *t*-BuOK, THF-C$_5$H$_{12}$, -85°

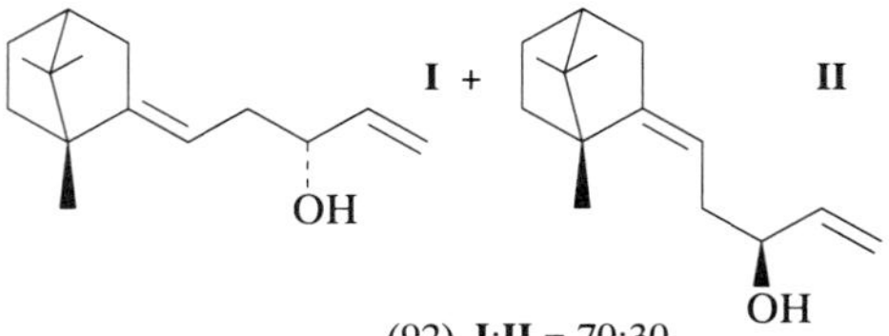

**I** + **II**

(92)  **I:II** = 70:30

173

172

Table III. REARRANGEMENTS OF ALLYL ETHERS (*Continued*)

| Ether | Conditions | Product(s) and Yield(s) (%) | Refs. |
|---|---|---|---|
| 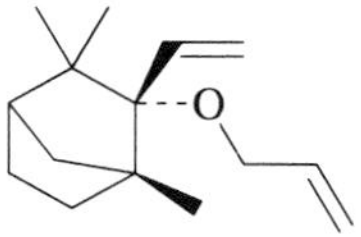 | *t*-BuLi, *t*-BuOK, HMPA-THF-$C_5H_{12}$, -85° | 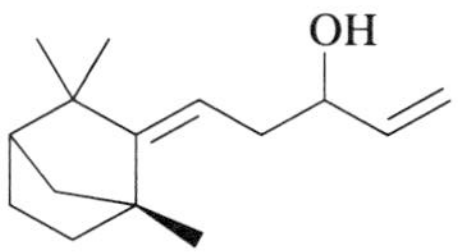 (74) | 173 |
| $C_{19}$ | *n*-BuLi, HMPA-$C_6H_{14}$, -85° | (51) 63% chirality transfer | 73 |

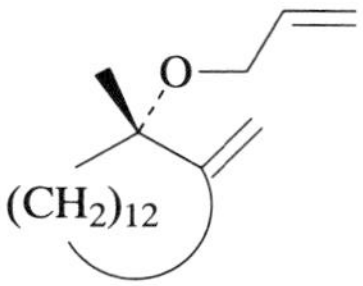

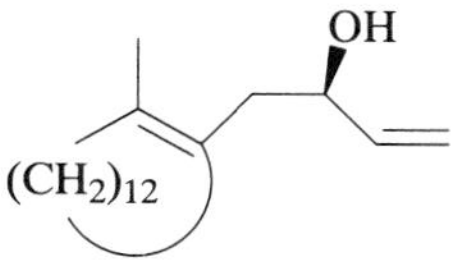

Table IV. REARRANGEMENTS OF PROPARGYL ETHERS

| Ether | Conditions | Product(s) and Yield(s) (%) | Refs. |
|---|---|---|---|

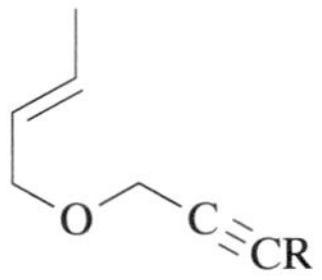

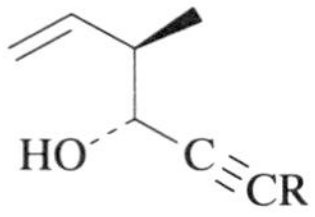

C$_7$

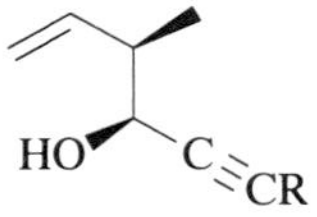

| | | |
|---|---|---|
| R = H | 93% E | *n*-BuLi, THF-C$_6$H$_{14}$ (1:1), -85°, 8 h | (72) **I:II** = 93:7 [99:1] |

| | | | |
|---|---|---|---|
| R = H | 98% Z | " | (76) **I:II** = 12:88 [10:90] |
| R = H | 93% E | THF-Et$_2$O (1:1) | (81) **I:II** = 89:11 |
| R = H | 98% Z | " | (58) **I:II** = 36:64 |

C$_8$
| R = Me | 93% E | " | (48) **I:II** = 85:15 |
| R = Me | 93% E | *n*-BuLi, THF-C$_6$H$_{14}$ (1:1), -85°, 8 h" | (78) **I:II** = 92:8 [99:1] |

| R = Me | 98% Z | " | (74) **I:II** = 2:98 [0:100] |
C$_{10}$
| R = TMS | 93% E | " | (72) **I:II** = 25:75 [27:73] |
| R = TMS | 98% Z | " | (74) **I:II** = 2:98 [0:100] |

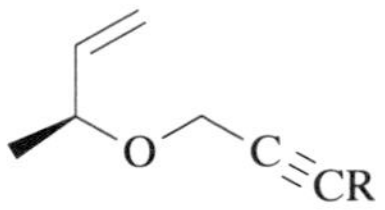

28, 29

| | | *n*-BuLi, THF, -85°, 6 h | 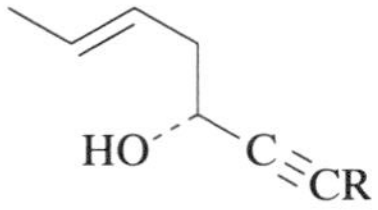 | 67 |
|---|---|---|---|---|

C$_7$
| R = H | 90% ee | | (−) 98% E, 84% ee |
| R = H | 94% ee | | (−) 95% E, 86% ee |
C$_8$
| R = Me | 90% ee | | (−) 97% E, 80% ee |
C$_{10}$
| R = TMS | 94% ee | | (−) 93% E, 76% ee |

174

Table IV. REARRANGEMENTS OF PROPARGYL ETHERS (*Continued*)

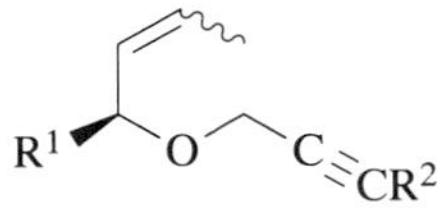

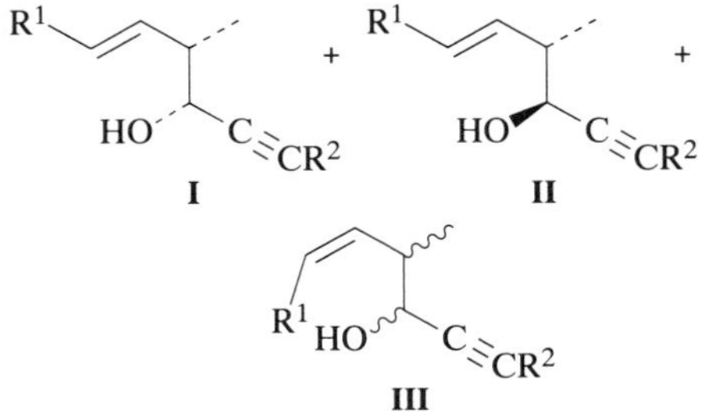

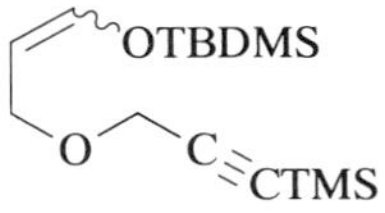

| Ether | Conditions | Product(s) and Yield(s) (%) | Refs. |
|---|---|---|---|

| | $R^1$ | $R^2$ | | | |
|---|---|---|---|---|---|
| $C_8$ | Me | H | 99% E | *n*-BuLi, THF, −78°, 6 h | (62) **I:II:III** = 91:9:0 | 28, 29 |
| | Me | H | 99% E | THF, $C_6H_{14}$ (1:1) | (62) **I:II:III** = 88:12:0 | 28, 29 |
| $C_9$ | Me | Me | >99% Z, 98% ee | *n*-BuLi, THF, −78° | (64) **I:II:III** = >99:1:0, 98% ee | 28, 29 |
| $C_{10}$ | *i*-Pr | H | 100% Z | *n*-BuLi, THF, −78° | (89) **I:II:III** = 91:9:0 | 81 |
| | *i*-Pr | H | 100% E | *n*-BuLi, THF, −78° | (63) **I:II:III** = 10:82:8 | 81 |
| $C_{11}$ | Me | TMS | 97% Z | *n*-BuLi, THF, −78°, 6 h | (64) **I:II:III** = 1:99:0 | 28, 29 |
| $C_{12}$ | BnOCH$_2$ | TMS | — | *n*-BuLi, THF, −78° | (90) | 127 |
| | Me | *n*-Bu | — | *n*-BuLi, THF, −78° | (96) | 125,126 |
| $C_{13}$ | *i*-Pr | TMS | 100% Z | *n*-BuLi, THF, −78° | (62) **I:II:III** = >98:2:0 | 81 |
| $C_{14}$ | *i*-Pr | TBDMSOCH$_2$ | Z, 92% ee | *n*-BuLi, THF, −78° | (85) 92% ee | 124 |

| | | | |
|---|---|---|---|
| $C_9$ | 65% E | *n*-BuLi, THF, −78°, 5 h | (79) **I:II** = 98:2 | 77 |
| | 95% E | | (66) **I:II** = 95:5 | |

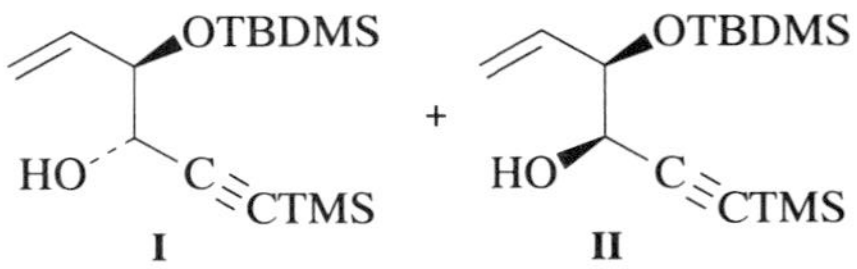

175

Table IV. REARRANGEMENTS OF PROPARGYL ETHERS (*Continued*)

| Ether | Conditions | Product(s) and Yield(s) (%) | Refs. |
|---|---|---|---|
| C₁₁ | | | |

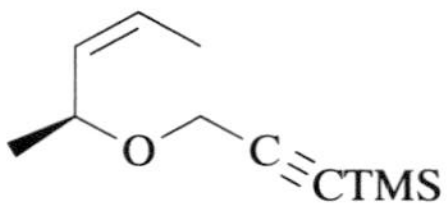

$C_{11}$

*n*-BuLi, THF, -85°, 6 h     (>64)     80

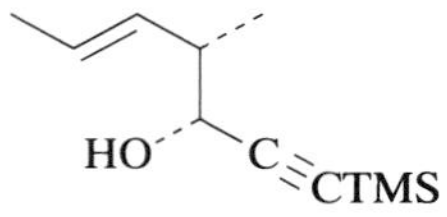

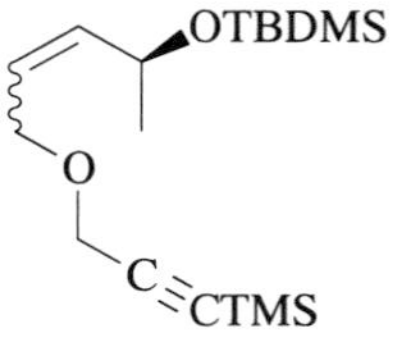

*n*-BuLi, THF, -78°     **I** +    **II** +    **III**     104, 105

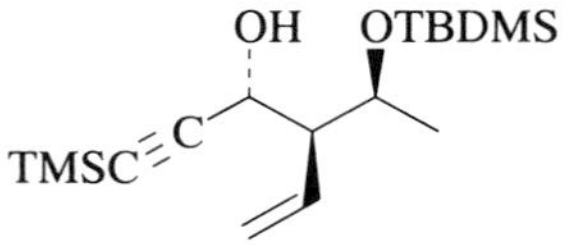

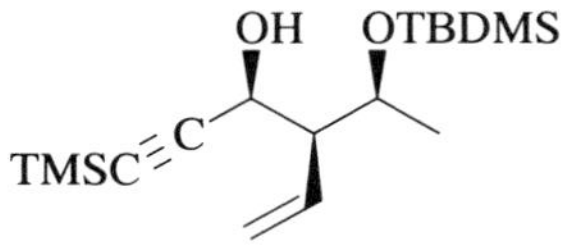

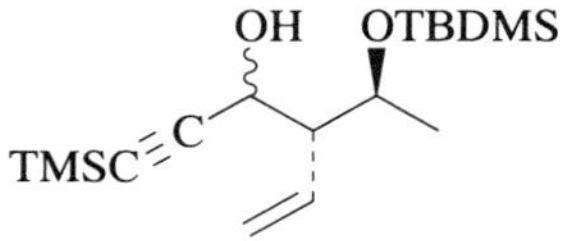

(77) **I:II:III** = 94:1:5
(93) **I:II:III** = 10:21:69

97% Z
97% E

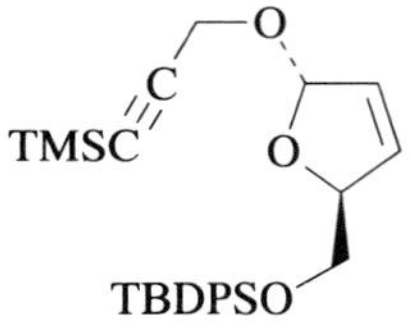

*n*-BuLi, THF, -78°     (73)     54

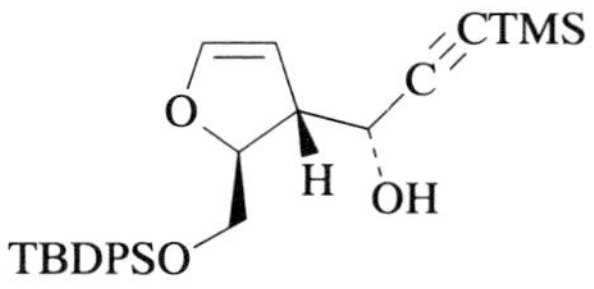

176

Table IV. REARRANGEMENTS OF PROPARGYL ETHERS (*Continued*)

| Ether | Conditions | Product(s) and Yield(s) (%) | Refs. |
|---|---|---|---|
| C$_{12}$  R = Me  100% Z | *n*-BuLi, THF, -78° | (62) **I:II:III** = 98:0:2 | 104, 105 |
| R = Me  100% E | " | (73) **I:II:III** = 7:77:16 | |
| C$_{14}$  R = TMS  100% Z | " | (77) **I:II:III** = >99:0:0 | |
| R = TMS  100% E | " | (86) **I:II:III** = 12:81:7 | |
| R = TMS  Z | -78 to 20° | (87) **I:II** = 100:0 | |
| R = TMS  E | " | (84) **I:II** = 17:83 | |
| C$_{12}$  R = TMS  98% E | *n*-BuLi, THF, -78°, 8 h | (65) **I:II** = 3:97 | 75 |
| C$_{15}$  R = Ph  E | | (85) **I:II** = <5:>95 | 174 |

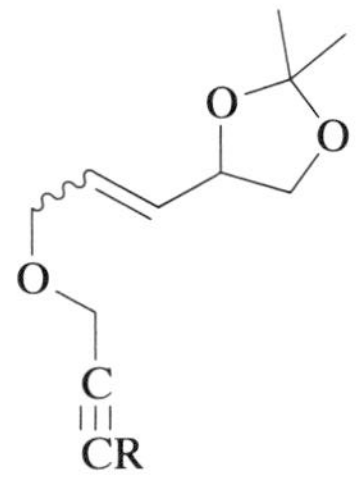

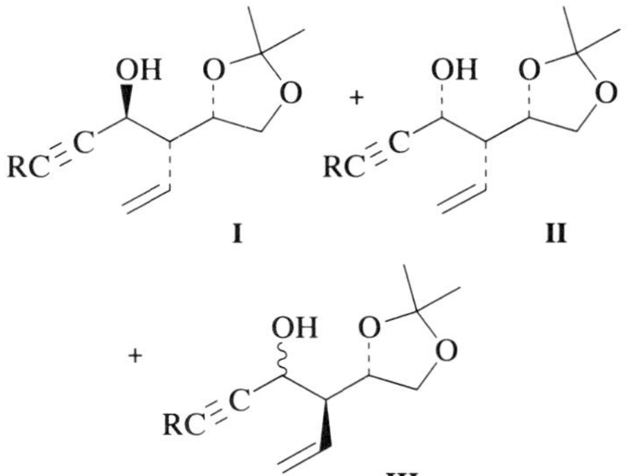

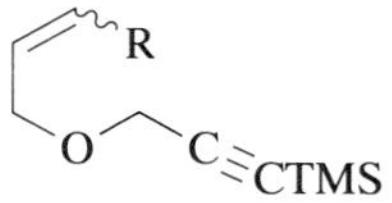

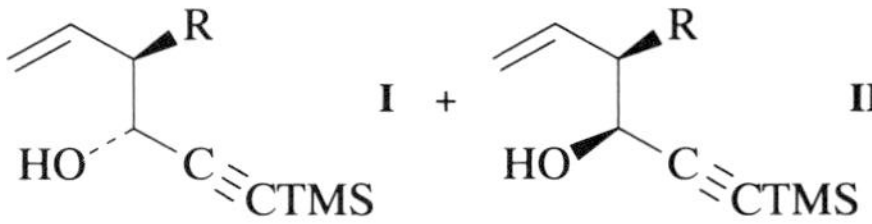

177

Table IV. REARRANGEMENTS OF PROPARGYL ETHERS (*Continued*)

| Ether | Conditions | Product(s) and Yield(s) (%) | Refs. |
|---|---|---|---|

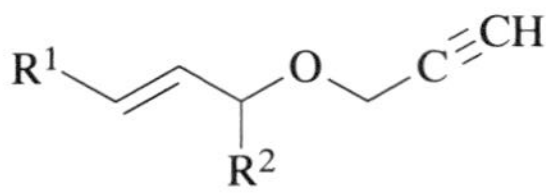

$C_9$  R¹ = TMS, R² = H    *n*-BuLi, THF, -78°, 8 h    (60)    33

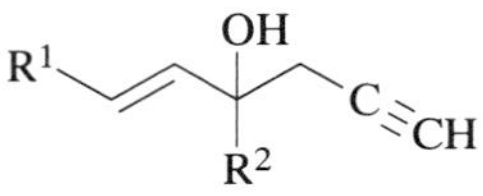

$C_{10}$ R¹ = TMS, R² = Me    (64)    33

$C_{12}$ R¹ = Ph, R² = H    (67)    174

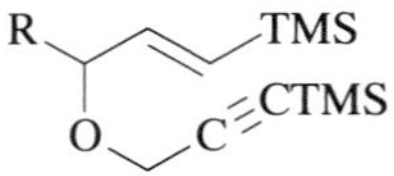

*n*-BuLi, THF, -78°

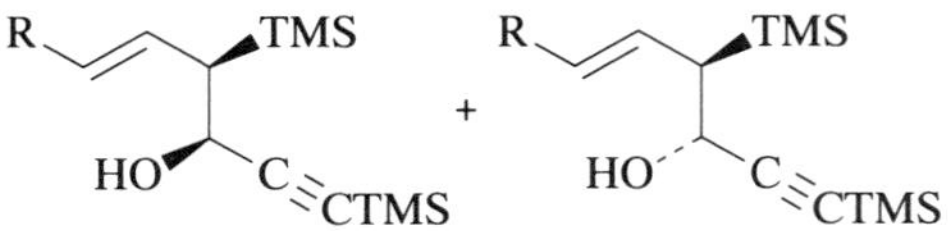

**I** + **II**

$C_{12}$ R = H    98% E    (93) **I:II** = 97:3

$C_{13}$ R = Me    95% E    (85) **I:II** = 4:96, 97% E

$C_{14}$ R = Et    100% E    (80) **I:II** = 3:97, 100% E

$C_{16}$ R = *n*-Bu    100% E    (61) **I:II** = 11:89, 100% E    75, 76

$C_{13}$    *n*-BuLi, THF, -78°    (>85)    124

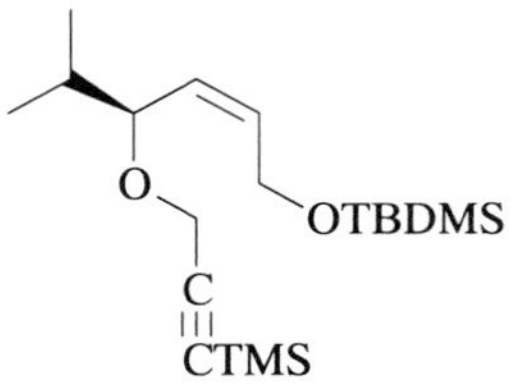

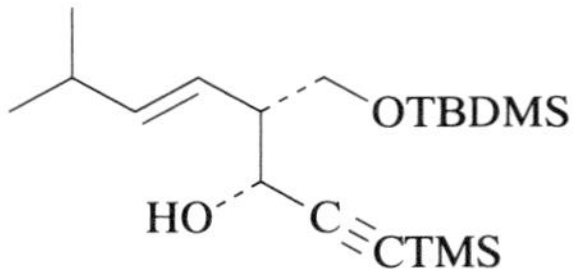

178

Table IV. REARRANGEMENTS OF PROPARGYL ETHERS (*Continued*)

| Ether | Conditions | Product(s) and Yield(s) (%) | Refs. |
|---|---|---|---|

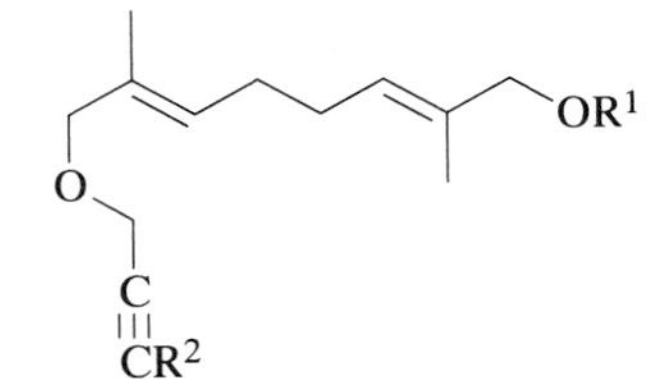

$C_{13}$, $C_{14}$

| $R^1$ | $R^2$ |
|---|---|
| H | H |
| MOM | H |
| MOM | $CH_2OH$ |
| MOM | $CH_2OH$ |
| MOM | $CH_2OH$ |
| TBS | $CH_2OH$ |
| TBS | $CH_2OH$ |

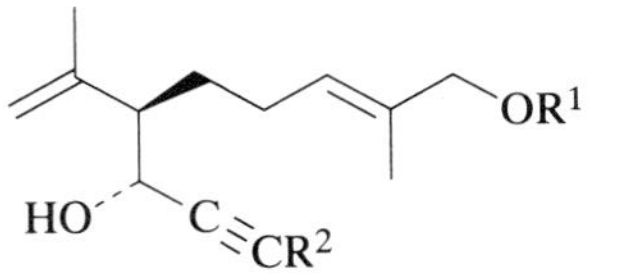

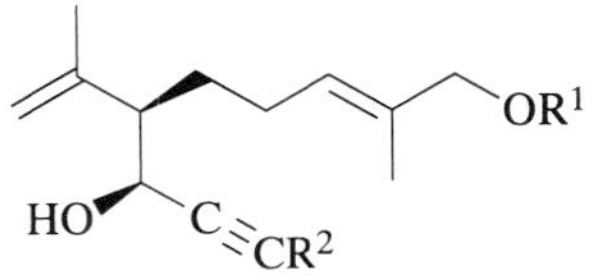

$n$-BuLi, THF, -85 to -15°

**I** + **II**

| Additive | | |
|---|---|---|
| TMEDA | (45) | **I:II** = 87:13 |
| — | (28) | **I:II** = 67:33 |
| — | (0) | |
| TMEDA | (63) | **I:II** = 67:33 |
| HMPA | (15) | **I:II** = 67:33 |
| — | (0) | |
| TMEDA | (57) | **I:II** = 55:45 |

42, 43

$C_{16}$

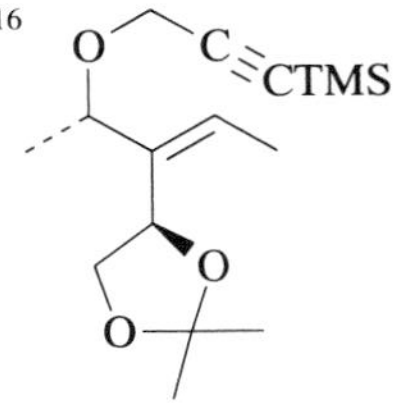

$n$-BuLi, THF, -78°, 5.5 h; to -20°, 20 h

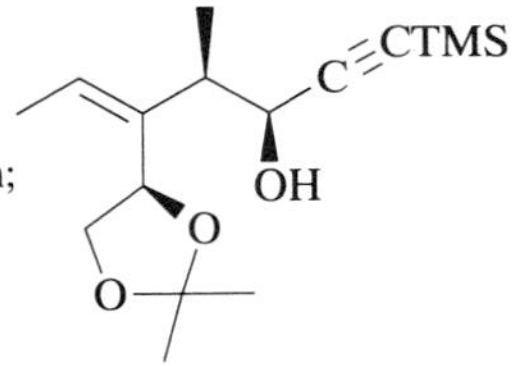

(—)

111

Table IV. REARRANGEMENTS OF PROPARGYL ETHERS (*Continued*)

| Ether | Conditions | Product(s) and Yield(s) (%) | Refs. |
|---|---|---|---|
| 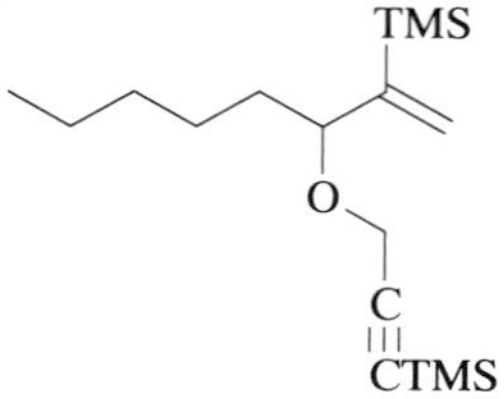 C₁₇ | *n*-BuLi, THF, -78°, 6 h | 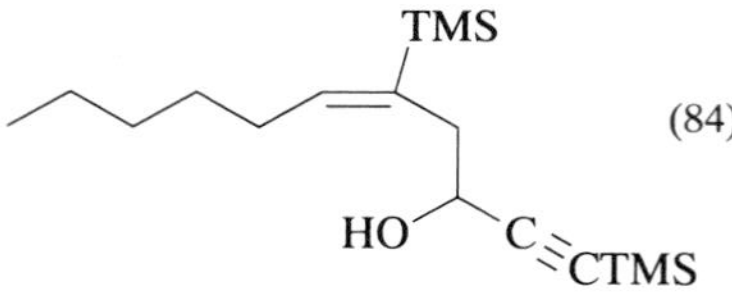 (84) >80% E | 71 |
| C₂₈ R = H R = TMS | *n*-BuLi, THF, -78° | I (75) II (quant) | 47 |
| 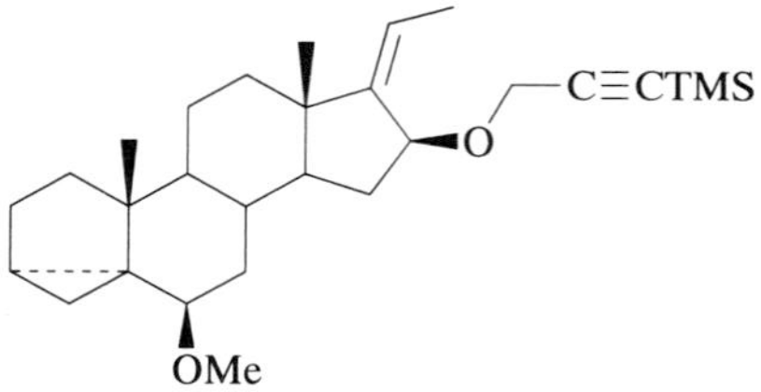 | *n*-BuLi, THF, -78° | II (85) | 138 |

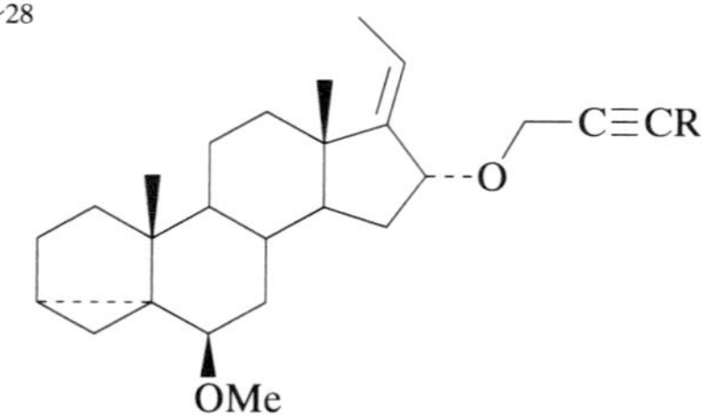

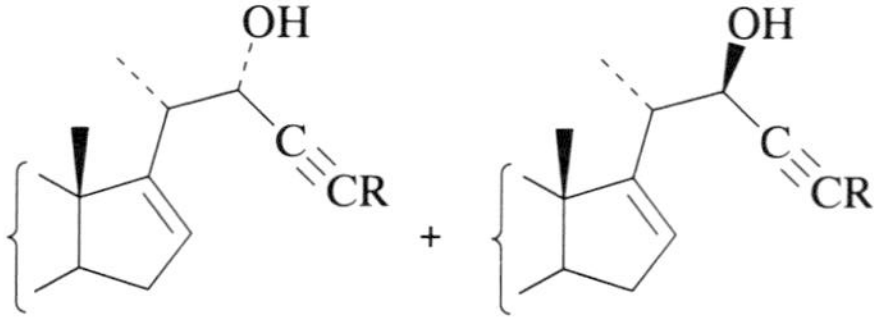

Table IV. REARRANGEMENTS OF PROPARGYL ETHERS (*Continued*)

| Ether | Conditions | Product(s) and Yield(s) (%) | Refs. |
|---|---|---|---|

C$_{29}$

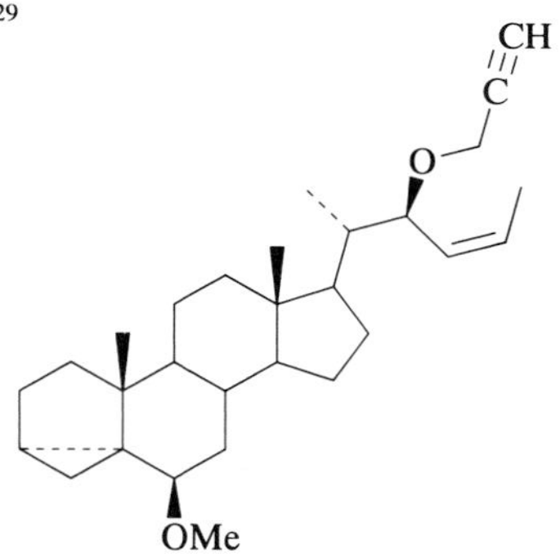

*n*-BuLi, THF,
-78 to 0°

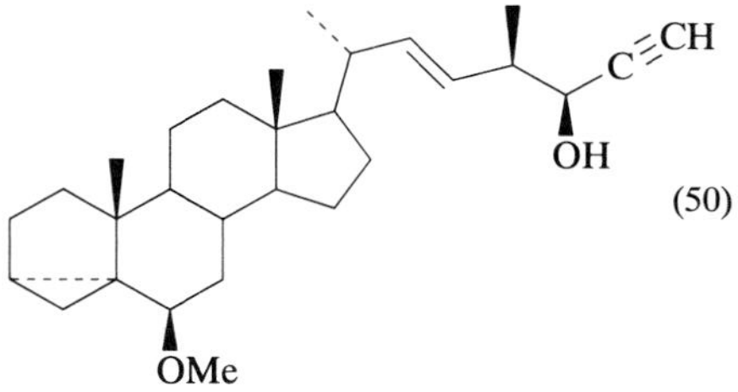

(50)    134

Table V. ENOLATE REARRANGEMENTS OF α-(ALLYLOXY)CARBONYL COMPOUNDS

| | Ether | Conditions | Product(s) and Yield(s) (%) | Refs. |
|---|---|---|---|---|
| $C_6$ 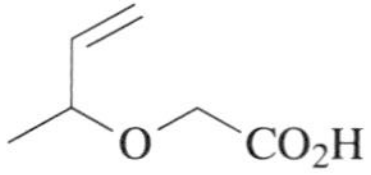 | (structure) 93% E, 95% Z | LDA, THF, -78° | 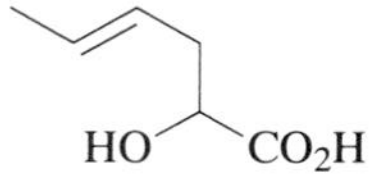 (80) 74% E | 58 |
| | (structure) | LDA, THF, -78° | 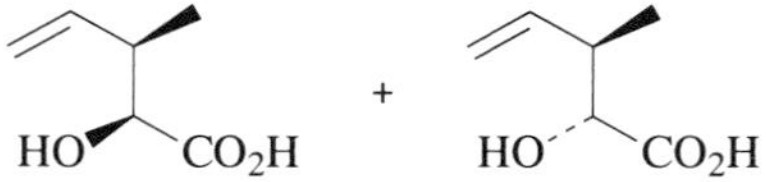 I + II<br>(60) **I:II** = 88:12 [92:8]<br>(73) **I:II** = 25:75 [21:79] | 58 |
| $C_7$<br>$C_8$ | R = H<br>R = Me | TMSOTf, Et₃N<br>CH₂Cl₂, rt | 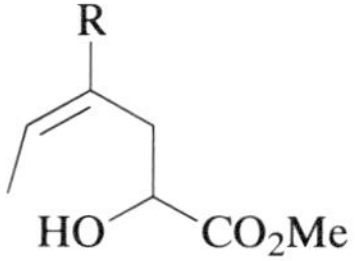 (93) 69% Z<br>(98) 69% Z | 164 |
| 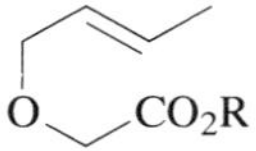 $C_7$<br><br>$C_9$ | R = Me 93% E<br>R = Me 93% Z<br>R = i-Pr 93% E | TMSOTf, Et₃N<br>CH₂Cl₂, rt | 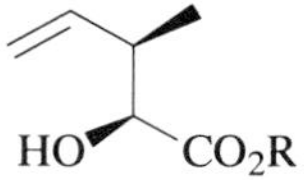 (83) 92% erythro<br>(86) 53% erythro<br>(86) 95% erythro | 164 |

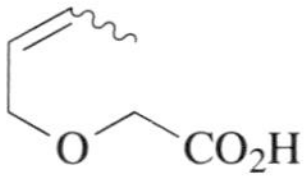

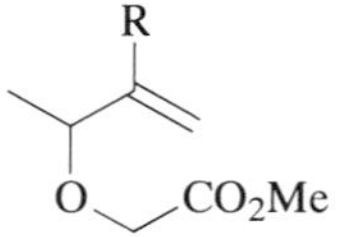

182

Table V. ENOLATE REARRANGEMENTS OF α-(ALLYLOXY)CARBONYL COMPOUNDS (*Continued*)

| Ether | Conditions | Product(s) and Yield(s) (%) | Refs. |
|---|---|---|---|

C₇

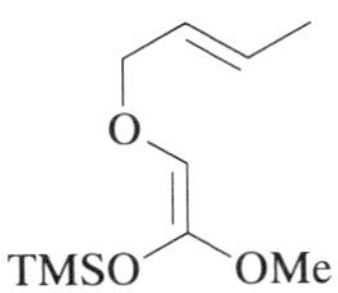

93% *E*

TMSOTf (20 mol %), CH₂Cl₂, -72°

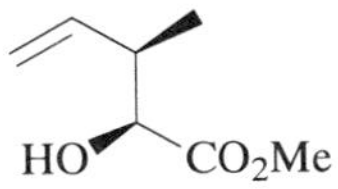

(72) 97% *erythro*

164

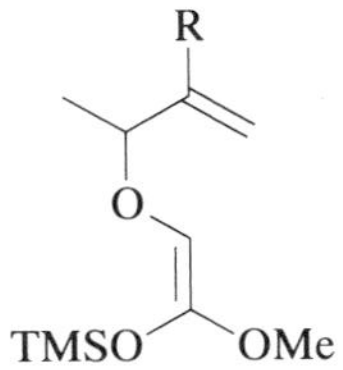

TMSOTf (20 mol %), CH₂Cl₂, -72°

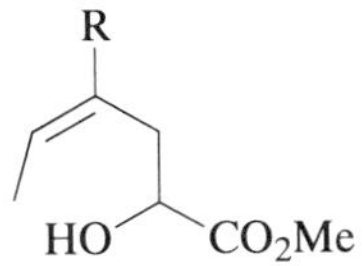

164

C₇ R = H
C₈ R = Me

(76) 81% *Z*
(78) 78% *Z*

C₇

65

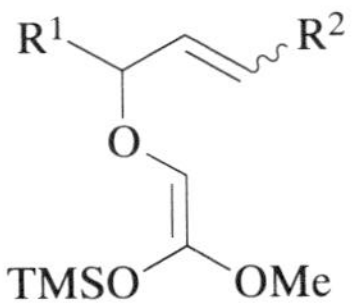

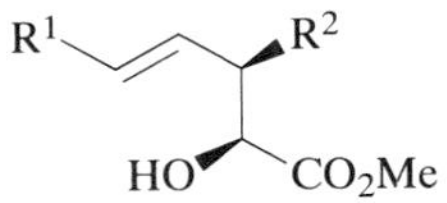

| R¹ | R² | | |
|---|---|---|---|
| Me | H | | |
| Me | H | | |
| H | Me, 93% *E* | TiCl₄, CH₂Cl₂, -50°, 5 h | (78) 54% *E* |
| H | Me, 93% *E* | SnCl₄, CH₂Cl₂, -50°, 5 h | (76) 59% *Z* |
| H | Me, 93% *Z* | TiCl₄, CH₂Cl₂, -50°, 5 h | (75) 91% *erythro* |
| H | Me, 93% *Z* | SnCl₄, CH₂Cl₂, -50°, 5 h | (73) 92% *erythro* |
| | | TiCl₄, CH₂Cl₂, -50°, 5 h | (75) 52% *erythro* |
| | | SnCl₄, CH₂Cl₂, -50°, 5 h | (63) 63% *erythro* |

183

Table V. ENOLATE REARRANGEMENTS OF α–(ALLYLOXY)CARBONYL COMPOUNDS (*Continued*)

| Ether | Conditions | Product(s) and Yield(s) (%) | Refs. |
|---|---|---|---|
| 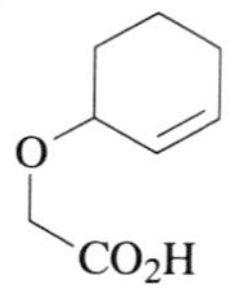 C₈ | LDA, THF, -78° | 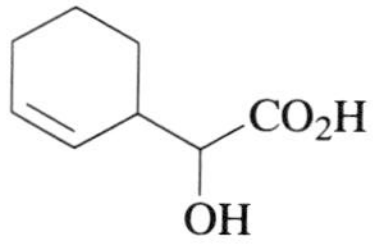 (78) | 58 |
| 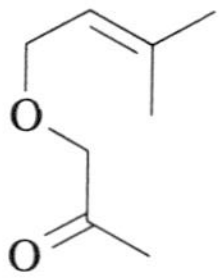 | KOBu-*t*, *t*-BuOH, THF, 0°, 40 min | 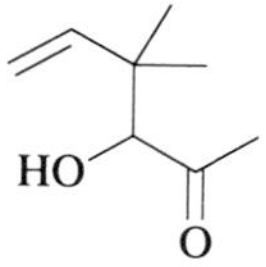 (64) | 55 |
| 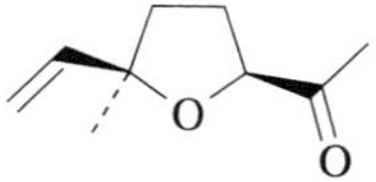 C₉ | KOBu-*t*, *t*-BuOH, 0°, 26 h | 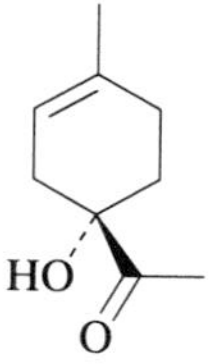 (22) | 55 |
| 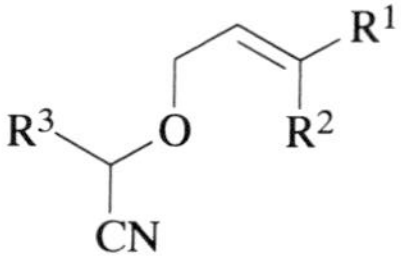 | LDA, THF, -78° | 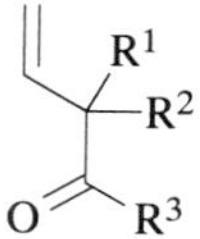 (76) (65) (60) | 35-37 |

|  | R¹ | R² | R³ |
|---|---|---|---|
| C₉ | H | H | *i*-Bu |
| C₁₀ | Me | H | *i*-Bu |
| C₁₁ | Me | Me | CH₂CH(OMe)Me |

184

Table V. ENOLATE REARRANGEMENTS OF α-(ALLYLOXY)CARBONYL COMPOUNDS (*Continued*)

| Ether | Conditions | Product(s) and Yield(s) (%) | Refs. |
|---|---|---|---|
| 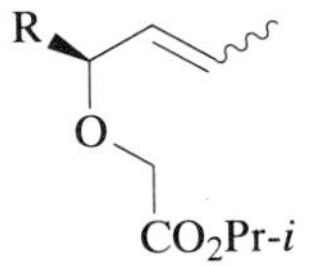 C$_9$ | LDA, THF, –100°, 1 h; to –20°, 18 h | 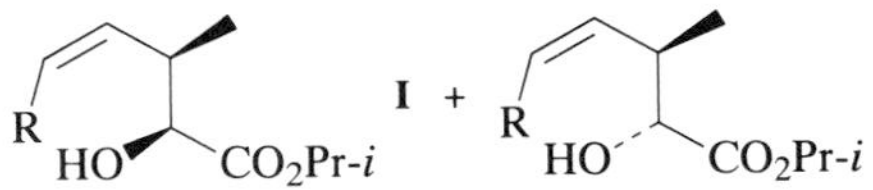 **I** + **II** | 69 |
| R | | Additive | |
| H   Z | | — | (20) **I:II** = 2:1 |
| H   Z | | Cp$_2$ZrCl$_2$ | (15) **I:II** = 7:1 |
| H   E | | — | (14) **I:II** = 4.5:1 |
| H   E | | Cp$_2$ZrCl$_2$ | (47) **I:II** = 45:1 |
| C$_{10}$   Me   E | | | (72) **I:II** = 4:1 |
| Me   E, S | | Cp$_2$ZrCl$_2$ | (91) **I:II** = 100*:1   *>96% ee, S, Z |
| C$_{12}$   i-Pr   E, R | | Cp$_2$ZrCl$_2$ | (26) **I:II** = 50*:1   *>96% ee, R, Z |
| C$_{13}$   n-Bu   E, R | | Cp$_2$ZrCl$_2$ | (81) **I:II** = 61*:1   *>96% ee, R, Z |
| C$_{17}$   n-C$_8$H$_{17}$   E, R | | Cp$_2$ZrCl$_2$ | (70) **I:II** = 90*:1   *>96% ee, R, Z |
| C$_{10}$ 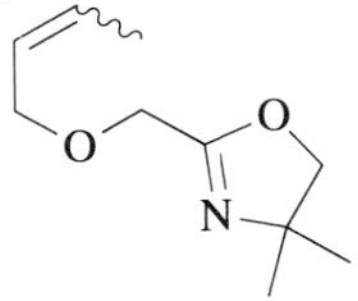 | *n*-BuLi, THF, –78° | 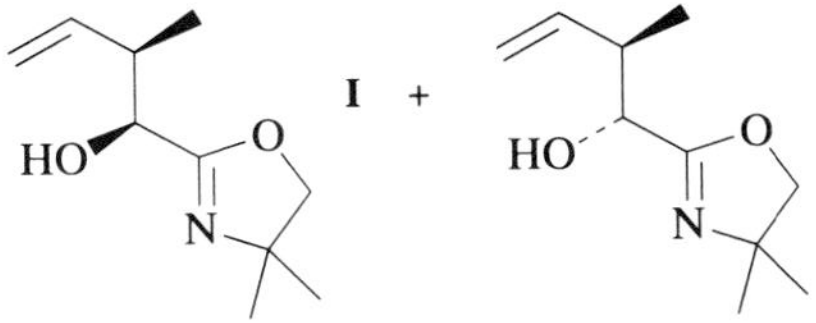 **I** + **II** | 74 |
| 93% E | LDA, THF, –78° | (98) **I:II** = 66:34 | |
| 93% E | LDA, THF, –78° | (80) **I:II** = 84:16 | |
| 94% Z | | (80) **I:II** = 28:78 | |
| 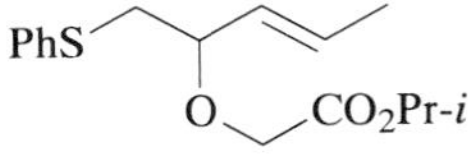 | LDA, THF, Cp$_2$ZrCl$_2$, –20° | 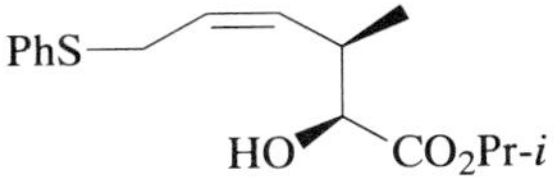 (67) | 83 |

Table V. ENOLATE REARRANGEMENTS OF α-(ALLYLOXY)CARBONYL COMPOUNDS (*Continued*)

| Ether | Conditions | Product(s) and Yield(s) (%) | Refs. |
|---|---|---|---|

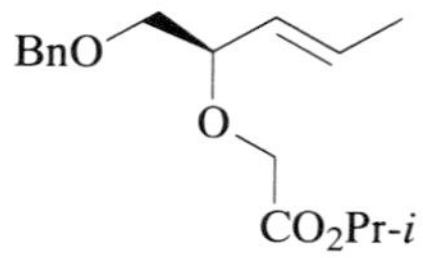

**Row 1:**

Ether: BnO— ...CO$_2$Pr-$i$ (79% ee)

Conditions: LDA, THF, Cp$_2$TiCl$_2$, -20°

Product(s): BnO— HO— CO$_2$Pr-$i$ (—) 79% ee

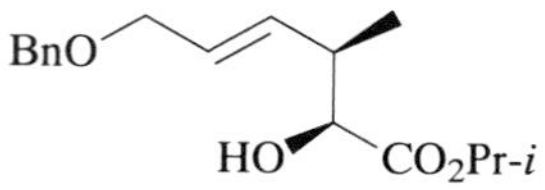

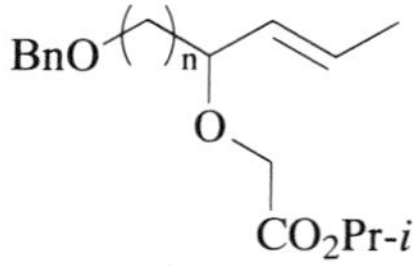

Refs.: 83

**Row 2:**

Ether: BnO—(...)$_n$ O— CO$_2$Pr-$i$

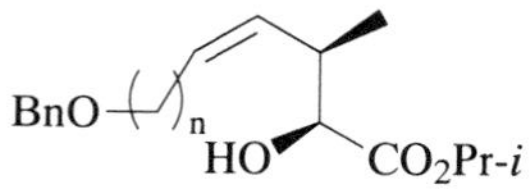

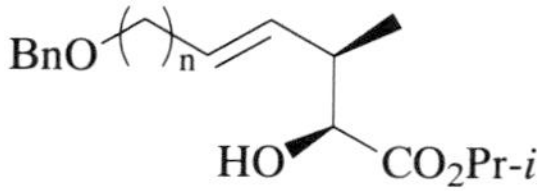

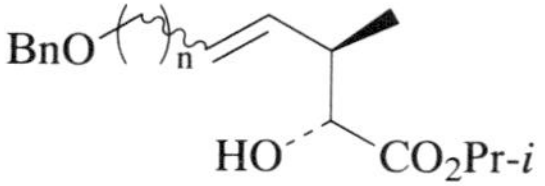

Conditions: LDA, THF, -20°

Product(s):

BnO—(...)$_n$ HO— CO$_2$Pr-$i$  **I** +

BnO—(...)$_n$ HO— CO$_2$Pr-$i$  **II** +

BnO—(...)$_n$ HO— CO$_2$Pr-$i$  **III**

Refs.: 83

| | | Additive | | |
|---|---|---|---|---|
| C$_{10}$ | n = 1 | — | (37) | **I:II:III** = 1:10:10 |
| | | Cp$_2$TiCl$_2$ | (72) | **I:II:III** = 1:58:1 |
| | | Cp$_2$ZrCl$_2$ | (53) | **I:II:III** = 1:1.5:0 |
| | | Cp$_2$HfCl$_2$ | (37) | **I:II:III** = 2:29:1 |
| C$_{11}$ | n = 2 | — | (56) | **I:II:III** = 1.5:3.8:1 |
| | | Cp$_2$TiCl$_2$ | (31) | **I:II:III** = 1:4:1.2 |
| | | Cp$_2$ZrCl$_2$ | (62) | **I:II:III** = 55:1:1 |
| | | Cp$_2$HfCl$_2$ | (58) | **I:II:III** = 21:1:1 |
| C$_{12}$ | n = 3 | — | (59) | **I:II:III** = 3.2:4.3:1 |
| | | Cp$_2$TiCl$_2$ | (37) | **I:II:III** = 1:2:1.7 |
| | | Cp$_2$ZrCl$_2$ | (57) | **I:II:III** = 54:2:1 |

Table V. ENOLATE REARRANGEMENTS OF α-(ALLYLOXY)CARBONYL COMPOUNDS (*Continued*)

| Ether | Conditions | Product(s) and Yield(s) (%) | Refs. |
|---|---|---|---|
| $C_{10}$ | LDA, THF, -85°<br>LHDS<br>LDA, HMPA-THF (23%)<br>KHDS, THF, -85°<br>LHDS, 0° | I +<br>(98) **I:II** = 96:4<br>(96) **I:II** = 96:4<br>(98) **I:II** = 96:4<br>(88) **I:II** = 96:4<br>(96) **I:II** = 96:4    **II** | 61 |
| $C_{11}$   R = H, 93% *E* | LDA, THF, -85°<br>LDA, THF, MgBr$_2$, -85°<br>KH, LDA<br>LHDS<br>LDA, THF, -85°<br>LDA, THF, Cp$_2$ZrCl$_2$, -85°<br>LDA | I + II <br>(—) **I:II** = 95*:5 *52% ee *R*<br>(—) **I:II** = 96*:4 *36% ee *R*<br>(—) **I:II** = 93*:7 *20% ee *R*<br>(—) **I:II** = 95*:5 *20% ee *R*<br>(—) **I:II** = 97*:3 *20% ee *S*<br>(—) **I:II** = 98*:2 *60% ee *S*<br>(—) **I:II** = 95*:5 *12% ee *S* | 61 |
| $C_{15}$   R = MEM, 93% *E* | | | |

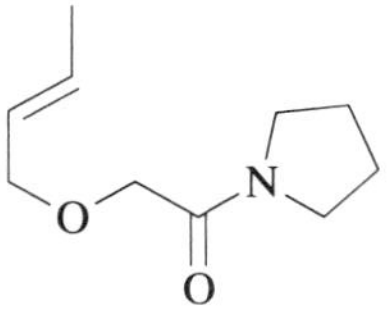

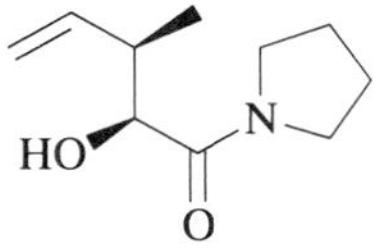

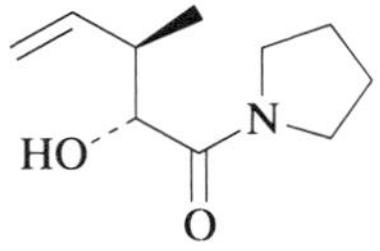

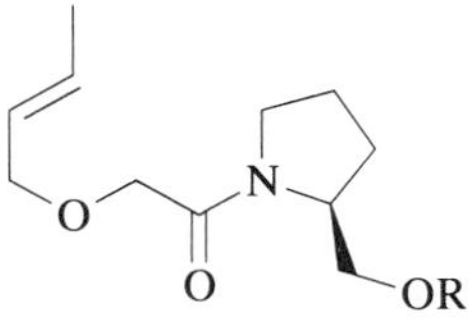

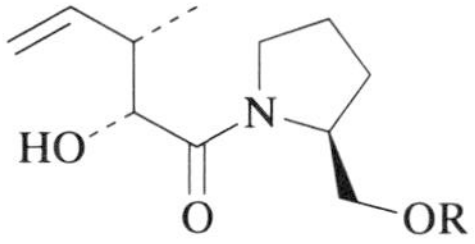

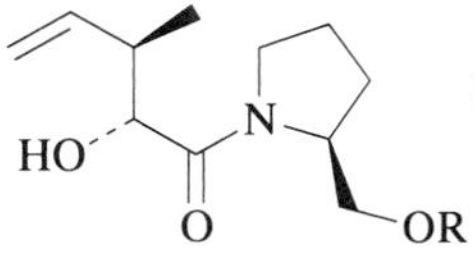

Table V. ENOLATE REARRANGEMENTS OF α–(ALLYLOXY)CARBONYL COMPOUNDS (*Continued*)

| Ether | Conditions | Product(s) and Yield(s) (%) | Refs. |
|---|---|---|---|

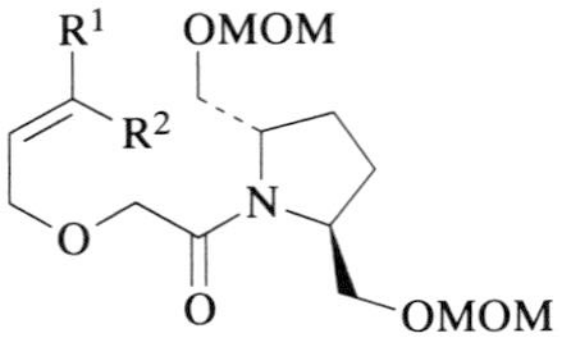

LDA, THF, -100°, 3 h; -70°, 3 h

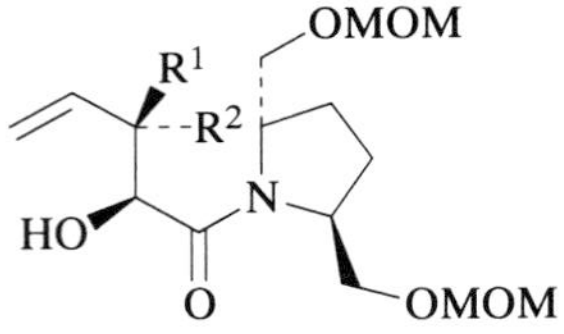

**I +**

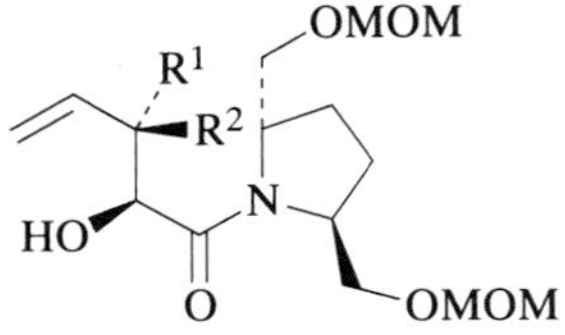

**II**

(61) 96% ee

| | R¹ | R² | Additive | | |
|---|---|---|---|---|---|
| C₁₁ | H | H | — | (61) 96% ee | |
| C₁₂ | Me | H | Cp₂ZrCl₂ | (43) **I:II** = 28*:1 | *96% ee |
| C₁₃ | Et | H | — | (75) **I:II** = 2*:1 | *92% ee |
| | Et | H | Cp₂ZrCl₂ | (65) **I:II** = 30*:1 | *96% ee |
| | Et | H | MgBr₂ | (13) **I:II** = 4*:1 | *82% ee |
| | H | Et | MgBr₂ | (33) **I:II** = 1*:3.5 | *90% ee |
| | H | Et | Cp₂ZrCl₂ | (7) **I:II** = 1*:1 | *92% ee |

C₁₁

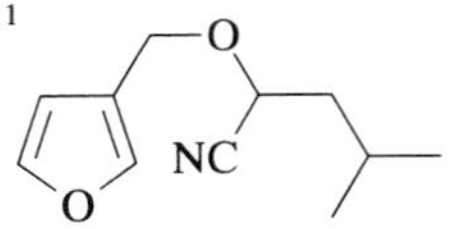

LDA, HMPA-THF (20%), -78°

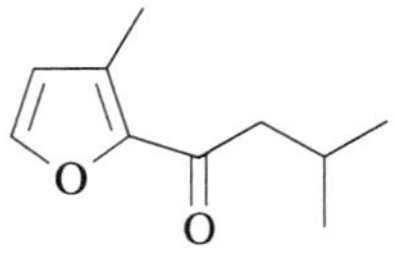

(60)

35-37

188

Table V. ENOLATE REARRANGEMENTS OF α–(ALLYLOXY)CARBONYL COMPOUNDS (*Continued*)

| Ether | Conditions | Product(s) and Yield(s) (%) | Refs. |
|---|---|---|---|
| C₁₂ | | | |
| 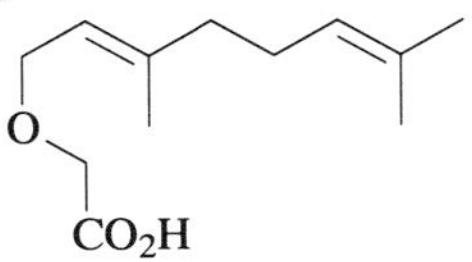 | LDA, THF, -78° | 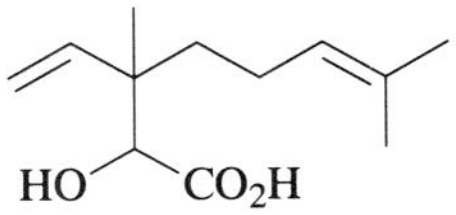 (63) | 58 |
| 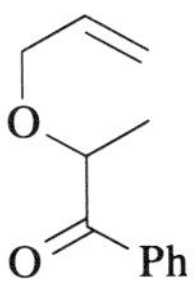 | LiH, MeOH, THF, 67° | 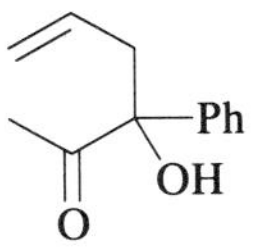 (20) | 60 |
| 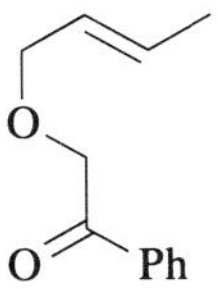 | LDA, HMPA-THF (1:4), -78° to rt | 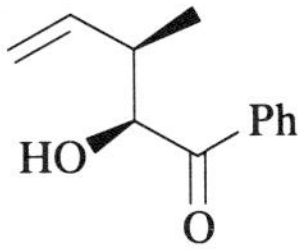 (45) ds 69% | 59 |
|  R = TMS  R = Br | LDA, THF, -78°  LDA, THF, -100 to -20° | (46) 98% Z  (31) Z:E >25:1 | 175 |

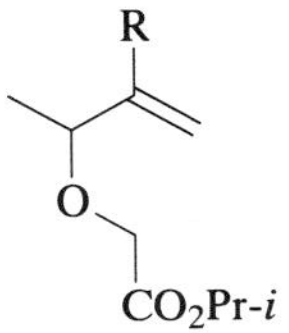

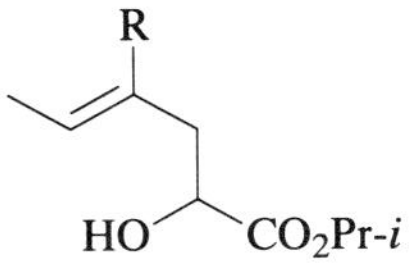

189

Table V. ENOLATE REARRANGEMENTS OF α-(ALLYLOXY)CARBONYL COMPOUNDS (*Continued*)

| Ether | Conditions | Product(s) and Yield(s) (%) | Refs. |
|---|---|---|---|
| 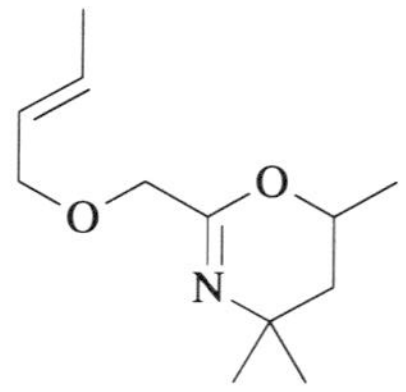 | n-BuLi, THF, -78°<br>LDA, THF, -78°<br>LDA, THF, -78° | 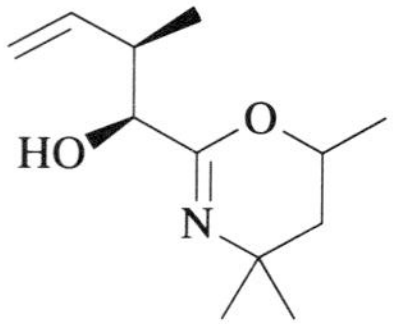 <br>I 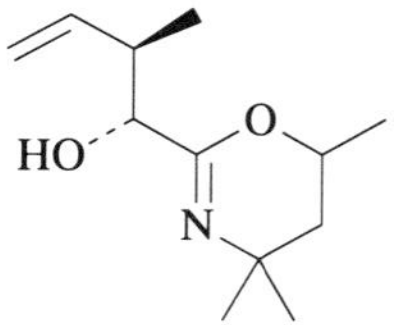 + II<br>(98) **I:II** = 98:2<br>(86) **I:II** = 96:4<br>(82) **I:II** = 65:35 | 74 |
| 95% *E*<br>95% *E*<br>98% *Z* | | | |
| 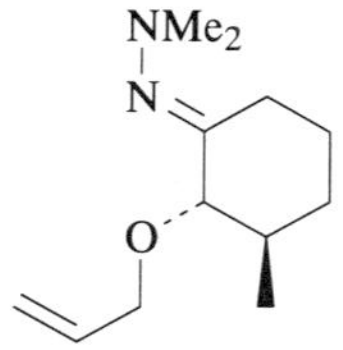 | KH, KOBu-*t*,<br>THF, rt, 10 h | 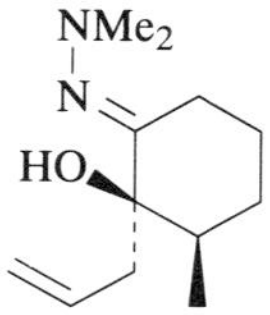 <br>(91) | 176 |
| 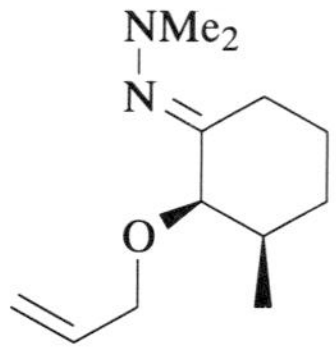 | KH, KOBu-*t*, THF,<br>reflux, 12 h | " (92) | 176 |

190

Table V. ENOLATE REARRANGEMENTS OF α-(ALLYLOXY)CARBONYL COMPOUNDS (*Continued*)

| Ether | Conditions | Product(s) and Yield(s) (%) | Refs. |
|---|---|---|---|
| <br><br>$C_{12}$, $C_{13}$<br><table><tr><td>$R^1$</td><td>$R^2$</td></tr><tr><td>H</td><td>H</td></tr><tr><td>Me</td><td>H</td></tr><tr><td>Me</td><td>H</td></tr><tr><td>H</td><td>Me</td></tr></table> | *n*-BuLi, THF, –78°<br>"<br>HMPA, THF<br>*n*-BuLi, THF, –78° | <br>(82) ds >100:1<br>(87) ds 68:1<br>(85) ds 8.8:1<br>(91) ds >100:1 | 72, 84 |
| <br><br>$C_{12}$  R = H<br>$C_{13}$  R = Me | *n*-BuLi, THF, –78°<br>MeMgBr, THF, 0°<br>*n*-BuLi, THF, –78°<br>*n*-BuLi, HMPA, THF, –78° | <br>(95) ds 1.8:1<br>(20) ds >100:1<br>(87) ds 2.2:1<br>(85) ds 1.2:1 | 72, 84 |

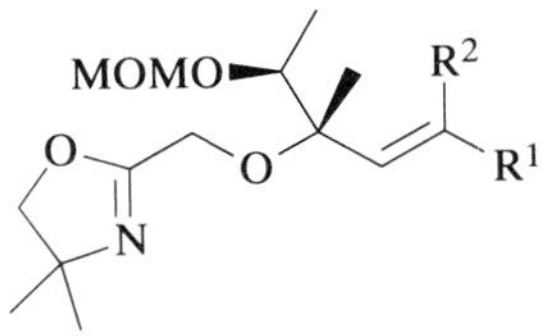

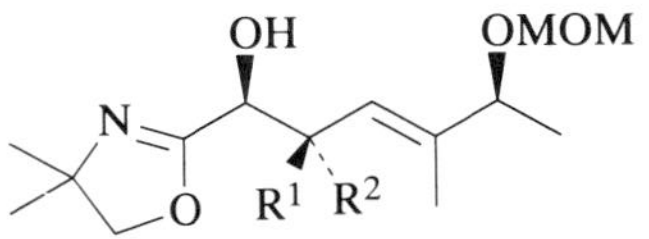

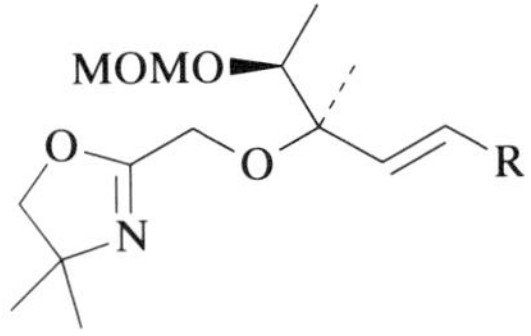

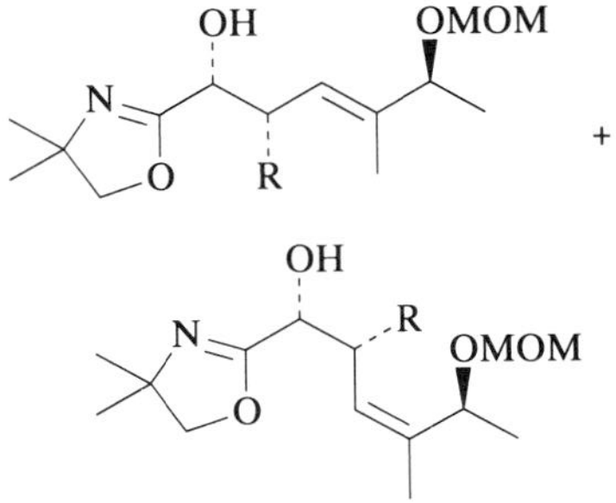

191

Table V. ENOLATE REARRANGEMENTS OF α-(ALLYLOXY)CARBONYL COMPOUNDS (*Continued*)

| Ether | Conditions | Product(s) and Yield(s) (%) | Refs. |
|---|---|---|---|

C₁₃

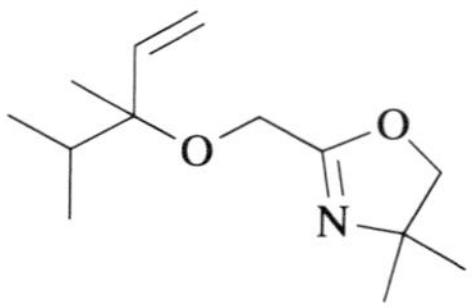

*n*-BuLi, THF, -78°

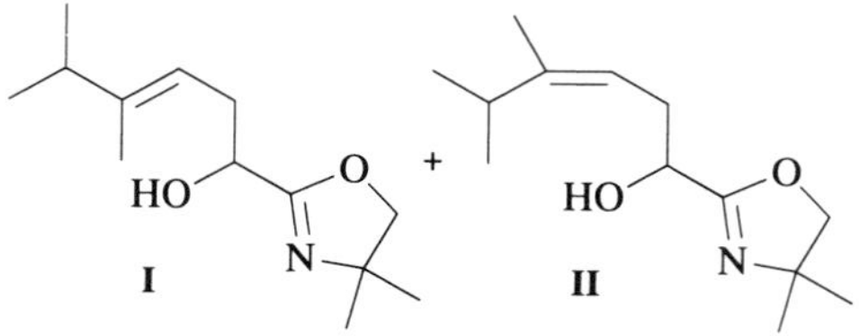

**I:II = 72:28**

72

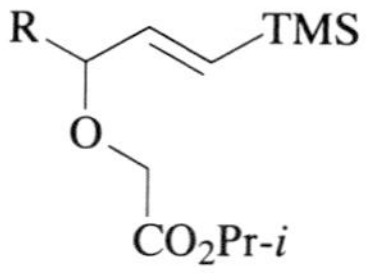

LDA, Cp₂ZrCl₂,
THF, -78 to -20°

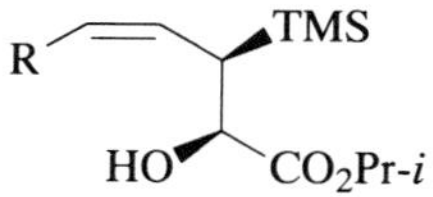

177

C₁₃  R = Et          (72)
C₁₄  R = *i*-Pr       (66)
C₁₇  R = C₆H₁₁       (30)

C₁₃

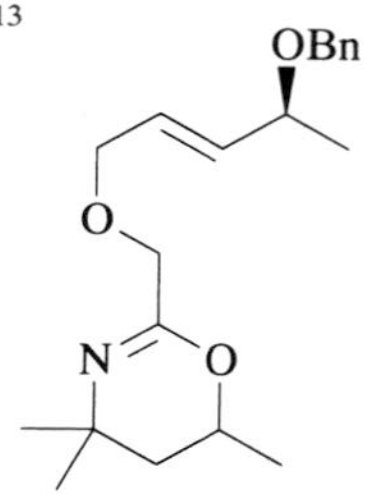

*n*-BuLi, THF, -85°

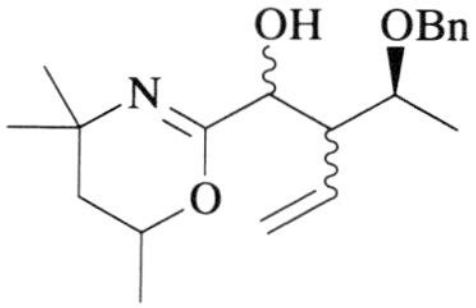

(98)
ds 62:22:3:13

178

Table V. ENOLATE REARRANGEMENTS OF α-(ALLYLOXY)CARBONYL COMPOUNDS (*Continued*)

| Ether | Conditions | Product(s) and Yield(s) (%) | Refs. |
|---|---|---|---|
| 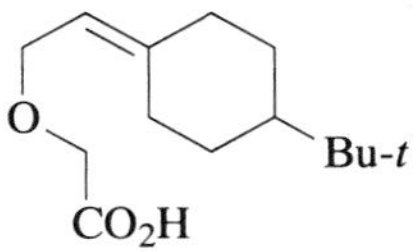 C₁₄ | *n*-BuLi, THF, -78° | 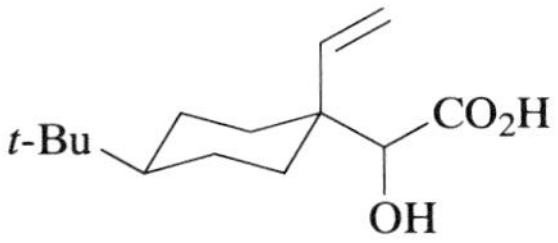 (63) | 58 |
| 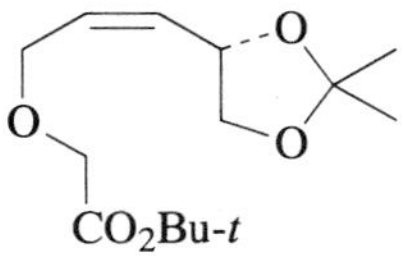 | LDA, THF, TMEDA -78°, 30 min; to -40°, 3 h | 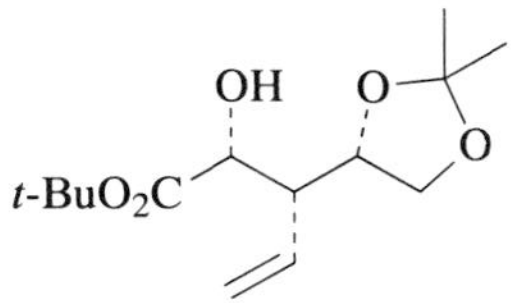 (40) >98% selective | 106 |
| C₁₅ | | | 96 |

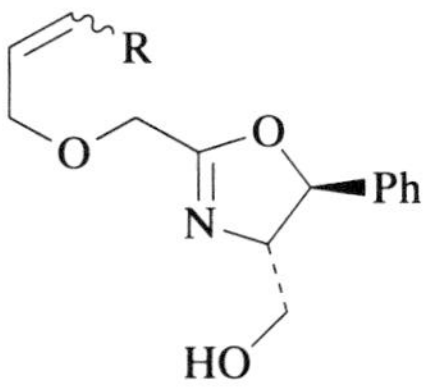

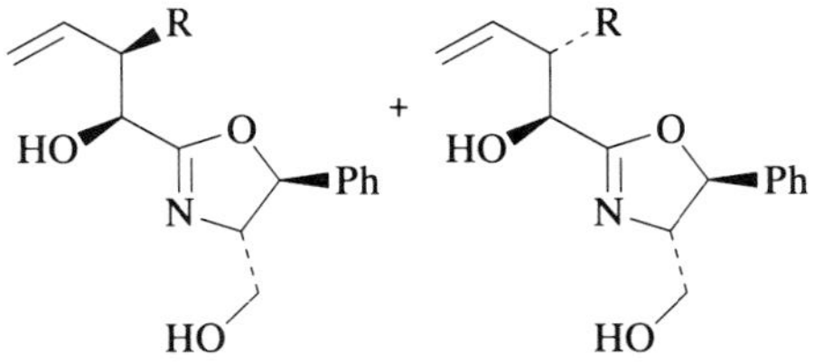

| | R | Conditions | Product(s) and Yield(s) (%) |
|---|---|---|---|
| C₁₄ | H | KH, THF, 20° | (−) 84% ee, *S* |
| | H | KH, 18-c-6, THF, -20° | (−) 96% ee, *R* |
| C₁₅ | Me, 93% *E* | KH, THF, 20° | (−) **I:II** = 41:59; **I**, 0% ee; **II**, 74% ee, *S* |
| | Me, 93% *Z* | KH, THF, 20° | (−) **I:II** = 43:57; **I**, 64% ee, *S*; **II**, 78% ee, *S* |
| | Me, 93% *E* | KH, 18-c-6, THF, -20° | (−) **I:II** = 46:54; **I**,86% ee, *R*; **II**, 84% ee, *R* |
| | Me, 93% *Z* | KH, 18-c-6, THF, -20° | (−) **I:II** = 54:46; **I**,86% ee, *R*; **II**, 86% ee, *R* |
| C₁₆ | Me₂ | KH, THF, 20° | (−) 56% ee, *S* |

193

Table V. ENOLATE REARRANGEMENTS OF α–(ALLYLOXY)CARBONYL COMPOUNDS (*Continued*)

| Ether | Conditions | Product(s) and Yield(s) (%) | Refs. |
|---|---|---|---|
| 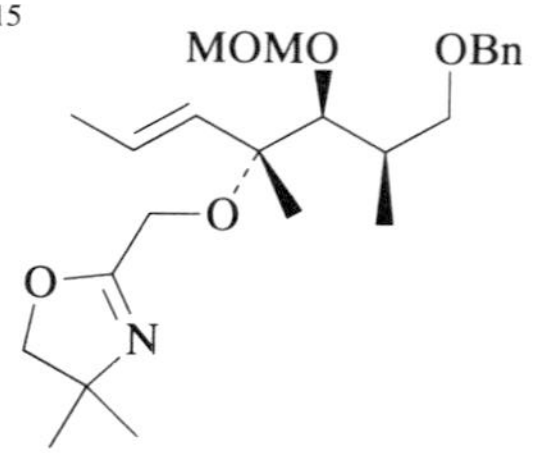 C₁₅ | LDA, THF, -78° | 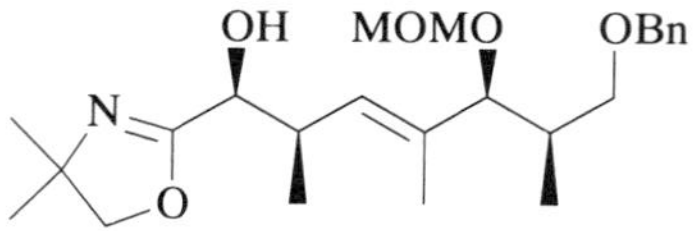 (95) ds 35:1 | 72, 84 |
| 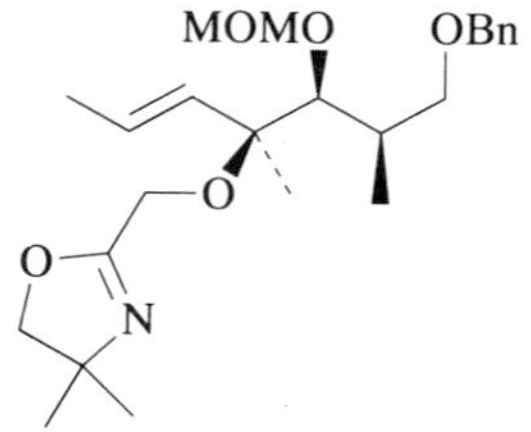 | LDA, THF, -78° | 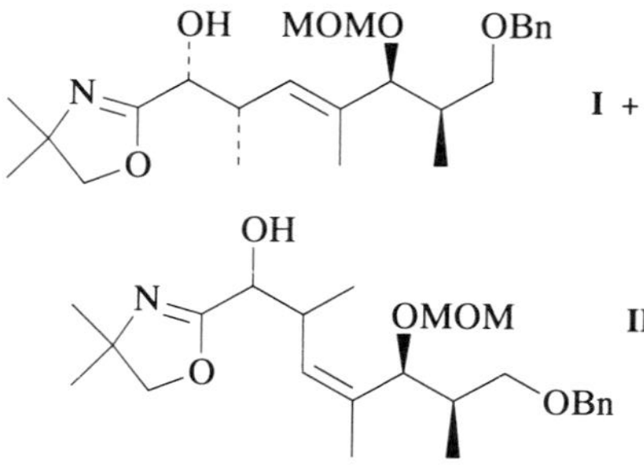 **I** + **II** (90) **I:II** = 1.7:1 | 72, 84 |

194

Table V. ENOLATE REARRANGEMENTS OF α-(ALLYLOXY)CARBONYL COMPOUNDS (*Continued*)

| Ether | Conditions | Product(s) and Yield(s) (%) | Refs. |
|---|---|---|---|
| | | | 95 |

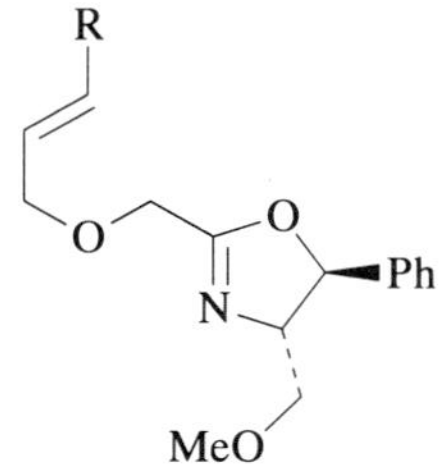

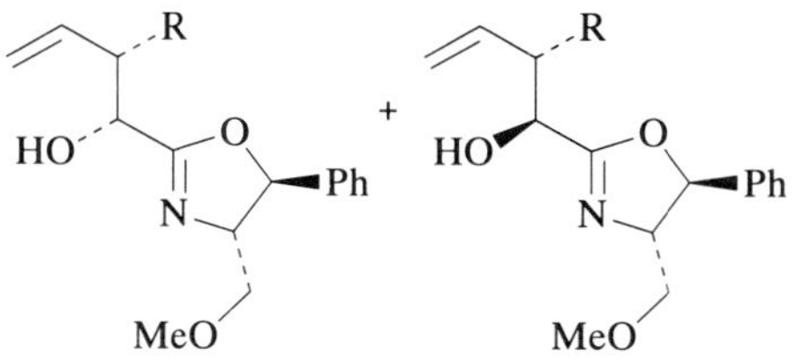

| | | R | Conditions | Product(s) and Yield(s) (%) |
|---|---|---|---|---|
| C15 | | H | *n*-BuLi, THF, –78° | (—) 38% ee, *R* |
| | | H | LDA, THF, –78° | (—) 14% ee, *R* |
| C16 | | Me | *n*-BuLi, THF, –78° | **I**:**II** = 90:10; **I**, 78% ee, *R*; **II**, 8% ee, *S* |
| | | Me | LDA, THF, –78° | (—) **I**:**II** = 84:16; **I**, 64% ee, *R*; **II**, 28% ee, *S* |
| C17 | | Me2 | *n*-BuLi, THF, –78° | (—) 75% ee, *R* |

195

Table V. Enolate Rearrangements of α-(Allyloxy)carbonyl Compounds (*Continued*)

| Ether | Conditions | Product(s) and Yield(s) (%) | Refs. |
|---|---|---|---|
| C₁₇ R = Me, C₁₈ R = Et, C₁₉ R = i-Pr | LDA, THF, -78 to 0°, 10 min | I + II <br> (60) **I:II** = 75:25 <br> (63) **I:II** = 95:5 <br> (78) **I:II** = 95:5 | 56 |
| C₁₇ R = Me, C₁₈ R = Et, C₁₉ R = i-Pr | LDA, THF, -78 to 0°, 10 min | **I** (56) <br> **I** (89) <br> **I + II** (66), **I:II** = 95:5 | 56 |

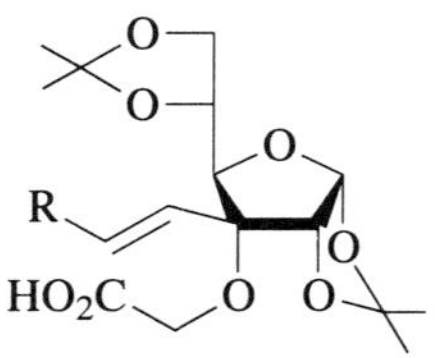

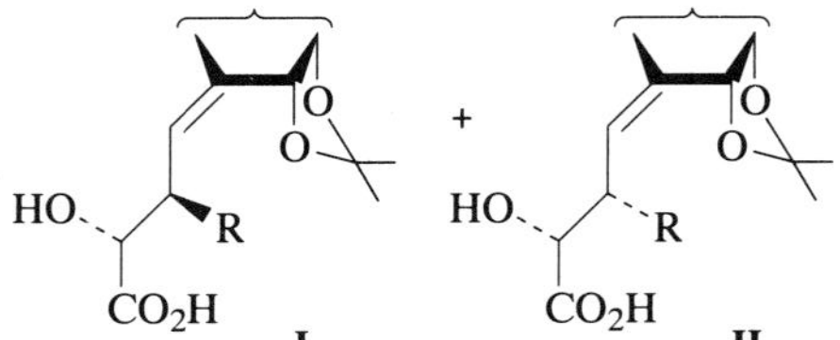

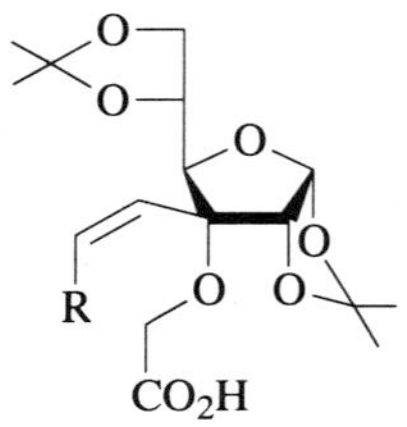

Table V. ENOLATE REARRANGEMENTS OF α–(ALLYLOXY)CARBONYL COMPOUNDS (*Continued*)

| Ether | Conditions | Product(s) and Yield(s) (%) | Refs. |
|---|---|---|---|
| C₁₇ | LDA, THF, -100°, 3 h; -70°, 3 h<br><br>LDA, THF, Cp₂ZrCl₂, -100°, 3 h; -70°, 3 h<br><br>LDA, THF, Cp₂ZrCl₂, -100°, 3 h; -20°, 3 h | <br>(83) **I:II** = 4:1; **I**; 33% ee<br><br>(25) **I:II** = 1:3; **II**, 87% ee<br><br>(70) **I:II** = 1:3; **II**, 41% ee | 97 |

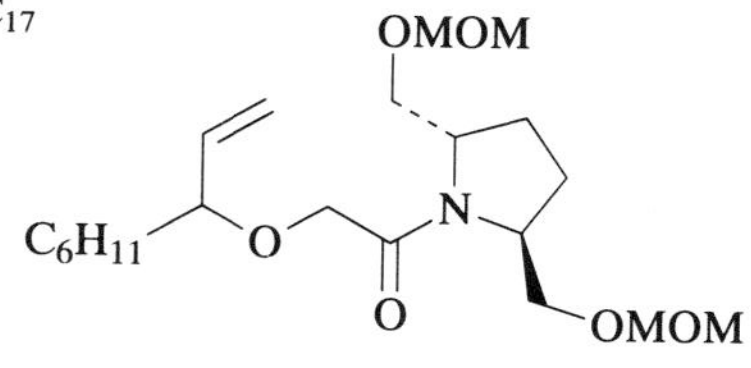

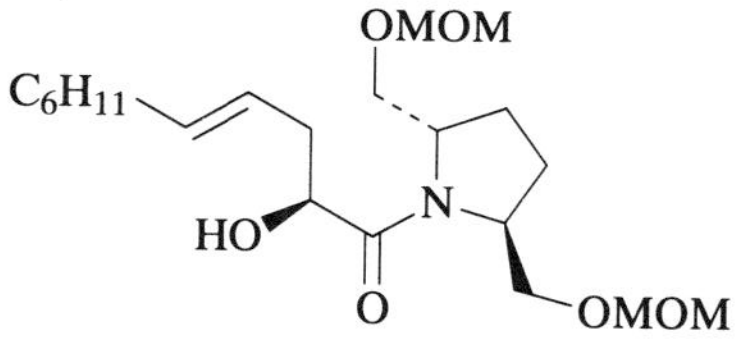

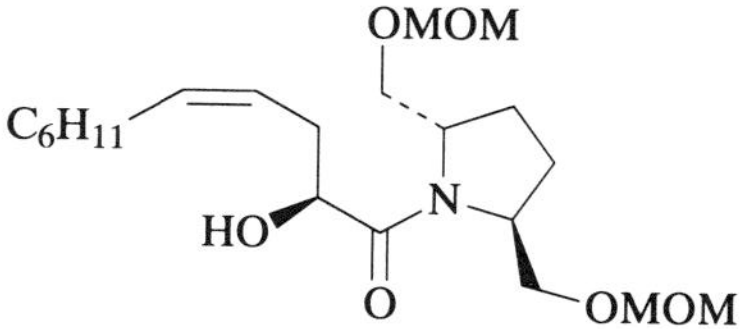

197

Table V. ENOLATE REARRANGEMENTS OF α–(ALLYLOXY)CARBONYL COMPOUNDS (*Continued*)

| Ether | Conditions | Product(s) and Yield(s) (%) | Refs. |
|---|---|---|---|
| (C₂₁, C₂₂, C₂₃, C₂₅) | | 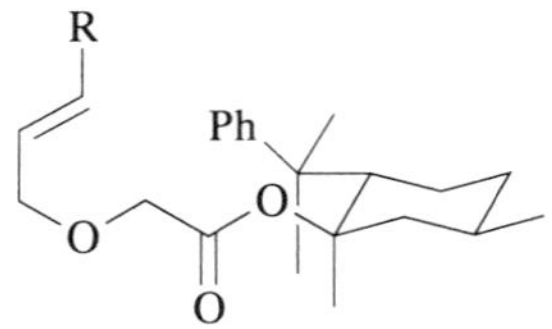 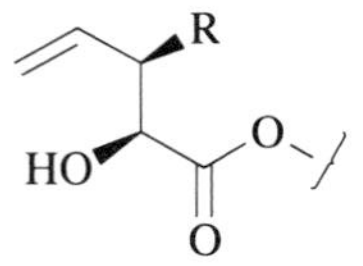 **I** 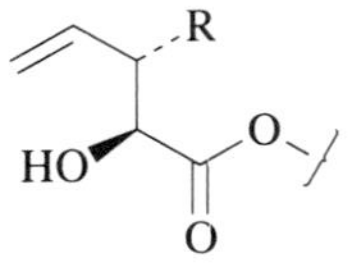 + **II** | 98 |
| R H (C₂₁) | LDA, THF, HMPA, –78° | (—) 11% ee, *S* | |
| Me (C₂₂) | LDA, THF, HMPA, –78° | (—) **I:II** = 90:10; **I**, 97% ee, *S* | |
| Me | LHDS, THF, HMPA, –78° | (—) **I:II** = 93:7; **I**, 96% ee, *S* | |
| Me | LTMP, THF, HMPA, –78° | (—) **I:II** = 92:8; **I**, 97% ee, *S* | |
| Me₂ (C₂₃) | LDA, THF, HMPA, –78° | (—) 96% ee, *S* | |
| *n*-Bu (C₂₅) | LDA, THF, HMPA, –78° | (—) **I:II** = 92:8; **I**, 95% ee, *S* | |
| 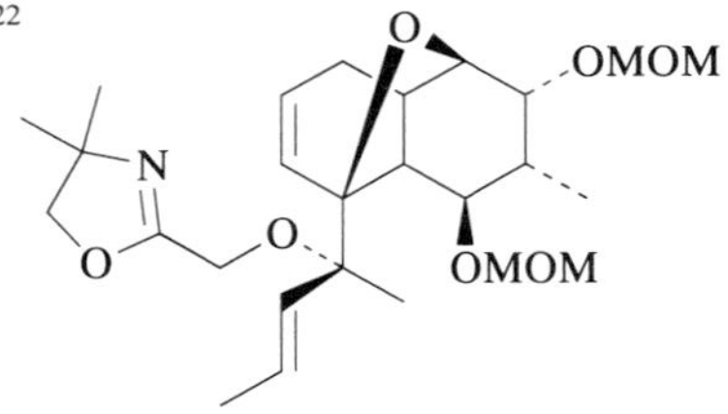 (C₂₂) | LDA, THF, –78° | 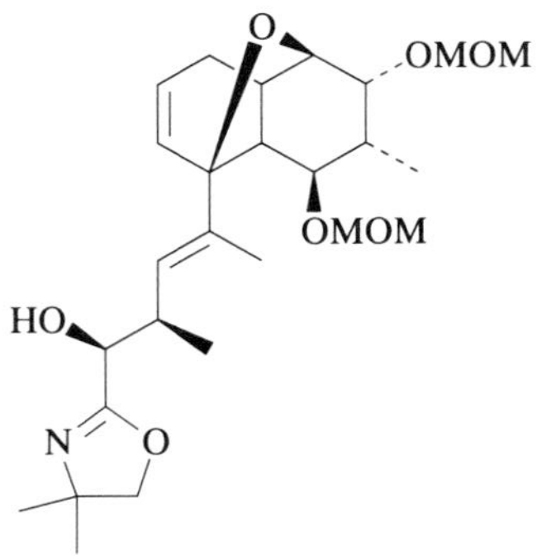 (94) | 131 |

198

Table V. ENOLATE REARRANGEMENTS OF α-(ALLYLOXY)CARBONYL COMPOUNDS (*Continued*)

| Ether | Conditions | Product(s) and Yield(s) (%) | Refs. |
|---|---|---|---|
| TMSO <br> 93% *E* <br> 93% *E* | TiCl$_4$, CH$_2$Cl$_2$, -50°, 5 h <br> SnCl$_4$, CH$_2$Cl$_2$, -50°, 5 h | **I** + **II** <br> (80) **I:II** = 85:15; **I**, 8% ee, *S* <br> (79) **I:II** = 93:7; **I**, 12% ee, *S* | 65 |
| 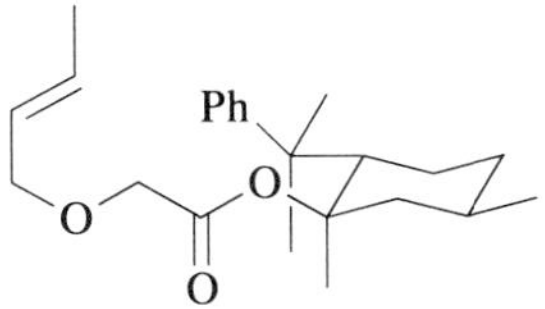 | LDA, THF, Cp$_2$ZrCl$_2$, -78° | **I + II** (75); **I:II** = 94:6; **I**, 91% ee, *S* <br> | 179 |
| C$_{24}$ | LDA, THF, -78° | (82) <br> | 85, 86 |

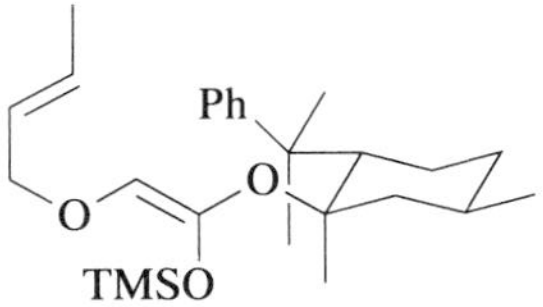

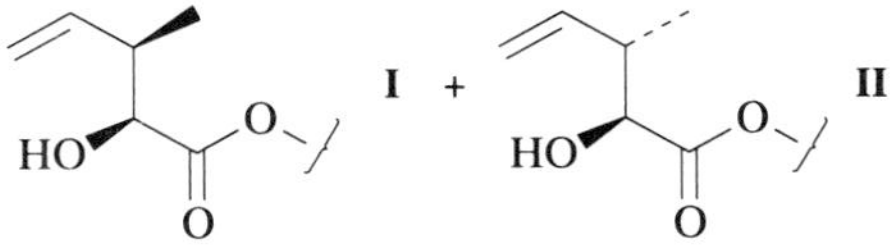

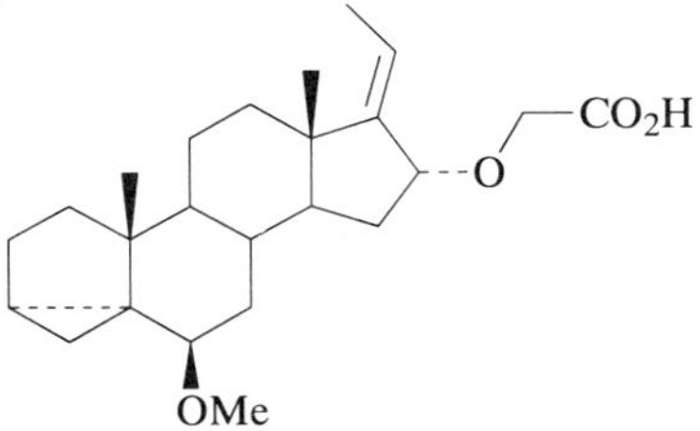

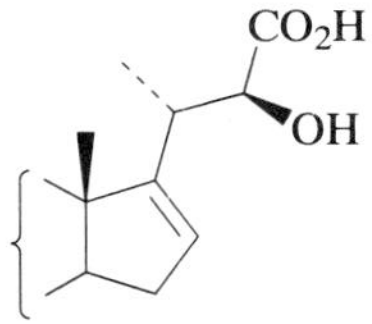

Table VI. [2,3]-WITTIG RING CONTRACTIONS

| Ether | Conditions | Product(s) and Yield(s) (%) | Refs. |
|---|---|---|---|
| 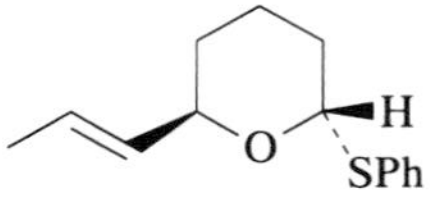 C₈ | LTBB, THF, -78 to 0°, 2 h or -78°, 2 h | 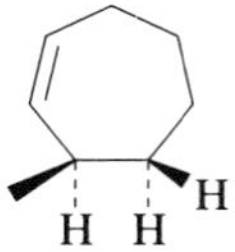 (45) 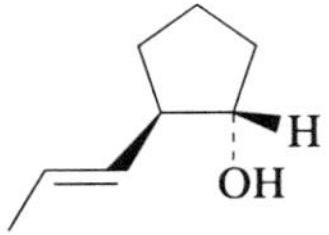 + (21) | 120 |
| C₁₀ | LTMP, THF, -78°, 5 h<br>LTMP, THF, 0°, 14 h<br>n-BuLi, THF, HMPA, 0°, 12 h<br>LBPEAS, THF, 0°, 24 h<br>LBPEAS, THF, 25°, 36 h<br>LBPEAS, THF, HMPA, 0°, 24 h | + <br>(10)<br>(78)<br>decomposition<br>(40) 25% ee *R,R*<br>(52) 25% ee *R,R*<br>(45) 25% ee *R,R* | 42-44 |
| 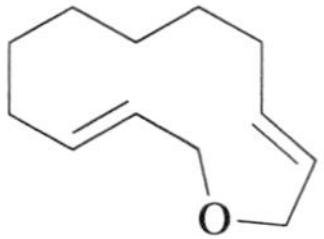 C₁₂ | t-BuLi, THF, -78°, 6-10 h | 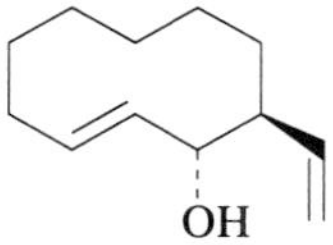 I (94) | 41 |
| 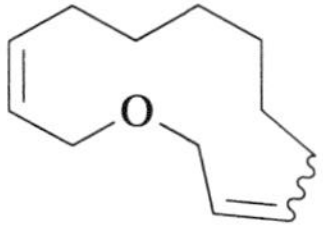 | t-BuLi, THF, -78°, 6-10 h | I 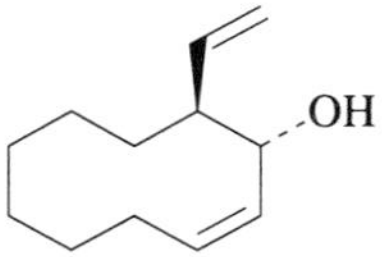 + II 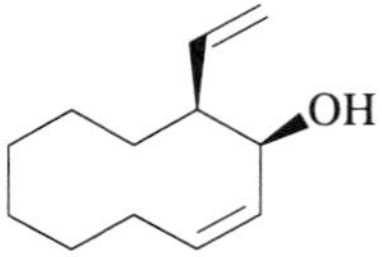 + III<br>(95) **I:II:III** = 9:50:41 | 41 |

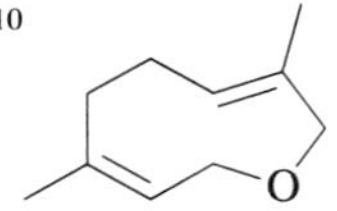

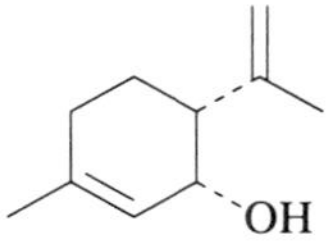

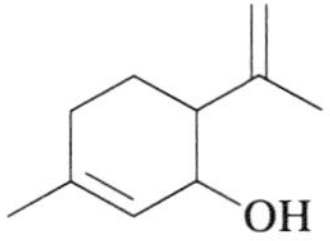

200

## Table VI. [2,3]-WITTIG RING CONTRACTIONS (*Continued*)

| Ether | Conditions | Product(s) and Yield(s) (%) | Refs. |
|---|---|---|---|
| 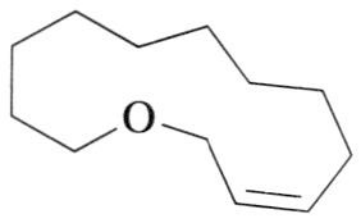 | *t*-BuLi, THF, <br> -78°, 6-10 h | 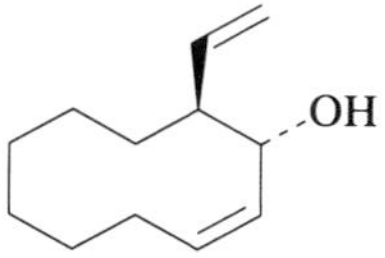 (96) | 41 |
| C<sub>14</sub> 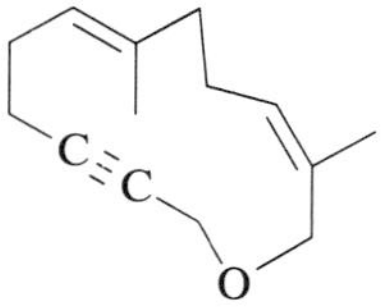 | LTMP, THF, -20° | 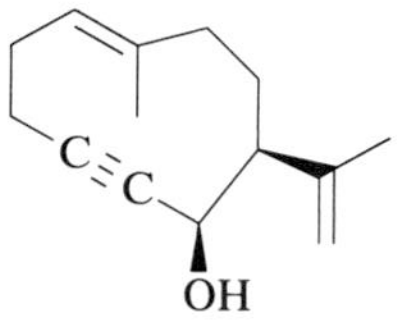 (—) | 78, 79 |
| 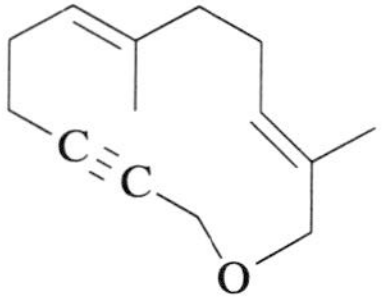 | *n*-BuLi, C<sub>5</sub>H<sub>12</sub>-THF, <br> (9:1), -78°, 2.5 h | 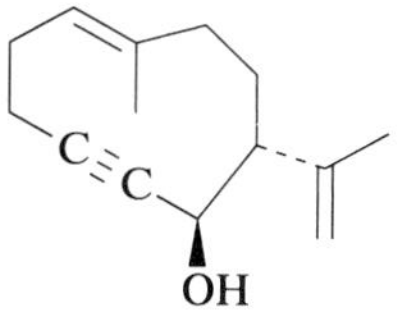 (92) | 78, 79 |
| | LBPEAS, THF, -70 to -15°, 45 min <br> LBPEAS, C<sub>5</sub>H<sub>12</sub>, THF, -25 to 0°, 90 min <br> LBPEAS, Et<sub>2</sub>O, -25 to -15°, 30 min <br> LBPEAS, THF, -20°, 60 min <br> LBPEAR, THF, -35 to -25°, 60 min <br> LBPEAR, THF, -20°, 40 min | (82) 69% ee <br> (68) 43% ee <br> (70) 9% ee <br> (82) 2% ee <br> (78) 70% ee <br> (75) 60% ee | 115-<br>117 |

201

Table VI. [2,3]-WITTIG RING CONTRACTIONS (*Continued*)

| Ether | Conditions | Product(s) and Yield(s) (%) | Refs. |
|---|---|---|---|

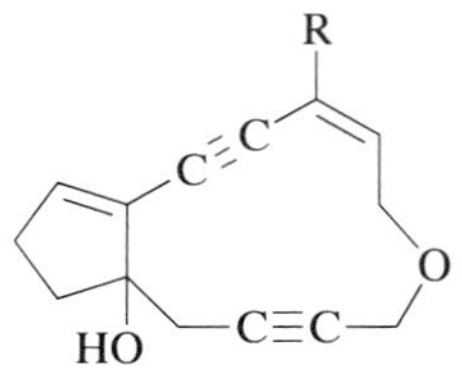

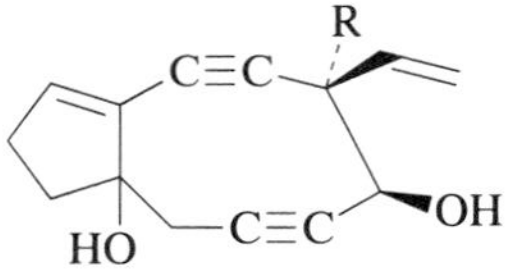

*n*-BuLi, HMPA, −78°   R = H   (<29)   146
*t*-BuLi, THF, −100°   R = Me   (62-66)   145

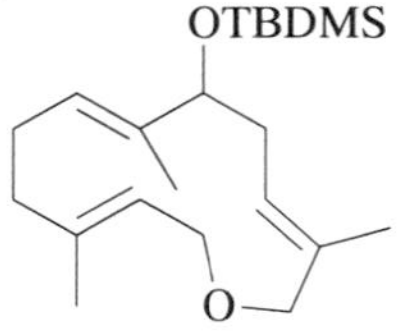

*t*-BuLi, Et₂O, −70°   25, 141

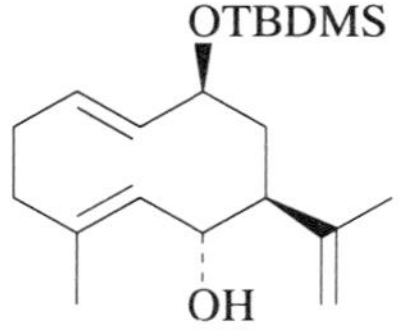

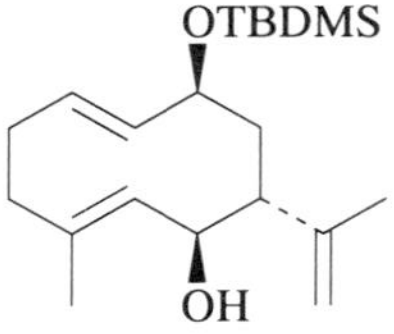

I + II + III (90)
I:II:III = 39:9:52

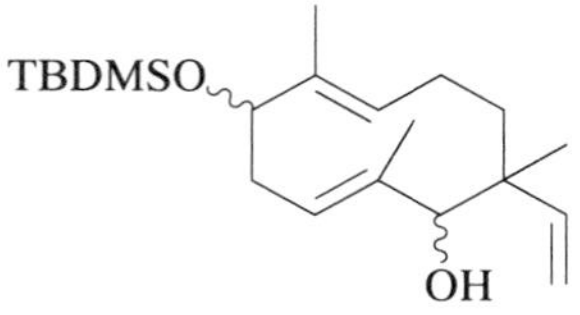

C₁₅

Table VI. [2,3]-WITTIG RING CONTRACTIONS (*Continued*)

| Ether | Conditions | Product(s) and Yield(s) (%) | Refs. |
|---|---|---|---|

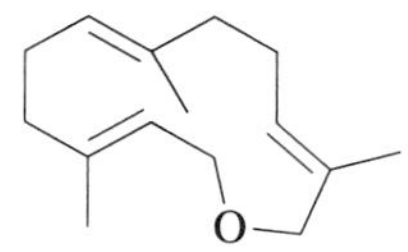

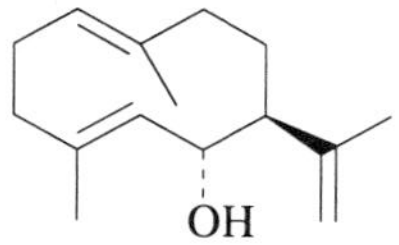

**I +**        **II**

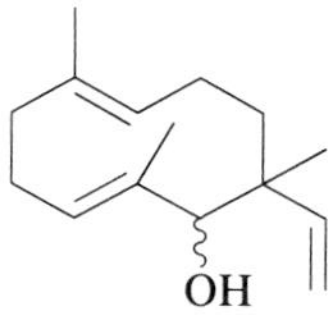

| | | | |
|---|---|---|---|
| | *t*-BuLi, Et$_2$O, -78°, 10 h to 0°, 30 min | **I + II** (90); **I:II** = 75:25 | 41 |
| | *t*-BuLi, C$_6$H$_{14}$, -70° | **I + II** (0) | 25, 141 |
| | *t*-BuLi, Et$_2$O, -70° | **I + II** (98); **I:II** = 75:25 | 25, 141 |
| | *t*-BuLi, THF, -70° | **I + II** (95); **I:II** = 55:45 | 25, 141 |
| | *t*-BuLi, THF-DME (3:1), -70° | **I + II** (—); **I:II** = 10:90 | 25, 141 |
| | *t*-BuLi, THF-TMEDA (3:1), -70° | **II** (—) | 25, 141 |
| | *t*-BuLi, THF-HMPA (3:1), -70° | **II** (—) | 25, 141 |

C$_{19}$

| 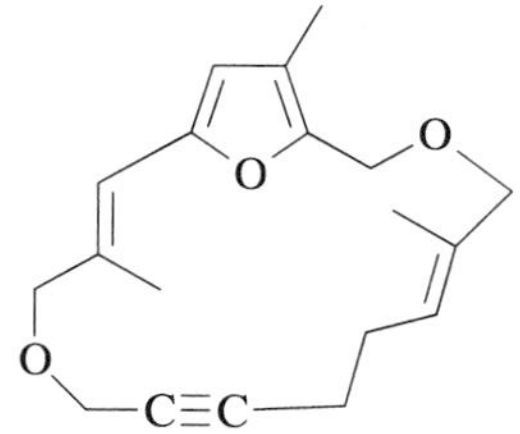 | LTMP, C$_6$H$_{14}$-THF, (10:1), -78 to -23° | 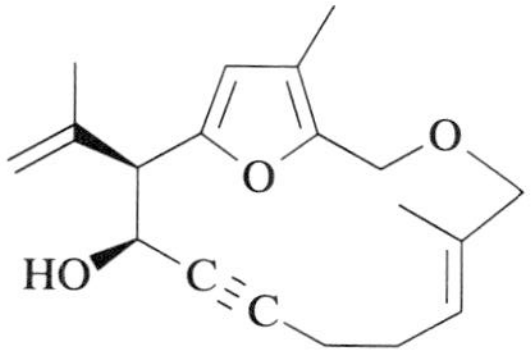 (12) | 142 |
|---|---|---|---|

203

Table VI. [2,3]-WITTIG RING CONTRACTIONS (Continued)

| Ether | Conditions | Product(s) and Yield(s) (%) | Refs. |
|---|---|---|---|
| | LBPEAS, THF, -55 to -10° | I + II, (56); **I:II** = 70:30, <5% ee | 115-117 |
| | LBPEAS, C$_5$H$_{12}$, -55 to -10° | I + II, (68); **I:II** = 60:40, 33% ee | 115-117 |
| | n-BuLi, THF, -78° | I + II, (80); **I:II** = 70:30 | 42, 43 |
| | n-BuLi, THF-C$_6$H$_{14}$ (1:10), -78° | I + II, (73); **I:II** = 29:71 | 42, 43 |
| | n-BuLi, THF-C$_6$H$_{14}$ (1:10), -20° | I + II, (78); **I:II** = 18:82 | 42, 43 |
| | n-BuLi, THF-HMPA (25:1), -78° | I + II, (76); **I:II** = 75:25 | 42, 43 |
| | n-BuLi, THF-HMPA (25:2), -78° | I + II, (49); **I:II** = 84:16 | 42, 43 |
| | n-BuLi, THF-HMPA (4:1), -78° | I + II, (45); **I:II** = 88:12 | 42, 43 |

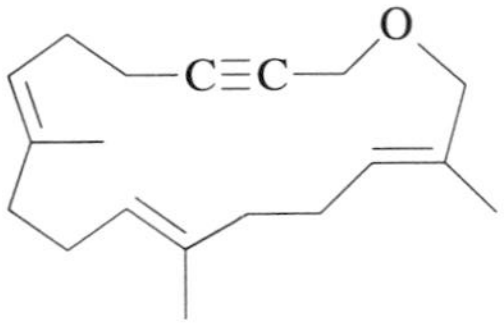

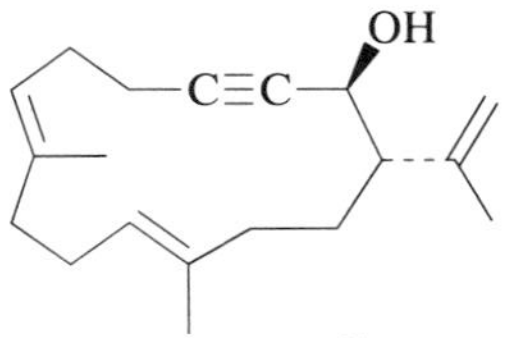

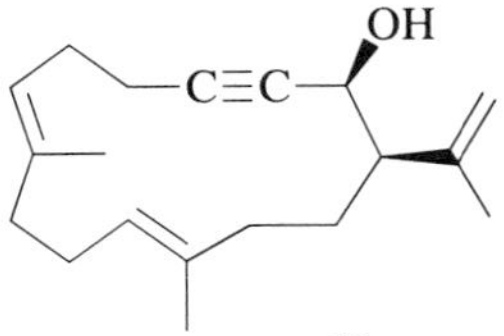

Table VI. [2,3]-WITTIG RING CONTRACTIONS (Continued)

| Ether | Conditions | Product(s) and Yield(s) (%) | Refs. |
|---|---|---|---|

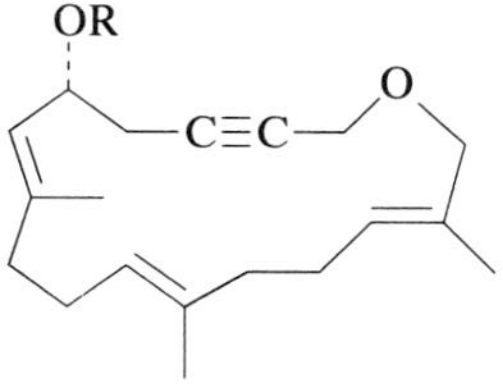

*n*-BuLi, TMEDA, C$_5$H$_{12}$, THF, -78°

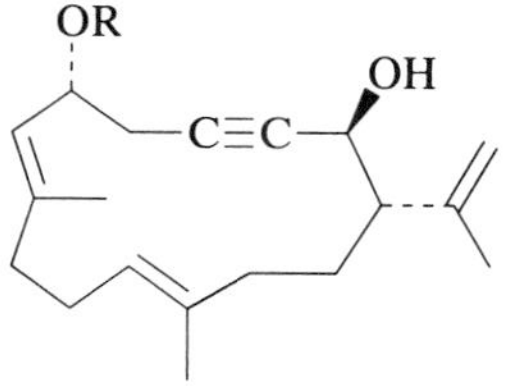

I + II

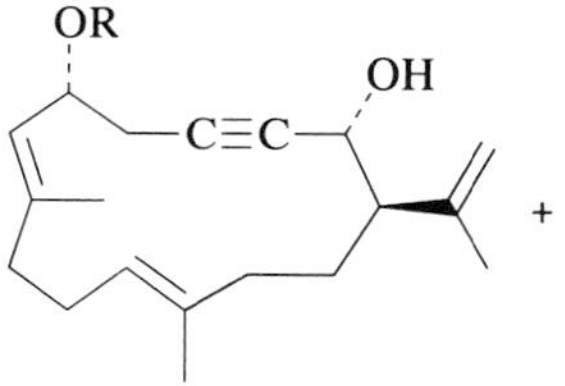

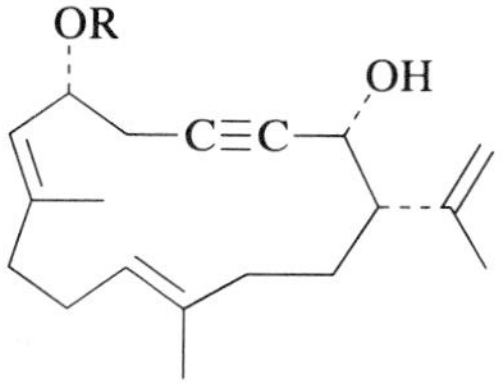

III + IV

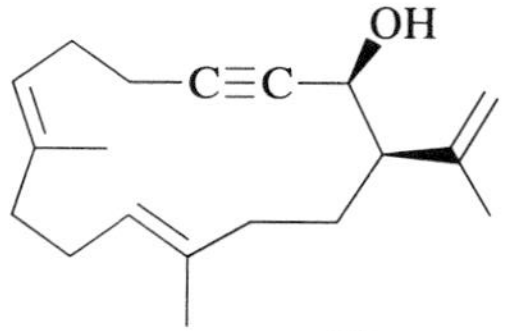

(71) **I:II:III:IV** = 20:29:39:12
(90) **I:II:III:IV** = 81:7:6:6
(94) **I:II:III:IV** = 74:8:14:4

| R | |
|---|---|
| H | |
| THP | |
| TBS | |

102,103
143,144

## REFERENCES

[1] Pine, S. H. *Org. React.* **1970**, *18*, 403.
[2] Wittig, G.; Lohman, L. *Justus Liebigs Ann. Chem.* **1942**, *550*, 260.
[3] Wittig, G. *Angew. Chem.* **1954**, 66, *10*.
[4] Cast, J.; Stevens, T. S.; Holmes, J. *J. Chem Soc.* **1960**, 3521.
[5] Schollkopf, U.; Fellenberger, K. *Justus Liebigs Ann. Chem.* **1966**, *80*, 698; Wittig, G.; Dösser, H.; Lorenz, I. *ibid.*, **1949**, *562*, 192.
[6] Makizumi, Y.; Notsumoto, S. *Tetrahedron Lett.* **1966**, 6393.
[7] Yamamoto, Y.; Oda, J.; Inouye, Y. *Tetrahedron Lett.* **1979**, *26*, 2411.
[8] Schollkopf, U. *Angew. Chem., Int. Ed. Engl.* **1970**, *9*, 76.
[9] Schollkopf, U.; Fellenberger, K.; Rizk, M. *Justus Liebigs Ann. Chem.* **1970**, *734*, 106.
[10] Baldwin, J. E.; Patrick, J. E. *J. Am. Chem. Soc.* **1971**, *93*, 3556.
[11] Still, W. C.; Mitra, A. *J. Am. Chem. Soc.* **1978**, *100*, 1927.
[12] Nakai, T.; Mikami, K.; Taya, S.; Fujita, Y. *J. Am. Chem. Soc.* **1981**, *103*, 6492.
[13] Nakai, T.; Mikami, K. *Chem. Rev.* **1986**, *86*, 885.
[14] Nakai, T.; Mikami, K.; Sayo, N. *Yuki Kagaku Kyokaishi* **1983**, *41*, 100 [*Chem. Abstr.* **1983**, *98*, 178323].
[15] Marshall, J. A. in *Comprehensive Organic Synthesis*, Vol. 3; Trost, B. M.; Fleming, I. Eds.; Pergamon: New York, 1991; p. 975.
[16] Mikami, K.; Nakai, T. *Synthesis,* **1991**, 594.
[17] Bruckner, R. in *Comprehensive Organic Synthesis*, Vol. 5; Trost, B. M.; Fleming, I. Eds.; Pergamon: New York, 1991; p. 813.
[18] Rautenstrauch, V. *J. Chem Soc., Chem. Commun.* **1970**, 4.
[19] Woodward, R. B.; Hoffmann, R. *The Conservation of Orbital Symmetry;* Academic: New York, 1970.
[20] Fukui, K. *Theory of Orientation and Stereoselection;* Springer-Verlag: Berlin, 1971.
[21] Felkin, H.; Frajerman, C. *Tetrahedron Lett.* **1977**, 3485, and references therein.
[22] Fuchs, B. in *Topics in Stereochemistry*, Vol. 10; Eliel, E. L.; Allinger, N. L., Eds.; Wiley: New York, 1978, p. 1.
[23] Trost, B. M.; Melvin Jr., A. *Sulfur Ylides;* Academic: New York, 1979; Chapter 7.
[24] Mikami, K.; Kimura, Y.; Kishi, N.; Nakai, T. *J. Org. Chem.* **1983**, *48*, 279.
[25] Takahashi, T.; Nemoto, H.; Kanda, Y.; Tsuji, J.; Fukazawa, Y.; Okajima, T.; Fujise, Y. *Tetrahedron,* **1989**, *43*, 5499.
[26] Mikami, K.; Nakai, T. in *Studies in Organic Chemistry (Physical Organic Chemistry 1986)*, Vol. 31; Kobayashi, M., Ed.; Elsevier: Amsterdam, 1987; p. 153.
[27] Mikami, K.; Uchida, T.; Hirano, T.; Wu, Y.-D.; Houk, K. N. *Tetrahedron,* **1994**, *50*, 5917; Wu, Y.-D.; Houk, K. N.; Marshall, J. A. *J. Org. Chem.* **1990**, *55*, 1421.
[28] Mikami, K.; Azuma, K.; Nakai, T. *Chem. Lett.* **1983**, 1379.
[29] Mikami, K.; Azuma, K.; Nakai, T. *Tetrahedron* **1984**, *40*, 2303.
[30] Broka, C. A.; Shen, T. *J. Am. Chem. Soc.* **1989**, *111*, 2981; Broka, C. A.; Hu, L.; Lee, W. J.; Shen, T. *Tetrahedron Lett.,* **1987**, *28*, 4993.
[31] Broka, C. A.; Lin, Y. T. *J. Org. Chem.* **1988**, *53*, 5876.
[32] Wada, M.; Fukui, A.; Nakamura, H.; Takei, H. *Chem. Lett.* **1977**, 557.
[33] Mikami, K.; Kishi, N.; Nakai, T. *Chem. Lett.* **1989**, 1683.
[34] Huche, M.; Cresson, P. *Tetrahedron Lett.* **1975**, 367.
[35] Cazes, B.; Julia, S. *Synth. Commun.* **1977**, *7*, 113.
[36] Cazes, B.; Julia, S. *Synth. Commun.* **1977**, *7*, 273.
[37] Cazes, B.; Julia, S. *Bull. Soc. Chim. Fr.* **1977**, 931.
[38] Marshall, J. A.; Robinson, E. D.; Zapata, A. *J. Org. Chem.* **1989**, *54*, 5854.
[39] Marshall, J. A.; Wang, X.-J. *J. Org. Chem.* **1990**, *55*, 1421.
[40] Ziegler, F. E. *Chem. Rev.* **1988**, *88*, 1423.
[41] Takahashi, T.; Nemoto, H.; Kanda, Y.; Tsuji, J.; Fujise, Y. *J. Org. Chem.* **1986**, *51*, 4316.
[42] Marshall, J. A.; Jenson, T. M.; DeHoff, B. S. *J. Org. Chem.* **1986**, *51*, 4319.

[43] Marshall, J. A.; Jenson, T. M.; DeHoff, B. S. *J. Org. Chem.* **1987**, *52*, 3860.

[44] Marshall, J. A.; Lebreton, J. *J. Org. Chem.* **1988**, *53*, 4108.

[45] Castedo, L.; Granja, J. R.; Mourino, A. *Tetrahedron Lett.* **1985**, *26*, 4959.

[46] Castedo, L.; Granja, J. R.; Mourino, A.; Pumar, M. C. *Synth. Commun.* **1987**, *17*, 251.

[47] Mikami, K.; Kawamoto, K.; Nakai, T. *Tetrahedron Lett.* **1985**, *26*, 5799.

[48] Nakai, T; Shirai, F. Unpublished results. *Cf.* Shirai, F. Masters Dissertation, Tokyo Institute of Technology, 1984.

[49] Sayo, N.; Kimura, Y.; Nakai, T. *Tetrahedron Lett.* **1982**, *23*, 795. *Cf.* Sayo, N., PhD Dissertation, Tokyo Institute of Technology, 1984.

[50] Paquette, L. A.; Sugimura, T. *J. Am. Chem. Soc.* **1986**, *108*, 3841.

[51] Sugimura, T.; Paquette, L. A. *J. Am. Chem. Soc.* **1987**, *109*, 3017.

[52] Fraser-Reid, B.; Dawe, R. D.; Tulshian, D. B. *Can. J. Chem.* **1979**, *57*, 1746.

[53] Tulshian, D. B.; Fraser-Reid, B. *J. Org. Chem.* **1984**, *49*, 518.

[54] Tomooka, K.; Watanabe, M.; Nakai, T. *Tetrahedron Lett.* **1990**, *31*, 7353.

[55] Thomas, A. F.; Dubini, R. *Helv. Chim. Acta* **1974**, *57*, 2084.

[56] Kakinuma, K.; Li, H.-Y. *Tetrahedron Lett.* **1989**, *30*, 4157.

[57] Schulte-Elte, K. H.; Rautenstrauch, V.; Ohloff, G. *Helv. Chim. Acta* **1971**, *54*, 1805.

[58] Nakai, T.; Mikami, K.; Taya, S.; Kimura, Y.; Mimura, T. *Tetrahedron Lett.* **1981**, *22*, 69.

[59] Takahashi, O.; Saka, T.; Mikami, K.; Nakai, T. *Chem. Lett.* **1986**, 1599.

[60] Koreeda, M.; Luengo, J. L. *J. Am. Chem. Soc.* **1985**, *107*, 5572.

[61] Mikami, K.; Takahashi, O.; Kasuga, T.; Nakai, T. *Chem. Lett.* **1985**, 1729.

[62] Takahashi, O.; Maeda, T.; Mikami, K.; Nakai, T. *Chem. Lett.* **1986**, 1355.

[63] Raucher, S.; Gustavson, L. M. *Tetrahedron Lett.* **1986**, *27*, 155.

[64] Kachinsky, J. L.; Salomone, R. G. *J. Org. Chem.* **1986**, *51*, 1393.

[65] Mikami, K.; Takahashi, O.; Fujimoto, K.; Nakai, T. *Synlett* **1991**, 629.

[66] Bartlett, P. A. *Tetrahedron* **1980**, *36*, 2.

[67] Sayo, N.; Shirai, F.; Nakai, T. *Chem. Lett.* **1984**, 255.

[68] Still, W. C.; McDonald III, J. H.; Collum, D. B.; Mitra, A. *Tetrahedron Lett.* **1979**, 593.

[69] Uchikawa, M.; Katsuki, T.; Yamaguchi, M. *Tetrahedron Lett.* **1986**, *27*, 4581.

[70] Mikami, K.; Kishi, N.; Nakai, T. *Chem. Lett.* **1982**, 1643.

[71] Kaye, A. D.; Pattenden, G.; Roberts, R. M. *Tetrahedron Lett.* **1986**, *27*, 2033.

[72] Wittman, M. D.; Kallmerten, J. *J. Org. Chem.* **1988**, *53*, 4631.

[73] Marshall, J. A.; Jenson, T. M. *J. Org. Chem.* **1984**, *49*, 1707.

[74] Mikami, K.; Fujimoto, K.; Nakai, T. *Tetrahedron Lett.* **1983**, *24*, 513.

[75] Mikami, K.; Maeda, T.; Nakai, T. *Tetrahedron Lett.* **1986**, *27*, 4189.

[76] Kishi, N.; Maeda, T.; Mikami, K.; Nakai, T. *Tetrahedron Lett.* **1992**, *48*, 4087.

[77] Nakai, E.; Nakai, T. *Tetrahedron Lett.* **1988**, *29*, 5409.

[78] Marshall, J. A.; Lebreton, J.; DeHoff, B. S.; Jenson, T. M. *J. Org. Chem.* **1987**, *52*, 3883.

[79] Marshall, J. A.; Lebreton, J.; DeHoff, B. S.; Jenson, T. M. *Tetrahedron Lett.* **1987**, *28*, 723.

[80] Sayo, N.; Azuma, K.; Mikami, K.; Nakai, T. *Tetrahedron Lett.* **1984**, *25*, 565.

[81] Tsai, D. J.-S.; Midland, M. M. *J. Org. Chem.* **1984**, *49*, 1842.

[82] Sayo, N.; Kitahara, E.; Nakai, T. *Chem. Lett.* **1984**, 259.

[83] Kuroda, S.; Sakaguchi, S.; Ikegami, S.; Hanamoto, T.; Katsuki, T.; Yamaguchi, M. *Tetrahedron Lett.* **1988**, *29*, 4763.

[84] Balestra, M.; Wittman, M. D.; Kallmerten, J. *Tetrahedron Lett.* **1988**, *29*, 6905.

[85] Mikami, K.; Kawamoto, K.; Nakai, T. *Tetrahedron Lett.* **1986**, *27*, 4899.

[86] Koreeda, M.; Ricca, D. J. *J. Org. Chem.* **1986**, *51*, 4090.

[87] Trost, B. M.; Mao, M. K.-T.; Balkovec, J. M.; Buhlmayer, P. *J. Am. Chem. Soc.* **1986**, *108*, 4965.

[88] Castedo, L.; Mascarenas, J. L.; Mourino, A. *Tetrahedron Lett.* **1987**, *28*, 2099.

[89] Crimmins, M. T.; Gould, L. D. *J. Am. Chem. Soc.* **1987**, *109*, 6199.

[90] Eguchi, S.; Ebihara, K.; Morisaki, M. *Chem. Pharm. Bull.* **1988**, *36*, 4638.

[91] Balestra, M.; Kallmerten, J. *Tetrahedron Lett.* **1988**, *29*, 6901.

[92] Coutts, S. J.; Wittman, M. D.; Kallmerten, J. *Tetrahedron Lett.* **1990**, *31*, 4301.

[93] Coutts, S. J.; Kallmerten, J. *Tetrahedron Lett.* **1990**, *31*, 4305.
[94] Barrish, J. C.; Wovkulich, P. M.; Tang, P. C.; Batcho, A. D.; Uskokovic, M. R. *Tetrahedron Lett.* **1990**, *31*, 2235.
[95] Mikami, K.; Fujimoto, K.; Kasuga, T.; Nakai, T. *Tetrahedron Lett.* **1984**, *25*, 6011.
[96] Mikami, K.; Kasuga, T.; Fujimoto, K.; Nakai, T. *Tetrahedron Lett.* **1986**, *27*, 4185.
[97] Uchikawa, M.; Hanamoto, T.; Katsuki, T.; Yamaguchi, M. *Tetrahedron Lett.* **1986**, *27*, 4577.
[98] Takahashi, O.; Mikami, K.; Nakai, T. *Chem. Lett.* **1987**, 69.
[99] Uemura, M.; Nishimura, H.; Hayashi, Y. *J. Organomet. Chem.* **1989**, *376*, C3.
[100] Uemura, M.; Nishimura, H.; Minami, T.; Hayashi, Y. *J. Am. Chem. Soc.* **1991**, *113*, 5402.
[101] Brocard, J.; Mahmoudi, M.; Pelenski, L.; Maciejewski, L. *Tetrahedron Lett.* **1989**, *30*, 2549.
[102] Marshall, J. A.; Robinson, E. D.; Lebreton, J. *Tetrahedron Lett.* **1988**, *29*, 3547.
[103] Marshall, J. A.; Robinson, E. D. *Tetrahedron Lett.* **1989**, *30*, 1055.
[104] Nakai, E.; Nakai, T. *Tetrahedron Lett.* **1988**, *29*, 4587.
[105] Bruckner, R. *Chem. Ber.* **1989**, *122*, 193.
[106] Bruckner, R. *Chem. Ber.* **1989**, *122*, 703.
[107] Priepke, H.; Bruckner, R. *Chem. Ber.* **1990**, *123*, 153.
[108] Bruckner, R.; Priepke, H. *Angew. Chem., Int. Ed. Engl.* **1988**, *27*, 278.
[109] Priepke, H.; Bruckner, R.; Harms, K. *Chem. Ber.* **1990**, *123*, 555.
[110] Mori, K.; Mori, H. *Tetrahedron* **1985**, *41*, 5487.
[111] Scheuplein, S. W.; Kusche, A.; Bruckner, R.; Harms, K. *Chem. Ber.* **1990**, *123*, 917.
[112] Paquette, L. A.; Wright, J.; Drtina, G. J.; Roberts, R. A. *J. Org. Chem.* **1987**, *52*, 2960.
[113] Wright, J.; Drtina, G. J.; Roberts, R. A.; Paquette, L. A. *J. Am. Chem. Soc.* **1988**, *110*, 5806.
[114] Tomooka, K.; Ishikawa, K.; Nakai, T. *Synlett* **1993**, 527.
[115] Marshall, J. A.; Lebreton, J. *Tetrahedron Lett.* **1987**, *28*, 3323.
[116] Marshall, J. A.; Lebreton, J. *J. Am. Chem. Soc.* **1988**, *110*, 2925.
[117] Marshall, J. A.; Lebreton, J. *J. Org. Chem.* **1988**, *53*, 4108.
[118] Marshall, J. A.; Wang, X.-J. *J. Org. Chem.* **1990**, *55*, 2995.
[119] Marshall, J. A.; Wang, X.-J. *J. Org. Chem.* **1991**, *56*, 4913.
[120] Verner, E. J.; Cohen, T. *J. Am. Chem. Soc.* **1992**, *114*, 375.
[121] Tomooka, K.; Igarashi, T.; Watanabe, M.; Nakai, T. *Tetrahedron Lett.* **1992**, *33*, 5795.
[122] Sayo, N.; Nakai, E.; Nakai, T. *Chem. Lett.* **1985**, 1723.
[123] Oishi, T.; Nakata, T. *Acc. Chem. Res.* **1984**, *17*, 338.
[124] Midland, M. M.; Gabriel, J. *J. Org. Chem.* **1985**, *50*, 1143.
[125] Takano, S.; Sekiguchi, Y.; Ogasawara, K. *Heterocycles* **1989**, *29*, 445.
[126] Takano, S.; Sekiguchi, Y.; Ogasawara, K. *J. Chem Soc., Chem. Commun.* **1987**, 555.
[127] Takano, S.; Shimizaki, Y.; Sekiguchi, Y.; Ogasawara, K. *Chem. Lett.* **1988**, 2041.
[128] Tsai, D. J.-S.; Midland, M. M. *J. Am. Chem. Soc.* **1985**, *107*, 3915.
[129] Nakai, E.; Kitahara, E.; Sayo, N.; Ueno, Y.; Nakai, T. *Chem. Lett.* **1985**, 1725.
[130] Ikegami, S.; Katsuki, T.; Yamaguchi, M. *Tetrahedron Lett.* **1988**, *29*, 5285.
[131] Rossano, L. Y.; Plata, D. J.; Kallmerten, J. *J. Org. Chem.* **1988**, *53*, 5189.
[132] Piatak, D. M.; Wicha, *J. Chem. Rev.* **1978**, *78*, 199.
[133] Redpath, J.; Zeelen, F. *J. Chem. Soc. Rev.* **1983**, *12*, 75.
[134] Fujimoto, Y.; Ohhana, M.; Terasawa, T.; Ikakawa, N. *Tetrahedron Lett.* **1985**, *26*, 3239.
[135] Midland, M. M.; Kwon, Y. C. *Tetrahedron Lett.* **1985**, *26*, 5017.
[136] Midland, M. M.; Kwon, Y. C. *Tetrahedron Lett.* **1985**, *26*, 5013.
[137] Midland, M. M.; Kwon, Y. C. *Tetrahedron Lett.* **1985**, *26*, 5021.
[138] Mikami, K.; Kawamoto, K.; Nakai, T. *Chem. Lett.* **1985**, 1719.
[139] Lee, E.; Liu, Y.-T.; Solomon, P. H.; Nakanishi, K. *J. Am. Chem. Soc.* **1976**, *98*, 1634.
[140] Hayami, H.; Sato, M.; Kanemoto, S.; Morisawa, Y.; Oshima, K.; Nazaki, H. *J. Am. Chem. Soc.* **1983**, *105*, 4491.
[141] Takahashi, T.; Nemoto, H.; Kanda, Y.; Tsuji, J. *Heterocycles* **1987**, *25*, 139.
[142] Marshall, J. A.; Nelsen, D. J. *Tetrahedron Lett.* **1988**, *29*, 741.
[143] Marshall, J. A.; Robinson, E. D.; Adams, R. D. *Tetrahedron Lett.* **1988**, *29*, 4913.
[144] Marshall, J. A.; Robinson, E. D.; Lebreton, J. *J. Org. Chem.* **1990**, *55*, 227.

[145] Doi, T.; Takahashi, T. *J. Org. Chem.* **1991**, *56*, 3465.

[146] Wender, P. A.; McKinney, J. A.; Mukai, C. *J. Am. Chem. Soc.* **1990**, *112*, 5369.

[147] Nakai, T.; Mikami, K. *Kagaku no Ryoiki* **1982**, *36*, 661 [*Chem. Abstr.* **1982**, *96*, 16001].

[148] Ziegler, F. E. in *Comprehensive Organic Synthesis*, Vol. 6; Trost, B. M.; Fleming, I. Eds.; Pergamon: New York, 1991, p. 875.

[149] Mikami, K.; Taya, S.; Nakai, T.; Fujita, Y. *J. Org. Chem.* **1981**, *46*, 544.

[150] Mikami, K.; Kishi, N.; Nakai, T.; Fujita, Y. *Tetrahedron* **1986**, *42*, 2911.

[151] Mikami, K.; Kishi, N.; Nakai, T. *Chem. Lett.* **1981**, 1721.

[152] Saucy, G.; Marbet, R. *Helv. Chim. Acta* **1967**, *50*, 1158, 2291.

[153] Johnson, W. S.; Wertherman, L.; Bartlett, W. R.; Brocksom, T. J.; Li, T.; Faulkner, D. J.; Petersen, M. R. *J. Am. Chem. Soc.* **1970**, *92*, 741.

[154] Ireland, R. E.; Mueller, R. H.; Willard, A. K. *J. Am. Chem. Soc.* **1976**, *98*, 2868.

[155] Paquette, L. A. *Angew. Chem., Int. Ed. Engl.* **1990**, *29*, 609.

[156] Evans, D. A.; Golob, A. M. *J. Am. Chem. Soc.* **1975**, *97*, 4765.

[157] Evans, D. A.; Nelson, J. V. *J. Am. Chem. Soc.* **1980** *102*, 774.

[158] Mikami, K.; Nakai, T. *Chem. Lett.* **1982**, 1349.

[159] Mikami, K.; Kishi, N.; Nakai, T. *Chem. Lett.* **1982**, 1643.

[160] Mikami, K.; Kishi, N.; Nakai, T. *Tetrahedron Lett.* **1983**, *24*, 795.

[161] Kishi, N.; Mikami, K.; Nakai, T. *Tetrahedron* **1991**, *47*, 8111.

[162] Tomooka, K.; Wei, S.-Y.; Nakai, T. *Chem. Lett.* **1991**, 43.

[163] Wei, S.-Y.; Tomooka, K.; Nakai, T. *J. Org. Chem.* **1991**, *56*, 5973.

[164] Mikami, K.; Takahashi, O.; Tabei, T.; Nakai, T. *Tetrahedron Lett.* **1986**, *27*, 4511.

[165] Baldwin, J. E.; DeBernardis, J.; Patrick, J. E. *Tetrahedron Lett.* **1970**, 353.

[166] Morgans, Jr., D. J. *Tetrahedron Lett.* **1981**, *22*, 3721.

[167] Boger, D. L.; Mathvink, R. J. *J. Am. Chem. Soc.* **1990**, *112*, 4008.

[168] Mori, K.; Kuwahara, S. *Tetrahedron* **1982**, *38*, 521.

[169] Kruse, B.; Bruckner, R. *Chem. Ber.* **1989**, *122*, 2023.

[170] Oppolzer, W.; Stevenson, T. *Tetrahedron Lett.* **1986**, *27*, 1139.

[171] Kano, S.; Yokomatsu, T.; Nemoto, H.; Shibuya, S. *J. Am. Chem. Soc.* **1986**, *108*, 6746.

[172] Garbers, C. F.; Scott, F. *Tetrahedron Lett.* **1976**, 507.

[173] Keegan, D. S.; Midland, M. M.; Werley, R. T.; McLoughlin, J. I. *J. Org. Chem.* **1991**, *56*, 1185.

[174] Hayakawa, K.; Hayashida, A.; Kanematsu, K. *J. Chem. Soc., Chem. Commun.* **1988**, 1108.

[175] Ishikawa, A.; Uchiyama, H.; Katsuki, T.; Yamaguchi, M. *Tetrahedron Lett.* **1990**, *31*, 2415.

[176] Luengo, J. I.; Koreeda, M. *J. Org. Chem.* **1989**, *54*, 5415.

[177] Kuroda, S.; Katsuki, T.; Yamaguchi, M. *Tetrahedron Lett.* **1987**, *28*, 803.

[178] Mikami, K.; Kasuga, T.; Nakai, T. Unpublished results. *Cf.* Kasuga, T. Masters Dissertation, Tokyo Institute of Technology, 1986.

[179] Mikami, K.; Takahashi, O.; Nakai, T. Unpublished results. *Cf.* Takahashi, O. PhD Dissertation, Tokyo Institute of Technology, 1988.

# CHAPTER 3

# REDUCTIONS WITH SAMARIUM(II) IODIDE

Gary A. Molander

*Department of Chemistry and Biochemistry, University of Colorado at Boulder, Boulder, Colorado*

## CONTENTS

| | PAGE |
|---|---|
| ACKNOWLEDGMENTS . | 213 |
| INTRODUCTION . | 213 |
| MECHANISM AND STEREOCHEMISTRY . | 215 |
|     Reduction of Organic Halides, Sulfonates, and Sulfones | 215 |
|     Reductive Elimination/Fragmentation Reactions . | 219 |
|     Reduction of Aldehydes and Ketones . | 222 |
|     Reduction of Carboxylic Acids and Their Derivatives . | 224 |
|     Reductive Cleavage of $\alpha$-Heterosubstituted Carbonyl Compounds and Related Substrates . | 225 |
|     Reductive Cleavage of Cyclopropyl Ketones . | 228 |
|     Deoxygenation Reactions . | 229 |
|     Reduction of Nitrogen-Based Functional Groups . | 230 |
|     Reduction of Miscellaneous Functional Groups | 233 |
| SCOPE AND LIMITATIONS | 236 |
|     Reduction of Organic Halides, Sulfonates, and Sulfones | 236 |
|     Reductive Elimination/Fragmentation Reactions . | 239 |
|     Reduction of Aldehydes and Ketones | 241 |
|     Reduction of Carboxylic Acids and Their Derivatives . | 241 |
|     Reduction of Conjugated Carbonyl Substrates | 243 |
|     Reductive Cleavage of $\alpha$-Heterosubstituted Carbonyl Compounds and Related Substrates | 244 |
|     Reductive Cleavage of Cyclopropyl Ketones . | 248 |
|     Deoxygenation Reactions . | 249 |
|     Reduction of Nitrogen-Based Functional Groups . | 250 |
|     Reduction of Miscellaneous Functional Groups | 252 |
| COMPARISON WITH OTHER METHODS . | 254 |
|     Reduction of Organic Halides, Sulfonates, and Sulfones | 254 |
|     Reductive Elimination/Fragmentation Reactions . | 255 |
|     Reduction of Aldehydes and Ketones | 255 |
|     Reduction of Carboxylic Acids and Their Derivatives . | 255 |
|     Reduction of Conjugated Carbonyl Substrates | 255 |
|     Reductive Cleavage of $\alpha$-Heterosubstituted Carbonyl Compounds and Related Substrates | 256 |
|     Reductive Cleavage of Cyclopropyl Ketones . | 256 |

*Organic Reactions, Vol. 46,* Edited by Leo A. Paquette et al.
ISBN 0-471-08619-3   © 1994 Organic Reactions, Inc. Published by John Wiley & Sons, Inc.

Deoxygenation Reactions . . . . . . . . . . . . . . . 257
Reduction of Nitrogen-Based Functional Groups . . . . . . . . . 257
Reduction of Miscellaneous Functional Groups . . . . . . . . . 257
EXPERIMENTAL CONDITIONS . . . . . . . . . . . . . . 258
EXPERIMENTAL PROCEDURES . . . . . . . . . . . . . . . 261
Samarium(II) Iodide (Preparation of the Reducing Agent from Samarium Metal
Using 1,2-Diiodoethane as the Oxidant) . . . . . . . . . . 261
Samarium(II) Iodide (Preparation of the Reducing Agent from Samarium Metal
Using Diiodomethane as the Oxidant) . . . . . . . . . . . 261
Samarium(II) Iodide (Preparation of the Reducing Agent from Samarium Metal
Using Iodine as the Oxidant) . . . . . . . . . . . . . 261
Samarium(II) Iodide [Preparation of a Samarium Iodide Equivalent in Acetonitrile
Solvent] . . . . . . . . . . . . . . . . . . . 262
1-Phenyldec-1-yne (Generation of Alkylidenecarbenes from 1,1-Dibromoalk-1-enes) . 262
3-Methoxy-14,21-cyclo-19-norpregna-1,3,5(10)-trien-17-ol Acetate (Reductive
Desulfonylation Reactions) . . . . . . . . . . . . . 262
[6R-(6α[2S*(R*),3S*],8β(2S*,3S*,5S*),9β,10β)]-α-Ethyl-10-hydroxy-3,9-dimethyl-8-
[tetrahydro-3-(hydroxymethyl)-2-methoxy-5-methyl-2H-pyran-2-yl]-1,7-
dioxaspiro[5.5]undecane-2-ethanol [Deprotection of (2,2,2-Trichloroethoxy)methoxy
Ethers by Reductive β-Elimination] . . . . . . . . . . . 263
1,4-Diphenylbut-1-ene (Reductive Elimination of β-Hydroxy Imidazoyl Sulfones) . . 264
(R)-Benzoin (Enantioselective Reduction of Ketones) . . . . . . . . 264
(2S,4S,5R,1'S)-3-Oxazolidinecarboxylic Acid, 2-(3-Methoxy-1-methyl-3-oxopropyl)-
4-methyl-5-phenyl Methyl Ester and (2S,4S,5R,1'R)-3-Oxazolidinecarboxylic Acid, 2-
(3-Methoxy-1-methyl-3-oxopropyl)-4-methyl-5-phenyl Methyl Ester (Reduction of
α,β-Unsaturated Carbonyl Substrates) . . . . . . . . . . . 265
5-Iodo-1-phenyl-1-pentanone (Reduction of α-Acyloxy Ketone Substrates) . . . 265
1,8-Dichloro-11,11-dimethoxy-3-exo-hydroxytetracyclo[6.2.1.0^{2,7}.0^{4,10}]undec-5-en-9-one
(Reductive Cleavage of α-Halo Ketones) . . . . . . . . . . 266
(R)-Diisopropyl Malate (Reduction of α-Hydroxy Ester Substrates) . . . . . 266
(R)-2-Deoxy-3,4-O-(phenylmethylene)-D-erythro-pentanoic Acid, δ-Lactone
(Reduction of α-Hydroxy Aldonolactones) . . . . . . . . . . 267
(1'R*,3R*,4S*,4aR*,7S*,8aS*)-3,4,4a,7,8,8a-Hexahydro-7-hydroxy-4-methyl-3-(1'methyl-
prop-2'-enyl)-2(1H)-naphthalenone (Reduction of α-Alkoxy Ketone Substrates) . . 267
trans-Hexahydro-4a-hydroxy-8a-methyl-1,6(2H,5H)-naphthalenedione (Reduction of
α,β-Epoxy Ketones) . . . . . . . . . . . . . . . . 268
Diethyl [(2E)-4,8-Dimethyl-4-hydroxy-2,7-nonadien-1-yl]phosphonate (Reduction of
Vinyl Oxiranes) . . . . . . . . . . . . . . . . . 268
Cyclohexadec-5-enecarbonitrile (Reduction of α-Heterosubstituted Nitriles) . . . 269
N-Hydroxy 2-(tert-Butyldiphenylsiloxy)ethanamine (Reduction of Nitro Compounds to
Hydroxylamines) . . . . . . . . . . . . . . . . 270
[S-(R*,S*)]-[1-Methyl-2-oxo-2-[2-(1-pyrrolidinylcarbonyl)-1-pyrrolidinyl]ethyl]carbamic
Acid, Phenylmethyl Ester (Reduction of 2-Hydroxyimino Amides) . . . . . 270
(S)-(−)-α-Methylbenzylamine (Reductive Cleavage of the Nitrogen–Nitrogen Bond of
N-Aroylhydrazines) . . . . . . . . . . . . . . . . 271
A-Nor-9,10-secocholesta-5(10),8-dien-6-yn-11α-ol (Palladium-Promoted Reductive
Cleavage of Propargyl Carboxylates) . . . . . . . . . . . 271
TABULAR SURVEY . . . . . . . . . . . . . . . . . . 272
Table I. Reduction of Organic Halides, Sulfonates, and Sulfones . . . . . 273
Table II. Reductive Elimination/Fragmentation Reactions . . . . . . . 283
Table III. Reduction of Aldehydes and Ketones . . . . . . . . . 292
Table IV. Reduction of Carboxylic Acids and Their Derivatives . . . . . . 296
Table V. Reduction of Conjugated Carbonyl Substrates . . . . . . . 301

Table VI. Reductive Cleavage of $\alpha$-Heterosubstituted Carbonyl Compounds and
Related Substrates . . . . . . . . . . . . . . 307
Table VII. Reductive Cleavage of Cyclopropyl Ketones . . . . . . . 336
Table VIII. Deoxygenation Reactions . . . . . . . . . . . 338
Table IX. Reduction of Nitrogen-Based Functional Groups. . . . . . . 344
Table X. Reduction of Miscellaneous Functional Groups . . . . . . . 355
REFERENCES . . . . . . . . . . . . . . . . 365

## ACKNOWLEDGMENTS

I wish to express my warmest thanks to Professeur Janine Cossy of the École Supérieure de Physique et de Chimie Industrielles de Paris for graciously providing me with a visiting professorship that made the completion of this chapter possible. Additionally, I would like to thank the National Institutes of Health, who have steadfastly supported our own studies in this area of chemistry since their inception. Finally, I gratefully acknowledge all of my research colleagues who have contributed so much to the success of our program, and who have made the science we have performed a great joy to carry out.

## INTRODUCTION

Reducing agents (along with complementary oxidants) constitute the most fundamental class of reagents available to synthetic chemists for the conversion of organic substrates to desirable organic products. As a consequence of their central importance, the search for novel reducing agents remains the focus of intense exploration. Especially valued are those reagents that not only transform a broad range of diverse functional groups, but do so with a high degree of selectivity. Since the early 1980s, samarium(II) iodide ($SmI_2$) has been increasingly recognized as a reducing agent capable of meeting the intensifying demands of synthetic organic chemistry. Although the compound itself has been known for many years,[1,2] it was not until Kagan and co-workers developed a convenient synthesis of $SmI_2$ and outlined its general reactivity with common organic functional groups[3] that $SmI_2$ became of general interest and importance to synthetic organic chemists.

Samarium(II) iodide is a powerful one-electron reducing agent that can be prepared in moderate concentrations (0.1 M) in tetrahydrofuran (THF) by one of several different reactions from samarium metal.[3–8]

$$Sm \quad + \quad ICH_2CH_2I \quad \xrightarrow[\text{rt, 1 h}]{\text{THF}} \quad SmI_2 \quad + \quad CH_2=CH_2$$

$$Sm \quad + \quad ICH_2I \quad \xrightarrow[\text{rt, 1 h}]{\text{THF}} \quad SmI_2 \quad + \quad 0.5 \; CH_2=CH_2$$

$$Sm \;+\; I_2 \;\xrightarrow[\text{heat, 24 h}]{\text{THF}}\; SmI_2$$

$$Sm \;\xrightarrow[\text{MeCN}]{2\ Me_3SiCl,\ 2\ NaI}\; SmI_2 \;+\; Me_3SiSiMe_3 \;+\; 2\ NaCl$$

Deep blue solutions of $SmI_2$ are generated in virtually quantitative yields by these procedures, but the preparations using diiodomethane or diiodoethane as the oxidants would appear to be the easiest and most reliable methods. The reducing agent can be stored as a solution in THF for reasonably long periods of time, particularly when it is stabilized by a small amount of samarium metal. Alternatively, the solvent may be removed, providing $SmI_2(THF)_n$ powder. For synthetic purposes, $SmI_2$ is typically generated and utilized in situ, although THF solutions of $SmI_2$ are commercially available.

The reduction potential of $Sm^{+2}/Sm^{+3}$ as measured in water is $-1.55V$.[9] However, the reduction potential as well as the ability of $SmI_2$ to promote the reduction of diverse organic substrates varies widely according to the solvent and the presence of various additives.

Kagan's pioneering studies on the reaction of $SmI_2$ with organic substrates not only provided a general outline for its reactivity,[3] but also inspired an extraordinary number of subsequent studies with important ramifications for selective organic synthesis. As a consequence of these extensive studies, $SmI_2$ has emerged as one of the more useful reducing agents in synthetic organic chemistry. The complementary reactivity of $SmI_2$ as compared to the vast inventory of other available reducing agents constitutes one reason for its appeal. Another attraction of $SmI_2$ is its ready accessibility, either from commercial sources or by in situ preparation via one of the convenient methods outlined above. The high chemoselectivity exhibited by $SmI_2$ and the ability to change its reactivity (selectivity) rather dramatically based upon solvent effects further enhance its attractiveness. Finally, $SmI_2$ is easily handled by standard techniques for the manipulation of air-sensitive materials.

As testimony to its increasing importance, several review articles have appeared that focus on the utility of $SmI_2$ both as a reducing agent and as a reductive coupling agent in selective organic synthesis.[10-22] This review outlines some of the practical aspects of reactions employing $SmI_2$, but is strictly limited to functional group reductions. Thus those reactions promoted by $SmI_2$ that result in the formation of new carbon–carbon bonds are not included. The chapter is organized according to the type of organic functional groups involved, and topics covered include the reduction of organic halides, sulfonates,[3,23-30] and sulfones;[31-33] reductive elimination/fragmentation processes;[26-28,31,33-39] reduction of aldehydes and ketones;[3,6,25,40-45] reduction of carboxylic acids and their derivatives;[42,46-48] reduction of conjugated carbonyl substrates;[3,47,49-52] reductive cleavage of $\alpha$-heterosubstituted carbonyl substrates;[7,29,49,53-72] reduction of cyclopropyl ketones;[61,73,74] various deoxygenation reactions;[3,41,49,75-78] reduction of nitrogen-based functional groups;[42,46,49,79-84] and finally an outline of the reduction of miscellaneous functional groups[25,85-91] that cannot be appropriately assigned to the general classifications outlined above. In covering these topics, the literature has been surveyed through 1992.

## MECHANISM AND STEREOCHEMISTRY

Because $SmI_2$ is a one-electron reducing agent, the transformations carried out with it are single-electron transfer processes with mechanisms similar to those established with other one-electron reducing systems (including other low-valent metals and metal complexes, dissolving metal reductions, and electrochemical methods). For example, reduction of organic halides, reductive elimination processes, reductions of aldehydes and ketones, reductions of conjugated carbonyl systems, and reductions of $\alpha$-heterosubstituted carbonyl substrates undoubtedly follow along the lines of traditional mechanistic interpretations for such substrates, although there are some caveats that will be discussed in greater detail.

There are several attractive features of $SmI_2$ that render its use advantageous over many other similar reducing agents. The first is that samarium ions are excellent Lewis acids. Thus in reactions with polarized functional groups, complexation with either Sm(II) or Sm(III) can greatly facilitate electron transfer by lowering the energy of the lowest unoccupied molecular orbital of the substrate. Additionally, samarium(III) ions produced upon electron transfer can enhance the leaving group ability of those functional groups with which it can complex, again facilitating intermediate steps in mechanistic schemes involving displacement of a leaving group. Finally, although they are considered rather powerful reducing agents, samarium(II) species are reasonably stable in alcohols, in water, and, apparently, even in acidic and basic aqueous solutions. Consequently, for reaction processes that generate reactive intermediates requiring immediate quenching upon generation, the use of $SmI_2$ in the presence of protic solvent additives provides a means to produce these species and protonate them before they can react via unproductive pathways.

## Reduction of Organic Halides, Sulfonates, and Sulfones

Early studies on the reduction of alkyl halides to the corresponding alkanes by $SmI_2$ suggested that organosamariums were not involved as intermediates, but that radical processes were uniquely responsible for the observed products.[3,26] However, these results appear to have been a consequence of the solvent used. Organosamariums now appear to be well-established intermediates in the conversion of many alkyl halides to alkanes, particularly when hexamethylphosphoric triamide (HMPA) is used as a cosolvent.

In these initial studies, most alkyl halide reductions with $SmI_2$ were performed in THF solvent heated at reflux.[3] Quenching the reaction mixtures with

D$_2$O afforded nondeuterated alkanes, leading to the conclusion that organosamariums were not involved as intermediates and that the alkane products arose as a result of hydrogen atom abstraction from THF by alkyl radicals.[3,26]

Subsequent studies with various alkyl halide substrates performed in THF/HMPA provided a growing body of evidence that organosamariums are frequently generated as intermediates from alkyl halide precursors.[24,41,73,92] The addition of HMPA to SmI$_2$ affords a much more powerful reducing system, allowing reactions to be carried out under considerably milder conditions.[24,41] Consequently, when SmI$_2$ reductions of 2-bromoadamantane are carried out at room temperature in the presence of D$_2$O, the adamantane recovered is deuterated to the extent of 80%.[24] This is in stark contrast to the results obtained when the reactions are carried out in boiling THF. Prompted by the accumulating results and apparent discrepancies, a mechanistic study was undertaken to provide further clarification of the mechanism of alkyl halide reductions.[93-95] These investigations provided convincing evidence that organosamariums are indeed intermediates in such reductions, and suggested that the lack of deuterium incorporation in the initial study was a result of decomposition of the organometallic formed at the relatively high temperatures required for reductions carried out in the absence of HMPA.[92]

The mechanism for SmI$_2$ reductions of primary and secondary alkyl halides in THF/HMPA is now reasonably clear. Dissociative electron transfer[96] from SmI$_2$ to the alkyl halide initiates the process, generating a primary or secondary alkyl radical. Rapid reduction of this radical ($k = 10^6$ M$^{-1}$ s$^{-1}$)[94] provides an organosamarium intermediate that can be protonated, affording the alkane.

$$RX \xrightarrow[-X^-]{SmI_2} \left[ R \cdot \right] \xrightarrow{SmI_2} RSmI_2 \xrightarrow{H^+} RH$$

With the exception of substrates possessing special structural features, there is a loss of stereochemical integrity during reductions of chiral organic halides. Along these lines, the reduction of cyclopropyl halides by SmI$_2$ not only provides interesting insight into reductions of this special class of alkyl halides, but also into the properties of SmI$_2$ and organosamarium species as well.[92] Thus treatment of enantiomerically pure 1-bromo-1-methyl-2,2-diphenylcyclopropane with SmI$_2$ in THF/HMPA followed by a MeOD quench provides a 58% yield of racemic 1-methyl-2,2-diphenylcyclopropane in which no more than 15% deuterium has been incorporated. An independent synthesis of the presumed organosamarium intermediate reveals that this species reacts slowly with THF, and slowly racemizes as well.

(58%) <15% D          (22%) 100% D

On the other hand, enantiomerically pure 1-fluoro-1-iodo-2,2-diphenylcyclopropane reacts with $SmI_2$ in THF/HMPA to provide 55% of 1-fluoro-2,2-diphenylcyclopropane with 95% retention of configuration.[92] Fluoro-substituted cyclopropyl radicals have greater configurational stability than simple alkyl-substituted cyclopropyl radicals. The configurational lifetime is thus long enough that reaction with $SmI_2$ can occur prior to inversion, providing racemic product. Consequently, stereochemical fidelity is maintained in conversion of the iodide to the organosamarium intermediate. This intermediate organosamarium itself must exhibit high configurational stability as evidenced by the high degree of stereospecificity of the overall process. A minor byproduct of the reaction (18%) is 1,1-diphenylallene, presumably formed as a result of carbene formation from the intermediate organosamarium species by an $\alpha$-elimination process.

Finally, nonracemic 1-bromo-1-methoxymethyl-2,2-diphenylcyclopropane affords a 50% yield of racemic reduced product upon reaction with $SmI_2$ followed by protonation of the reaction mixture with methanol.[92] In addition, a 16% yield of 1-methylene-2,2-diphenylcyclopropane is generated as a result of $\beta$ elimination of the intermediate organosamarium species.

The mechanism of reduction of tertiary alkyl halides is a matter of some contention. Simple reduction of such substrates with samarium(II) iodide in THF/HMPA followed by a $D_2O$ quench provides no deuterium incorporation in the alkane product.[24,41,93] The implication is that the tertiary radicals formed by reduction of the tertiary halides are not reduced rapidly enough by $SmI_2$ to prevent radical–radical and/or radical–solvent reactions. However, studies have surfaced in which tertiary organosamariums may be intermediates formed from the corresponding halide.[27,73,92] Consequently, the issue of whether tertiary organosamariums can or cannot be formed by reduction of tertiary halides at this point remains unresolved.

Samarium(II) iodide can be utilized to reduce aryl halides and alkenyl halides in THF/HMPA, but it does so without the intermediacy of organosamariums. Thus both aryl radicals and alkenyl radicals undergo rapid hydrogen atom abstraction (from the solvent) or other relatively facile processes before they can be reduced to the corresponding organosamariums.[24,93-95,97,98]

$$ArX \xrightarrow[\substack{THF, HMPA \\ -X^-}]{SmI_2} [Ar \bullet] \xrightarrow{S-H} ArH$$

One apparent exception to this general phenomenon may be seen in the reduction of 1,1-dibromoalkenes, wherein the intermediacy of an alkenylsamarium species has been postulated. In this case, it is likely that the second electron transfer is facilitated because the anion thus formed is electronically stabilized by the remaining bromide.

$$\underset{R}{\overset{R}{\diagdown}}C=C\underset{Br}{\overset{Br}{\diagup}} \xrightarrow[-SmI_2Br]{SmI_2} \left[\underset{R}{\overset{R}{\diagdown}}C=C\underset{Br}{\overset{\bullet}{\diagup}}\right] \xrightarrow{SmI_2} \left[\underset{R}{\overset{R}{\diagdown}}C=C\underset{Br}{\overset{SmI_2}{\diagup}}\right] \xrightarrow{-SmI_2Br}$$

$$\left[\underset{R}{\overset{R}{\diagdown}}C=C:\right] \longrightarrow RC\equiv CR$$

Allylic and benzylic halides are rapidly reduced by $SmI_2$, but the major products are homocoupled dimers. Although it was originally suggested that these products arise from coupling of the respective radicals,[3,26] more recent evidence suggests that the bibenzyl and 1,5-hexadiene products are generated by reaction of intermediate organosamariums with the benzylic or allylic halide precursors, respectively.[94]

Although alkyl tosylates are reduced by $SmI_2$ to the corresponding alkanes,[3] experimental evidence suggests that the tosylates are first converted to iodoalkanes, and that the iodides are reduced to the hydrocarbons.[26] Thus addition of sodium iodide to $SmI_2$-promoted reductions of alkyl tosylates enhances the rate of reduction, and, even in the absence of sodium iodide, alkyl iodides are detected among the products of reaction between $SmI_2$ and alkyl tosylates.

The reduction of sulfones undoubtedly follows along the same lines as those of alkyl halides.[37] Single-electron transfer from $SmI_2$ to the sulfone provides a radical anion, which dissociates to the alkyl radical and the sulfinate anion.[99] Reduction of the alkyl radical provides an organosamarium intermediate, which is protonated to provide the alkane. Alkenyl sulfones presumably generate alkenyl radicals, which are quenched by hydrogen abstraction from the solvent before they can be reduced further to the organosamarium species.

$$RSO_2Ar \xrightarrow{SmI_2} \left[RSO_2Ar\right]^{-\bullet} \xrightarrow{-ArSO_2^-} [R \bullet] \xrightarrow{SmI_2} [RSmI_2] \xrightarrow{H^+} RH$$

## Reductive Elimination/Fragmentation Reactions

A variety of appropriately functionalized 1,2-disubstituted alkanes undergo elimination reactions by initial formation of the organosamarium followed by rapid $\beta$ elimination.

X = halide, SO$_2$Ar

Y = OH, OR, O$_2$CR, SO$_2$Ar

This reductive elimination process has been utilized as a mechanistic probe for intermediate organosamariums,[18,73,92] as a means to deprotect protected alcohols and amines,[35,36] and also as a synthesis of alkenes.[26-28,31,33-37] As expected, there appears to be little difference in the rate of reaction or yields obtained in elimination of diastereomeric cyclic substrates wherein the two substituents to be eliminated are either *cis* or *trans* to one another.[33] Acyclic substrates generally provide diastereomeric mixtures of olefinic products. However, in the reductive fragmentation of cyclic $\beta$-halogeno tetrahydrofurans, high levels of selectivity for the $E$ isomer have been achieved regardless of the stereochemistry of the starting material.[34] This is in contrast to the same process carried out with sodium metal, which provides different mixtures of $E$ and $Z$ olefinic isomers depending on the stereochemistry of the starting material.

Interestingly, the analogous cleavage of cyclic $\beta$-halo tetrahydropyrans with SmI$_2$ is less useful than that of its sodium-promoted counterpart. The SmI$_2$-promoted reactions provide mixtures of $E$ and $Z$ olefin products that appear to arise from a common intermediate from diastereomeric halide precursors.[34]

In general these 1,2-reductive elimination processes are reasonably straight-forward, proceeding without interference from other mechanistic pathways. However, one particular substitution pattern has been discovered in which reactions of the intermediate radical species intervene in the desired process, providing unexpected products.[37] Thus in the attempted reductive elimination of β-hydroxy imidazoyl sulfones the initial radical formed upon single electron transfer from $SmI_2$ to the sulfone forms a homoallylic radical that undergoes 3-*exo* ring closure, forming a benzylically stabilized cyclopropylcarbinyl radical. This radical can fragment, forming an isomeric homoallylic radical that is stabilized by the adjacent hydroxy group. Hydrogen atom abstraction from the solvent affords the observed product of the reaction.

Diverse fragmentation reactions resulting in the cleavage of carbon–carbon bonds with concomitant formation of enolates can be promoted by $SmI_2$. As in the β-elimination processes, the fragmentation reactions appear to transpire via intermediate organosamariums, and some of these would appear to involve the generation of tertiary organosamariums.[27,38,39]

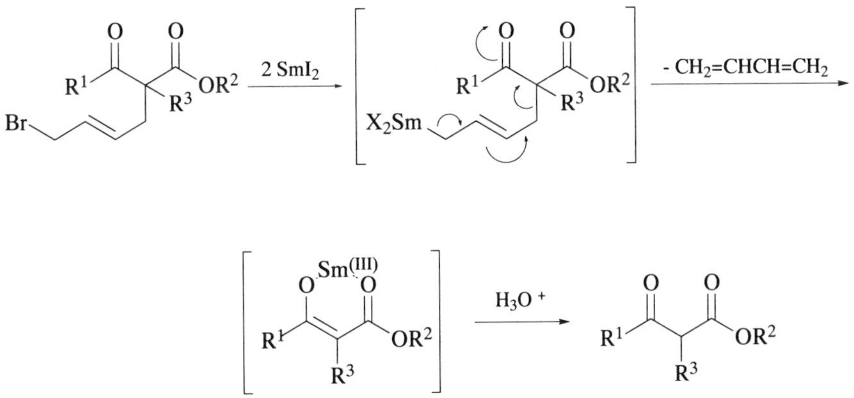

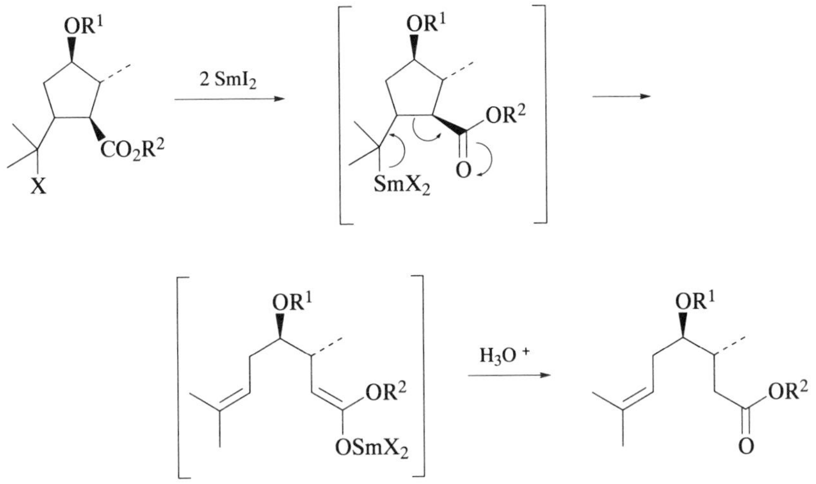

Reductive elimination reactions involving cyclopropylogous $\alpha$-bromoketones are particularly interesting.[28] In these systems, near-complete stereospecificity is observed in the fragmentations. Ambiguities in the point of electron transfer from $SmI_2$ to the substrates make mechanistic interpretations extremely difficult. However, the fact that radical reducing agents (such as $Bu_3SnH$) render the process nonstereospecific rule out a radical fragmentation and suggest a fully concerted reaction.

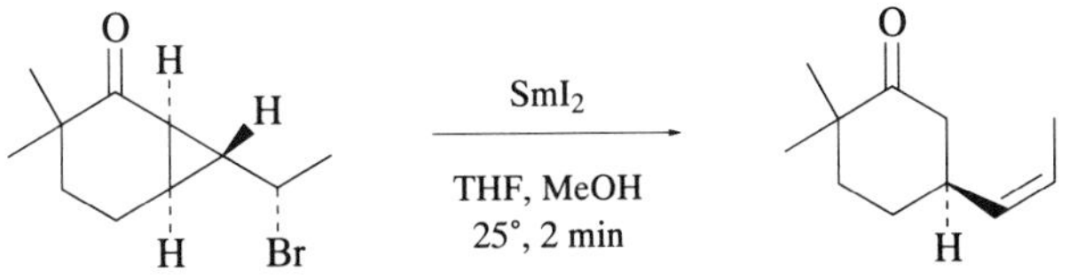

## Reduction of Aldehydes and Ketones

The scope of carbonyl reductions with $SmI_2$ is not nearly so well delineated as that of Bouveault–Blanc or dissolving metal reductions of aldehydes and ketones. However, it seems likely that the same general mechanistic and stereochemical considerations are in play in all of these various protocols. Nevertheless, some general cautions may be in order concerning the mechanisms outlined below. First, the dimer model of alkali metal ketyls has been arbitrarily extended to that of samarium ketyls, even though there is no direct experimental evidence for (or against) such species. Second, the timing of electron transfer and proton transfer events in the reduction of carbonyl substrates by $SmI_2$ has not been investigated, and thus the mechanisms proposed are simply logical possibilities derived from other, more firmly established, processes.

Extensive studies on carbonyl reduction with a variety of different electron transfer agents have led to a reasonably clear picture of the mechanism of these transformations.[100–102] Single-electron transfer to the aldehyde or ketone generates a ketyl radical anion that can form dimeric or polymeric ion pairs (Eq. 1).

(Eq. 1)

The nature and importance of the various species present in solution depend on the metal, the structure of the substrate, the solvent, the concentration, and the temperature. In the absence of good proton donors, collapse to a pinacol (Eq. 1) or disproportionation to a molecule of enolate and a molecule of alkoxide are the major pathways open to these ion pairs (Eq. 2). However, in the presence of a pro-

(Eq. 2)

ton source (as most $SmI_2$-promoted reduction reactions of aldehydes and ketones are performed), two additional pathways become available. The first is a rapid disproportionation of the ketyl dimers to a molecule of alcohol and a molecule of ketone (Eq. 3). A second pathway invokes reversible protonation of the ketyl

$$
\left[ \begin{array}{c} \text{Sm(III)} \\ \diagdown \text{O} \diagdown \diagdown \text{O} \diagup \\ \text{Sm(III)} \end{array} \right] \xrightarrow{\text{H}^+} \left[ \begin{array}{c} \diagdown \text{—OSm(III)} \\ \text{H} \end{array} \right] + \diagup \text{=O} + \text{Sm(III)}
$$

(Eq. 3)

(monomer or dimer) to provide a carbinol radical. This undergoes a second electron transfer, generating a hydroxyalkyl carbanion that is protonated to afford the observed alcohol product (Eq. 4). The extent to which these diverse pathways

$$
\left[ \overset{\cdot}{\diagdown}\text{—OSm(III)} \right] \xrightarrow{\text{H}^+} \left[ \overset{\cdot}{\diagdown}\text{—OH} \right] \xrightarrow{\text{SmI}_2} \left[ \overset{-}{\diagdown}\text{—OH} \right] \xrightarrow{\text{H}^+} \diagdown\text{—OH} \atop \text{H}
$$

(Eq. 4)

are followed is postulated to be a function of the acidity of the proton donor, the nature of the metal used for the reduction, and to some extent the structure of the substrate ketone or aldehyde. Because nothing is known of the structure or aggregation of samarium(III) ketyls, and the relative rates of the various processes outlined above are also unknown, any attempt to define more precisely the pathway traversed in SmI$_2$-mediated reductions would be highly speculative.

Very few details of the stereochemical issues involved in SmI$_2$-promoted carbonyl reductions have been elaborated. Those described, however, are consistent with the notion that the stereochemistry is determined by the relative conformational stabilities of ketyl intermediates. Cyclohexanones typically lead to equatorial alcohols.[41,45] For example, 4-*tert*-butylcyclohexanone can be reduced under optimized conditions to a 93:7 mixture of *trans*- and *cis*-4-*tert*-butylcyclohexanols. Only very modest "anti-Cram" stereochemical control can be achieved in the reduction of chiral acyclic ketones.[40,41] Calculations have provided theoretical models and rationalizations for these experimental observations.[103] Finally, limited success has been achieved in attempts to induce asymmetry in the reduction of achiral ketones by using chiral protonating agents.[43]

Conjugated aldehydes and ketones provide mixtures of 1,2- and 1,4-reduction products. However, $\alpha,\beta$-unsaturated esters and amides undergo selective conjugate addition with SmI$_2$. Mechanisms for the reduction of conjugated carbonyl systems would appear to be reasonably straightforward.[3,50,51,104,105] Many of the same factors involved in the mechanism of carbonyl reduction as outlined above apply to the $\alpha,\beta$-unsaturated systems as well.

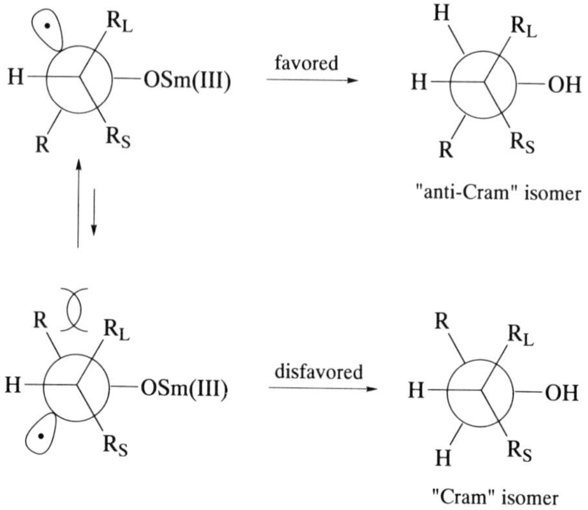

"anti-Cram" isomer

"Cram" isomer

## Reduction of Carboxylic Acids and Their Derivatives

Carboxylic acids and their derivatives are not readily reduced by single-electron transfer reductants,[106] and thus it is somewhat surprising that conditions have been developed that permit the reduction of these functional groups with $SmI_2$.[42,46,47] Under either acidic or basic conditions, the reduction of a variety of substituted aromatic acyl derivatives is relatively rapid.

Aromatic sodium carboxylates, carboxylic acids, carboxylic acid esters, and carboxylic acid amides react with $SmI_2$ under aqueous basic conditions to provide the corresponding benzylic alcohols.[42,46,47] The precise nature of the reducing agents formed under these conditions and definitive mechanisms for the reactions are somewhat speculative because no detailed studies have been performed. For example, it is conceivable that $SmI_2$ reacts with the bases in these reactions, producing much more powerful reducing agents [$SmX_2$, X = OH, OR, $NH_2$, or perhaps even "ate" complexes of samarium(II)]. Although there is some doubt about the nature of the reducing agent, the timing of the various electron transfer and protonation steps are likely to be analogous to those of the ketone and aldehyde reductions outlined previously.

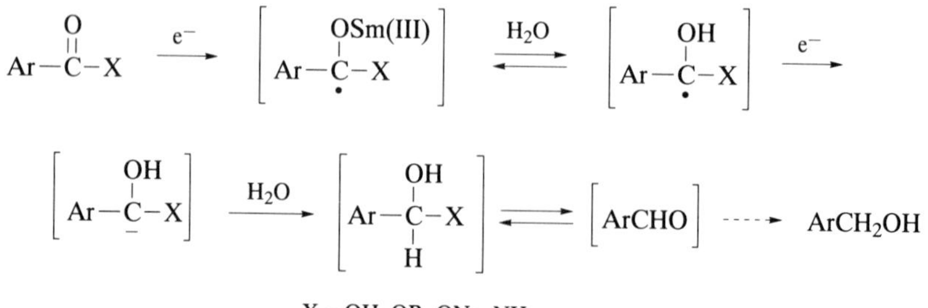

X = OH, OR, ONa, $NH_2$

The presence of Lewis acidic samarium(III) ions could facilitate departure of the leaving group (X) in the tetrahedral intermediate leading to the aldehyde. From this point, reduction to the alcohol proceeds as described above for aldehyde and ketone substrates.

Under acidic conditions aromatic carboxylic acids and esters are converted to alcohols efficiently in very good yields.[46] Under such conditions electron transfer to the carbonyl system may be facilitated by protonation of the carbonyl group, rendering the system more easily reduced, owing to a lowering of the lowest unoccupied molecular orbital of the substrate. Subsequent electron transfer followed by proton transfer generates a tetrahedral intermediate, which can be protonated to ease departure of the leaving group. Decomposition of the tetrahedral intermediate again generates an aldehyde, which can be reduced to the benzyl alcohol.

$$X = OH, OR$$

The addition of $SmI_2$ to primary amides in the presence of phosphoric acid results in the production of aromatic aldehydes. This might be explained by a series of sequential proton transfers and electron transfers that result in the formation of an intermediate carbinolamine. Under the acidic conditions of the reaction, this carbinolamine may form a stable salt that resists reduction to the alcohol.[107] Decomposition of the ethanolamine salt upon aqueous workup thus provides the observed aldehyde products of the reaction.

## Reductive Cleavage of α-Heterosubstituted Carbonyl Compounds and Related Substrates

The most widely applied reduction process promoted by $SmI_2$ is the reductive cleavage of α-heterosubstituted carbonyl substrates.[7,29,49,53–72] A wide variety of substituents can serve as leaving groups in this process, including hydroxy groups.[53–57]

$$\text{RCHXCOY} \xrightarrow[\text{H}^+]{2\,\text{SmI}_2} \text{RCH}_2\text{COY}$$

X = OH, OR, O$_2$CR, halide, OSiR$_3$, OSO$_2$Ar, SAr, S(O)Ar, SO$_2$Ar, OP(O)(OR)$_2$

Y = R, OR

Two limiting mechanisms can be envisioned for such transformations. In the first, initial electron transfer to the carbonyl generates a ketyl. This ketyl reacts immediately with a proton source, generating an $\alpha$-hydroxy radical. Reduction of this radical to the anion with rapid $\beta$ elimination affords an enol that tautomerizes to the observed carbonyl product.[53] It is likely that the oxophilicity and Lewis acidity of samarium(III) ions generated in the reduction are responsible for the success of this reaction with the wide variety of substituents, in many cases facilitating the departure of otherwise reluctant leaving groups.[108]

The second limiting mechanism involves initial dissociative electron transfer to an easily reducible $\alpha$ substituent. Subsequent reduction by a second equivalent of SmI$_2$ generates an enolate that undergoes protonation leading to the observed carbonyl product.[53] This type of mechanistic pathway is undoubtedly followed in substrates such as $\alpha$-halo esters, wherein the halogen is far more easily reduced than the ester moiety, and is also an excellent leaving group.

$\alpha,\beta$-Epoxy ketones and $\alpha,\beta$-epoxy esters comprise another subclass of $\alpha$-heterosubstituted carbonyl substrates in which the mechanism for reductive cleavage must closely resemble that of other compounds wherein ketyl formation initiates the reaction.[67-69] The same fundamental mechanistic pathway is followed thereafter, resulting in the production of aldol products in which stereogenicity at the $\beta$ carbon has remained intact throughout the process. Care must be taken in reactions with less easily reduced epoxy ester substrates because Lewis acid promoted ring opening of the epoxides competes with the reduction process if unoptimized solvent conditions are employed.[67] Nevertheless, this reaction provides an excellent entry to enantiomerically enriched $\alpha$-unsubstituted aldol-type products that are sometimes difficult to access by more conventional means.[67-69]

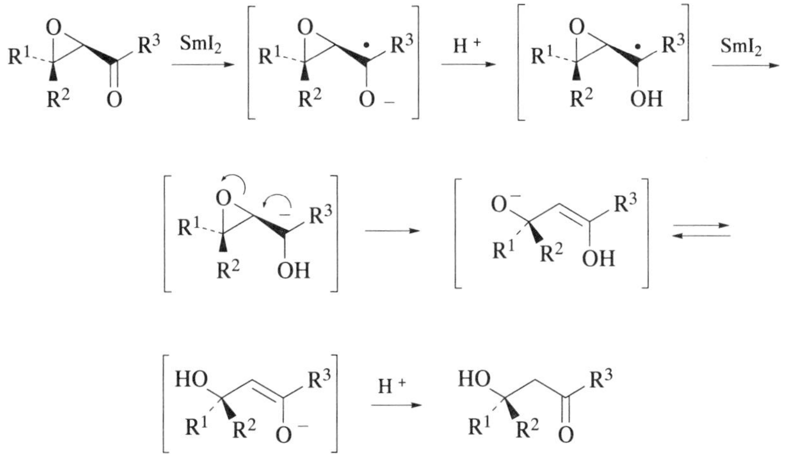

In addition to the diverse array of $\alpha$ substituents that can be reductively cleaved from carbonyl systems, other electron acceptors can be utilized for reductive cleavage processes. For example, $\alpha$-heterosubstituted nitriles have also proven to be useful substrates for the reaction.[71,72] Mechanisms similar to those of analogous $\alpha$-heterosubstituted carbonyl substrates are undoubtedly in effect.

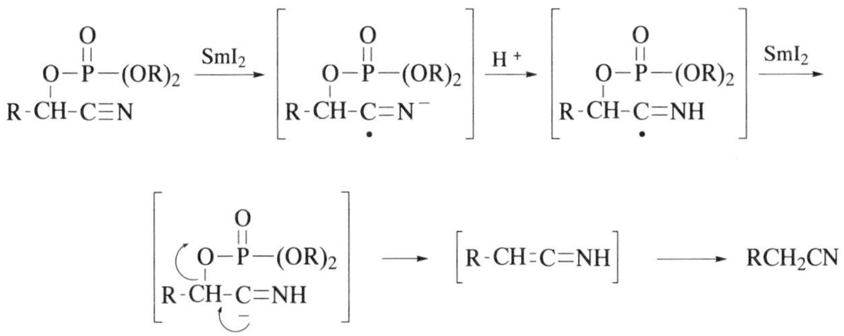

Finally, vinylogous epoxy carbonyl compounds and related vinyloxiranes also undergo facile reductive epoxide ring opening with $SmI_2$, providing $\delta$-hydroxy-β,γ-unsaturated carbonyl substrates.[67,70] In order to react efficiently, the substituent on the olefin in these systems must be a reasonably good electron-withdrawing group. This lowers the energy of the lowest unoccupied molecular orbital of the system, facilitating electron transfer in the crucial first step of the process. When electron-rich epoxy olefins are subjected to the reaction, yields fall dramatically and substantial deoxygenation byproducts contaminate the reaction mixture (see below).[70] The key feature of these reductions with electron-deficient unsaturated epoxides is that a conjugated dienolate (or related species) is generated as an intermediate in the reaction. This is kinetically protonated, generating a nonconjugated olefin that resists isomerization under the rather mild reaction conditions.[70]

Starting from enantiomerically enriched epoxides, the overall process provides an excellent means to generate enantiomerically enriched allylic alcohols, although in some instances mixtures of diastereomeric alkenes are produced as a result of protonation of the dienolate.[70]

## Reductive Cleavage of Cyclopropyl Ketones

The reductive cleavage of cyclopropyl ketones[61,73,74] has many similarities to that of the reduction of α-heterosubstituted carbonyl substrates. In both cases the process is initiated by single-electron transfer to the carbonyl, producing a samarium(III) ketyl. When a proton source is present,[61,73] the ketyl is protonated and strain induced ring fragmentation of the cyclopropylcarbinyl radical thus generated affords a homoallylic radical. Electron transfer followed by proton transfer and protonation provides the observed products of the reaction.

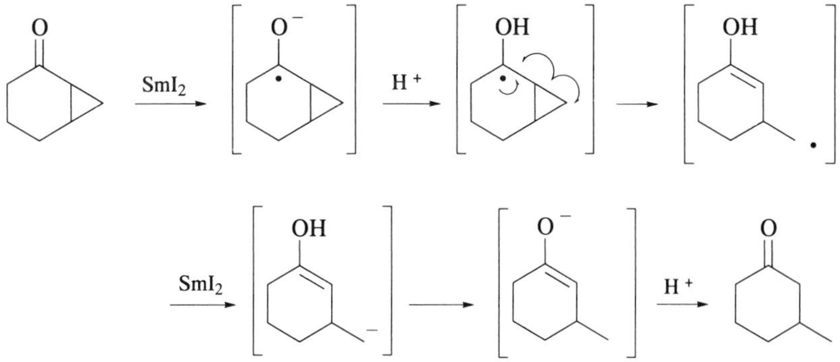

The mechanism and intermediates involved in the absence of a proton source are less clear. The latter conditions can be used in intramolecular carbon–carbon bond formation employing a judiciously placed sidechain on the ring possessing an unsaturated radical acceptor.[74] In the absence of such radicophiles, the homoallylic radical that is generated is presumably reduced to an organosamarium intermediate. However, trapping of such organosamariums with added electrophiles has not been observed. Indeed, when reactions conducted under aprotic conditions are quenched with added electrophiles, it is the samarium enolate that reacts.[74] Although no explanation has been provided for this observation, it is possible that the organosamarium is quenched by enolization of the starting ketone. The yields in these reactions wherein alkylsamariums would be generated are always less than 40%.

## Deoxygenation Reactions

In contrast to $\alpha,\beta$-epoxy ketones and vinylogous epoxy ketones in which the system is activated by an electron-deficient functional group, simple alkyl-substituted epoxides undergo a regiospecific but nonstereospecific deoxygenation, providing the corresponding alkenes as mixtures of diastereomers.[3,75] One suggested mechanism for this process invokes a Lewis acid promoted ring opening of the epoxide, providing a mixture of iodohydrins or related species.[75] Reductive elimination of the intermediate thus generated provides the alkenes as mixtures of diastereomers.

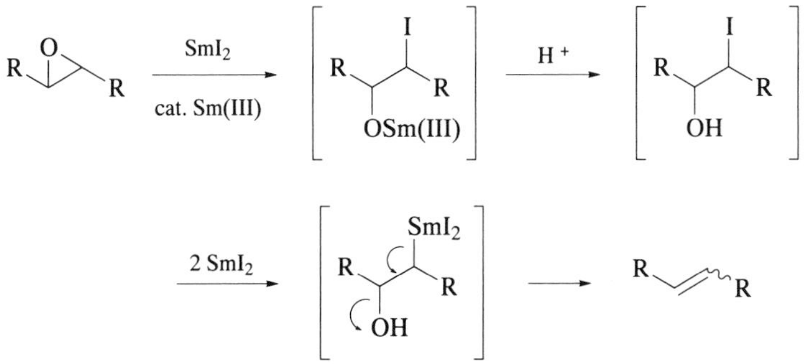

A second mechanistic scheme invokes direct reduction of the epoxide by samarium(II) in analogy to deoxygenation reactions promoted by a number of other low valent metal reductants.[109–113] By this pathway, intermediate radicals and subsequently anions are formed that eventually suffer $\beta$ elimination as well. By either mechanistic pathway, it is probable that the presence of samarium(III) ions facilitates departure of the leaving group in the $\beta$-elimination process.

Although sulfoxides, phosphine oxides, nitrogen oxides, and arsine oxides all undergo deoxygenation with $SmI_2$, there is little information on the mechanisms of these processes.[41,49,76] Furthermore, no enantiomerically enriched phosphine oxides have been utilized in the reaction to determine if the process can be achieved stereospecifically.

### Reduction of Nitrogen-Based Functional Groups

Reduction of nitroalkanes and aromatic nitro compounds can be stopped at the hydroxylamine stage or carried all the way to the amine, depending on the reaction conditions.[49,79,80] With four equivalents of $SmI_2$ in THF/MeOH for short periods of time (minutes), the hydroxylamines are produced. The mechanism is undoubtedly the same as that proposed for dissolving metal reduction of the same substrates.[104] A series of electron transfers from $SmI_2$ followed by proton transfers ultimately generates the hydroxylamine. Samarium(III) ions generated in the reaction mixture can be envisioned to facilitate departure of a hydroxy leaving group from one of the intermediates.

With six equivalents of $SmI_2$ over a period of several hours, the amines can be generated directly from the nitro compounds through the hydroxylamines by a mechanism involving electron transfer to the weak nitrogen–oxygen bond in the final stages of the reaction (vide infra).

$$\text{RNHOH} \xrightarrow[\text{H}^+]{\text{2 SmI}_2} \text{RNH}_2$$

Nitriles are resistant to $SmI_2$ in THF/MeOH at ambient temperatures,[3,80] but aromatic nitriles do react under acidic or basic reaction conditions to provide the corresponding amines in excellent yields.[46] Under acidic conditions the process may be facilitated by protonation of the nitriles, which serves to activate the nitrile toward reduction by $SmI_2$. The mechanism for reduction could follow the classic electron transfer/proton transfer scheme.

$$\text{RC}{\equiv}\text{N} \underset{}{\overset{\text{H}^+}{\rightleftharpoons}} \left[\text{RC}{\equiv}\overset{+}{\text{N}}\text{H}\right] \xrightarrow{\text{SmI}_2} \left[\text{R}\overset{\bullet}{\text{C}}{=}\text{NH}\right] \xrightarrow{\text{SmI}_2} \left[\text{R}\overset{-}{\text{C}}{=}\text{NH}\right]$$

$$\xrightarrow{\text{H}^+} \left[\begin{array}{c}\text{R}\\ \diagdown \\ \phantom{x}\text{C}{=}\text{NH}\\ \diagup \\ \text{H}\end{array}\right] \xrightarrow{\text{H}^+} \left[\begin{array}{c}\text{R}\\ \diagdown \\ \phantom{x}\overset{+}{\text{C}}{=}\text{NH}_2\\ \diagup \\ \text{H}\end{array}\right] \xrightarrow{\text{SmI}_2} \left[\text{R}\overset{\bullet}{\text{C}}\text{HNH}_2\right]$$

$$\xrightarrow{\text{SmI}_2} \left[\text{R}\overset{-}{\text{C}}\text{HNH}_2\right] \xrightarrow{\text{H}^+} \text{RCH}_2\text{NH}_2$$

Alternatively, it has been suggested that phosphorylated imines, formed by addition of phosphoric acid to the nitriles, may be the intitial substrates for the reaction.[46]

$$\text{RC}{\equiv}\text{N} \xrightarrow{\text{H}_3\text{PO}_4} \left[\begin{array}{c}\text{R}{-}\text{C}{=}\text{NH}\\ |\\ \text{O}{-}\text{P}{-}\text{(OR)}_2\\ \|\\ \text{O}\end{array}\right] \xrightarrow{\text{SmI}_2} \left[\begin{array}{c}\text{R}{-}\overset{\bullet}{\text{C}}{-}\overset{-}{\text{NH}}\\ |\\ \text{O}{-}\text{P}{-}\text{(OR)}_2\\ \|\\ \text{O}\end{array}\right] \xrightarrow{\text{H}^+}$$

$$\left[\begin{array}{c}\text{R}{-}\overset{\bullet}{\text{C}}{-}\text{NH}_2\\ |\\ \text{O}{-}\text{P}{-}\text{(OR)}_2\\ \|\\ \text{O}\end{array}\right] \xrightarrow{\text{SmI}_2} \left[\begin{array}{c}\text{R}{-}\overset{-}{\text{C}}{-}\text{NH}_2\\ |\\ \text{O}{-}\text{P}{-}\text{(OR)}_2\\ \|\\ \text{O}\end{array}\right] \xrightarrow{\text{H}^+} \left[\begin{array}{c}\text{R}{-}\text{CHNH}_2\\ |\\ \text{O}{-}\text{P}{-}\text{(OR)}_2\\ \|\\ \text{O}\end{array}\right]$$

$$\underset{}{\overset{\text{H}^+}{\rightleftharpoons}} \left[\begin{array}{c}\text{R}{-}\text{CH}{-}\text{NH}_2\\ |\\ (\text{O}{-}\text{P}{-}\text{(OR)}_2\\ \| \\ {+}\,\overset{-}{\text{O}}\text{H}\end{array}\right] \longrightarrow \left[\begin{array}{c}\text{R}\\ \diagdown \\ \phantom{x}\overset{+}{\text{C}}{=}\text{NH}_2\\ \diagup \\ \text{H}\end{array}\right] \xrightarrow{\text{SmI}_2}$$

$$\left[\text{R}\overset{\bullet}{\text{C}}\text{HNH}_2\right] \xrightarrow{\text{SmI}_2} \left[\text{R}\overset{-}{\text{C}}\text{HNH}_2\right] \xrightarrow{\text{H}^+} \text{RCH}_2\text{NH}_2$$

The species formed by addition of sodium hydroxide to $SmI_2$ is apparently a much more powerful reductant than $SmI_2$ itself. As proof of this, aryl chlorides are also extensively reduced during the reduction of aromatic nitriles with $SmI_2$ under basic conditions.[46] Thus, whatever the actual reductant is, it is also capable of reducing the nitrile very rapidly under these conditions. Samarium(III) ions formed during the reaction may ease electron transfer to the imine intermediate formed during the reaction.

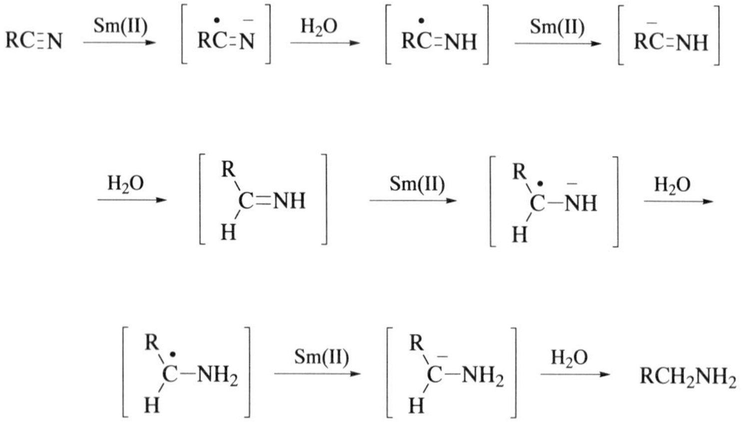

In direct analogy to the reduction of ketones, imines can be reduced with $SmI_2$ in THF/MeOH.[80,81] The mechanism of these reactions is undoubtedly analogous to that of ketones as well.

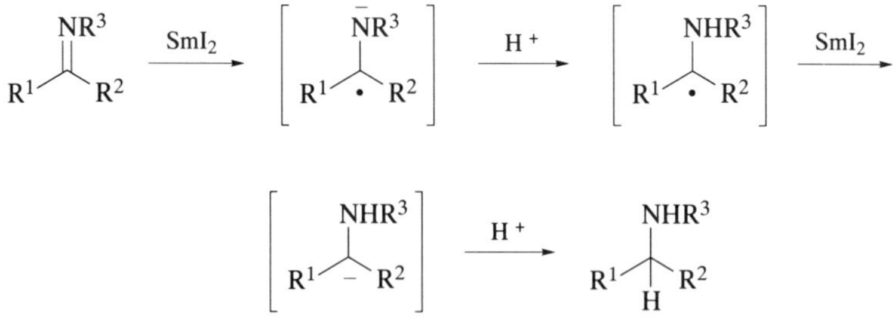

The mechanism for reduction of oximes to imines by $SmI_2$[42,80,82] can be envisioned as being similar to that proposed for the electrochemical reduction of oximes.[104,105] Further reduction of the imine thus generated to the amine follows the mechanism outlined above. A series of chiral 2-hydroximino amides of $\alpha$-amino acids has been reduced with fair to good diastereoselectivities in the protonation of the amine $\alpha$-carbanion.[82] There has been no rationalization for the sense of asymmetric induction observed.

Azo compounds react with two equivalents of $SmI_2$ in THF/MeOH to form hydrazines,[49] and these intermediates (and related acyl hydrazines) are reductively cleaved to amines with an excess of $SmI_2$.[80,84]

$$RCH=NOH \xrightarrow{SmI_2} \left[ R\overset{\bullet}{C}H-\overset{-}{N}OH \right] \xrightarrow{H^+} \left[ R\overset{\bullet}{C}H-NHOH \right] \xrightarrow{SmI_2}$$

$$\left[ \underset{H}{RCH-N-OH} \right] \longrightarrow \left[ RCH=NH \right] \dashrightarrow RCH_2NH_2$$

$$RN=NR \xrightarrow{SmI_2} \left[ \overset{\bullet}{R}N-\overset{-}{N}R \right] \xrightarrow{H^+} \left[ \overset{\bullet}{R}N-NHR \right] \xrightarrow{SmI_2}$$

$$\left[ \overset{-}{R}N-NHR \right] \xrightarrow{H^+} \left[ RHN-NHR \right] \xrightarrow{SmI_2}$$

$$\left[ \overset{\bullet}{R}NH \atop + \atop \overset{-}{R}NH \right] \begin{array}{l} \xrightarrow{SmI_2} \left[ \overset{-}{R}NH \right] \xrightarrow{H^+} RNH_2 \\[3em] \xrightarrow{H^+} RNH_2 \end{array}$$

## Reduction of Miscellaneous Functional Groups

As pointed out in some of the preceding examples, heteroatom–heteroatom bonds are readily cleaved by $SmI_2$. Thus peroxides also suffer rapid reductive bond cleavage, and bicyclic peroxides produce diols.[85,86]

$$ROOR \xrightarrow{SmI_2} \begin{cases} RO\bullet \xrightarrow{SmI_2} RO^- \xrightarrow{H^+} ROH \\[1em] RO^- \xrightarrow{H^+} ROH \end{cases}$$

The nitrogen–oxygen bond of isoxazoles suffers the same fate, but in this case the enol oxygen is tautomerized to a ketone, and the imine equilibrates to an enamine in the final product.[25]

Halo phosphine oxides and halo phosphine sulfides undergo reduction to the corresponding phosphines upon reduction by $SmI_2$.[87] The mechanism undoubtedly mimics that of reduction of alkyl halides by $SmI_2$, proceeding through radical and presumably anionic intermediates. In reactions with enantiomerically enriched substrates, substantial racemization occurs in the transformation from starting material to products. It is unclear whether this occurs as a result of epimerization of the intermediate radical or the anion. However, analogous anionic phosphorus complexes have been reported to maintain their stereochemistry.

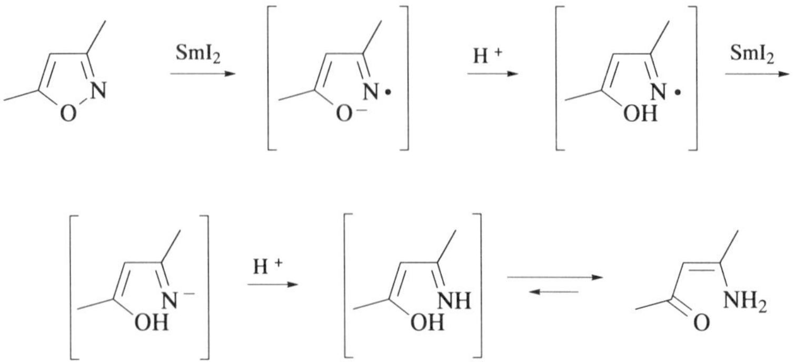

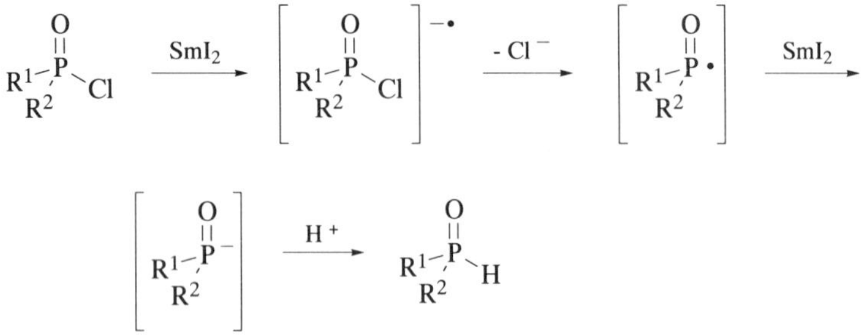

Samarium(II) iodide serves as the stoichiometric reductant in the palladium(0)-catalyzed reduction of allylic acetates.[88] In the process, the $\pi$-allylpalladium species initially formed is reduced by $SmI_2$, forming an allyl anion with regeneration of the palladium(0) catalyst. The allylic anion is subsequently quenched by isopropanol that is present in the reaction mixture. As a consequence of this mechanism, mixtures of regioisomeric and stereoisomeric alkenes are generally formed as products of the reactions.

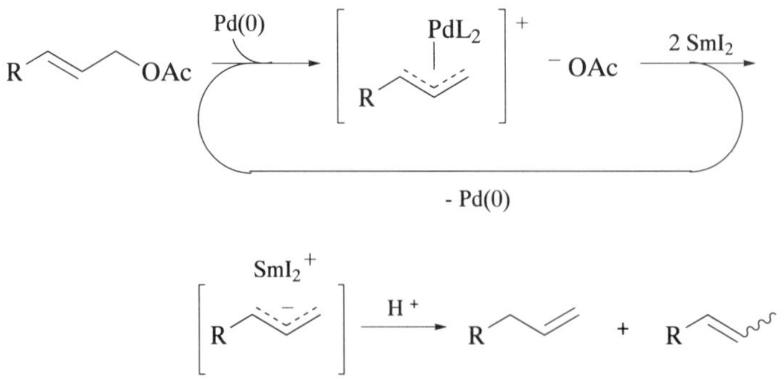

In an analogous process, propargyl acetates react under similar conditions to form allenes or alkynes.[89] The ratio of the two isomeric products is a function of the structure of the substrate as well as the protonating agent used in the reactions. Sterically bulky protonating agents typically provide a greater proportion of allene in these reactions, and the preference for allene formation is also enhanced in tertiary propargyl acetates relative to that of secondary and primary propargyl acetates.

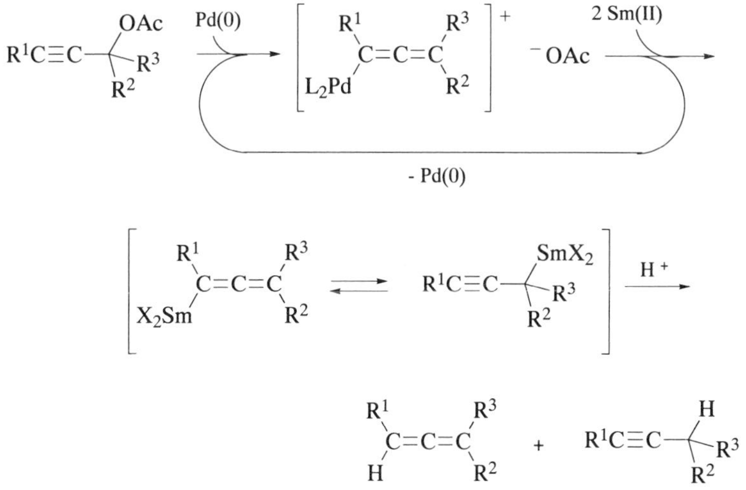

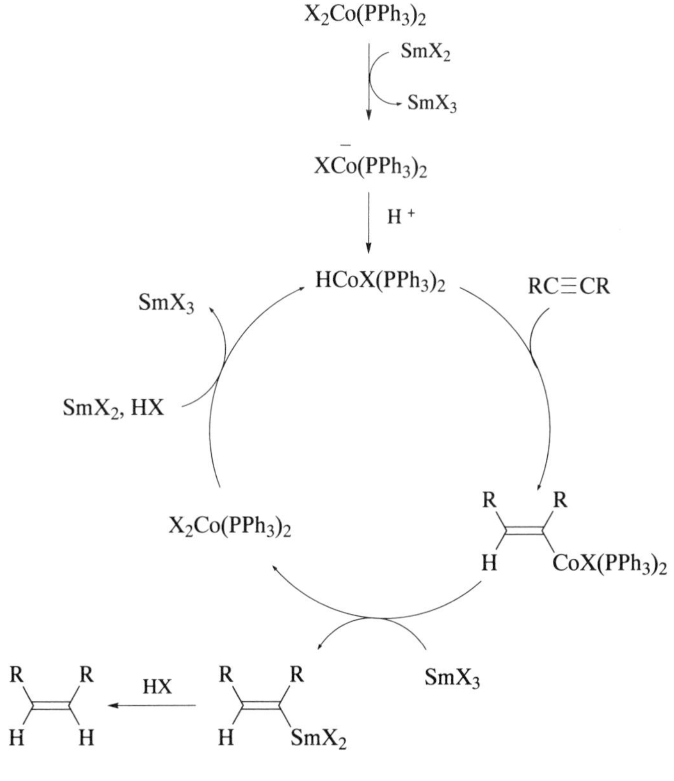

Samarium(II) iodide can also be utilized as the stoichiometric reductant in transition metal catalyzed reductions of alkynes to produce *cis*-alkenes.[91] Although few mechanistic details exist for these transformations, transition metal hydrides are suspected as the reactive reducing agents in the reactions. Thus $SmI_2$ may serve as a reductant for $CoCl_2 \cdot 4PPh_3$, which becomes protonated to generate $HCoCl(PPh_3)_2$, the active catalyst for the process. *cis*-Addition of the reactive hydride to the alkyne followed by transmetalation with $SmX_3$ may provide the corresponding alkenylsamarium and $X_2Co(PPh_3)_2$. The *cis* alkenes can be derived from the alkenylsamarium species by protonation, and the active catalyst would be regenerated from $X_2Co(PPh_3)$ by reduction with $SmI_2$ and subsequent protonation.

Alternatively, the alkenylsamarium species could be derived from the alkenyl-cobalt species by a reductive transmetalation process, generating the anionic cobalt complex. Simple protonation of the latter species would again regenerate the active catalyst.

## SCOPE AND LIMITATIONS

### Reduction of Organic Halides, Sulfonates, and Sulfones

In the presence of HMPA, $SmI_2$ appears to be a remarkably versatile and powerful reagent for the conversion of a variety of alkyl halides to the corresponding alkanes.[24] Primary, secondary, and even tertiary alkyl halides are reduced rapidly under very mild conditions. Although alkyl iodides and bromides react within minutes at room temperature, chloroalkanes require heating for several hours in THF for complete reaction. Selectivity for the reduction of alkyl bromides or alkyl iodides in the presence of chloroalkanes has not been specifically addressed. However, on the basis of the relative rates outlined above[24] and related data from intermolecular Barbier-type reactions,[3] high chemoselectivity in the reduction of bromochloroalkanes and chloroiodoalkanes can be anticipated. Although fluoroalkanes have apparently not been employed as substrates, it is likely that they are resistant to reduction under reasonable experimental conditions.

Several different functional groups can be tolerated under conditions required for alkyl halide reduction. These functional groups include alcohols, aromatic rings, ethers, and for most substitution patterns alkenes and esters.[24] Suitably disposed alkenes can trap radicals generated from alkyl halide precursors, and similarly esters (and other carboxylic acid derivatives) can undergo nucleophilic acyl substitution if five- or six-membered rings can result. The procedure for alkyl halide reduction is certainly not effective for substrates possessing more easily reducible functional groups such as aldehydes, ketones, and conjugated carbonyl groups. Reduction of these functional groups or Barbier-type reactions compete with the alkyl halide reduction. Attempted reductions of alkyl halides in the presence of still other readily reduced functional groups or reactive electrophiles should be carried out with care.

Certain structural classes of alkyl halides may pose problems as well. For example, attempted reduction of geminal dihalides may lead to unwanted side reactions owing to carbene formation,[23,29,114] and alkyl halides possessing a leaving group $\beta$ to the halide will undergo conversion to alkenes via a reductive $\beta$-elimination process.[26-28,31,33-39]

Alkyl tosylates can be reduced to the corresponding alkanes with $SmI_2$.[3] The scope of this transformation has not been fully explored. It has been suggested that the mechanism for this process involves an initial Finkelstein-type conversion of the tosylate to the corresponding alkyl iodide, which subsequently undergoes reductive cleavage. If this is the case, then the reduction will be restricted to alkyl tosylates that undergo facile conversion to the corresponding iodide under the reaction conditions.

Owing to the intervention of radical intermediates in the reduction of alkyl halides, an obvious limitation to $SmI_2$-promoted reductions is the loss of stereochemistry that will result in nearly all cases when chiral alkyl halide substrates are reduced. One exception is the retention observed in reduction of some cyclopropyl halides, in which configurational stability is conferred on the intermediates of the reduction process.[92]

(55%) 95% retention

Aryl iodides, aryl bromides, and even aryl chlorides are all reduced effectively by $SmI_2$ in the presence of HMPA.[24] Although somewhat less well studied, alkenyl halides also appear to be suitable substrates for reduction reactions. The intermediacy of rapidly inverting vinyl radicals would appear to prevent stereospecific reduction of diastereomeric alkenyl halides.

Neither allylic halides nor benzylic halides are effectively reduced to the corresponding alkanes with $SmI_2$ in THF. In both sets of substrates, extensive reductive dimerization occurs.[3] However, these substrates have not been subjected to reductions employing $SmI_2$ in HMPA, and perhaps under these conditions a clean reduction can be achieved.[115]

Alkyl phenyl sulfones and alkenyl tert-butyl sulfones are reductively cleaved by $SmI_2$ in the presence of HMPA, providing the corresponding hydrocarbons.[33] These reactions can be acccomplished in spite of the fact that diaryl sulfones and dialkyl sulfones have been reported to be converted to the corresponding sulfides under nearly the same reaction conditions.[41,76] Perhaps even more surprising is the fact that the alkenyl sulfones can be reductively cleaved without interference from competitive conjugate reduction. Although further studies will be required to outline the full scope of this process, it is clear that the procedure is fairly chemoselective (tolerating alcohols, esters, benzyl ethers, and aromatic systems), and thus may be of reasonable synthetic utility.

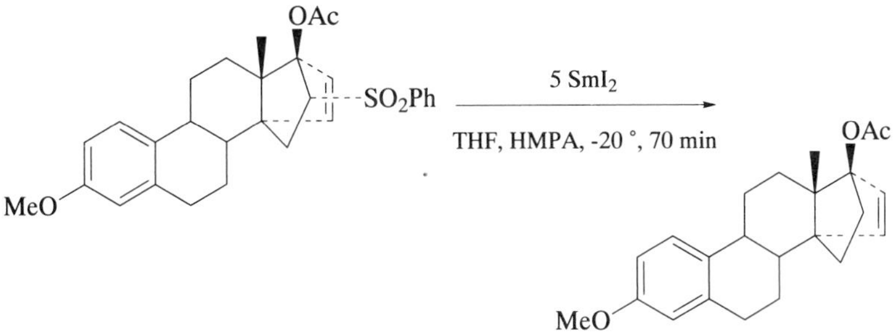

(70%)

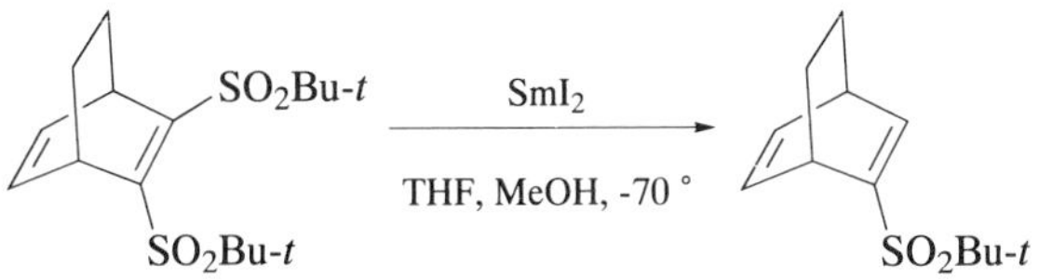

(85%)

## Reductive Elimination/Fragmentation Reactions

Reductive cleavage of $\beta$-halo ethers and $\beta$-acyloxy halides is a general reaction, but tertiary alkyl halides have not been employed as substrates. Although only modest effort has been devoted to this reaction as a synthetic tool, substrates examined to date make it clear that ketones, esters, alcohols, and alkynes do not react under conditions required for the $\beta$-elimination processes. Reactions proceed under mild conditions using both chloride and bromide precursors.[26,31,34-36] The procedure would thus appear to be a reliable, chemoselective method for the deprotection of appropriately protected ($\beta$-haloethyl-substituted) alcohols and amines in highly functionalized molecules,[35,36] and for the construction of olefins from suitable geminally substituted substrates. Additionally, the reductive fragmentation reaction can be utilized as a mechanistic tool to probe for the intermediacy of organosamarium intermediates in processes promoted by SmI$_2$.[18,73] 1,2-Dihaloalkenes and 1,2-dihaloarenes have not been subjected to reduction by SmI$_2$. However, because alkenyl and aryl radicals generally undergo hydrogen atom abstraction from the solvent more rapidly than they are reduced to the anion,[24,93-95,97,98] it is not likely that such substrates would lead to alkynes or benzynes, respectively.

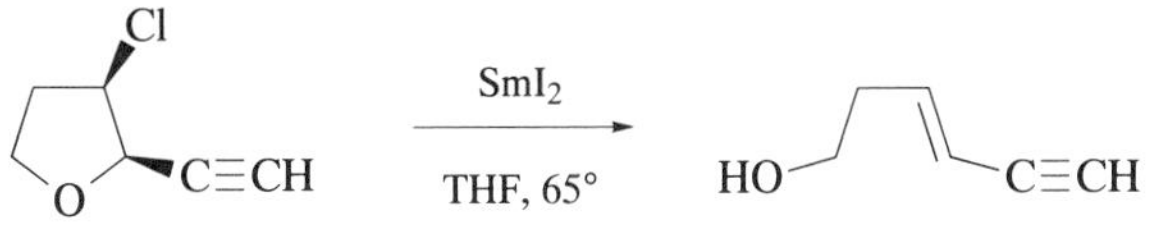

(75%) >99% E

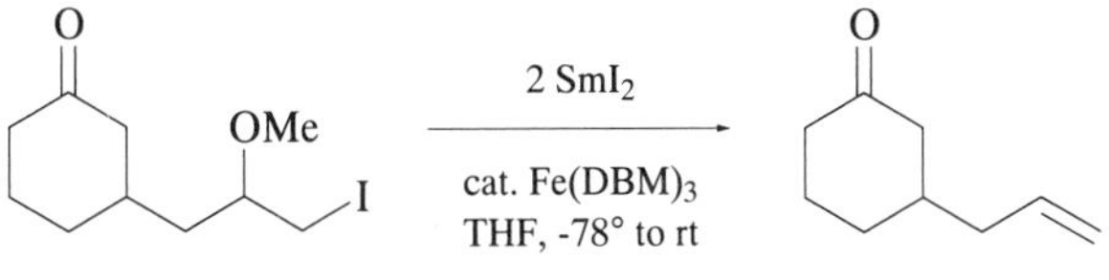

(81%)

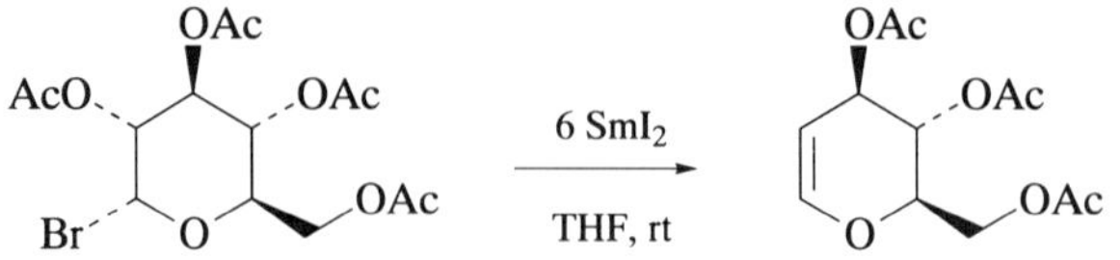

Only two $\alpha,\beta$-disulfones have been reduced by $SmI_2$, but yields are high in both cases, and thus the expectation is that a diverse range of such substrates might undergo reductive elimination.[33] There are some reservations in this expectation, however, because of the results obtained in the reductive elimination of $\beta$-hydroxy sulfones. Thus although reductive cleavage of alkyl imidazolyl sulfones in this series proceeds very effectively, alkyl phenyl sulfones subjected to identical reaction conditions lead only to the recovery of starting material or to desulfonated alcohol.[33,37]

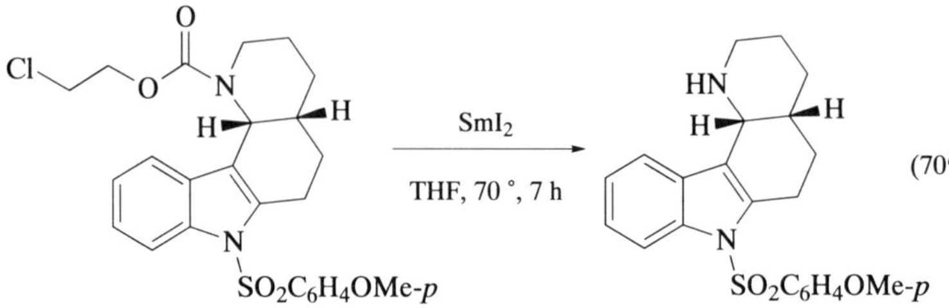

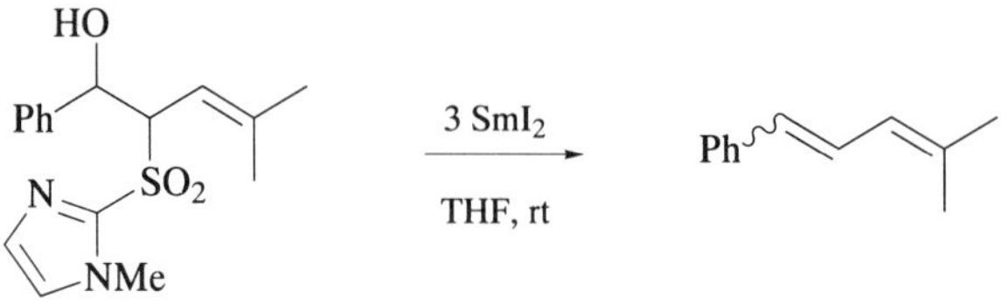

The reductive elimination of enolates is an intriguing reaction that has seen only preliminary development with $SmI_2$ utilized as the reducing agent to induce the fragmentation. In certain cases it has the potential to generate stereocontrolled, highly functionalized acyclic structural units from readily available cyclic precursors.[27,28,38,39]

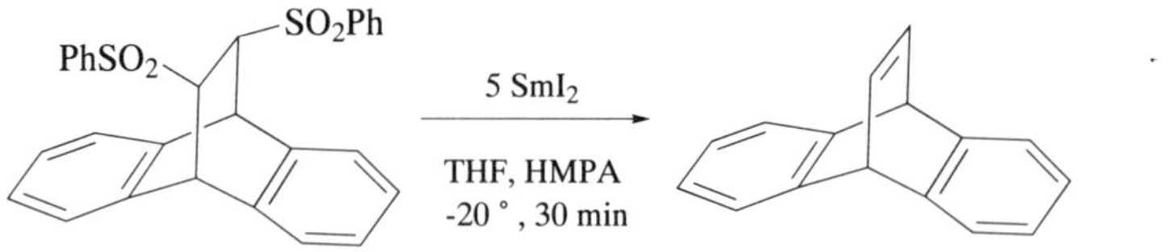

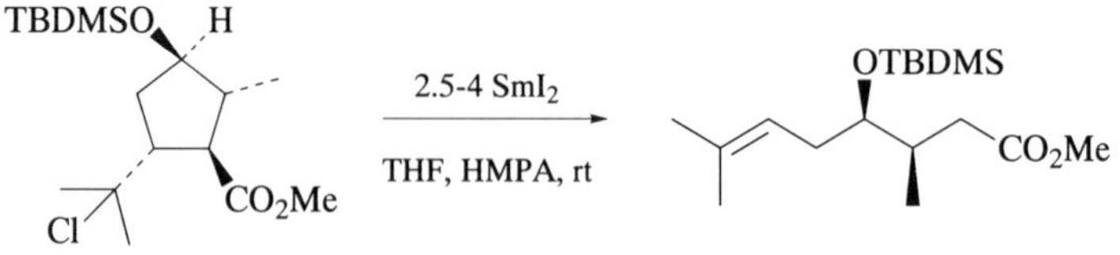

## Reduction of Aldehydes and Ketones

The reduction of aldehydes and ketones by $SmI_2$ has not been explored in great detail. In the few examples that are reported, the reaction works reasonably well on some simple substrates.[3] Competitive rate studies indicate that $SmI_2$ is highly chemoselective for the reduction of aldehydes in the presence of ketones.[3] Additionally, ketones can be reduced in the presence of esters[45] and, to some extent, $\alpha,\beta$-unsaturated esters.[44] Chemoselectivity in the presence of halides is difficult to achieve because of competitive Barbier-type reactions, although alkyl chlorides will survive the reaction conditions if HMPA is not employed as a cosolvent. The ability of other reducible functional groups to remain intact during carbonyl reduction has yet to be submitted to close scrutiny.

Diastereoselectivity in the reduction of aldehydes and ketones with $SmI_2$ in most cases is not high,[40,41] although in specific instances preference for the equatorial alcohol in six-membered ring systems can be quite respectable.[41,45] Although reasonably high enantioselectivities have been achieved in the reduction of benzil to benzoin using $SmI_2$ as a reductant along with chiral protonating agents,[43] this approach to asymmetric synthesis is unlikely to be competitive with other available methods.

## Reduction of Carboxylic Acids and Their Derivatives

Carboxylic acids, carboxylic acid amides, and carboxylic acid esters are quite resistant to reduction by $SmI_2$ in THF or even in THF/HMPA. However, under acidic (aqueous phosphoric acid) or basic (e.g., aqueous sodium hydroxide) conditions, reduction of aromatic carboxylic acids and their derivatives can be effected within seconds.[42,46,47] Under basic conditions the reduction of simple, unfunctionalized aliphatic carboxylic acids also proceeds in good-to-excellent yields, but the acidic reaction protocol appears applicable only to aromatic carboxylic acids. There is no information on the scope of the reaction with regard to the nature of the alcohol-derived portion of the ester, but for those esters that possess alkyl units that can form reasonably stable radicals, the electron transfer nature of the reaction may lead to the intervention of radical cleavage processes leading to undesired byproducts.[116,117] Furthermore, because of the powerfully reducing reaction conditions, only a very limited array of functionality can be tolerated in these transformations.

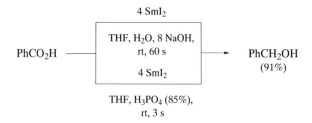

As might be expected, based upon the reduction of carboxylic acids in basic media, sodium benzoate is also readily reduced in aqueous $SmI_2$, providing a good yield of benzyl alcohol.[47] Anhydrides and carboxylic acid halides, however, are poor substrates for $SmI_2$-based reduction under either anhydrous or aqueous conditions. Under anhydrous conditions the acyl anion intermediate generated upon initial reduction of the acid halide by $SmI_2$ is acylated, leading to the generation of $\alpha$-diketones and $\alpha$-ketols.[48] Even under aqueous conditions substantial dimerization occurs, and the process is therefore not synthetically useful.

Methyl benzoate can be reduced under acidic or basic conditions to provide benzyl alcohol,[42,46] but this reduction for nonaromatic esters has not been examined, and the reaction is not likely to be general.

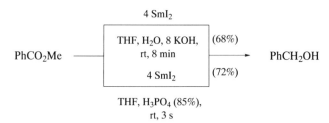

Curiously, aromatic carboxamides are reduced to benzyl alcohols under basic conditions.[42] The reaction under acidic conditions provides the corresponding benzaldehydes in excellent yield.[46] Unfortunately, the reaction in both cases appears limited to primary amides. Secondary amides, hydrazides, and hydroxamic acids give mixtures of products in modest yields. Consequently, this process, too, appears of quite limited synthetic value.

## Reduction of Conjugated Carbonyl Substrates

Reduction of $\alpha,\beta$-unsaturated aldehydes and ketones with $SmI_2$ invariably leads to mixtures of 1,2- and 1,4-addition products, and to products in which both the olefin and the carbonyl group have been reduced.[3] In some cases, extensive polymerization is observed as well. Consequently, with the exception of benzoquinone and perhaps its analogs,[49] clean reductions in such systems seem difficult to achieve.

On the other hand, the reduction of conjugated esters and amides by $SmI_2$ is a quite useful process, proceeding within minutes in N,N-dimethylacetamide with an added proton source.[3,50-52] The nature of the solvent is reported to be critical for success in these conjugate reductions, although there is some controversy surrounding this point. Thus HMPA is reported to lead to extensive reductive dimerization of the unsaturated systems in one study,[51] an observation that is disputed in a second investigation.[50] With the exception of tetrasubstituted systems, all substitution patterns about the olefin have been examined in conjugate reductions of unsaturated esters and amides, and all proceed in reasonable yields. Little information is available about other functional groups that might be compatible with the reactions.

Conjugated carboxylic acids and anhydrides can also be reduced to the saturated derivatives in the presence of $SmI_2$.[50] For these substrates, THF/HMPA appears to be the solvent of choice. Di-, tri-, and tetrasubstituted olefins have all been examined, and modest to excellent yields are maintained throughout the series.

## Reductive Cleavage of $\alpha$-Heterosubstituted Carbonyl Compounds and Related Substrates

The most broadly applied reduction processes promoted by $SmI_2$ are the reductive cleavage reactions of various $\alpha$-heterosubstituted carbonyl compounds. These reactions are general for both the types of $\alpha$-heterosubstituents that may be employed, as well as for the carbonyl substrates that undergo the transformation. For example, the reductive cleavage of both $\alpha$-halo ketones and $\alpha$-halo esters with $SmI_2$ can be effected under extremely mild conditions, affording the dehalogenated products in virtually quantitative yields.[8,29,49,53] Both bromide and chloride precursors have been utilized for the reaction, and ketones, acyclic esters, and lactone substrates have been employed successfully as well. The facility with which $\alpha$-heterosubstituents are cleaved permits their selective removal in the presence of other alkyl halides,[29] and it is likely that the process can be carried out in the presence of most other functional groups as well.

One of the unique features of $SmI_2$ is its ability to effect the reduction of $\alpha$-hydroxy carbonyl substrates.[53–55,57] Several different reaction protocols have been developed for this process, and it can be utilized most effectively with $\alpha$-hydroxy esters and $\alpha$-hydroxy lactones. Under optimized reaction conditions these transformations transpire at room temperature within hours, and yields are uniformly high.[54,55] The reductive cleavage of $\alpha$-hydroxy ketone substrates is less extensively established,[53,57] and the best reaction protocol has perhaps not yet been developed. Nevertheless, it is obvious that this transformation can be applied to most systems, and it is likely to prove general.

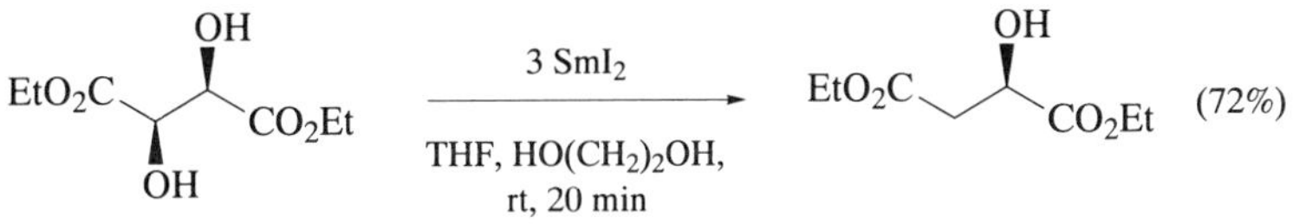

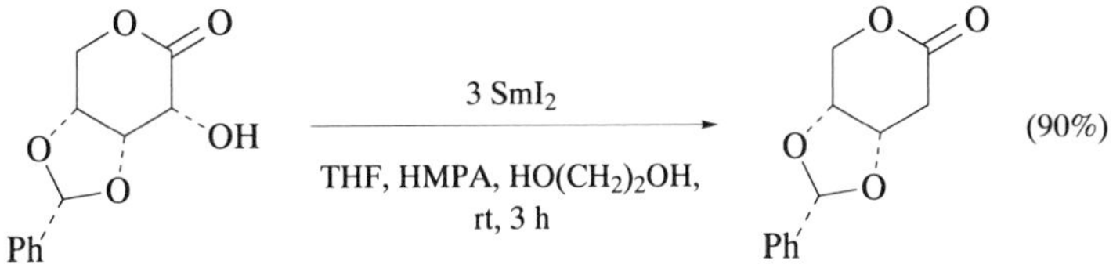

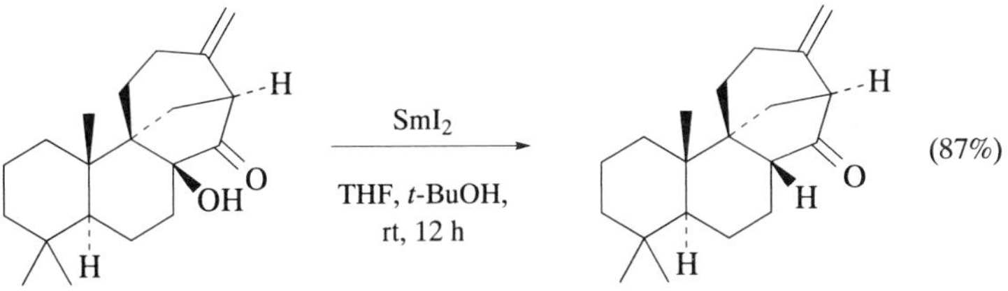

In a similar fashion, $\alpha$-alkoxy, $\alpha$-silyloxy, $\alpha$-carboalkoxy, and $\alpha$-tosyloxy carbonyl substrates are all rapidly reduced by $SmI_2$, and several different reaction protocols have been developed to carry out these transformations.[53-55,58-66] With simple ketone substrates bearing such $\alpha$ substituents, reactions are nearly always carried out in THF at $-78°$ and are complete within minutes, providing nearly quantitative yields of the desired products.[53,59-64,66] Because of the facility with which these reactions proceed, most functional groups (e.g., isolated ketones, esters, halides, and nitriles) are compatible with the extraordinarily mild reduction conditions required. $\alpha$-Heterosubstituted ester and lactone substrates are more difficult to reduce, and thus reactions are typically performed at ambient temperatures in THF/HMPA.[53-55,58] Nevertheless, the desired materials are isolated in excellent yields, and in terms of both the $\alpha$-heterosubstituent and the structure of the ester, the process is highly general.

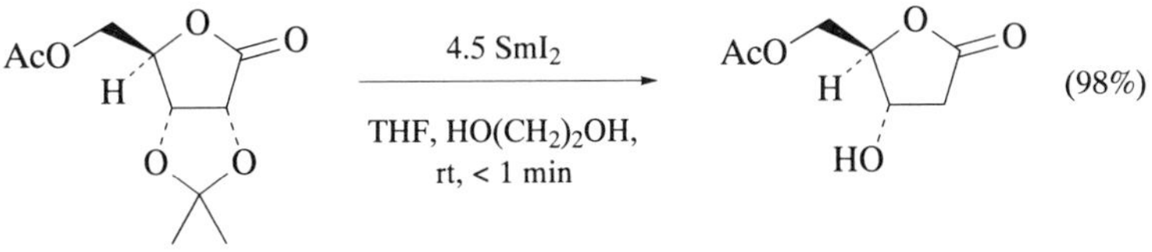

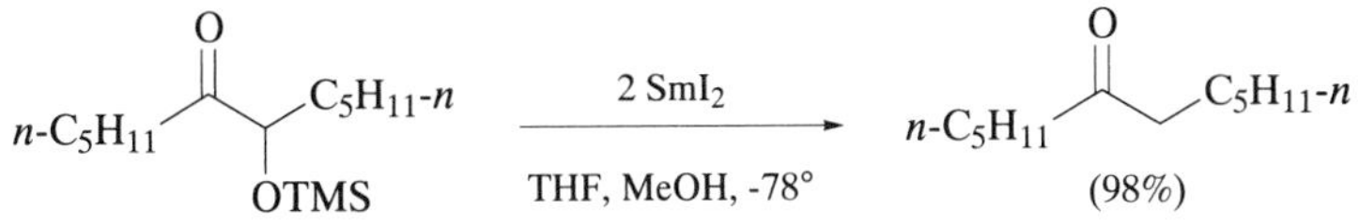

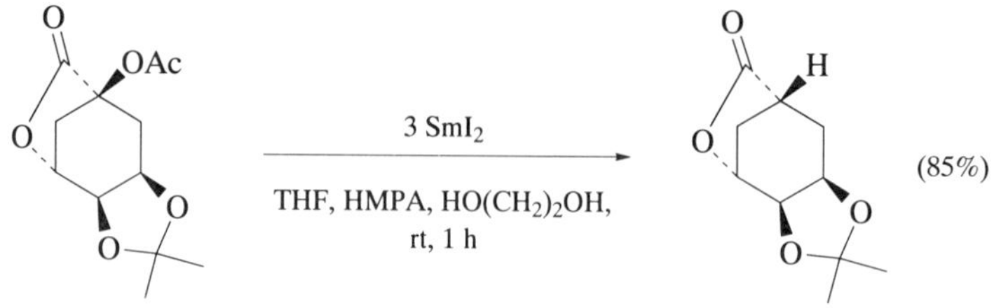

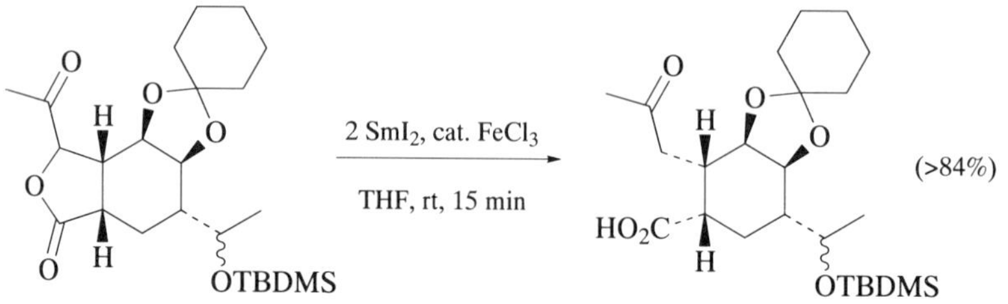

(85%)

(>84%)

Product mixtures are often observed when $\alpha,\beta$-diheterosubstituted lactones are reduced with $SmI_2$.[55,65] For such substrates, the enolate that initially forms after reductive cleavage of the $\alpha$-heterosubstituent may be protonated or may lead to rapid $\beta$ elimination. The $\beta$ elimination is usually prevented when substrates possess a $\beta$-alkoxy substituent and when reactions are carried out in THF/ ethylene glycol mixtures.[55] However, the unsaturated lactones are synthesized predominantly, if not exclusively, when reductions are carried out with $\beta$-acyloxy substrates in THF alone or in THF with added pivalic acid or acetic acid.[55,65]

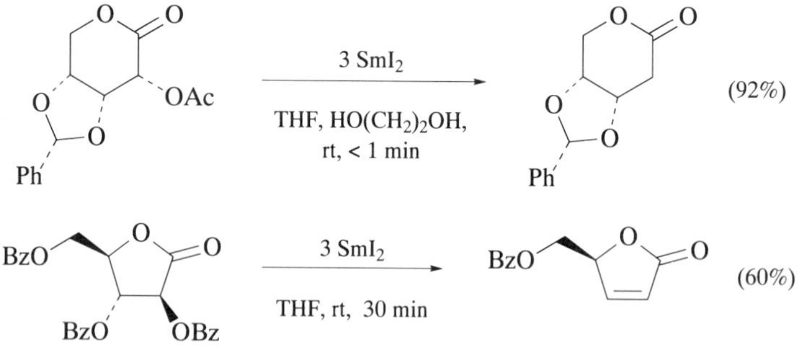

(92%)

(60%)

Only three $\alpha$-thiosubstituted ketones have been subjected to reduction by $SmI_2$, but the $\alpha$-thiophenyl ether, $\alpha$-phenyl sulfoxide, and $\alpha$-phenyl sulfone of cyclohexanone are all reduced in reasonable yields (64–88%).[53] This provides some confidence that the process can be generalized in more diverse systems.

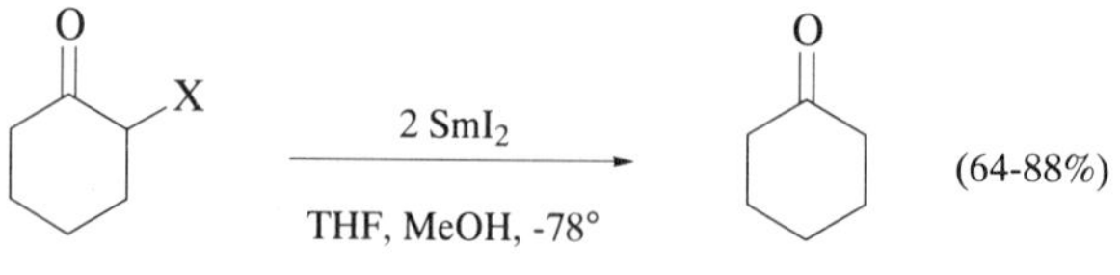

(64-88%)

X = SPh, S(O)Ph, SO₂Ph

Reductive cleavage reactions of $\alpha$-heterosubstituted carbonyl substrates have been extended to $\alpha,\beta$-epoxy carbonyl substrates.[67-69] Reactions proceed rapidly at very low temperatures in THF solvent ($-90°$) for epoxy ketones,[69] but epoxy esters appear to react best in THF/HMPA/$N,N$-dimethylaminoethanol solvent systems at room temperature.[67,68] Diverse systems have been examined in which both substitution and orientation about the epoxy carbonyl centers vary, and excellent yields are realized in virtually every case. Conditions required for reduction of epoxy ketones permit most functional groups (e.g., isolated ketones, esters, halides, and amines) to be present in substrates of interest, but the situation with less easily reduced epoxy esters has not been established. Of some significance is the fact that nonracemic $\alpha,\beta$-epoxy carbonyl substrates can be reduced to aldol-type products with little or no epimerization through a retroaldol-aldol sequence.[67-69] This provides a novel route to $\alpha$-unsubstituted, enantiomerically enriched aldol products that are typically difficult to access by more traditional means.

Vinylogous epoxy carbonyl compounds are also extremely reactive with $SmI_2$, providing the $\delta$-hydroxy-$\beta,\gamma$-unsaturated carbonyl products in excellent yields.[67,70] In fact, unsaturated epoxides bearing cyano groups, sulfones, phosphonates, and thioesters on the olefin work as well as ketones and esters in activating the substrate toward reduction. Unactivated epoxy alkenes undergo much slower reactions, and mixtures of diastereomeric alkenes result.[70] In virtually every successful reductive cleavage reaction, kinetic protonation of the intermediate dienolate (or analog) leads to exclusive formation of the nonconjugated olefin, and under the mild reaction conditions little, if any, conjugated material is produced. All substitution patterns about the epoxide and the alkene units in these systems work quite well, although stereoisomeric mixtures of unsaturated products are often isolated. As with the simple $\alpha,\beta$-epoxy ketones, the use of enantiomerically enriched substrates provides a facile route to nonracemic allylic alcohols with complete retention of stereochemistry.[70]

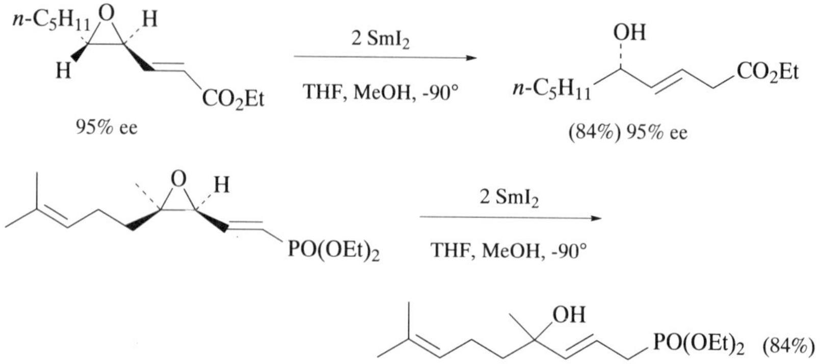

Cyanohydrin reactions carried out with lithium cyanide in the presence of diethyl phosphorocyanidate generate intermediate $\alpha$-cyanophosphates. These $\alpha$-heterosubstituted nitriles are efficiently reduced by $SmI_2$ in THF/methanol, providing the unsubstituted nitriles.[71] Kinetic protonation of the intermediate anion generated is evident in these substrates, because allylic and propargylic $\alpha$-cyanophosphates are cleaved to provide exclusively the $\beta,\gamma$-unsaturated nitriles.

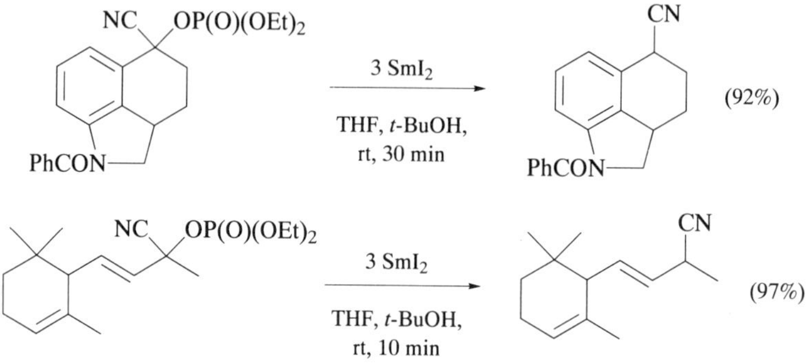

## Reductive Cleavage of Cyclopropyl Ketones

Stereoelectronically controlled radical ring opening of cyclopropyl ketones by $SmI_2$-promoted reduction of acyl cyclopropanes occurs in modest to good yields.[61,73,74] The number of examples tested is quite limited at this point, but the process would appear to have some utility. Yields are higher when the reaction is carried out under protic conditions, in which case both the organosamarium and samarium enolate intermediates generated in the process are immediately protonated.[61,73] Esters, alkenes, and methoxyethoxymethyl ethers can withstand the reaction conditions; however, heterosubstituents $\alpha$ to the carbonyl are reduced at the same or more rapid rates than the cyclopropyl ketone.[61,73,74]

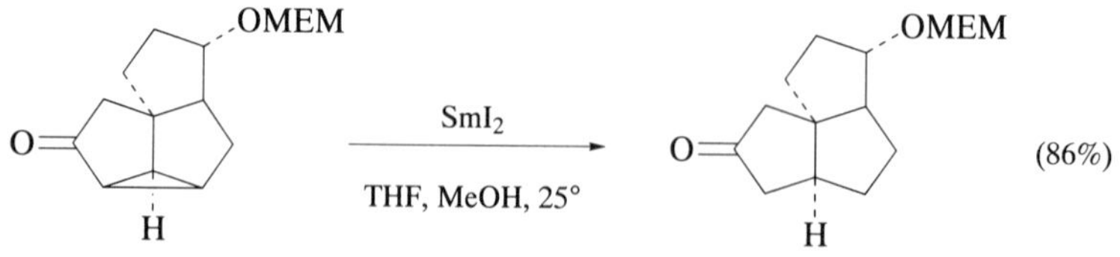

## Deoxygenation Reactions

Samarium(II) iodide effects the deoxygenation of a variety of functional groups, and thus has proven itself synthetically useful for such transformations. Epoxides, for example, can be reduced effectively to the corresponding alkenes with $SmI_2$. In THF/alcohol solvent mixtures, Lewis acid promoted epoxide rearrangements compete with the deoxygenation, and as a consequence mixtures of alkene, ketone, and alcohol can result.[3] On the other hand, when reductions are carried out in HMPA/$N,N$-dimethylaminoethanol, the Lewis acid promoted epoxide ring opening is inhibited, and clean conversion to the alkenes results.[75] Although deoxygenations carried out under such conditions are completely regiospecific, they are not stereospecific. The deoxygenation thus affords both diastereomeric alkenes. These deoxygenations can be carried out in the presence of esters, but the tolerance of other functional groups has not been examined.

The $SmI_2$-promoted reduction of sulfoxides to sulfides is an extraordinarily rapid and facile process when carried out in THF/HMPA,[41,76] but it is much slower in THF alone.[3] Dialkyl, diaryl, and aralkyl sulfoxides are all converted to the corresponding sulfides in virtually quantitative yields, and the reaction can be carried out in the presence of ester and even ketone groups.[41,76]

The remarkable effects of HMPA on $SmI_2$-promoted deoxygenation reactions is also demonstrated in reductions of sulfones. Thus, although studies indicate that sulfones are resistant to reduction by $SmI_2$ carried out in THF, reactions performed in THF/HMPA demonstrate that diaryl sulfones can be reduced to sulfides in nearly quantitative yields.[41,76] Dialkyl sulfones are converted in only modest yields even with use of the THF/HMPA solvent system. For aryl alkyl sulfones, reductive cleavage may be a competitive process (see above).[33]

$N$-Oxides, triphenylphosphine oxide, and triphenylarsine oxide are all effectively deoxygenated with $SmI_2$ under relatively mild reaction conditions.[41,49,76] With the exception of the phosphine oxides, it does not appear that HMPA is necessary for these conversions because the reactions can be carried out efficiently in THF alone.

(98%)

## Reduction of Nitrogen-Based Functional Groups

Selective reduction of several nitrogen-based functional groups by $SmI_2$ has been examined, and useful procedures have emanated from these studies. For example, the reduction of nitro groups can be effectively controlled to provide either hydroxylamines or amines, depending upon the stoichiometry and duration of the reaction. In the presence of four equivalents of $SmI_2$ for short periods of time (less than five minutes), both nitroalkanes and nitroaromatics are reduced to hydroxylamines in modest to excellent yields.[79] With six to eight equivalents of $SmI_2$ over a period of hours, the corresponding amines are synthesized in equally good yields.[49,79,80] Amines, nitriles, tert-butyldiphenylsilyl ethers, and some acetals generally remain intact during such reactions, but in one case an ester functionality led to a complex mixture of reaction products.[79] Both reductions to hydroxylamines and to amines would appear to be quite general in scope and of synthetic utility for a wide range of substrates.

Aromatic nitriles are reportedly unreactive toward $SmI_2$ in THF/MeOH solvent mixtures,[80] but by using either acidic (aqueous phosphoric acid) or basic (aqueous potassium hydroxide) reaction conditions these substrates can be effectively converted to the corresponding amines.[46] In aromatic nitrile substrates containing an aromatic chloride, both the nitrile and the halide are reduced by $SmI_2$ under basic conditions. Reduction of the halide can be prevented, however, when acidic conditions are used. These transformations appear quite limited not only because of the lack of chemoselectivity, but also because alkyl nitriles provide only modest yields of the desired amines[46] or are completely unreactive.[80]

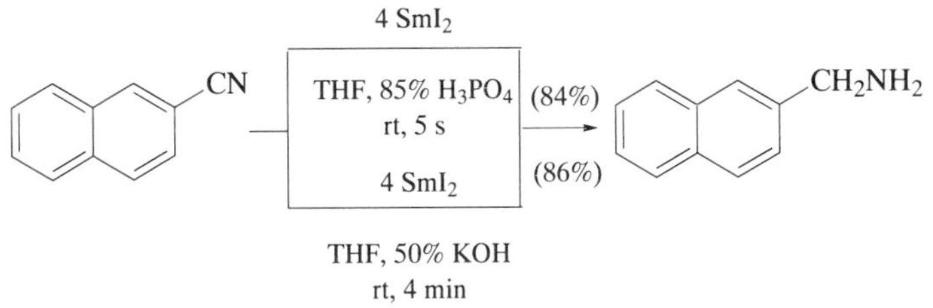

Only three imines have been employed as substrates in $SmI_2$-promoted reductions.[80,81] Yields in these cases are high, but with such a limited database it is difficult to extrapolate the generality of such transformations.

$$\text{PhCH=NPh} \xrightarrow[\text{THF, MeOH, rt, 24 h}]{2.5\ \text{SmI}_2} \text{PhCH}_2\text{NHPh} \quad (80\%)$$

Although oximes as a general class have not been examined for their ability to take part in $SmI_2$-mediated reductions,[42,80] 2-hydroximino amides of $\alpha$-amino acids have been extensively investigated as precursors to dipeptides.[82] Reductions of chiral amino acid precursors to the desired amines occur in good yields, and with fair to good diastereoselectivities (1,4-asymmetric induction).

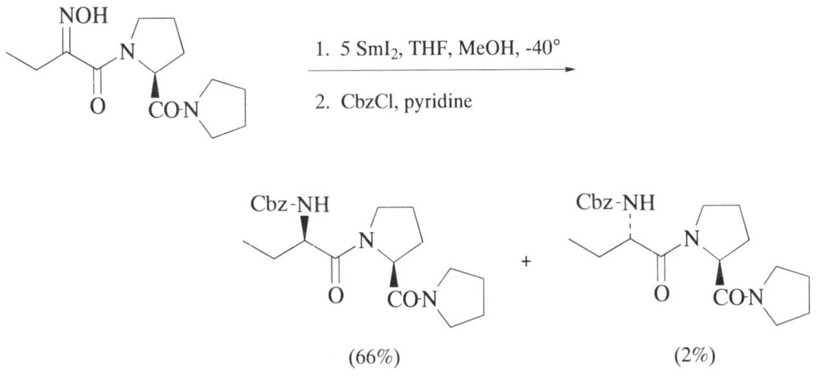

A single azo compound, azobenzene, has been reduced by $SmI_2$. With two equivalents of $SmI_2$, diphenylhydrazine is isolated,[49] but with eight equivalents of $SmI_2$ the hydrazine intermediate initially generated is reductively cleaved in modest yield to provide two equivalents of aniline.[80]

Although hydrazines have not been generally examined for their susceptibility to $SmI_2$-promoted reductive cleavage, indications are that such reactions are slow (four days at ambient temperature) and rather inefficient.[80] In contrast, *N*-acylhydrazines are rather good substrates for this reaction.[84] The reductive cleavage takes place within 30 minutes and provides acceptable yields of the desired

amines. This transformation is useful for the cleavage of enantiomerically enriched *N*-acylhydrazines, affording optically active amines.

89% ee                                         (72%) 89% ee

## Reduction of Miscellaneous Functional Groups

Various other functional groups, some in very limited samplings, have been
subjected to reduction by $SmI_2$. In addition, $SmI_2$ has been utilized as a stoichiometric reductant in processes catalyzed by transition metals. A number of these
transformations are outlined in this section.

Peroxides are quite efficiently reduced by $SmI_2$ in THF.[85,86] The process has
been utilized to release the diol functionality in bicyclic peroxides. The reaction
can be carried out in the presence of alcohols, alkenes, and primary alkyl tosylates.

The heteroatom–heteroatom bond in isoxazoles can also be cleaved efficiently
by $SmI_2$.[25] Although esters survive the process intact, aldehydes and benzylic
halides located off the isoxazole ring appear to be reduced in preference to the
nitrogen-oxygen bond.

A reasonably broad array of chloro phosphine oxides and chloro phosphine
sulfides have been subjected to reduction with $SmI_2$.[87] Reaction of the chloride
proceeds smoothly in THF at room temperature to provide the corresponding P-H
compound, without interference from deoxygenation of the phosphorus–oxygen
bond in the final product. In some cases, reductive dimerization is a minor side
reaction. Phosphine oxides are generally reduced more quickly than their phosphine sulfide analogs, and increasing the number of alkoxy groups attached to
phosphorus dramatically increases reaction times required for complete reduction
and markedly decreases the yield of the desired products as well. Somewhat unex-
pectedly, the reduction occurs with substantial loss of stereochemistry at an asymmetric phosphorus center. Consequently, it does not seem likely that the process
will be useful for the reduction of enantiomerically enriched starting materials.

$$\underset{\underset{EtO}{Ph}}{\overset{\overset{S}{\overset{||}{P}}}{\diagdown}}Cl \xrightarrow[\text{THF, rt, 48 h}]{2 \ SmI_2} \underset{\underset{EtO}{Ph}}{\overset{\overset{S}{\overset{||}{P}}}{\diagdown}}H$$

Allylic acetates are reduced to alkenes in high yields with $SmI_2$ and 2-propanol in the presence of a catalytic amount of palladium(0) catalyst.[88] Unfortunately, mixtures of regioisomeric and stereoisomeric alkenes are generated in these reactions, detracting from the synthetic utility of the process.

$$\xrightarrow[\text{THF, } i\text{-PrOH, 65°, 8 h}]{2 \ SmI_2, \ Pd(PPh_3)_4 \ (5 \ mol \ \%)}$$

(81%) racemic

In contrast to the palladium(0)-catalyzed reduction of allylic acetates, conditions for reductive cleavage of propargyl $\gamma$-lactones have been developed to the point whereby some control for the synthesis of allenes can be achieved.[89] Thus tertiary propargylic carboxylates usually provide allenes exclusively in the reaction with $SmI_2$-palladium(0). Secondary carboxylates and $\gamma$-lactones provide mixtures of allene and alkyne products, with allenes typically predominating. On the other hand, primary carboxylates often lead to product mixtures in which the alkyne products predominate.

$$n\text{-}C_8H_{17}C\equiv C \xrightarrow[\text{2. } CH_2N_2]{\substack{\text{1. } 2.5 \ SmI_2, \ Pd(PPh_3)_4 \ (\ 5 \ mol\%), \\ \text{THF, } i\text{-PrCHOHPr-}i, \\ -40°, \ 2 \ h}}$$

$$\underset{H}{\overset{n\text{-}C_8H_{17}}{\diagdown}}C=C=\underset{(CH_2)_2CO_2Me}{\overset{H}{\diagup}}C \qquad (66\%)$$

Some success has been achieved in accomplishing selective reduction reactions of alkynes in which $SmI_2$ is used as the stoichiometric reductant in THF, and cobalt, nickel, or iron salts are employed as catalysts.[91] The method has been touted as a means to prevent overreduction of the *cis*-alkene intermediates that commonly plague transition metal catalyzed hydrogenation reactions. Under optimized reaction conditions, diphenylacetylene is selectively reduced to *cis*-stilbene in 99% yield, and the selectivity for this product with respect to *trans*-stilbene and 1,2-diphenylethane is $>99:1$. In two different alkyne substrates, reasonably high selectivity for the $E$ isomer is observed when HMPA is added to the reaction mixture. Only a limited number of substrates have been examined in these hydrogenation studies, and the one functionalized molecule studied (a propargyl alcohol) leads to a mixture of alkene, allene, and alkyne. Thus although the method may be of use for hydrocarbon substrates, considerable development will be required before it can be viewed as a general method for the stereoselective reduction of alkynes.

$$PhC{\equiv}CPh \quad \xrightarrow[\text{THF, AcOH, rt, 30 min}]{2\ SmI_2,\ CoCl_2{\cdot}4\ PPh_3\ (3\ mol\%)} \quad \substack{Ph \diagup Ph \\ H \diagdown H} \quad (>99\%)$$

$$PhC{\equiv}CPh \quad \xrightarrow[\text{THF, HMPA, }i\text{-PrOH, rt, 30 min}]{2\ SmI_2,\ NiCl_2{\cdot}4\ PPh_3\ (3\ mol\%)} \quad \substack{Ph \diagup H \\ H \diagdown Ph} \quad (81\%)$$

## COMPARISON WITH OTHER METHODS

Given the very fundamental nature of the processes outlined in this chapter and the numerous other methods that have been developed to effect these same transformations, it would be impossible to compare $SmI_2$ with every other reducing agent for each individual class of compounds in a succinct manner. However, an attempt is made to compare some of the more useful reductions achieved by $SmI_2$ to those of other more popular methods.

### Reduction of Organic Halides, Sulfonates, and Sulfones

There are scores of reducing agents, including many low-valent metals or metal complexes, which have proven effective for the reduction of organic halides. As might be expected, there are significant areas where $SmI_2$ is effective and those where other reducing agents certainly surpass its ability. Samarium(II) iodide cannot be employed to reduce allylic halides or benzylic halides, and chiral alkyl halides will undergo nonstereospecific reduction. However, $SmI_2$ is a versatile reagent that can be utilized to reduce both $sp^3$ (primary, secondary, and tertiary) and $sp^2$-hybridized organic halides with reasonable efficiency. Perhaps one of the most useful characteristics of the reagent is that these reductions can be performed in the presence of unprotected alcohols and, in general, esters. This, plus the ability to vary the solvent to increase at will the reactivity and selectivity of the reducing agent, makes it likely that $SmI_2$ can be utilized for reduction of organic halides in the presence of many other functional groups as well.

The direct reduction of alkyl sulfonates to alkanes by $SmI_2$ appears promising, but because only a single example has been examined[2] extrapolation to other systems is premature. Samarium(II) iodide may not be as general or as selective as the more traditional hydride reducing agents in this respect.[118]

Samarium(II) iodide has shown tremendous promise in the reduction of $\alpha,\beta$-unsaturated sulfones to the corresponding alkenes without significant interference from conjugate reduction of the double bond.[32,33] In this regard it is a considerable improvement over other reducing agents, including sodium amalgam and sodium dithionite.[118] The reduction of alkyl aryl sulfones is also competitive with dissolving metal reductions and sodium amalgam in alcohol as a means to cleave the sulfone moiety, although the scope of the reaction with regard to the tolerance of various functional groups is of some concern[33] and reduction to sulfides may also compete.[41,76]

## Reductive Elimination/Fragmentation Reactions

The $SmI_2$-mediated reductive elimination of appropriately functionalized 1,2-disubstituted substrates is competitive with many other methods previously developed for the same transformation.[118] Drawbacks to the $SmI_2$ route are that it is not stereospecific, although in some cases it is more diastereoselective than sodium-promoted processes.[34] It may also be more tolerant of different functional groups (e.g., unprotected alcohols, esters)[31,35] than other methods. Furthermore, in direct comparison with sodium amalgam, $SmI_2$ has displayed enhanced efficiency for the reduction of $\beta$-hydroxy sulfones,[37] and thus it may be used as an alternative reagent in the conventional Julia sequence for the synthesis of alkenes.

A unique process that deserves further study is the fragmentation of $\gamma$-halo carbonyl substrates by $SmI_2$.[27,28,38,39] Although the full scope of this reaction is unknown at the present time, this distinctive transformation cannot be effected by other reductants like tri-$n$-butyltin hydride/azobis(isobutyronitrile) or zinc in acetic acid.[27] The $SmI_2$ promoted process thus certainly deserves further study.

## Reduction of Aldehydes and Ketones

As a general rule, $SmI_2$ will not compete with the numerous hydride reducing agents that have been developed for the chemoselective and stereoselective reduction of acyclic aldehydes and ketones.[118] However, in limited cases it may provide advantages over other single electron transfer agents that might normally be utilized for reductions. Samarium(II) iodide is generally more chemoselective than these other reagents, and in reductions of chiral cyclohexanones, higher selectivity for the equatorial alcohol isomer may be expected under optimized conditions using $SmI_2$.[41,45] This latter aspect of the chemistry bears further exploration.

## Reduction of Carboxylic Acids and Their Derivatives

Although conditions for the $SmI_2$-mediated reduction of carboxylic acids and various derivatives have been developed, by and large these methods are not particularly useful for the synthesis of alcohols.[42,46,48] In general the reaction conditions are not amenable to the incorporation of other functionality in substrates of interest, and for the most part the method is restricted to aromatic carboxylic acids and carboxylic acid derivatives. Consquently, it is unlikely that $SmI_2$-promoted reductions of carboxylic acids and their derivatives will compete with those of more established, general methods.[118]

## Reduction of Conjugated Carbonyl Substrates

Although $SmI_2$ provides mixtures of products in the reduction of conjugated aldehydes and ketones, it has proven to be quite general and selective for the conjugate reduction of $\alpha,\beta$-unsaturated carboxylic acids and their derivatives (esters and amides).[50,51] Perhaps most useful among these is the reduction of the unsaturated carboxylic acids themselves,[50] because relatively few general methods for this conversion exist.[118] This reactivity pattern is useful for the direct reduction of unsaturated carboxylic acids because it circumvents synthetic designs requiring conjugate reduction through an ester where inefficient protection–deprotection schemes are necessitated.

### Reductive Cleavage of α-Heterosubstituted Carbonyl Compounds and Related Substrates

The reduction of α-heterosubstituted carbonyl compounds and related substrates has long been recognized as a useful synthetic transformation, and many different reagents have been developed to realize this conversion.[53,118] Among the more widely utilized methods are zinc metal and chromium(II) ion-induced reductions, which require the use of acidic media for extended reaction times, often at elevated temperatures. Only a limited array of functionality can be tolerated under such conditions. Other methods (e.g., dissolving metal reductions, phosphorus-based or silicon-based reducing agents) either lack chemoselectivity or have not been thoroughly investigated to ascertain the scope of their capabilities. Thus $SmI_2$ appears to be among the more versatile and useful of the reagents available for this transformation.

Of notable interest is the ability of $SmI_2$ to cleave α-hydroxy carbonyl substrates by a reductive process.[53–57,118] Because it is one of only a handful of reagents for this particular transformation and because the $SmI_2$-mediated method appears to be mechanistically distinct from the others, it is likely to find considerable use in this transformation.

Likewise the reductive cleavage of α,β-epoxy ketones and α,β-epoxy esters with $SmI_2$ is a highly useful tool for the construction of aldol-type products.[67–69] Although several other methods have been developed for the same process,[69,118] $SmI_2$ is certainly competitive, if not superior, to these with respect to yields, versatility, and selectivity.

Among the reagents reported to cleave vinyloxiranes in a reductive process, $SmI_2$ may be unique in terms of its specificity and overall capabilities.[67,72,118] The observed regioselectivity of the final products, along with the generality of the process for a host of activated substrates, combine to make $SmI_2$ the reagent of choice for such conversions.

Reductive cleavage of cyano phosphates with $SmI_2$ provides a facile entry to the corresponding nitriles.[71,72] Previous attempts at this and related conversions with other reducing agents resulted either in complete failure or in methods with rather severe limitations.[72,119] Thus $SmI_2$ fills a useful niche, permitting the rather direct formation of one-carbon homologated nitriles from aldehyde and ketone precursors.

### Reductive Cleavage of Cyclopropyl Ketones

As a general rule the reductive cleavage of cyclopropyl ketones has seen little development as a synthetic method for the synthesis of open-chain ketones.[74] Nevertheless, of the few strategies that have been developed for such processes (e.g., photochemical and radical methods), conversions mediated by $SmI_2$ would appear to be competitive based upon the limited number of substrates examined.

## Deoxygenation Reactions

Although $SmI_2$ is reasonably effective for the deoxygenation of epoxides to produce the corresponding alkenes,[3,75] the lack of stereospecificity engendered in this process makes it of limited utility relative to the numerous methods exhibiting this feature.[118]

A number of methods have been developed for the deoxygenation of sulfoxides and N-oxides, but only a handful of methods are available for the deoxygenation of sulfones and phosphorus oxides. Consequently, the ability of $SmI_2$ to deoxygenate these functional groups clearly represents a potentially significant procedure.[75] If the limited number of substrates subjected to these deoxygenations could be expanded, providing a better idea of the scope of the reaction, $SmI_2$ could well prove to be the reagent of choice for such operations.

## Reduction of Nitrogen-Based Functional Groups

There are notable areas of success in $SmI_2$-mediated reductions of functional groups involving nitrogen. One example is the reduction of nitro groups, wherein appropriate control of stoichiometry and reaction conditions permits the isolation of either hydroxylamines or amines.[79] This method is fully comparable to alternative procedures for other representative syntheses of alkyl hydroxylamines and alkyl amines from nitro precursors.

On the other hand, the reduction of aromatic nitriles to amines must be performed under rather harsh (acidic or basic) reaction conditions.[46] Consequently, there is little tolerance for other functionality. Furthermore, the method is not applicable to alkyl nitriles, and thus the countless other reagents that can be utilized for the synthesis of amines from nitriles are probably best used for this transformation.[118]

Only three imines have been reduced to the corresponding amines with $SmI_2$,[80,81] and with such a limited database it is risky to speculate on the scope of this reaction in relation to other methods for the same transformation.

The enhanced chemoselectivity of $SmI_2$ relative to that of lithium aluminum hydride, sodium borohydride, and borane has made it an attractive alternative to these more traditional reagents for the diastereoselective reduction of 2-hydroximino amides.[82] It is likely that with further development more uses of the reductant with similar substrates will be possible.

Although hydrazines in general react slowly with $SmI_2$, acyl hydrazines are excellent substrates for the reagent.[84] Because of the lack of studies wherein functional group compatibility and other factors are available, direct comparison with other potential reducing agents is difficult. However, $SmI_2$ appears to be an excellent reagent for this type of reductive cleavage reaction.

## Reduction of Miscellaneous Functional Groups

Limited data are available on the reduction of other functional groups with $SmI_2$. In reactions with bicyclic peroxides, the reagent appears comparable to zinc/acetic acid for the reductive cleavage, and may hold an advantage in systems with acid-sensitive arrays of functionality.[85]

The reduction of halo phosphine oxides and halo phosphine sulfides holds some promise as a general method for the synthesis of phosphine oxides and phosphine sulfides, but unfortunately will not compete with routes that provide these materials in enantiomerically pure form.[87]

Similarly, the palladium(0)-catalyzed reductive cleavage of allylic acetates[88] and propargyl acetates[89] is somewhat limited in scope given the mixtures of products that are typically generated. Nevertheless, the $SmI_2$-mediated process would appear comparable in many respects to those transformations using ammonium formate, borohydride, trialkyltin hydride, and other reducing reagents.

There is some promise for the use of transition metal catalyzed hydrogenations of alkynes to stereodefined alkenes using $SmI_2$ as a stoichiometric reductant.[91] Although innumerable other catalysts and synthetic methods have been developed for this transformation,[118] the method using $SmI_2$ has demonstrable selectivity for alkene formation, and under suitable reaction conditions both the *cis-* and *trans-*olefinic isomers can be generated with considerable selectivity. However, substantial development of the method will be required before it can be considered a general method for the selective hydrogenation of alkynes to alkenes.

## EXPERIMENTAL CONDITIONS

Pure samarium(II) iodide is a deep blue, air-sensitive compound. Consequently, all manipulations involving this material must be carried out in an inert atmosphere (e.g., argon or nitrogen). However, $SmI_2$ is not so air-sensitive that glovebox techniques or even Schlenk-type glassware must be utilized, and thus normal benchtop techniques for the handling of air-sensitive materials suffice. Somewhat surprisingly, $SmI_2$ does not react appreciably with water over several hours, and is clearly even less reactive toward other protic solvents (e.g., alcohols). Consequently, these may be added as cosolvents to reactions of $SmI_2$ in the presence of various organic substrates with little or no detrimental effect to the reducing agent, and in fact may be necessary for the efficacy of the reactions themselves.

Most often solutions of $SmI_2$ in THF are utilized to effect the desired transformations. Although solutions of $SmI_2$ in THF are commercially available, it is very easy to prepare the reagent in situ. As far as can be determined, yields in the published preparations are virtually quantitative and thus these methods provide a rapid and convenient source of the reducing agent. Perhaps the most convenient method is the oxidation of samarium metal with diiodomethane.[6] The oxidant in this case is a liquid that can be injected into a rapidly stirring slurry of samarium metal in THF. Another efficient method for the preparation of $SmI_2$ employs 1,2-diiodoethane as the oxidant.[3-5] The only drawbacks to this procedure are that the oxidant must be purified to some extent before use, and the fact that the oxidant is a solid. This makes its introduction to the slurry of samarium metal somewhat less convenient than the diiodomethane route. The other methods for preparation of $SmI_2$ (using iodine[8] or trimethylsilyl chloride/sodium iodide[9] as the oxidant) suffer from either extremely long reaction times or require less useful solvent systems and are therefore much less generally utilized. Samarium(II) iodide can be stored as a solution in THF for reasonably long periods of time, particularly when it is stabilized by a small amount of samarium metal.[3-5] Alternatively, the solvent may be removed altogether, providing $SmI_2(THF)_n$ powder.[3-5]

The $SmI_2$ prepared by these methods can be characterized in solution in several different ways, including techniques involving absorption spectroscopy, magnetic susceptibility measurements, titration of samarium ions with $N,N,N',N'$-tetramethylethylenediamine, potentiometric titrations of iodide ion, and acidometric titration and reaction of iodine, which measures the reductive capability of the solutions.[3,4] The last method serves for the rapid and convenient determination of the molar concentration of $SmI_2$ in, for example, THF solution.

Because $SmI_2$ is a one-electron reducing agent, synthetic conversions that require a net two-electron reduction necessitate the addition of two molar equivalents of $SmI_2$. Similarly, four-electron conversions require four molar equivalents, and so forth. In published experimental procedures it is not unusual to find that an excess of $SmI_2$ is used in certain difficult conversions, although in general it is sufficient to utilize a stoichiometric amount of the reagent or a slight excess to effect complete reaction.

The vast majority of reactions employing $SmI_2$ are carried out in THF, and because $SmI_2$ is conveniently generated in this solvent this makes the in situ preparations particularly useful. However, there is a dramatic solvent effect on the ability of $SmI_2$ to reduce various functional groups.[24,41] In general polar aprotic solvents such as acetonitrile, $N,N'$-dimethylpropyleneurea, tetramethylurea, $N,N$-dimethylformamide, $N,N$-dimethylacetamide, 1-methyl-2-pyrrolidinone, $N,N,N',N'$-tetramethylethylenediamine, and in particular, HMPA greatly enhance the reducing power of $SmI_2$. The precise reason for this dramatic activation is unknown, but could be attributed to deaggregation of $SmI_2$ (the solution structure of $SmI_2$ in THF is unknown) or $f$-orbital perturbation owing to ligand field effects in the presence of the strong donor ligands, raising the energy of the highest occupied molecular orbital (electron-donating orbital) and thereby increasing the samarium(II)/samarium(III) cell electromotive force.[120] In fact, a combination of these two effects might be responsible. Whatever the reason, reports in the literature reveal that the reduction potentials of several lanthanides are strongly dependent on the associated ligand(s).[121-124] As an example, a negative shift of 0.86 V in the reduction potential of europium(III) is observed upon addition of 5% dimethyl sulfoxide by volume as cosolvent in propylene carbonate.[124,125] Similar effects are observed for the reduction potentials of samarium(III) and ytterbium(III).

Some explanation for the unique nature of the interaction of HMPA with lanthanide ions and its effect on the reduction potential may be derived from the X-ray crystal structures of various lanthanide(III) ion–HMPA complexes. Crystal structures of such complexes reveal that in all cases the HMPA molecules are coordinated via their oxygen atoms. For example, reaction of ytterbium(III) chloride with excess HMPA in THF produces $YbCl_3(HMPA)_3$ in quantitative yield.[126] The X-ray crystal structure determination reveals that HMPA molecules occupy the two axial positions and an equatorial position in the distorted octahedron of this complex. Remarkably, further studies reveal that the maximum number of HMPA ligands able to bind to any of the lanthanide(III) perchlorates is six.[127] Thus, in all cases (Ln = La − Lu), treatment of $LnCl_3$ with three equivalents of silver perchlorate followed by addition of excess HMPA results in the isolation of crystals that correspond to complexes with the stoichiometry $[Ln(HMPA)_6](ClO_4)_3$. These results are in stark contrast to studies using dimethyl sulfoxide or $N,N$-dimethylacetamide complexes, in which the number of ligands decreases with decreasing ionic radius of the lanthanides.[128,129] The singular ability of HMPA to activate Sm(II) may thus be attributed to its unique binding and stabilization of Sm(III) cations generated as a result of electron transfer.

The only drawback to the unique effect of HMPA activation of $SmI_2$ is that HMPA is a potent carcinogen, and thus must be handled with extreme care. However, unlike many of their main group and transition metal counterparts, inorganic lanthanide complexes themselves are generally classified as nontoxic when introduced orally, and only modestly toxic when injected.[18] Although toxicity may vary to some extent based on the ligands attached to the metal, in nearly all cases the lanthanide complexes are converted to hydroxides immediately upon ingestion, and thus are believed to be poorly absorbed in the digestive tract.

Another important factor in the efficacy of many $SmI_2$-promoted reductions is the presence of a proton source in the reaction mixture. Often, yields of the desired products are greatly enhanced when reactive intermediates such as enolates, alkoxides, or organosamariums are quenched immediately by in situ proton sources such as water, aliphatic alcohols, carboxylic acids, glycols, or amino alcohols in the reaction mixture. Furthermore, in some cases the presence or absence of a proton source may be the major determinant in mechanistic pathways followed by reactive intermediates. For example, treatment of aldehydes and ketones with $SmI_2$ in the presence of protic solvents generally leads to clean reduction to the corresponding alcohols. In the absence of such additives, ketyl dimerization occurs, leading to the efficient production of pinacols.[18,22,100,101]

Iron(III) salts have been found to catalyze various carbon−carbon bond-forming reactions promoted by $SmI_2$ (e.g., Barbier-type reactions).[3,18,21,22] However, such catalysts have only rarely been utilized for selective reduction reactions. Additionally, some highly specific reactions have been developed in which $SmI_2$ serves as a stoichiometric reductant for processes promoted by palladium(0)[88−90] or transition metal catalysts.[91] As a general rule, though, the potential role of catalysts in $SmI_2$-mediated reactions has not been explored.

Finally, workup procedures for $SmI_2$-promoted reactions may vary considerably depending on the stability of the desired products. If the organic products of the $SmI_2$-promoted reactions are stable to aqueous acids, aqueous hydrochloric acid can be utilized to quench the reaction. With this protocol the samarium salts generated as a result of the reaction are water soluble and are easily removed in the aqueous layer. For acid-sensitive organic products, mildly basic solutions or pH 7-8 buffers may be utilized to quench the reaction mixtures. In such cases the samarium salts are typically insoluble, but do form a suspension in the aqueous phase from which the desired organic product can be extracted by standard experimental procedures.

## EXPERIMENTAL PROCEDURES

$$Sm \ + \ ICH_2CH_2I \ \longrightarrow \ SmI_2 \ + \ CH_2=CH_2$$

**Samarium(II) Iodide (Preparation of the Reducing Agent from Samarium Metal Using 1,2-Diiodoethane as the Oxidant).**[3-5] The synthesis of samarium(II) iodide was performed under a nitrogen atmosphere. In a standard procedure, 1.504 g (10 mmol) of samarium powder was placed in a Schlenk tube. A 50-mL tetrahydrofuran solution of 1,2-diiodoethane (1.410 g, 5 mmol) was slowly added. The reactants were vigorously stirred with a magnetic stirrer. After a short induction period, a deep blue-green color appeared. After one hour, a 0.1 M solution of samarium(II) iodide in tetrahydrofuran was obtained. Titrations showed that the yield of the reaction was quantitative. Such solutions can be stored for long periods of time without a decrease in Sm(II) concentration if kept under an inert atmosphere and in the presence of a small amount of samarium metal.

$$Sm \ + \ ICH_2I \ \longrightarrow \ SmI_2 \ + \ 0.5 \, CH_2=CH_2$$

**Samarium(II) Iodide (Preparation of the Reducing Agent from Samarium Metal Using Diiodomethane as the Oxidant).**[6] Samarium metal powder (0.15 g, 1 mmol) was added under a flow of argon to an oven-dried round-bottomed flask containing a magnetic stirring bar and a septum inlet. The flask and the samarium metal had been flame-dried and cooled under a stream of argon. Tetrahydrofuran (10 mL) was added. The vigorously stirred slurry of samarium metal and tetrahydrofuran was cooled to 0°, and neat diiodomethane (0.228 g, 0.85 mmol) was added. The resulting green slurry was stirred at 0° for 15 minutes, then allowed to warm to room temperature and vigorously stirred for an additional hour. The resulting solution of samarium(II) iodide was a deep blue color.

$$Sm \ + \ I_2 \ \longrightarrow \ SmI_2$$

**Samarium(II) Iodide (Preparation of the Reducing Agent from Samarium Metal Using Iodine as the Oxidant).**[7] Iodine (5.1 g, 20 mmol) was added with stirring to a mixture of 40-mesh samarium powder (3.3 g, 22 mmol) and dry tetrahydrofuran (200 mL) under argon. The initial mildly exothermic reaction subsided in several minutes to form a yellow suspension of $SmI_3$. The mixture was then heated at reflux with stirring. The color of the suspension gradually turned from yellow to green and finally to an intense blue-green. Heating at reflux overnight provided a 0.1 M solution of $SmI_2$.

$$\text{Sm} \ + \ 2\,\text{Me}_3\text{SiCl} \ + \ 2\,\text{NaI} \ \longrightarrow \ \text{SmI}_2 \ + \ \text{Me}_3\text{SiSiMe}_3 \ + \ 2\,\text{NaCl}$$

**Samarium(II) Iodide (Preparation of a Samarium Iodide Equivalent in Acetonitrile Solvent).[8]**   To a solution of sodium iodide (0.9 g, 6 mmol) in dry acetonitrile (20 mL) was added chlorotrimethylsilane (0.76 mL, 6 mmol) followed by samarium powder (2 mmol) under a nitrogen atmosphere at room temperature. The samarium gradually reacted and the color of the solution turned to deep green, indicating the production of a samarium(II) iodide equivalent.

**1-Phenyldec-1-yne (Generation of Alkylidenecarbenes from 1,1-Dibromoalk-1-enes).[23]**   A solution of 1,2-diiodoethane (3.55 mmol) in benzene (32 mL) [Caution: Potent Carcinogen] and hexamethylphosphoric triamide (3.6 mL) [Caution: Potent Carcinogen] was added to 40-mesh samarium powder (5.32 mmol) under nitrogen. Gentle heating was required to initiate the reaction. The reaction mixture was stirred for 5 days at room temperature to afford a purple solution of samarium(II) iodide in benzene–hexamethylphosphoric triamide [Caution: Potent Carcinogen]. The concentration of $\text{SmI}_2$ was determined by titration using iodine under nitrogen according to the method of Imamoto and Ono.[8] To the purple solution of samarium(II) iodide thus generated (4 mL of a 0.094 mol $\text{dm}^{-3}$ solution, 0.38 mmol) was added a solution of 1,1-dibromo-2-phenyldec-1-ene in benzene [Caution: Potent Carcinogen] at room temperature under nitrogen. After stirring for 10 minutes, the mixture was quenched with dilute hydrochloric acid, and then extracted with diethyl ether. The organic layer was dried and concentrated to afford a crude mixture that was purified by preparative TLC to afford the title compound in 67% yield, along with 8% of 2-phenyl-1-decene.

**3-Methoxy-14,21-cyclo-19-norpregna-1,3,5(10)-trien-17-ol Acetate (Reductive Desulfonylation Reactions).[33]**   Under an atmosphere of argon, 0.494 g of (16$\alpha$,17$\alpha$)-3-methoxy-16-(phenylsulfonyl)-14,21-cyclo-19-norpregna-1,3,5(10)-trien-17-ol acetate was dissolved in 50 mL of a freshly prepared solution of samarium(II) iodide–tetrahydrofuran (approximately 0.1 M). This mixture was cooled to −20° under stirring and hexamethylphosphoric triamide (4 mL) [**Caution: Potent Carcinogen**] was added dropwise by syringe, whereupon the color of the solution changed from blue to purple. After 90 minutes, the reaction was terminated with aqueous ammonium chloride (5 mL). Most of the tetrahydrofuran was removed in vacuo by rotary evaporation. The product was precipitated by addition of cold hydrochloric acid (0.5 M) and isolated by suction–filtration. The solid residue was dissolved in ethyl acetate, and the resulting organic phase was washed with aqueous sodium thiosulfate solution, followed by brine, and then dried over anhydrous sodium sulfate. Chromatography of the crude product on silica gel (hexane-ethyl acetate, 9:1) provided 0.308 g (87%) of the title compound, mp 119–120° (acetone-hexane).

**[6R-(6α[2S\*(R\*),3S\*],8β(2S\*,3S\*,5S\*),9β,10β)]-α-Ethyl-10-hydroxy-3,9-dimethyl-8-[tetrahydro-3-(hydroxymethyl)-2-methoxy-5-methyl-2H-pyran-2-yl]-1,7-dioxaspiro[5.5]undecane-2-ethanol [Deprotection of (2,2,2-Trichloroethoxy)methoxy Ethers by Reductive β-Elimination].**[35] [6R-(6α[2S\*(R\*),3S\*],8β(2S\*,3S\*,5S\*),9β,10β)]-α-Ethyl-10-[(2,2,2-trichloroethoxy)methoxy]-3,9-dimethyl-8-[tetrahydro-3-(hydroxymethyl)-2-methoxy-5-methyl-2H-pyran-2-yl]-1,7-dioxaspiro[5.5]undecane-2-ethanol (6.2 mg, 10.2 μmol) was azeotropically dried with two 1-mL portions of toluene and was subsequently dissolved in 0.8 mL of tetrahydrofuran. Freshly prepared samarium(II) iodide (71.2 mmol, 0.72 mL, 0.10 M in tetrahydrofuran) was introduced in one portion, affording a dark blue solution. After 35 minutes at ambient temperature, the reaction was diluted with 15 mL of diethyl ether and was extracted with 10 mL of saturated aqueous potassium carbonate. The aqueous extract was washed with 10 mL of ethyl acetate, and the combined organic layers were washed successively with 15 mL of saturated aqueous sodium sulfite and 15 mL of brine, dried (anhydrous sodium sulfate), filtered, and concentrated. Purification of the residue by flash chromatography (1 × 18 cm, linear gradient of 60–80% ethyl acetate/hexane) yielded 3.2 mg (71%) of the title compound as a clear oil. This material proved identical in all respects (¹H NMR, optical rotation, TLC, GC coinjection, mass spec) with natural material.

**1,4-Diphenylbut-1-ene (Reductive Elimination of β-Hydroxy Imidazoyl Sulfones).**[37] To a stirred solution of samarium(II) iodide (1.5 mmol) in tetrahydrofuran (12 mL) was rapidly added a solution of β-[(1-methyl-1H-imidazol-2-yl)sulfonyl]-α-phenylbenzenebutanol (0.185 g, 0.5 mmol) in tetrahydrofuran (6 mL) under an argon atmosphere. After 15 minutes at room temperature the reaction was still blue because of the excess of $SmI_2$ utilized. The reaction mixture was then poured into a 10% solution of sodium thiosulfate (20 mL) and extracted with ethyl acetate. The residue was chromatographed over silica gel (hexane–ethyl acetate, 99:1) to provide 0.085 g (82%) of the title compound as an 8:1 mixture of E/Z olefins. Satisfactory spectral data (IR, $^1$H NMR, $^{13}$C NMR, and mass spectral data) were obtained for the product, the spectra matching that of material previously described in the literature.

**(R)-Benzoin (Enantioselective Reduction of Ketones).**[43] Benzil (30 mg, 0.14 mmol) and quinidine (93 mg, 0.28 mmol) were dissolved in tetrahydrofuran (1.8 mL), and to this solution hexamethylphosphoric triamide (0.1 mL) [**Caution: Potent Carcinogen**] and then samarium(II) iodide–tetrahydrofuran (0.1 mol dm$^{-3}$, 2.8 mL, 0.28 mmol) were added under an atmosphere of argon. After stirring for 30 minutes at room temperature, hydrochloric acid (0.1 mol dm$^{-3}$, 5 mL) was added and the mixture was extracted with benzene [**Caution: Potent Carcinogen**]. The organic phase was washed with brine, aqueous sodium thiosulfate, brine, 3 mol dm$^{-3}$ hydrochloric acid (20 mL), and brine successively, and dried over anhydrous magnesium sulfate. Benzene was removed and the residue was adsorbed on a silica gel column (Wako gel C-300, 7 g, 1.80 × 8 cm) and eluted with benzene (400 mL) [**Caution: Potent Carcinogen**]. The eluate was concentrated and the residue (18.2 mg) was analyzed by HPLC using a chiral column (Chiralcel OD; hexane:2-propanol, 9:1). Benzil (22.4%), benzyl phenyl ketone (trace), and benzoin (77.5%) were detected, and the enantiomers of benzoin were completely separated. Under these reaction conditions the benzoin was generated in 56.2% ee (R isomer predominating).

**(2S,4S,5R,1′S)-3-Oxazolidinecarboxylic Acid, 2-(3-Methoxy-1-methyl-3-oxopropyl)-4-methyl-5-phenyl Methyl Ester and (2S,4S,5R,1′R)-3-Oxazolidinecarboxylic Acid, 2-(3-Methoxy-1-methyl-3-oxopropyl)-4-methyl-5-phenyl Methyl Ester (Reduction of α,β-Unsaturated Carbonyl Substrates).[52]** A solution of (2S,4S,5R)-3-oxazolidinecarboxylic acid, 2-(3-methoxy-1-methyl-3-oxo-1-propenyl)-4-methyl-5-phenyl methyl ester (56.5 mg, 0.18 mmol of an 88:12 mixture of E:Z isomers) in tetrahydrofuran/water (5/1) was treated with a 0.1 M tetrahydrofuran solution of samarium(II) iodide (8.8 mL, 0.88 mmol) at room temperature and under nitrogen until a persistent blue color was obtained. The mixture was extracted with diethyl ether. The organic extracts were dried over anhydrous sodium sulfate and filtered, and the solvent was evaporated under reduced pressure. The crude product was purified by flash chromatography (n-hexane/ethyl acetate, 75/25) to provide the title compounds in 33% yield as a 9:1 mixture of 1′S:1′R isomers. $^1$H NMR (CDCl$_3$, 200 MHz): $\delta$ 0.82 (3H, d, $J$ = 7.1 Hz), 1.05 (3H, d, $J$ = 6.1 Hz), 2.25–2.70 (2H, m), 2.85 (1H, m), 3.71 (3H, s), 3.76 (3H, s), 4.81 (1H, m), 5.04 (1H, d, $J$ = 2.8 Hz), 5.14 (1H, d, $J$ = 6.0 Hz), 7.25–7.46 (5H, m). $^{13}$C NMR (CDCl$_3$, 50 MHz): $\delta$ 12.74, 16.06, 32.7, 37.7, 51.58, 52.51, 56.34, 80.556, 91.12, 125.9, 127.7, 128.2.

**5-Iodo-1-phenyl-1-pentanone (Reduction of α-Acyloxy Ketone Substrates).[53]** To a slurry of samarium powder (0.32 g, 2.1 mmol) in 2 mL of tetrahydrofuran at room temperature was added a solution of 1,2-diiodoethane (0.56 g, 2 mmol) in 2 mL of tetrahydrofuran. The resultant olive-green slurry was stirred at ambient temperature for 1 hour, after which time the resulting dark blue slurry of samarium(II) iodide that had formed was cooled to −78° and treated with a solution of 2-acetoxy-5-iodo-1-phenyl-1-pentanone (0.35 g, 1 mmol) in 1 mL of methanol and 2 mL of tetrahydrofuran. The resultant brown mixture was stirred for 10 minutes at −78°, warmed to room temperature, and then poured into saturated aqueous potassium carbonate. The aqueous phase was extracted with diethyl ether (5 × 10 mL) and the combined extracts were dried (anhydrous magnesium sulfate). Evaporation of the solvent left a solid that was recrystallized from diethyl ether to afford 0.24 g (87%) of the title compound, mp 72–73°. IR (CCl$_4$): 1690 cm$^{-1}$. $^1$H NMR (CCl$_4$): $\delta$ 7.9 (m, 2H), 7.4 (m, 3H), 3.2 (t, $J$ = 6 Hz, 2H), 2.9 (t, $J$ = 7.5 Hz, 2 H), 1.8 (m, 4H). $^{13}$C NMR: $\delta$ 199.26,

136.63, 132.90, 128.21 (2C), 127.84 (2C), 37.09, 32.85, 24.92, 6.12. Exact mass spectral analysis, calcd for $C_{11}H_{13}IO$, 288.0012; found, 288.0011.

**1,8-Dichloro-11,11-dimethoxy-3-*exo*-hydroxytetracyclo[6.2.1.0²,⁷.0⁴,¹⁰]** **undec-5-en-9-one (Reductive Cleavage of α-Halo Ketones).**[29] A solution of 1,8,10-trichloro-11,11-dimethoxy-3-*exo*-hydroxytetracyclo[6.2.1.0²,⁷.0⁴,¹⁰]undec-5-en-9-one (0.34 g, 1 mmol) in a tetrahydrofuran–methanol solution (3 mL, 2:1) was added to a solution of samarium(II) iodide (2.8 equiv, 6 mL) at −78°. The reaction mixture was allowed to warm to room temperature and stirred overnight (16 hours). The reaction mixture was worked up by pouring it into saturated aqueous potassium carbonate (25 mL) and extracting the resultant mixture with ethyl acetate (50 mL). The organic layer was washed with water and brine and dried (anhydrous magnesium sulfate). The residue obtained after removal of solvent under reduced pressure was applied to a silica gel column. Elution with 10% ethyl acetate–hexane furnished the title compound (218 mg, 68%), mp 148–149°. IR (KBr): 3526, 2951, 1764, 1215, 1087, 933, 787 cm⁻¹. ¹H NMR (CDCl₃): δ 2.87 (ddd, $J$ = 8.8, 4.5, 2.4 Hz, 1H), 2.95 (d, $J$ = 9.3 Hz, 1H), 3.30–3.22 (m, 1H), 3.32 (dd, $J$ = 5.7, 1.4 Hz, 1H), 3.37 (dd, $J$ = 8.1, 2.4 Hz, 1H), 3.55 (s, 3H), 3.70 (s, 3H), 3.98 (dd, $J$ = 9.3, 1.8 Hz, 1H), 5.91 (ddd, $J$ = 8.3, 5.7, 1.1 Hz), 6.32 (ddd, $J$ = 8.3, 7.1, 1.44 Hz, 1H). ¹³C NMR: δ 197.49, 137.81, 127.52, 99.15, 85.17, 83.56, 78.30, 62.58, 52.09, 49.57, 48.66, 46.75. Anal. Calcd for $C_{13}H_{14}O_4Cl_2$: C, 51.31; H, 4.64; Cl, 23.00. Found: C, 51.24; H, 4.61; Cl, 22.77.

**(R)-Diisopropyl Malate (Reduction of α-Hydroxy Ester Substrates).**[54] To a mixture of (R,R)-diisopropyl tartrate (148 mg, 0.63 mmol) and a samarium(II) iodide–tetrahydrofuran solution (1.9 mmol, 0.1 mol dm⁻³, 19 mL) was added dropwise a solution of ethylene glycol (0.5 mL) in tetrahydrofuran (19 mL) over a period of 30 minutes at room temperature. After stirring for an additional 30 minutes, the reaction mixture was exposed to air to quench the excess samarium(II) iodide. Ethylene glycol (0.57 mL), silica gel (approximately 3 g), and hexane (10 mL) were added and the mixture was stirred for 10 minutes. Chromatographic purification (silica gel, hexane/ethyl acetate, 3:1) provided the title compound (137 mg, 99%) as an oil.

**(R)-2-Deoxy-3,4-O-(phenylmethylene)-D-erythro-pentanoic Acid, δ-Lactone (Reduction of α-Hydroxy Aldonolactones).**[55]  To a solution of (R)-3,4-O-(phenylmethylene)-D-ribonic acid, δ-lactone (236 mg, 1.0 mmol), anhydrous ethylene glycol (650 μL, 12 equiv), and hexamethylphosphoric triamide (1.5 mL) **[Caution: Potent Carcinogen]** in tetrahydrofuran (10 mL) was added dropwise a solution of 0.1 M samarium(II) iodide in tetrahydrofuran (30 mL, 3 mmol, 3 equiv) at room temperature under argon. After stirring for 3 hours, the mixture was quenched with saturated aqueous sodium bicarbonate, then extracted with ethyl acetate. The organic layer was washed with aqueous sodium thiosulfate, brine, and water, and then dried. The volatiles were removed and the residue was chromatographed to afford 199 mg (90%) of the title compound, mp 138–139°. $[\alpha]_D = -167.9°$ (c 1.47). Characterization was accomplished by comparison with known physical constants from the literature. Additionally, a correct microanalysis was obtained, and the compound was further characterized by 300 MHz $^1$H NMR.

**(1'R*,3R*,4S*,4aR*,7S*,8aS*)-3,4,4a,7,8,8α-Hexahydro-7-hydroxy-4-methyl-3-(1'-methylprop-2'-enyl)-2(1H)-naphthalenone (Reduction of α-Alkoxy Ketone Substrates).**[63]  To a slurry of samarium powder (2.6 g, 17.2 mmol) in 18 mL of tetrahydrofuran at 25° was added 1,2-diiodoethane (4.34 g, 17.21 mmol) in 18 mL of tetrahydrofuran via cannula over 15 minutes. The transfer was completed with 5 mL of tetrahydrofuran. A dark blue-green color developed. After being stirred at 25° for 1 hour the solution was cooled to −78° and (1R*,1'R*,3R*,4S*,4aR*,7S*,8aS*)-1,7-epoxy-3,4,4a,7,8,8a-hexahydro-4-methyl-3-(1'-methylprop-2'-enyl)-2(1H)-naphthalenone (1.7 g, 7.3 mmol) in 24 mL of tetrahydrofuran was added over 10 minutes via cannula. The transfer was completed with 5 mL of tetrahydrofuran. The reaction mixture was stirred for 30 minutes, followed by addition of 30 mL of saturated aqueous potassium carbonate. The reaction mixture was warmed to 25°, the organic layer was separated, and the aqueous layer was extracted with four 100-mL portions of diethyl ether. The organic extracts were combined, washed with saturated aqueous sodium chloride, dried over anhydrous sodium sulfate, filtered, and concentrated in vacuo to yield 1.71 g of crude title compound that was utilized without further purification in a subsequent experiment. A sample of the material was purified by preparative TLC (25% ethyl acetate/hexanes) for spectral analysis. $^1$H NMR (500 MHz, CDCl$_3$): $\delta$ 1.2 [d, $J$ = 8 Hz, 6H, CH$_3$ (1″,4)], 1.6 (m, 2H), 1.75 (m, 1H), 1.95 (m, 1H), 2.05 (s, 1H), 2.3 (m, 2H), 2.55 (m, 1H), 4.3 [br s, 1H, H(7)], 4.95 [d, $J$ = 7 Hz, 1H, H(3′)], 5.0 [d, $J$ = 14 Hz, 1H, H(3′)], 5.9 [m, 3H, H(5,6,2′)]. IR (neat): 3420, 2960, 1700 cm$^{-1}$. LRMS (EI) $m/z$ 234 (M$^+$), 216 (M$^+$ −H$_2$O), 201 (M$^+$ −CH$_5$), 166.

***trans*-Hexahydro-4a-hydroxy-8a-methyl-1,6(2*H*,5*H*)-naphthalenedione (Reduction of $\alpha,\beta$-Epoxy Ketones).[69]** To a slurry of samarium powder (0.32 g, 2.1 mmol) in 2 mL of tetrahydrofuran at room temperature was added a solution of 1,2-diiodoethane (0.56 g, 2 mmol) in 2 mL of tetrahydrofuran. The resultant olive-green slurry was stirred at ambient temperature for 1 hour, after which time the resulting dark blue slurry of samarium(II) iodide that had formed was cooled to −90° and treated with a solution of (1a$\alpha$,4a$\alpha$,8a$R$*)-tetrahydro-4a-methyl-(1a$H$)-naphth[1,8a-$b$]oxirene-2,5(3$H$,6$H$)-dione (0.19g, 1.00 mmol) in 1 mL of methanol and 2 mL of tetrahydrofuran. The resultant brown mixture was stirred for 5 minutes at −90°, quenched at this temperature by the addition of saturated aqueous potassium carbonate or pH 8 phosphate buffer, and then warmed to room temperature. The aqueous phase was extracted with diethyl ether (5 × 10 mL), and the combined organic extracts were dried (anhydrous magnesium sulfate). Evaporation of the solvent left a solid that was recrystallized from diethyl ether to afford 0.14 g (76%) of the title compound, mp 186–187°. IR(CHCl$_3$/DMSO): 3330, 1700 cm$^{-1}$. $^1$H NMR (CDCl$_3$): $\delta$ 2.9–1.5 (m, 13H), 1.3 (s, 3H). $^{13}$C NMR (CDCl$_3$/DMSO-$d_6$): $\delta$ 213.59, 209.74, 79.04, 50.56, 49.87, 36.95, 35.93, 32.18, 27.35, 19.92, 19.79. Exact mass spectral analysis, calcd for C$_{11}$H$_{16}$O$_3$, 196.1099; found, 196.1105.

**Diethyl [(2E)-4,8-Dimethyl-4-hydroxy-2,7-nonadien-1-yl]phosphonate (Reduction of Vinyl Oxiranes).[70]** To a slurry of samarium powder (0.32 g, 2.1 mmol) in 2 mL of tetrahydrofuran at room temperature under argon was added a solution of 1,2-diiodoethane (0.56 g, 2 mmol) in 2 mL of tetrahydrofuran. The resultant olive-green slurry was stirred at ambient temperature for 1 hour, after which time the resulting dark blue slurry of samarium(II) iodide was cooled to $-90°$ and treated with a solution of diethyl [(3$R$*,4$S$*)-(1$E$)-4,8-dimethyl-3,4-epoxy-1,7-nonadien-1-yl]phosphonate (0.233 g, 0.077 mmol) in 2 mL of tetrahydrofuran and 1 mL of methanol. The resultant brown reaction mixture was stirred for 5 minutes at $-90°$, quenched at this temperature by the addition of pH 8 phosphate buffer, and then warmed to room temperature. The aqueous phase was extracted with diethyl ether (5 × 3 mL), the combined extracts were dried (anhydrous magnesium sulfate/potassium carbonate or anhydrous sodium sulfate), and the volatiles were removed in vacuo. The remaining oil was kugelrohr distilled (bp 100°, 0.1 mm Hg) to provide 0.197 g (84%) of the title compound. IR (neat): 3400, 1250, 1040 $cm^{-1}$. $^1$H NMR: $\delta$ 5.7 (m, 2H), 5.1 (m 1H), 4.1 (m, 4H), 2.64 (d, $J$ = 6 Hz, 1H), 2.54 (d, $J$ = 6 Hz, 1H), 2.0 (m, 2H), 1.8–1.2 (m, 19H). $^{13}$C NMR: $\delta$ 142.33, 131.06, 124.19, 115.91, 72.17, 61.68, 42.08, 31.10, 28.30, 27.38, 25.31, 22.45, 17.29, 16.14. Exact mass spectral analysis, calcd for $C_{15}H_{30}O_4P$, 305.1882; found, 305.1904.

**Cyclohexadec-5-enecarbonitrile (Reduction of $\alpha$-Heterosubstituted Nitriles).**[72] Cyclohexadec-5-enone (0.5 mmol) was stirred with $O,O'$-diethyl phosphorocyanidate (245 mg, 1.5 mmol) and lithium cyanide (24.5 mg, 1.5 mmol) in 10 mL of tetrahydrofuran for 10–30 minutes at room temperature. Water (10 mL) was added, and the mixture was extracted with ethyl acetate–hexane (1:1, 50 mL). The extract was washed with brine (2 × 20 mL), dried (anhydrous magnesium sulfate), and evaporated under reduced pressure. A solution of the crude cyanophosphate thus formed and *tert*-butanol (37 mg, 0.5 mmol) in 5 mL of tetrahydrofuran was added to a solution of samarium(II) iodide, prepared from samarium metal (345 mg, 2.3 mmol) and 1,2-diiodoethane (413 mg, 1.5 mmol) in 10 mL of tetrahydrofuran at room temperature. The reaction mixture was quenched by addition of 10% hydrochloric acid (10 mL) and extracted with diethyl ether (2 × 50 mL). The extracts were washed with 5% sodium thiosulfate (10 mL), water (10 mL) and brine (10 mL) and dried (anhydrous magnesium sulfate). After removal of the solvent, the residue was purified by column chromatography (benzene–hexane, 1:1, **[Caution: Potent Carcinogen]**) to provide the title compound (97%) as a colorless oil, bp 156° (2 mm Hg). IR (film): 2240 $cm^{-1}$. $^1$H NMR: $\delta$ 1.1–1.7 (m, 22H), 2.2 (br s, 4H), 2.5 (m, 1H), 5.3 (m, 2H). Mass spectrum, $m/z$ 247 (M$^+$). HRMS calcd for $C_{17}H_{29}N$, 247.2298; found, 247.2299. Anal. Calcd for $C_{17}H_{29}N$: C, 82.85; H, 11.82; N, 5.66. Found: C, 82.51; H, 11.97; N, 5.72.

$$\text{TDBMSO} \diagdown \diagup \text{NO}_2 \longrightarrow \text{TDBMSO} \diagdown \diagup \text{NHOH}$$

**N-Hydroxy 2-(*tert*-Butyldiphenylsiloxy)ethanamine (Reduction of Nitro Compounds to Hydroxylamines).[79]** To a solution of freshly prepared samarium(II) iodide (4 mmol) in 30 mL of tetrahydrofuran was rapidly added a solution of 2-(*tert*-butyldiphenylsiloxy)-1-nitroethane (1 mmol) in a 2:1 mixture of tetrahydrofuran/methanol (6 mL). The reaction mixture was stirred at room temperature for 3 minutes, poured into a 10% solution of sodium thiosulfate (30 mL) and extracted with ethyl acetate several times. The residue was chromatographed over silica gel (ethyl acetate) to provide the title compound in 79% yield, mp 68–69°. IR (CHCl$_3$) 3580, 3260, 2920, 1470, 1425, 1110 cm$^{-1}$. $^1$H NMR (300 MHz, CDCl$_3$): δ 7.70 (m, 4H), 7.43 (m, 6H), 6.32 (br, 2H), 3.85 (t, $J$ = 5.1 Hz, 2H), 3.08 (t, $J$ = 5.1 Hz, 2H), 1.09 (s, 9H). $^{13}$C NMR (60 MHz, CDCl$_3$): δ 135.53, 133.38, 129.72, 127.73, 60.02, 55.61, 26.84, 19.20. Mass spectrum: *m/z* 298 (M$^+$-17, 1). Anal. Calcd for C$_{18}$H$_{25}$NO$_2$Si: C, 68.52; H, 7.99; N, 4.44; Found: C, 68.79; H, 8.03; N, 4.30.

**[S-(R*,S*)]-[1-Methyl-2-oxo-2-[2-(1-pyrrolidinylcarbonyl)-1-pyrrolidinyl] ethyl]-Carbamic Acid, Phenylmethyl Ester (Reduction of 2-Hydroxyimino Amides).[82]** To a solution of (*S*)-1-[2-(hydroxyimino)-1-oxopropyl]-2-(1-pyrrolidinylcarbonyl)pyrrolidine (51.0 mg, 0.20 mmol) in methanol (10 mL) and tetrahydrofuran (4.0 mL) was added a solution of samarium(II) iodide in tetrahydrofuran (10 mL, 0.1 M, 10 mmol) under argon at −40°. The resulting mixture was stirred for 1 hour. The reaction mixture was quenched with a mixture of pH 7 buffer (2.5 mL) and methanol (2.5 mL) at −40°. The reaction mixture was poured into 10% aqueous potassium carbonate solution and extracted with dichloromethane. The organic layer was dried over anhydrous magnesium sulfate and the solvent was removed under reduced pressure. Dichloromethane (10 mL) was added to the residue, and carbobenzoxy chloride (0.04 mL, 0.3 mmol) and pyridine (0.03 mL, 0.3 mmol) were added to this solution. After stirring for 1 hour at 0°, the reaction mixture was quenched with brine and extracted with dichloromethane. The organic layer was dried over anhydrous magnesium sulfate. Subsequently, the solvent was removed and the crude product was purified by silica gel TLC (ethyl acetate) providing the title compound (60.8 mg, 82% yield).

**(S)-(−)-α-Methylbenzylamine (Reductive Cleavage of the Nitrogen–Nitrogen Bond of N-Aroylhydrazines).**[84] To (S)-(−)-1-phenyl-1-(2-benzoyl-hydrazino)ethane (0.40 g, 1.66 mmol, 89% ee) in methanol (7 mL) was added rapidly dropwise a solution of samarium(II) iodide (3.5 mmol, 70 mL of a 0.05 M solution in tetrahydrofuran). After complete addition, the reaction was allowed to stir for 30 minutes. The reaction was then concentrated on a rotary evaporator, and to the resulting residue was added 1 M hydrochloric acid (15 mL). The aqueous layer was extracted with diethyl ether (8 × 25 mL). The aqueous layer was made basic to litmus by the addition of 3 M sodium hydroxide and then was extracted with diethyl ether (8 × 25 mL). The combined ether extracts were diluted with pentane (1:1) and dried over a small amount of anhydrous magnesium sulfate. Concentration of the diethyl ether/pentane solution on a rotary evaporator provided the title compound (0.144 g, 72%) as a colorless oil, $[\alpha]_D^{20} = -37.1°$ (c 1.33, $C_6H_6$). ${}^1$H NMR (CDCl$_3$): δ 1.40 (d, $J_{HH}$ = 6.3 Hz, 3H, CH$_3$), 1.70 (br, 2H, NH), 4.14 (q, $J_{HH}$ = 6.3 Hz, 1H, CH), 7.25 (m, 1H, Ph), 7.45 (m, 4H, Ph). The enantiomeric purity of the product was determined to be 89% ee using capillary GC methods [J & W Cyclodex B column, 80°, isothermal, R isomer retention time 20.57 minutes; S isomer retention time 21.33 minutes].

**A-Nor-9,10-secocholestα-5(10),8-dien-6-yn-11α-ol (Palladium-Promoted Reductive Cleavage of Propargyl Carboxylates).**[90] To a suspension of samarium powder (0.451 g, 3.0 mmol) in dry tetrahydrofuran (5 mL) was added a solution of 1,2-diiodoethane (0.724 g, 2.57 mmol) in tetrahydrofuran (5 mL) under argon at room temperature via cannula. After stirring for 1 hour, a deep blue solution was obtained, and a solution of A-nor-9α,11α-oxido-9,10-secocholesta-5(10)-en-6-yn-8β-yl benzoate (0.251 g, 0.51 mmol) and tetrakis(triphenyl-phosphine)palladium(0) (0.018 g, 3 mol %) in tetrahydrofuran (7 mL) was added via cannula. The deep blue color persisted, and the solution was stirred for one hour. Water (5 mL) was added, and the mixture was stirred until it became yellow. Solid sodium carbonate was added to separate the layers, the entire mixture was extracted with diethyl ether ($2 \times 25$ mL), and the organic layers were combined and dried. Concentration gave a dark orange oil that was subjected to HPLC purification (10% ethyl acetate/hexanes) to provide 0.168 g (89%) of the title compound as a viscous oil. $^1$H NMR (CDCl$_3$): $\delta$ 0.72 (s, 3H, C$_{18}$CH$_3$), 0.85 (overlapping d, $J = 6.7$ Hz, 6 H, C$_{26,27}$2CH$_3$), 0.95 (d, $J = 6.2$ Hz, 3H, C$_{21}$CH$_3$), 1.84 (s, 3H, C$_{19}$CH$_3$), 4.41 (ddd, $J = 1.9, 7.0, 3.1$ Hz, 1H), 5.91 (dd, $J = 3.1, 3.1$ Hz, 1H).

## TABULAR SURVEY

Tables I–X are organized in the sequence used in the Scope and Limitations section. Literature coverage through 1992 is as exhaustive as possible, using both computer scanning services and hand searches. Unspecified yields are denoted by (−).

Tables I–X are ordered by increasing carbon number of the basic structural unit of the educt, omitting the carbon count of, for example, protecting groups or the alcohol portion of a carboxylic ester.

Abbreviations used in all of the tables are as follows:

| | |
|---|---|
| Ac | acetyl |
| Bn | benzyl |
| Bz | benzoyl |
| C$_6$H$_{11}$ | cyclohexyl |
| Cbz | carbobenzyloxy |
| DBM | dibenzoylmethanato |
| DMA | N,N-dimethylacetamide |
| DMAE | N,N-dimethylaminoethanol |
| DMF | N,N-dimethylformamide |
| DMPU | N,N′-dimethylpropyleneurea |
| ee | enantiomeric excess |
| HMPA | hexamethylphosphoric triamide |
| MEM | methoxyethoxymethyl |
| MOM | methoxymethyl |
| NMP | 1-methyl-2-pyrrolidinone |
| rt | room temperature |
| SEM | 2-(trimethylsilyl)ethoxymethyl |
| TBDMS | tert-butyldimethylsilyl |
| TBDPS | tert-butyldiphenylsilyl |
| THF | tetrahydrofuran |
| TMEDA | N,N,N′,N′-tetramethylethylenediamine |
| TMS | trimethylsilyl |
| TMU | tetramethylurea |
| Tol | tolyl |
| Tr | triphenylmethyl |
| Ts | p-toluenesulfonyl |

Table I. REDUCTION OF ORGANIC HALIDES, SULFONATES, AND SULFONES

### A. Organic Halides and Sulfonates

| Substrate | Conditions | Product(s) and Yield(s) (%) | Refs. |
|---|---|---|---|
| C₂ <br><br> $R^1, R^2$ on dibromoalkene (Br, Br) | | $R^1C{\equiv}CR^2$ + $\underset{\mathbf{I}}{}$ + $\underset{\mathbf{II}}{\begin{array}{c}R^1\quad H\\ \diagup\diagdown\\ R^2\quad Br\end{array}}$ + $\underset{\mathbf{III}}{\begin{array}{c}R^1\quad H\\ \diagup\diagdown\\ R^2\quad H\end{array}}$ | |
| $R^1 = C_6H_{11}; R^2 = H$ | 2.5 SmI₂, C₆H₆, HMPA, rt, 10 min | $\mathbf{I}$ (86) | 23 |
| $R^1 = C_6H_{13}; R^2 = H$ | " | $\mathbf{I}$ (90) | 23 |
| $R^1 = C_8H_{17}; R^2 = H$ | " | $\mathbf{I}$ (74) | 23 |
| $R^1 = Me; R^2 =$ 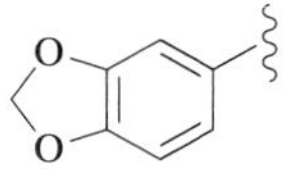 (benzodioxole substituent) | " | $\mathbf{I}$ (41) | 23 |
| $R^1 = C_8H_{17}; R^2 = Ph$ | 8 SmI₂, THF, HMPA, rt | $\mathbf{I}$ (27) + $\mathbf{II}$ (39) + $\mathbf{III}$ (6) | 23 |
| | 2.5 SmI₂, C₆H₆, HMPA, rt, 10 min | $\mathbf{I}$ (67) + $\mathbf{II}$ (8) | 23 |
| C₅ <br><br> (prenyl bromide structure) Br | 1 SmI₂, THF, rt, 1 d | 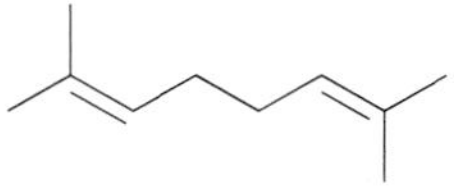 (42) 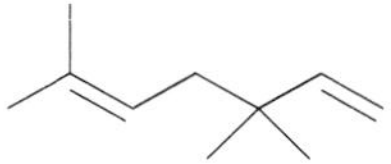 + (25)  + (6) | 3 |

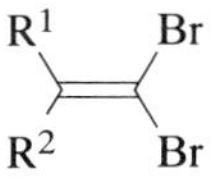

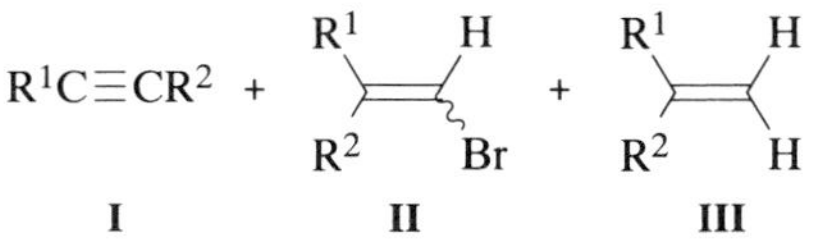

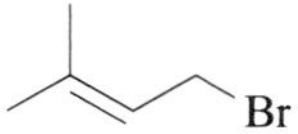

Table I. REDUCTION OF ORGANIC HALIDES, SULFONATES, AND SULFONES (*Continued*)

| Substrate | Conditions | Product(s) and Yield(s) (%) | Refs. |
|---|---|---|---|
| C₆ <br><br> Br-⟨C₆H₄⟩-OMe | 2.5 SmI₂, THF, HMPA, *i*-PrOH, rt, 8 h | 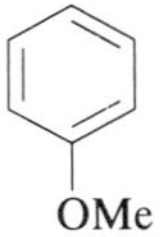 (82) | 24 |
| isoxazole-CH₂X (X = Cl) | 2 SmI₂, THF, rt, 2 d | **II** (54-60) | 25 |
| isoxazole-CH₂X (X = Br) | 2 SmI₂, THF, rt, 3 h | **I** (36) + **II** (47) | 25 |
| C₇ <br><br> Br-⟨C₆H₄⟩-CH₂OAc | 2.5 SmI₂, THF, HMPA, rt, 2 h | ⟨C₆H₅⟩-CH₂OAc (97) | 24 |
| PhCH₂Cl | 1 SmI₂, THF, rt, 1.5 h | PhCH₂CH₂Ph (67) | 3 |
| PhCH₂Cl | 1 SmI₂, THF, rt, 20 min | " (82) | 3 |
| C₈ <br><br> Ph-CH=CH-Br | SmI₂, THF | 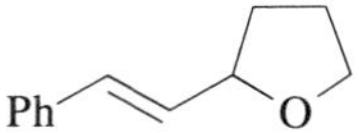 (22) | 26 |
| Ph-CH=CH₂ | 2.5 SmI₂, THF, HMPA, rt, 20 min | Ph (>95) | 24 |

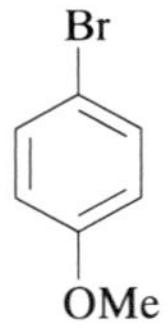

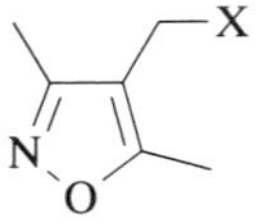

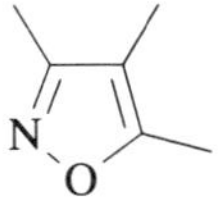

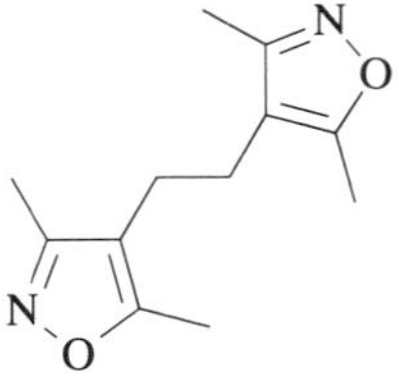

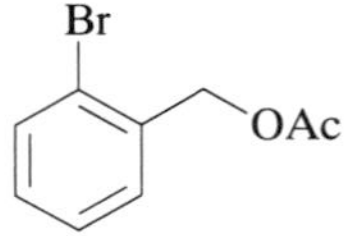

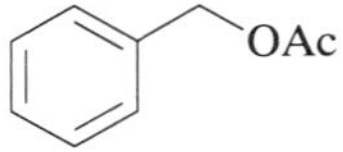

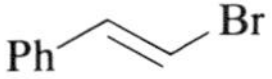

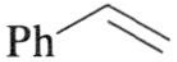

**I +** (two isoxazole rings connected by CH₂CH₂) **II**

274

Table I. REDUCTION OF ORGANIC HALIDES, SULFONATES, AND SULFONES (*Continued*)

| Substrate | Conditions | Product(s) and Yield(s) (%) | Refs. |
|---|---|---|---|
| 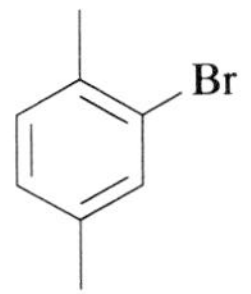 (image: 2-bromo-1,4-dimethylbenzene, with Br) | 2.5 SmI$_2$, THF, HMPA, *i*-PrOH, rt, 15 h |  (image: p-xylene) (84) | 24 |
| C$_9$ 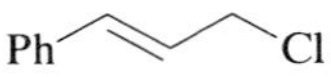 (image: Ph-CH=CH-CH$_2$-Cl, cinnamyl chloride) | SmI$_2$, THF, rt, 30 min | 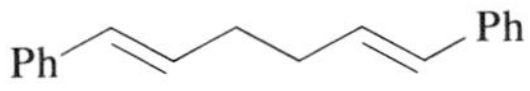 (images) **I** (51) 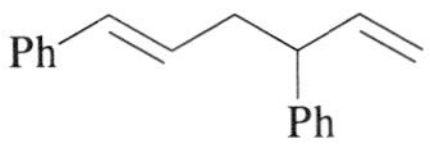 + **II** (23) 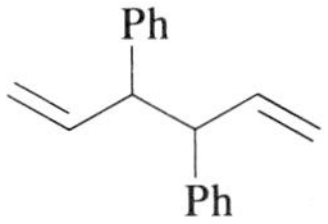 + **III** (7) | 3 |
| 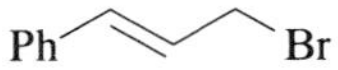 C$_{10}$ (image: Ph-CH=CH-CH$_2$-Br) | 1 SmI$_2$, THF, rt, 5 min | **I** (55) + **II** (21) + **III** (6) | 3 |
| 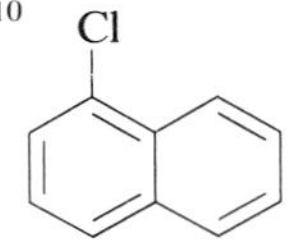 (image: 1-chloronaphthalene, Cl) | 2.5 SmI$_2$, THF, HMPA, 15 min | 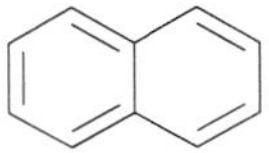 (image: naphthalene) **I** (>95) | 24 |
| 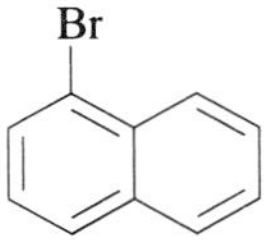 (image: 1-bromonaphthalene, Br) | 2.5 SmI$_2$, THF, HMPA, D$_2$O, rt, 5 min | **I** (98) | 24 |

275

Table I. REDUCTION OF ORGANIC HALIDES, SULFONATES, AND SULFONES (*Continued*)

| Substrate | Conditions | Product(s) and Yield(s) (%) | Refs. |
|---|---|---|---|
| 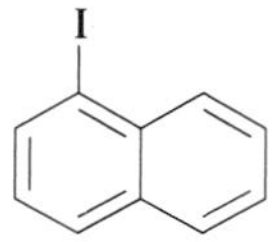 | 2.5 SmI₂, THF, HMPA, rt, 1 min | **I** (>95) | 24 |
| 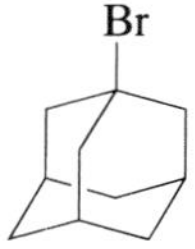 | 2.5 SmI₂, MeCN, HMPA, *i*-PrOH, rt, 10 min | 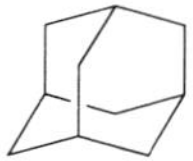 **I** (>95) | 24 |
| 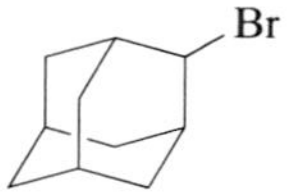 | 2.5 SmI₂, THF, HMPA, D₂O, rt, 10 min | **I** (20) 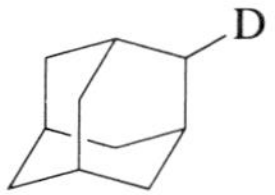 + **D** (80) | 24 |
| 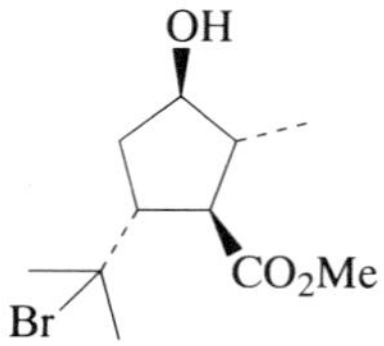 | 2.5-4 SmI₂, THF, HMPA, rt | 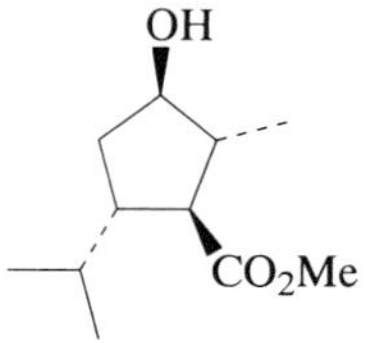 **I** (39) 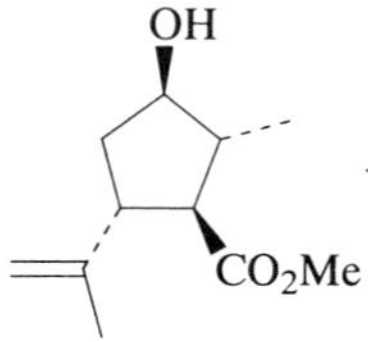 + **II** (17) 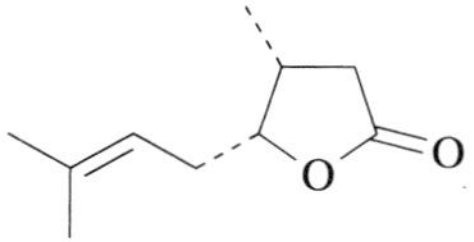 + **III** (15) | 27 |

276

| Substrate | Conditions | Product(s) and Yield(s) (%) | Refs. |
|---|---|---|---|
| 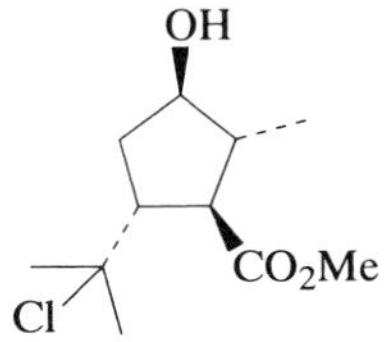 | 2.5-4 SmI$_2$, THF, HMPA, rt | **I** (37) + **III** (24) | 27 |
| $n$-C$_{10}$H$_{21}$X | | $n$-C$_{10}$H$_{22}$    **I** | |
| X = Cl | 2.5 SmI$_2$, THF, HMPA, $i$-PrOH, 60°, 8 h | **I** (>95) | 24 |
| X = Br | 2.5 SmI$_2$, THF, HMPA, $i$-PrOH, rt, 10 min | **I** (>95) | 24 |
| X = I | 2.5 SmI$_2$, THF, HMPA, $i$-PrOH, rt, 5 min | **I** (>95) | 24 |
| C$_{11}$ | 12 SmI$_2$, THF, MeOH, −78°, 16 h | | 29 |

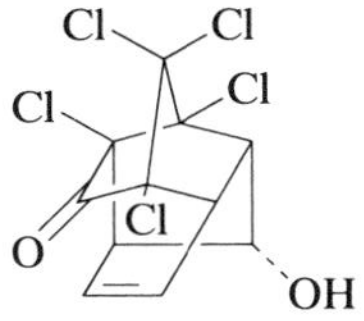

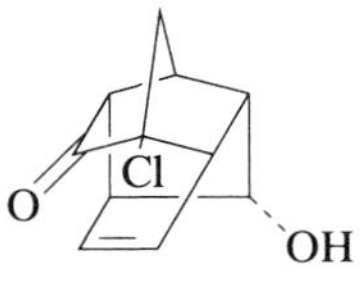

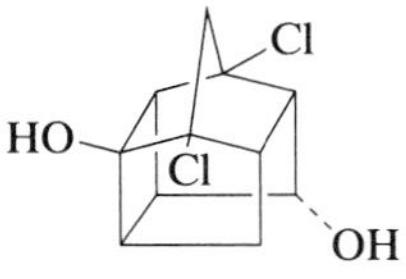

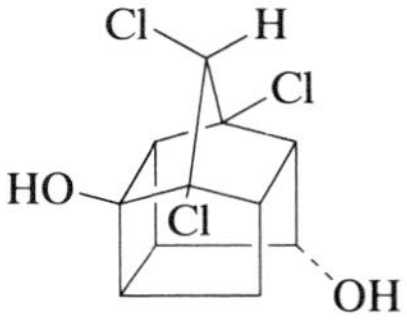

277

Table I. REDUCTION OF ORGANIC HALIDES, SULFONATES, AND SULFONES (*Continued*)

| Substrate | Conditions | Product(s) and Yield(s) (%) | Refs. |
|---|---|---|---|
| 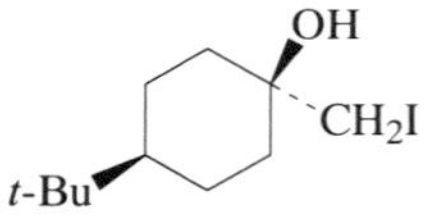 $C_{12}$ (structure: cyclohexane with OH, CH₂I, *t*-Bu) | SmI₂, THF, HMPA, rt, 5 min | 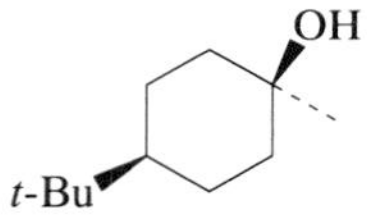 (structure: cyclohexane with OH, *t*-Bu) (100) | 30 |
| (4-bromobiphenyl, Br) | 2.5 SmI₂, THF, HMPA, rt, 1 h | (biphenyl) (>95) | 24 |
| 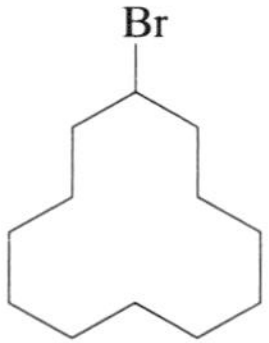 (bromocyclododecane, Br) | 2.5 SmI₂, THF, HMPA, *i*-PrOH, rt, 10 min | 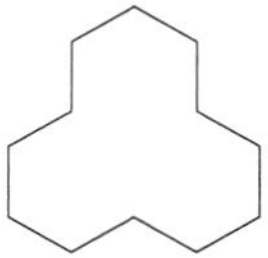 I (>95) | 24 |
| 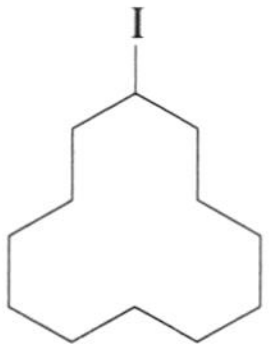 (iodocyclododecane, I) | 2.5 SmI₂, THF, HMPA, *i*-PrOH, rt, 10 min | I (>95) | 24 |
| *n*-C₁₂H₂₅X | | *n*-C₁₂H₂₆  I | |
| X = Cl | 2 SmI₂, THF, MeOH, 65°, 2 d | I (0) | 3 |
| X = Br | " | I (82) | 3 |
| X = I | 2 SmI₂, THF, MeOH, 65°, 6 h | I (95) | 3 |
| X = OTs | 2 SmI₂, THF, MeOH, 65°, 12 h | I (76) | 3 |

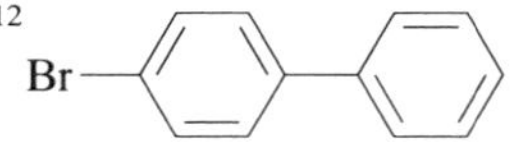

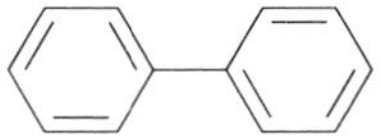

278

Table I. REDUCTION OF ORGANIC HALIDES, SULFONATES, AND SULFONES (*Continued*)

| Substrate | Conditions | Product(s) and Yield(s) (%) | Refs. |
|---|---|---|---|
| C$_{15}$ 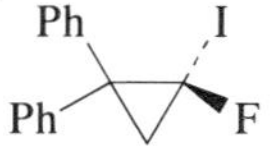 | 2 SmI$_2$, THF, HMPA, rt, 2 h | 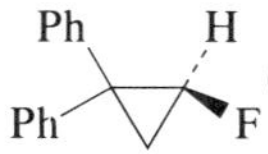 (55) 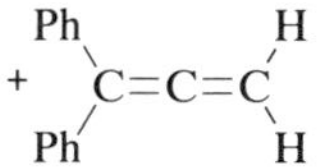 + (18) 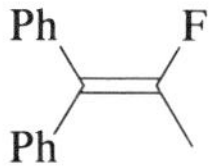 + (1) | 92 |
| C$_{16}$ 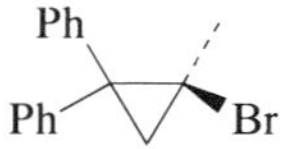 | 2 SmI$_2$, THF, HMPA, rt, 2 h | 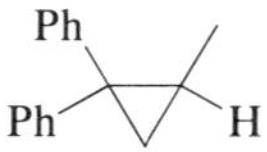 (58) 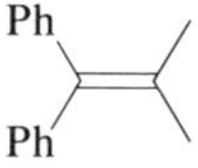 + (22) 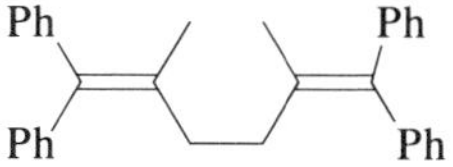 + (4) | 92 |
| 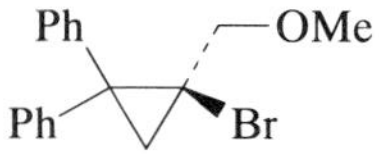 | 2 SmI$_2$, THF, HMPA, rt, 2 h | 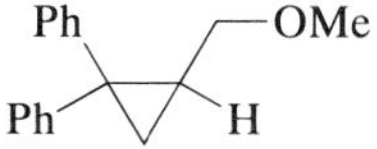 (50) 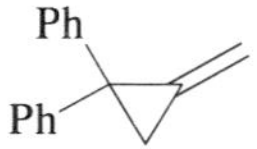 + (16) 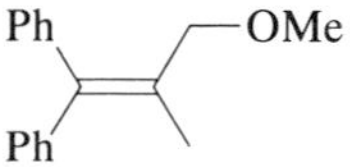 + (2) | 92 |

279

Table I. REDUCTION OF ORGANIC HALIDES, SULFONATES, AND SULFONES (*Continued*)

| Substrate | Conditions | Product(s) and Yield(s) (%) | Refs. |
|---|---|---|---|
| C$_{27}$ 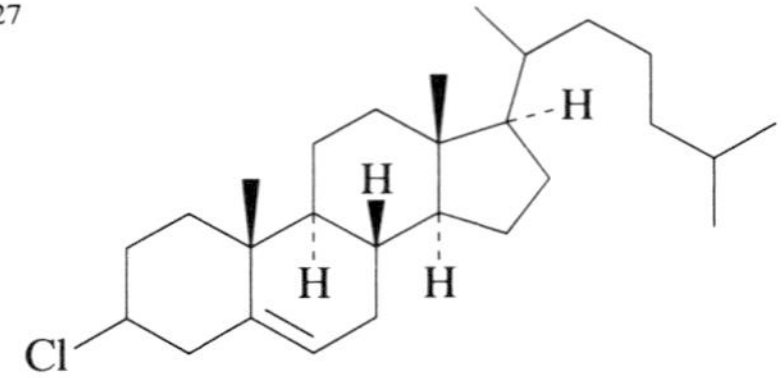 | 2.5 SmI$_2$, THF, HMPA, *i*-PrOH, rt, 3 h | 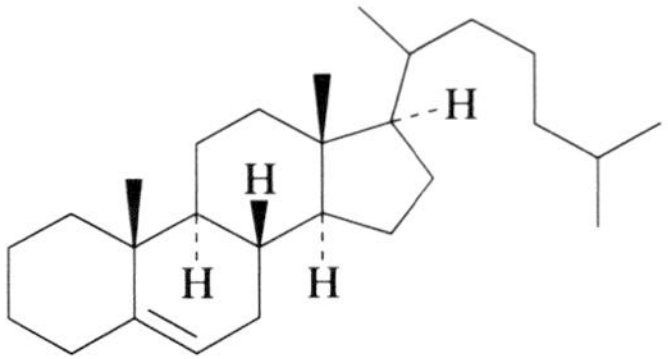 (99) | 24 |
| *B. Organic Sulfones* | | | |
| C$_6$ 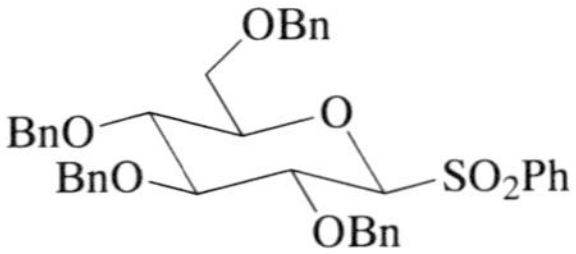 | SmI$_2$, THF, HMPA | 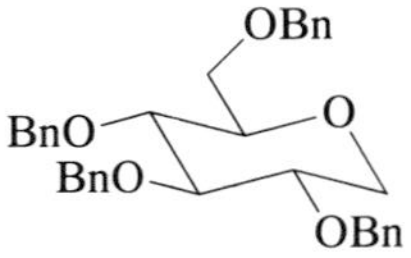 (77) 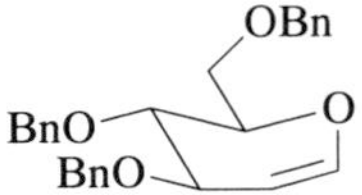 + (—) | 31 |
| C$_7$ 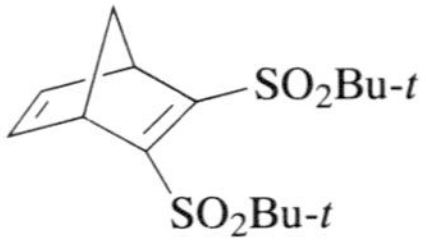 | SmI$_2$, THF, MeOH, -70° | 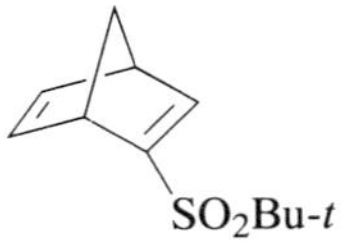 (100) | 32 |
| C$_8$ 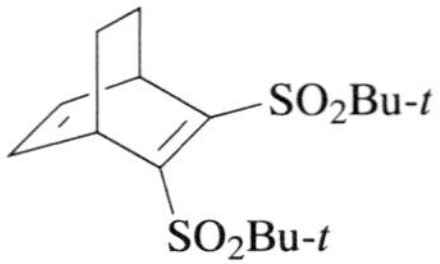 | SmI$_2$, THF, MeOH, -70° | 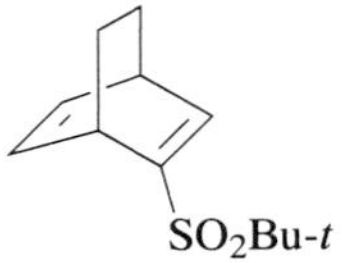 (85) | 32 |

Table I. REDUCTION OF ORGANIC HALIDES, SULFONATES, AND SULFONES (*Continued*)

| Substrate | Conditions | Product(s) and Yield(s) (%) | Refs. |
|---|---|---|---|
| C<sub>16</sub> | | | |
| 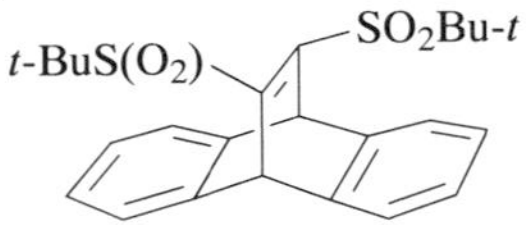 t-BuS(O<sub>2</sub>) / SO<sub>2</sub>Bu-t | SmI<sub>2</sub>, THF, MeOH, -70° | 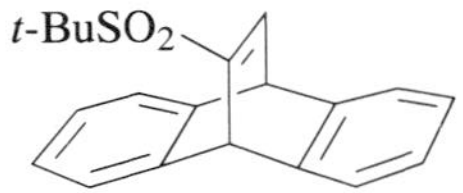 t-BuSO<sub>2</sub> (96) | 32 |
| 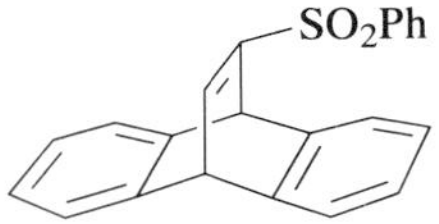 SO<sub>2</sub>Ph | 5 SmI<sub>2</sub>, THF, HMPA, -20°, 30 min | 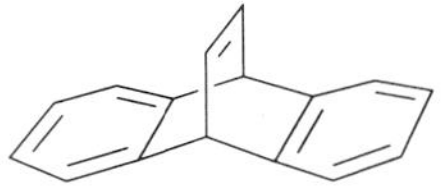 (77) | 33 |
| 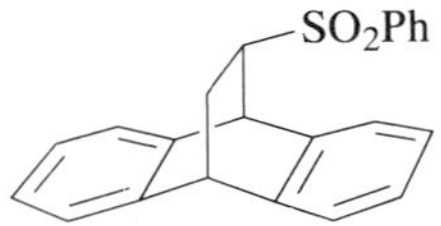 SO<sub>2</sub>Ph | 5 SmI<sub>2</sub>, THF, HMPA, -20°, 30 min | 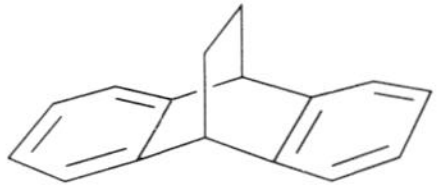 (74) | 33 |
| C<sub>18</sub> | | | |
| 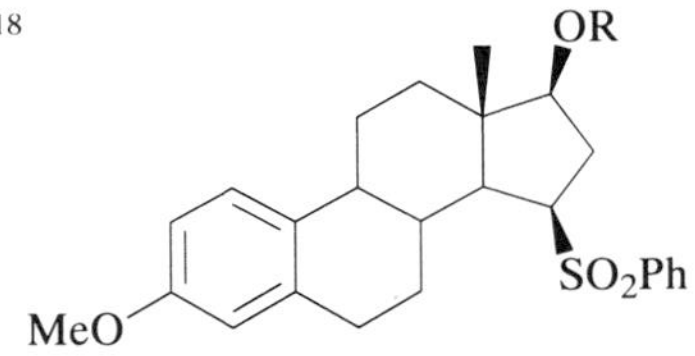 MeO, OR, SO<sub>2</sub>Ph | | 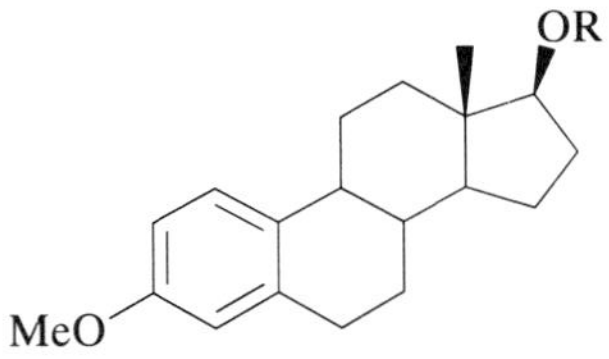 MeO, OR  **I** | |
| R = OH | 5 SmI<sub>2</sub>, THF, HMPA, 22°, 90 min | **I** (50) | 33 |
| R = OAc | 5 SmI<sub>2</sub>, THF, HMPA, -20°, 90 min | **I** (52) | 33 |

Table I. REDUCTION OF ORGANIC HALIDES, SULFONATES, AND SULFONES (*Continued*)

| Substrate | Conditions | Product(s) and Yield(s) (%) | Refs. |
|---|---|---|---|
| C$_{19}$ | | | |
| 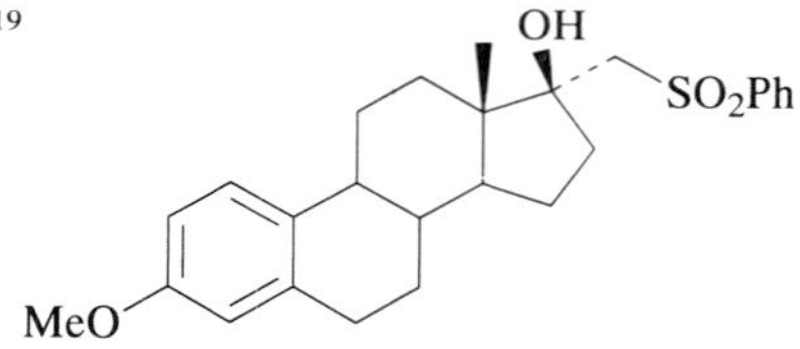 | 5 SmI$_2$, THF, HMPA, 22°, 60 min | 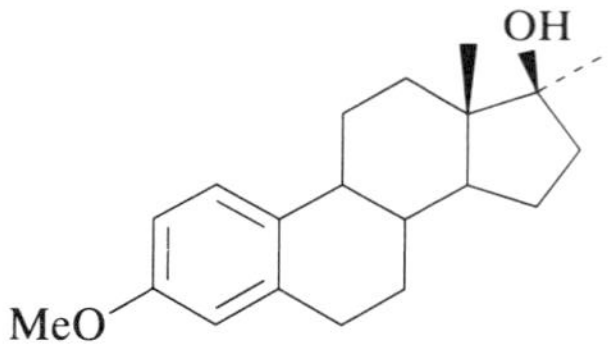 (53) | 33 |
| 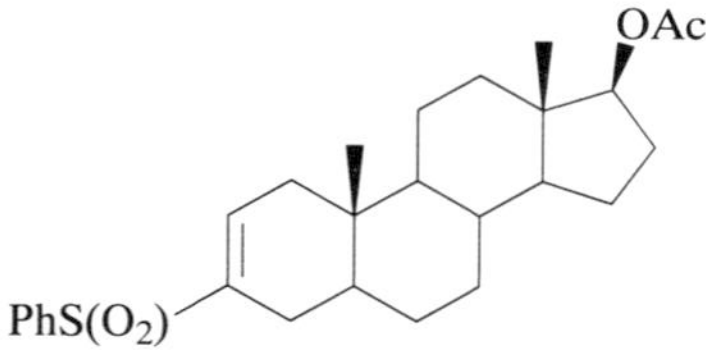 | 5 SmI$_2$, THF, HMPA, -20°, 90 min | 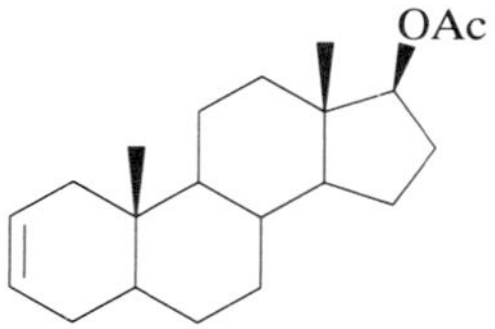 (68) | 33 |
| C$_{20}$ | | | |
| 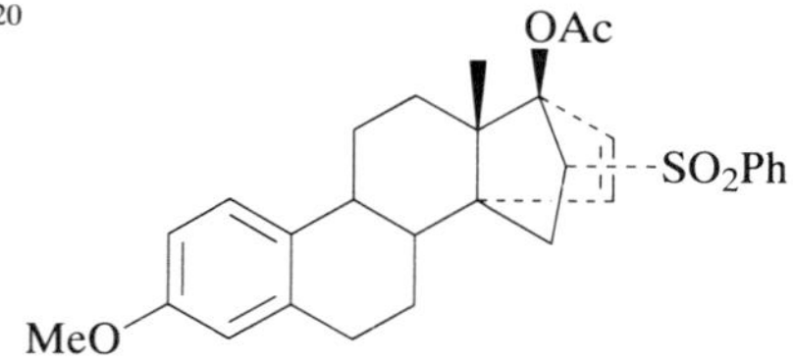 | 5 SmI$_2$, THF, HMPA, -20°, 70 min | 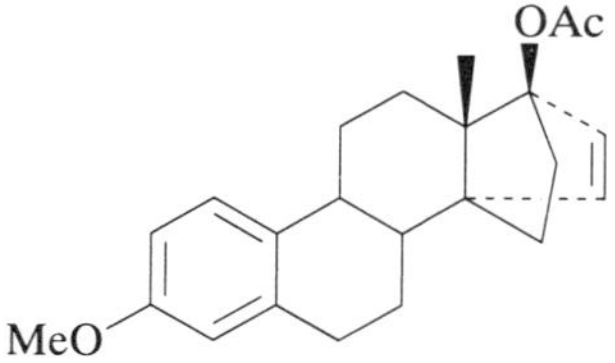 (70) | 33 |
| 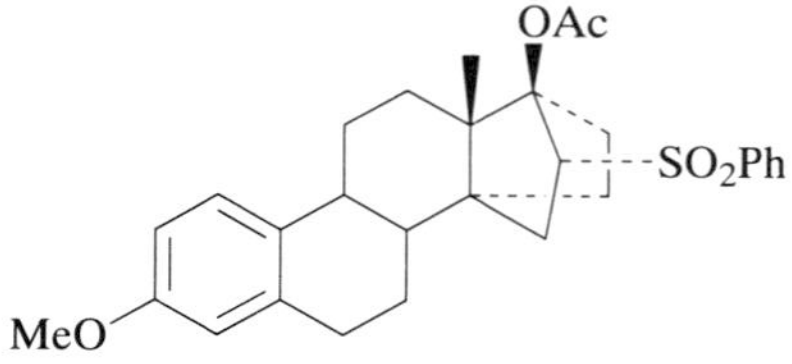 | 5 SmI$_2$, THF, HMPA, -20°, 90 min | 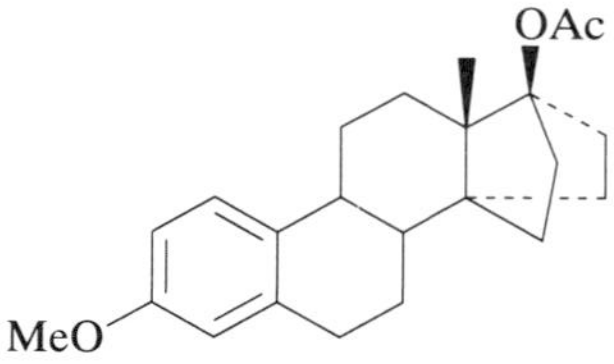 (87) | 33 |

282

## Table II. REDUCTIVE ELIMINATION/FRAGMENTATION REACTIONS

*A. β-Halo Ethers*

Product structure: R—CH=CH—CH₂CH₂—OH (**I**)

Substrate structure (C₄): 3-chlorotetrahydrofuran bearing R group

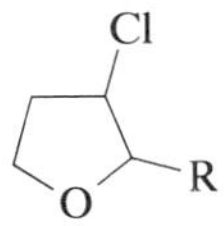

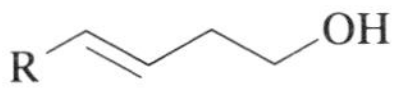

| Substrate | Conditions | Product(s) and Yield(s) (%) | Refs. |
|---|---|---|---|
| C₄ | | | |
| R = D (*cis + trans*) | SmI$_2$, THF, 65° | **I** 51 : 49 *Z/E* | 34 |
| R = Me (*cis*) | " | **I** (75-90), >95% *E* | 34 |
| R = Me (*trans*) | " | **I** (75-90), >95% *E* | 34 |
| R = C≡CH (*cis*) | " | **I** (75), >99% *E* | 34 |
| R = CH=CH$_2$ (*cis*) | " | **I** (84), >97% *E* | 34 |
| R = Et (66 : 34 *cis/trans*) | SmI$_2$, THF, 65°, 76 h | **I** (95), >98% *E* | 34 |
|  | SmI$_2$, THF, HMPA, 65°, 9 h | **I** (92) *E* | 34 |
|  | SmI$_2$, THF, DMPU, 65°, 5 h | **I** (96), >93% *E* | 34 |
| R = CH$_2$CH=CH$_2$ (2 : 1 *cis/trans*) | SmI$_2$, THF, 65° | **I** (93), >97% *E* | 34 |
| R = Ph (*trans*) | " | **I** (95), >97% *E* | 34 |
| R = *p*-Tol (*trans*) | " | **I** (100), >99% *E* | 34 |
| R = C≡CC$_5$H$_{11}$-*n* (*cis*) | SmI$_2$, THF, 65°, 22 h | **I** (84), >99% *E* | 34 |
|  | SmI$_2$, THF, DMPU, 65°, 3 h | **I** (83), >94% *E* | 34 |
| R = C≡CPh (*cis*) | SmI$_2$, THF, 65° | **I** (75), >99% *E* | 34 |

## Table II. REDUCTIVE ELIMINATION/FRAGMENTATION REACTIONS (Continued)

| Substrate | Conditions | Product(s) and Yield(s) (%) | Refs. |
|---|---|---|---|
| C₅ <br><br> Br-CH₂-(tetrahydrofuran) | SmI₂, THF | ⌇⌇⌇—OH (50) | 26 |
| 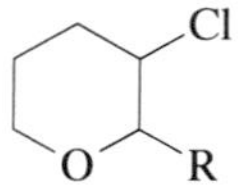 (Cl, R tetrahydropyran) | | 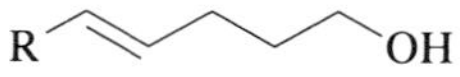 R—CH=CH-CH₂-CH₂-OH, I | |
| R = D (74 : 26 *trans/cis*) | SmI₂, THF, 65° | **I** 53 : 47 *E/Z* | 34 |
| R = Et (*cis*) | " | **I** (85-90), 76 : 24 *Z/E* | 34 |
| R = Et (*trans*) | " | **I** (85-90), 72 : 28 *Z/E* | 34 |
| R = Et (*cis + trans*) | SmI₂, THF, DMPU, 65° | **I** 13 : 87 *Z/E* | 34 |
| R = *i*-Pr (85 : 15 *trans/cis*) | SmI₂, THF, 65° | **I** (85-90), 79 : 21 *Z/E* | 34 |
| R = C≡CPr-*n* (70 : 30 *trans/cis*) | " | **I** (79), >99% *Z* | 34 |
| R = C≡CC₅H₁₁-*n* (80 : 20 *trans/cis*) | " | **I** (93), >97% *Z* | 34 |
| R = C≡CC₅H₁₁-*n* (83 : 17 *trans/cis*) | SmI₂, THF, DMPU, 65° | **I** (90), 63 : 37 *Z/E* | 34 |
| C₉ <br><br> (cyclohexanone with OMe, I, side chain) | 2 SmI₂, cat. Fe(DBM)₂, THF, -78 to 0° | (cyclohexanone with allyl) (69) | 73 |

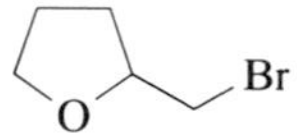

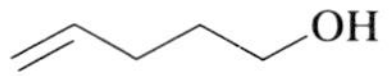

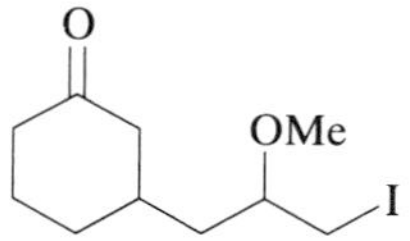

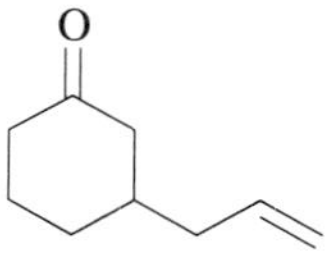

284

Table II. REDUCTIVE ELIMINATION/FRAGMENTATION REACTIONS (*Continued*)

| Substrate | Conditions | Product(s) and Yield(s) (%) | Refs. |
|---|---|---|---|
| C$_{23}$ 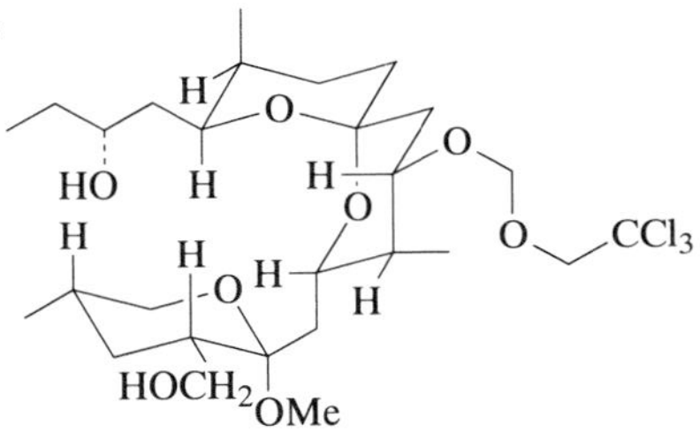 | 7 SmI$_2$, THF, rt, 35 min | 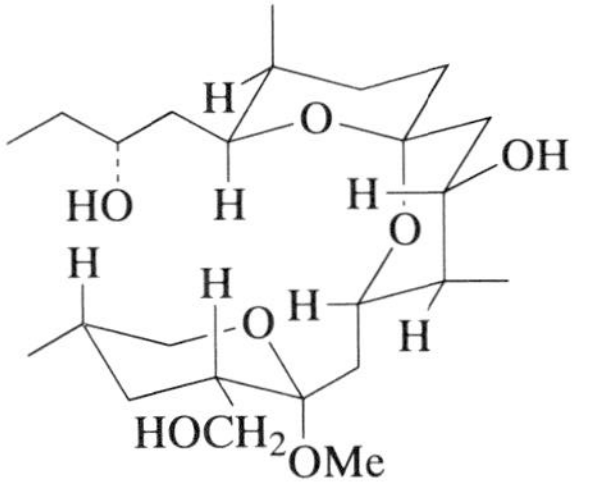 (71) | 35 |

*B. β-Carboalkoxy Halides and Related Substrates*

| | | | |
|---|---|---|---|
| C$_6$ 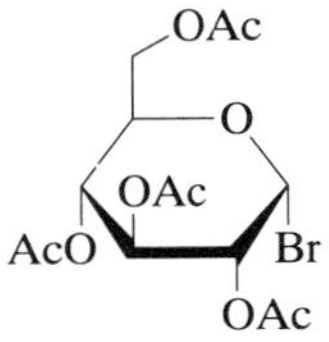 | 6 SmI$_2$, THF, rt | 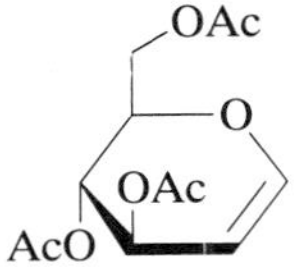 (90) | 31 |
| C$_{15}$ 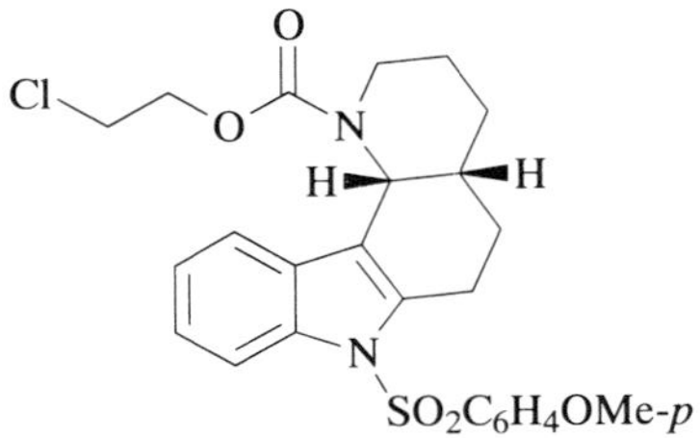 | SmI$_2$, THF, 70°, 7 h | 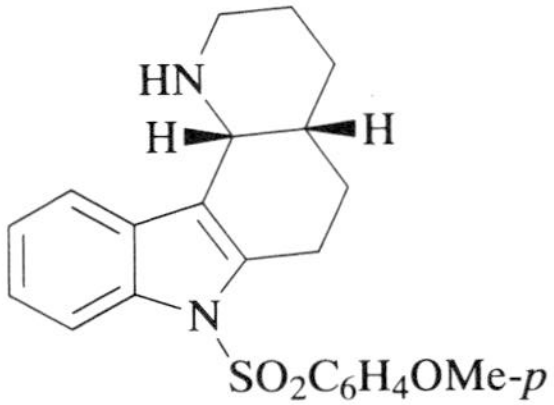 (70) | 36 |

285

Table II. REDUCTIVE ELIMINATION/FRAGMENTATION REACTIONS (*Continued*)

| Substrate | Conditions | Product(s) and Yield(s) (%) | Refs. |
|---|---|---|---|

*C. β-Hydroxy Sulfones*

$C_2$

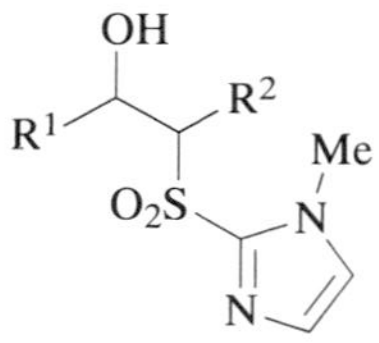

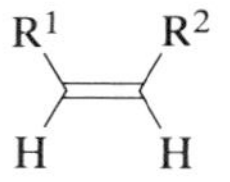

| $R^1$ | $R^2$ | | | |
|---|---|---|---|---|
| Ph | CH=CMe$_2$ | 3 SmI$_2$, THF, rt | **I** (87), 5 : 1 *E/Z* | 37 |
| (*E*)-CH=CHPh | CH=CMe$_2$ | " | **I** (78), *E,E* isomer only | 37 |
| CH=CMe$_2$ | (CH$_2$)$_2$Ph | " | **I** (82), 5 : 1 *E/Z* | 37 |
| Ph | (CH$_2$)$_2$Ph | 3 SmI$_2$, THF, rt, 15 min | **I** (82), 8 : 1 *E/Z* | 37 |
| (CH$_2$)$_2$Ph | (CH$_2$)$_2$Ph | 3 SmI$_2$, THF, rt | **I** (55), 3 : 1 *E/Z* | 37 |
| (*E*)-CH=CHPh | (CH$_2$)$_2$Ph | " | **I** (20), 9 : 2 *E/Z* 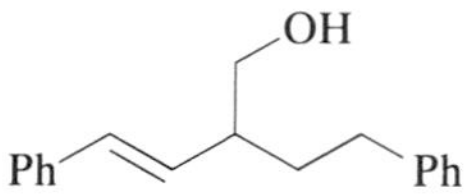 + | 37 |

| 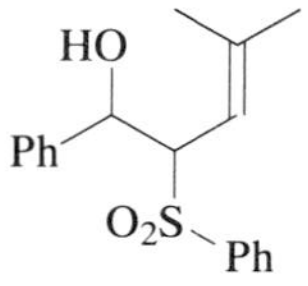 | 3 SmI$_2$, THF, rt, 15 min | 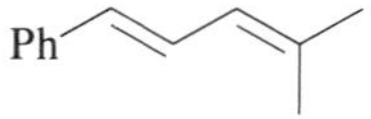 (trace) | 37 |

Table II. REDUCTIVE ELIMINATION/FRAGMENTATION REACTIONS (*Continued*)

| Substrate | Conditions | Product(s) and Yield(s) (%) | Refs. |
|---|---|---|---|
| *D. α,β-Disulfones* | | | |
| C16 | 5 SmI2, THF, HMPA, -20°, 30 min | 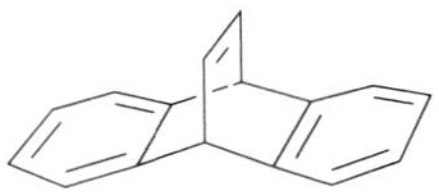 (91) | 33 |
| | 5 SmI2, THF, HMPA, -20°, 30 min | " (83) | 33 |

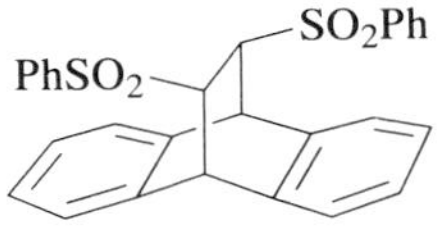

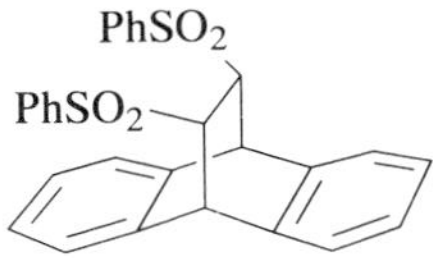

*E. γ-Halo Carbonyl Compounds and Related Substrates*

C9

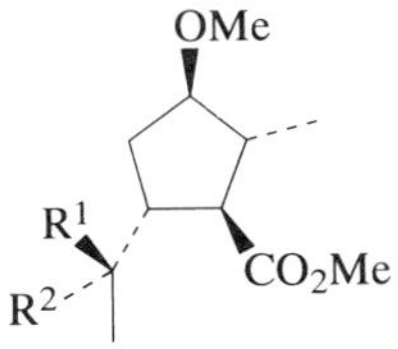

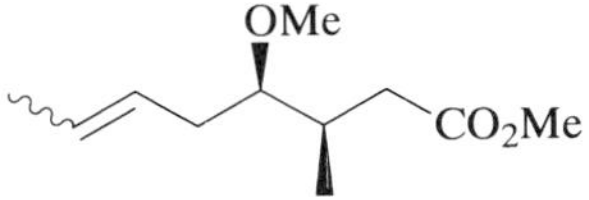

**I** + **II**

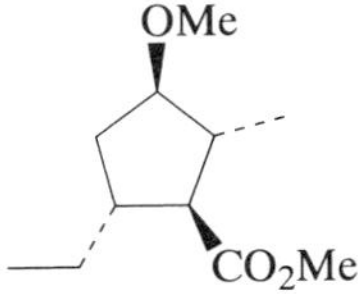

| R¹ | R² | | | |
|---|---|---|---|---|
| H | Cl | 2.5-4 SmI2, THF, HMPA, rt | **I** (21) + **II** (17) | 27 |
| Cl | H | " | **I** (29) + **II** (12) | 27 |
| H | Br | " | **I** (64) | 27 |
| Br | H | " | **I** (66) | 27 |

287

Table II. REDUCTIVE ELIMINATION/FRAGMENTATION REACTIONS (*Continued*)

| Substrate | Conditions | Product(s) and Yield(s) (%) | Refs. |
|---|---|---|---|
| 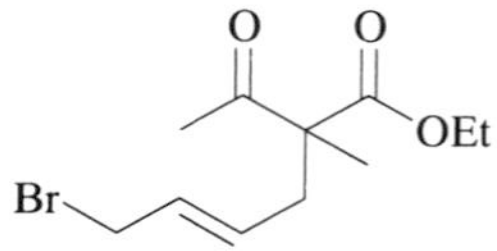 | 2 SmI$_2$, THF, -78° | 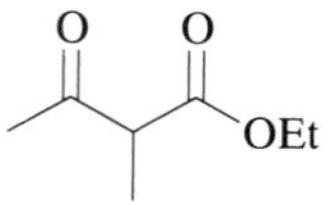 (—) | 38 |
| 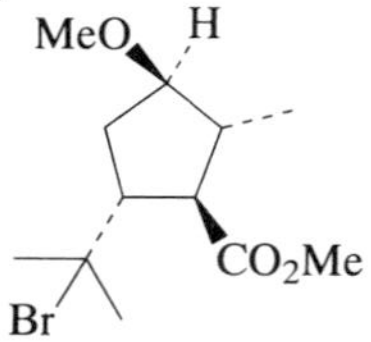 | 2 SmI$_2$, THF, -78° | 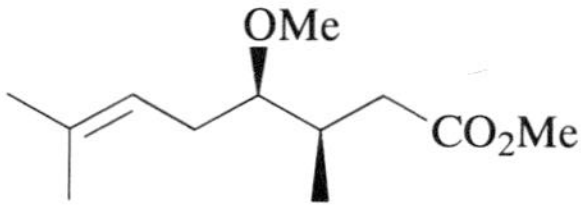 **I** (48) 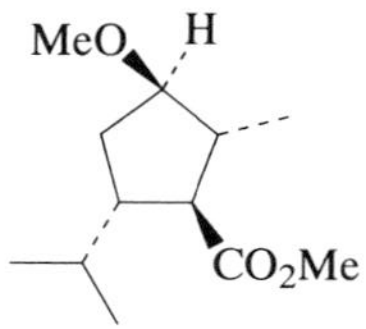 + (17) 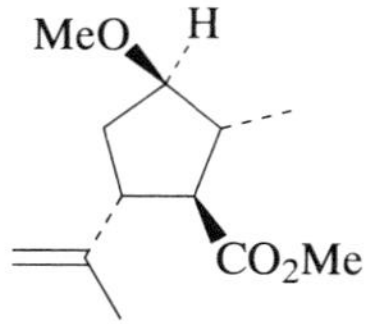 + (14) | 27 |
| 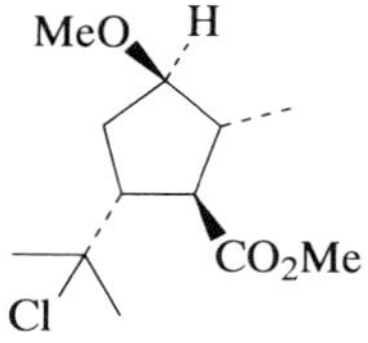 | 2 SmI$_2$, THF, -78° | **I** (77) | 27 |

288

Table II. REDUCTIVE ELIMINATION/FRAGMENTATION REACTIONS (*Continued*)

| Substrate | Conditions | Product(s) and Yield(s) (%) | Refs. |
|---|---|---|---|
| 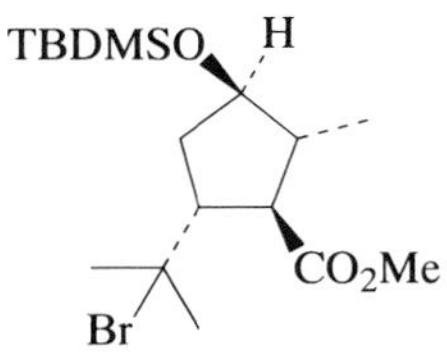 TBDMSO $\overset{H}{\cdots}$ cyclopentane with CO$_2$Me and C(CH$_3$)$_2$Br substituents | 2 SmI$_2$, THF, –78° | 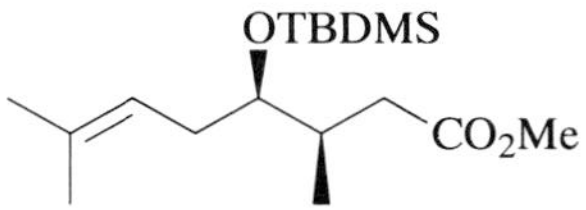 OTBDMS chain structure with CO$_2$Me, **I** (55) 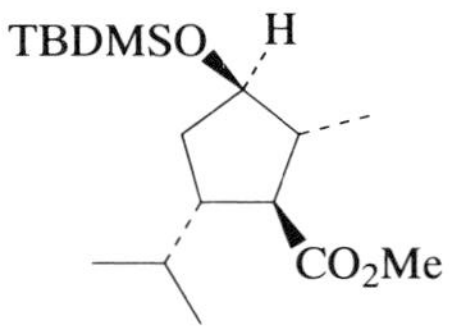 + cyclopentane with TBDMSO, CO$_2$Me, isopropyl (17) 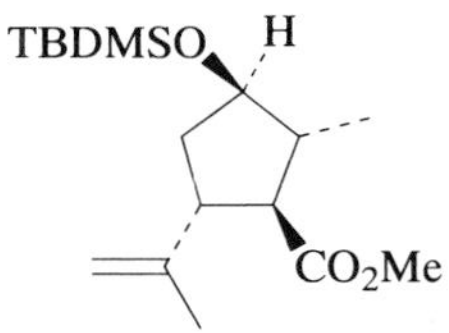 + cyclopentane with TBDMSO, CO$_2$Me, isopropenyl (10) | 27 |
| 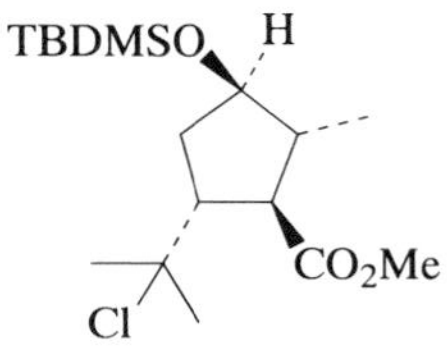 TBDMSO $\overset{H}{\cdots}$ cyclopentane with CO$_2$Me and C(CH$_3$)$_2$Cl substituents | 2 SmI$_2$, THF, –78° | **I** (78) | 27 |

289

Table II. REDUCTIVE ELIMINATION/FRAGMENTATION REACTIONS (*Continued*)

| Substrate | Conditions | Product(s) and Yield(s) (%) | Refs. |
|---|---|---|---|
| 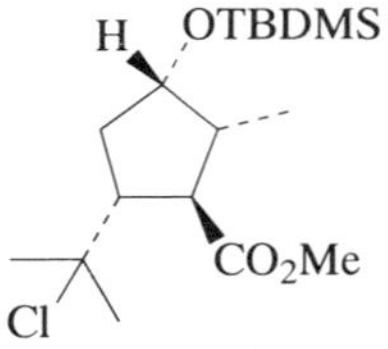 C11 | 2 SmI$_2$, THF, -78° | 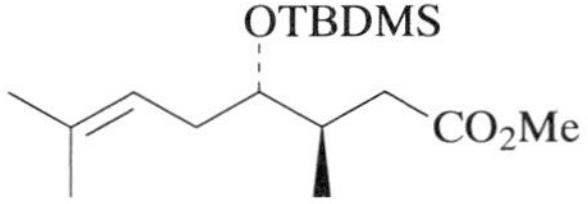 OTBDMS CO$_2$Me (60) | 27 |
| 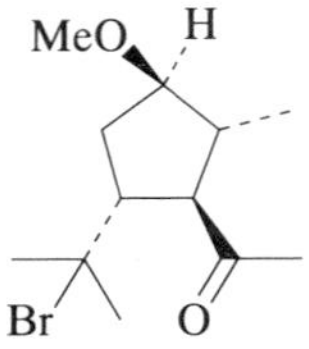 | 2 SmI$_2$, THF, -78° | 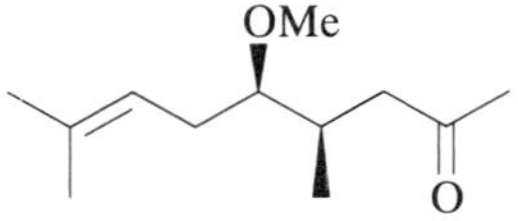 I (43) | 27 |
| 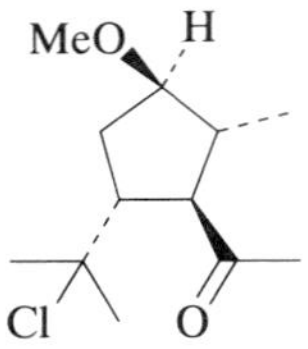 | 2 SmI$_2$, THF, -78° | I (18) | 27 |

290

Table II. REDUCTIVE ELIMINATION/FRAGMENTATION REACTIONS (*Continued*)

| Substrate | Conditions | Product(s) and Yield(s) (%) | Refs. |
|---|---|---|---|
| 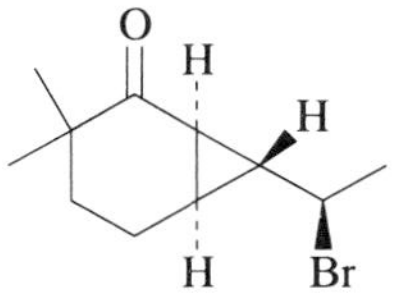 | SmI$_2$, THF, MeOH, 25°, 2 min | 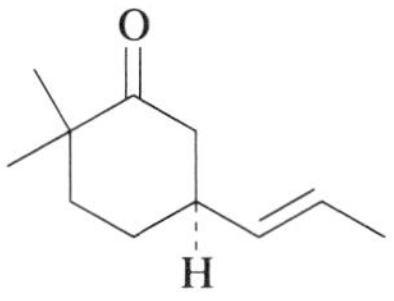 **I** (83) 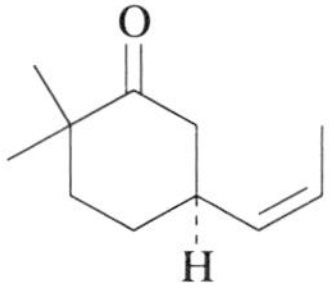 + **II** (2) | 28 |
| 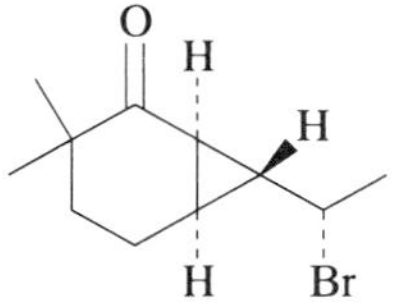 | SmI$_2$, THF, MeOH, 25°, 2 min | **I** (1) + **II** (75) | 28 |
| C$_{14}$ 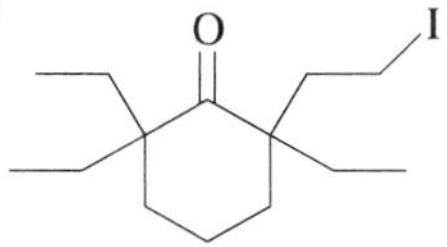 | SmI$_2$, THF | 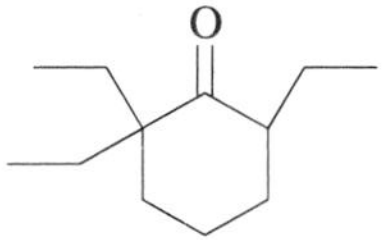 (—) | 39 |
| | *F. Miscellaneous* | | |
| C$_{21}$ 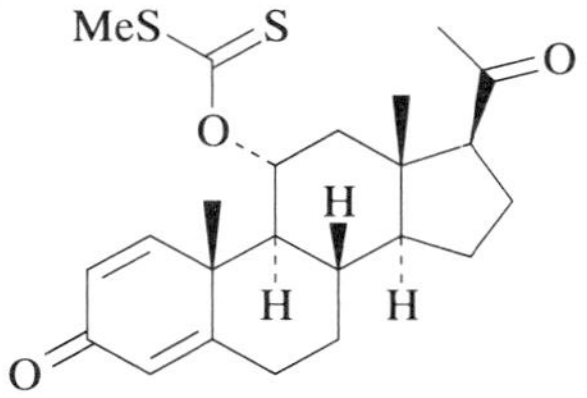 | THF, rt, 5 min | 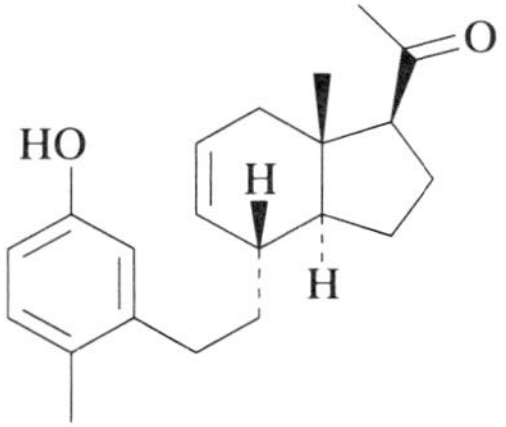 (88) | 36 |

291

Table III. REDUCTION OF ALDEHYDES AND KETONES

| Substrate | | Conditions | Product(s) and Yield(s) (%) | Refs. |
|---|---|---|---|---|
| 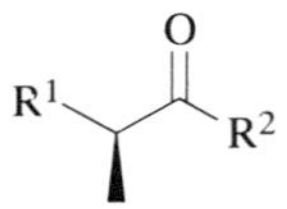 C3 | | | 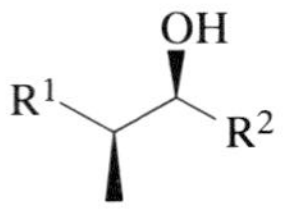 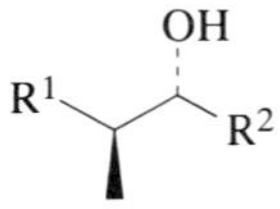 **I** + **II** | |
| R$^1$ | R$^2$ | | | |
| Ph | Me | SmI$_2$, THF, H$_2$O | **I** (7) + **II** (4) | 40 |
| | | SmI$_2$, THF, HMPA, t-BuCO$_2$H, rt, 3 min | **I** (55) + **II** (35) | 41 |
| | | SmI$_2$, THF, HMPA, MeOH, rt, 5 min | **I** (59) + **II** (23) | 41 |
| | | SmI$_2$, THF, HMPA, n-Bu$_3$SnH, 65°, 2 h | **I** (17) + **II** (66) | 41 |
| C$_6$H$_{11}$ | Me | SmI$_2$, THF, H$_2$O | **I** (9) + **II** (5) | 40 |
| C$_6$H$_{11}$ | Et | SmI$_2$, THF, H$_2$O | **I** (25) + **II** (18) | 40 |
| Ph | t-Bu | SmI$_2$, THF, HMPA, t-BuCO$_2$H, rt, 3 min | **I** (16) + **II** (76) | 41 |
| Ph | t-Bu | SmI$_2$, THF, HMPA, n-Bu$_3$SnH, 65°, 4 h | **I** (7) + **II** (51) | 41 |
| Ph | (CH$_2$)$_2$Ph | SmI$_2$, THF, HMPA, t-BuCO$_2$H, rt, 3 min | **I** (53) + **II** (46) | 41 |
| Ph | (CH$_2$)$_2$Ph | SmI$_2$, THF, HMPA, n-Bu$_3$SnH, 65°, 2 h | **I** (19) + **II** (80) | 41 |

Table III. REDUCTION OF ALDEHYDES AND KETONES (*Continued*)

| Substrate | Conditions | Product(s) and Yield(s) (%) | Refs. |
|---|---|---|---|
| C$_8$ <br><br> (acetophenone structure) | 2 SmI$_2$, THF, MeOH, rt, 1 d | (1-phenylethanol structure) (80) | 3 |
| $n$-C$_7$H$_{15}$CHO | 2 SmI$_2$, THF, MeOH, rt, 1 d | $n$-C$_7$H$_{15}$CH$_2$OH (99) | 3 |
| $n$-C$_6$H$_{13}$COMe | 2 SmI$_2$, THF, MeOH, rt, 1 d | $n$-C$_6$H$_{13}$CHOHMe **I** (12) | 3 |
| | 2 SmI$_2$, THF, H$_2$O, rt, 1 d | **I** (64) | 3 |
| C$_{10}$ <br><br> (4-t-Bu-cyclohexanone structure) | SmI$_2$, THF, HMPA, $t$-BuCO$_2$H | **I** (74) + **II** (26) | 41 |
| | SmI$_2$, THF, HMPA, $n$-Bu$_3$SnH | **I** (93) + **II** (7) | 41 |

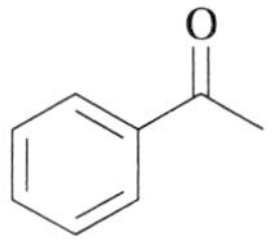

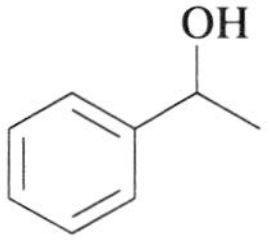

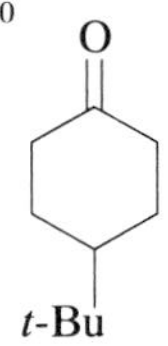

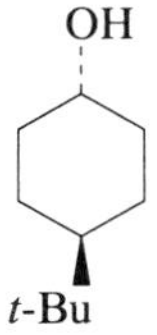

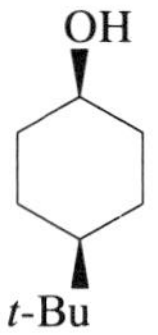

(product structures: **I** = cis cyclohexanol with t-Bu, **II** = trans cyclohexanol with t-Bu) **I** + **II**

Table III.  REDUCTION OF ALDEHYDES AND KETONES (*Continued*)

| Substrate | Conditions | Product(s) and Yield(s) (%) | Refs. |
|---|---|---|---|
| C$_{13}$ 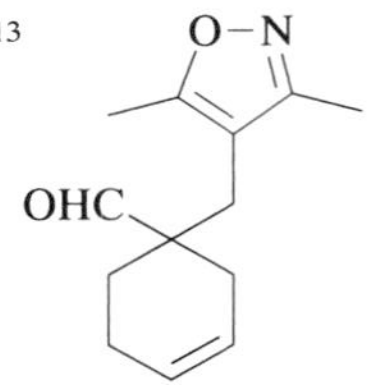 <br> OHC—(structure with O–N dimethylisoxazole ring attached to cyclohexene) | 2 SmI$_2$, THF, rt, 24 h | 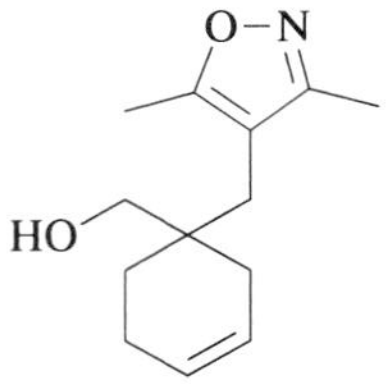 HO—(structure with O–N dimethylisoxazole ring attached to cyclohexene)  (53) | 25 |
| C$_{14}$ <br> PhCOCOPh | 4 SmI$_2$, THF, MeOH, 6-8 LiNH$_2$, rt, 25 min | PhCHOHCHOHPh (10) + PhCH$_2$CHOHPh (5) + PhCH$_2$COPh (45) + PhCH$_2$OH (13) | 42 |
|  | 2 SmI$_2$, THF, HMPA, 2 quinidine, rt, 30 min | 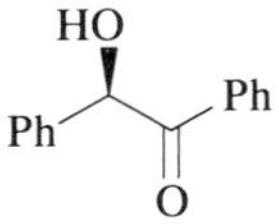 (structure: HO, Ph, Ph, O)  (77.5) 56.2 % ee | 43 |
|  | 2 SmI$_2$, THF, quinidine, rt | " (—) 18.8% ee | 43 |
|  | 2 SmI$_2$, THF, quinine, rt | " (—) 9.4% ee | 43 |
|  | 2 SmI$_2$, THF, cinchonidine, rt | " (—) 11.3% ee | 43 |
|  | 2 SmI$_2$, THF, N-[N,N-dimethyl-(S)-phenylalanyl]-(S)-1-phenylethylamine, rt | " (—) 15.7% ee | 43 |
|  | 2 SmI$_2$, THF, diethyl L-(+)-tartrate, rt | " (—) 7.1% ee | 43 |

Table III. REDUCTION OF ALDEHYDES AND KETONES (*Continued*)

| Substrate | Conditions | Product(s) and Yield(s) (%) | Refs. |
|---|---|---|---|
| PhCOCHOHPh | 4 SmI$_2$, THF, MeOH, 6-8 LiNH$_2$, rt, 3 s | PhCHOHCHOHPh (36) + PhCH$_2$CHOHPh (40) + PhCH$_2$COPh (tr) + PhCH$_2$OH (tr) | 42 |
| C$_{15}$ 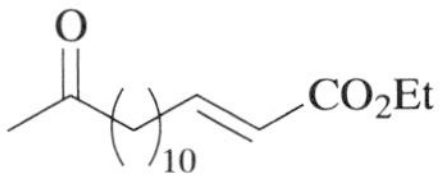 | 4 SmI$_2$, THF, HMPA, 65°, 1 h | 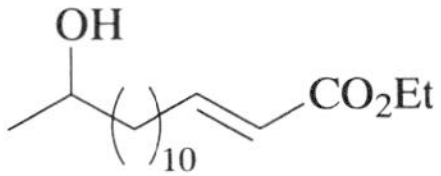 (40) | 44 |
| C$_{19}$ 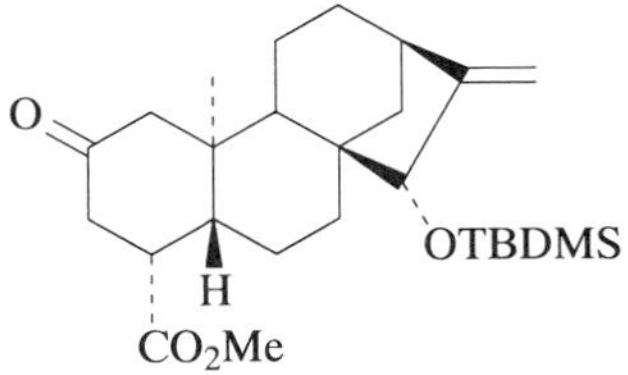 | SmI$_2$, THF, H$_2$O, rt, 10 min | 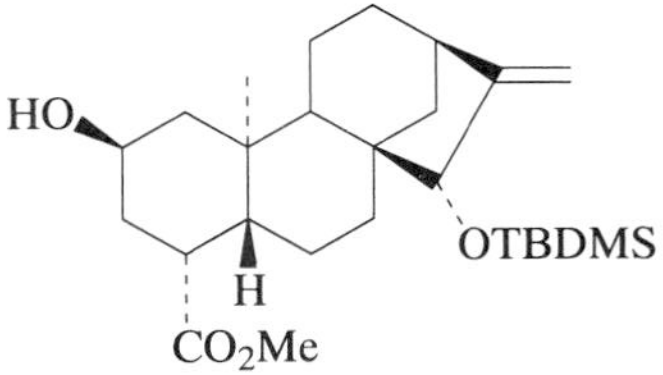 (90) 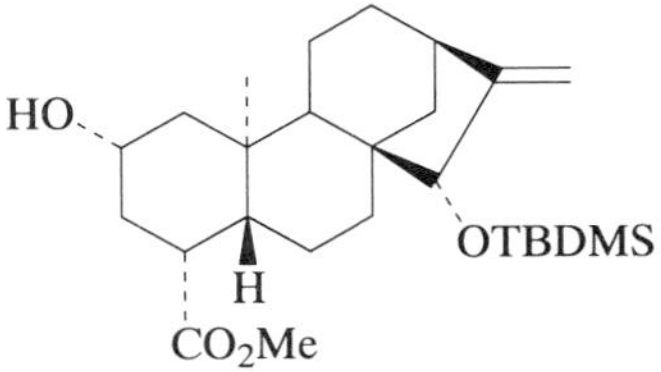 + (7) | 45 |

295

Table IV. REDUCTION OF CARBOXYLIC ACIDS AND THEIR DERIVATIVES

| Substrate | Conditions | Product(s) and Yield(s) (%) | Refs. |
|---|---|---|---|
| | | A. *Carboxylic Acids* | |
| $C_6$ | | | |
| 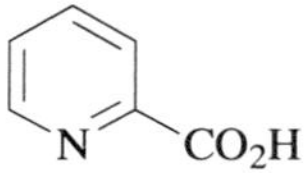 2-pyridyl-$CO_2H$ | 2 $SmI_2$, THF, $H_3PO_4$ (85%), rt, 3 s | 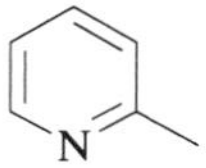 2-methylpyridine (43) | 46 |
| 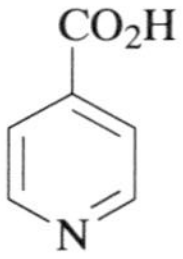 $CO_2H$ 4-pyridyl | 2 $SmI_2$, THF, $H_3PO_4$ (85%), rt, 3 s | 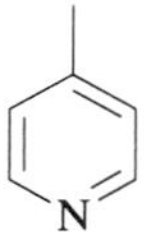 4-methylpyridine (48) | 46 |
| $n$-$C_5H_{11}CO_2H$ | 4 $SmI_2$, THF, $H_2O$, 8 NaOH, rt, 271 s | $n$-$C_5H_{11}CH_2OH$ (61) | 47 |
| $C_7$ | | | |
| $o$-$ClC_6H_4CO_2H$ | 4 $SmI_2$, THF, $H_3PO_4$ (85%), rt, 3 s | $o$-$ClC_6H_4CH_2OH$ (97) | 46 |
| $m$-$ClC_6H_4CO_2H$ | " | $m$-$ClC_6H_4CH_2OH$ (96) | 46 |
| $p$-$ClC_6H_4CO_2H$ | " | $p$-$ClC_6H_4CH_2OH$ (94) | 46 |
| $PhCO_2H$ | 4 $SmI_2$, THF, 10 h | $PhCH_2OH$ **I** (0) | 47 |
| | 4 $SmI_2$, THF, $H_2O$, 8 NaOH, rt, 60 s | **I** (91) | 47 |
| | 4 $SmI_2$, THF, $H_2O$, 10 $NH_3$, rt, 3 s | **I** (41) | 47 |
| | 4 $SmI_2$, THF, $H_2O$, 8 $LiNH_2$, rt, 60 s | **I** (87) | 47 |
| | 4 $SmI_2$, THF, $H_3PO_4$ (85%), rt, 3 s | **I** (91) | 46 |

296

Table IV. REDUCTION OF CARBOXYLIC ACIDS AND THEIR DERIVATIVES (*Continued*)

| Substrate | Conditions | Product(s) and Yield(s) (%) | Refs. |
|---|---|---|---|
| $C_6H_{11}CO_2H$ | 4 SmI$_2$, THF, H$_2$O, 8 NaOH, 58 s | $C_6H_{11}CH_2OH$ (78) | 47 |
| $n$-$C_6H_{13}CO_2H$ | 4 SmI$_2$, THF, H$_2$O, 8 NaOH, 291 s | $n$-$C_6H_{13}CH_2OH$ (57) | 47 |
| **C$_8$** | | | |
| $PhCH_2CO_2H$ | 4 SmI$_2$, THF, H$_2$O, 8 NaOH, rt, 82 s | $PhCH_2CH_2OH$ (73) | 47 |
| $o$-$MeC_6H_4CO_2H$ | 4 SmI$_2$, THF, H$_3$PO$_4$ (85%), rt, 4 s | $o$-$MeC_6H_4CO_2H$ (91) | 46 |
| $p$-$MeC_6H_4CO_2H$ | 4 SmI$_2$, THF, H$_3$PO$_4$ (85%), rt, 4 s | $p$-$MeC_6H_4CO_2H$ (95) | 46 |
| $n$-BuCH(Et)CO$_2$H | 4 SmI$_2$, THF, H$_2$O, 8 NaOH, rt, 10 s | $n$-BuCH(Et)CH$_2$OH (94) | 47 |
| *B.  Carboxylic Acid Salts* | | | |
| $PhCO_2Na$ | 4 SmI$_2$, THF, H$_2$O, rt, 60 s | $PhCH_2OH$ (92) | 47 |
| *C.  Carboxylic Acid Chlorides* | | | |
| **C$_1$** | | | |
| $Ph_2NCOCl$ | 2 THF, 6 H$_2$O, rt, 30 s | $Ph_2NCHO$ (84) | 48 |
| **C$_7$** | | | |
| $PhCOCl$ | 4 SmI$_2$, THF, H$_3$PO$_4$ (85%), rt, 3 s | $PhCH_2OH$ (31) + $PhCH_2CHOHPh$ (14) + $PhCHOHCHOHPh$ (40) + $PhCHOHCOPh$ (tr) | 46 |
| **C$_8$** | | | |
| $PhCH_2COCl$ | 2 SmI$_2$, THF, 6 H$_2$O, rt, 30 s | $PhCH_2CH_2OH$ (22) | 48 |

297

Table IV. REDUCTION OF CARBOXYLIC ACIDS AND THEIR DERIVATIVES (*Continued*)

| Substrate | Conditions | Product(s) and Yield(s) (%) | Refs. |
|---|---|---|---|
| | *D. Carboxylic Acid Derivatives* | | |
| $C_{14}$ | | | |
| $(PhCO)_2O$ | 4 SmI$_2$, THF, MeOH, 8 LiOMe, rt, 46 min | PhCH$_2$OH **I** (12) + PhCHO **II** (38) + PhCO$_2$Me **III** (52) | 42 |
| | 4 SmI$_2$, THF, MeOH, 8 KOH, rt, 7.6 min | **I** (50) + **II** (14) + **III** (16) | 42 |
| | 4 SmI$_2$, THF, H$_3$PO$_4$ (85%), rt, 3 s | **I** (46) + **II** (13) + PhCH$_2$CHOHPh (22) + PhCHOHCHOHPh (9) | 46 |
| | *E. Carboxylic Acid Esters* | | |
| $C_7$ | | | |
| PhCO$_2$Me | 4 SmI$_2$, THF, MeOH, 12 LiNH$_2$, rt, 14 min | PhCH$_2$OH **I** (64) | 42 |
| | 4 SmI$_2$, THF, MeOH, 8 LiOMe, rt, 27 min | **I** (59) | 42 |
| | 4 SmI$_2$, THF, MeOH, 8 KOH, rt, 8 min | **I** (68) | 42 |
| | 4 SmI$_2$, THF, H$_3$PO$_4$ (85%), rt, 3 s | **I** (72) | 46 |
| | *F. Carboxylic Acid Amides and Derivatives* | | |
| $C_6$ | | | |
| ![pyridine-2-carboxamide](N=/CONH_2) | 8 SmI$_2$, THF, H$_3$PO$_4$ (85%), rt, 2 s | ![2-methylpyridine] (88) | 46 |

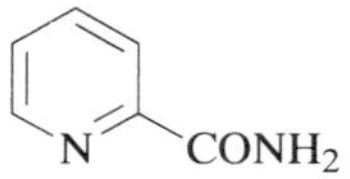

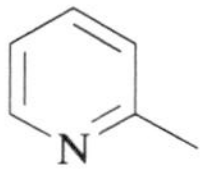

Table IV. REDUCTION OF CARBOXYLIC ACIDS AND THEIR DERIVATIVES (*Continued*)

| Substrate | Conditions | Product(s) and Yield(s) (%) | Refs. |
|---|---|---|---|
| 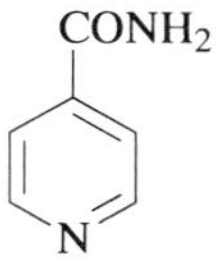 CONH$_2$ (4-pyridyl) | 8 SmI$_2$, THF, H$_3$PO$_4$ (85%), rt, 3 s | 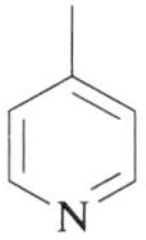 (67) | 46 |
| **C$_7$** | | | |
| $o$-HOC$_6$H$_4$CONH$_2$ | 4 SmI$_2$, THF, H$_3$PO$_4$ (85%), rt, 3 s | $o$-HOC$_6$H$_4$CHO  **I** (97) | 46 |
| $o$-HOC$_6$H$_4$CONHOH | " | **I** (59)  +  $o$-HOC$_6$H$_4$CONH$_2$  **II** (40) | 46 |
| $o$-HOC$_6$H$_4$CONHNH$_2$ | 4 SmI$_2$, THF, MeOH H$_3$PO$_4$ (85%), rt, 3 s | **I** (48) + **II** (50) | 46 |
| $o$-ClC$_6$H$_4$CONH$_2$ | " | $o$-ClC$_6$H$_4$CHO      (>99) | 46 |
| $m$-ClC$_6$H$_4$CONH$_2$ | " | $m$-ClC$_6$H$_4$CHO      (>99) | 46 |
| $p$-ClC$_6$H$_4$CONH$_2$ | " | $p$-ClC$_6$H$_4$CHO      (>99) | 46 |
| PhCONH$_2$ | 8 SmI$_2$, THF, H$_2$O, rt, 90 min | PhCH$_2$OH  **I** (63) +  PhCH$_2$NH$_2$  **II** (1) | 46 |
| | 4 SmI$_2$, THF, MeOH 12 LiNH$_2$, rt, 180 s | **I** (81) + **II** (8) | 42 |
| | 4 SmI$_2$, THF, MeOH 8 LiOMe, rt, 510 s | **I** (72) + **II** (4) | 42 |
| | 4 SmI$_2$, THF, MeOH 8 KOH, rt, 123 s | **I** (82) + **II** (8) | 42 |
| | 4 SmI$_2$, THF, H$_3$PO$_4$ (85%), rt, 3 s | PhCHO  **I** (>99) | 46 |
| PhCONHOH | " | **I** (91) +  PhCONH$_2$  **II** (6) | 46 |

299

Table IV. REDUCTION OF CARBOXYLIC ACIDS AND THEIR DERIVATIVES (*Continued*)

| Substrate | Conditions | Product(s) and Yield(s) (%) | Refs. |
|---|---|---|---|
| PhCONHNH$_2$ | 4 SmI$_2$, THF, H$_3$PO$_4$ (85%), rt, 3 s | **I** (81) + **II** (17) | 46 |
| PhCONHPh | 4 SmI$_2$, THF, MeOH, 12 LiNH$_2$, rt, 60 s | PhCH$_2$OH **I** (31) + PhCH$_2$NHPh **II** (20) + PhNH$_2$ **III** (28) | 42 |
| | 4 SmI$_2$, THF, H$_3$PO$_4$ (85%), rt, 3 s | **I** (4) + **II** (66) + PhCHO (23) | 46 |
| C$_8$ | " | (>99) | 46 |
| PhCH$_2$CONH$_2$ | 4 SmI$_2$, THF, MeOH, H$_2$O, 6 LiNH$_2$, rt, 68 s | PhCH$_2$CH$_2$OH (56) + PhCH$_2$CH$_2$NH$_2$ (2) | 42 |
| | 4 SmI$_2$, THF, H$_3$PO$_4$ (85%), rt, 90 s | PhCH$_2$CHO (14) | 46 |
| o-MeC$_6$H$_4$CONH$_2$ | 4 SmI$_2$, THF, H$_3$PO$_4$ (85%), rt, 4 s | o-MeC$_6$H$_4$CHO (90) | 46 |
| m-MeC$_6$H$_4$CONH$_2$ | " | m-MeC$_6$H$_4$CHO (74) | 46 |
| p-MeC$_6$H$_4$CONH$_2$ | " | p-MeC$_6$H$_4$CHO (91) | 46 |
| C$_9$ Ph(CH$_2$)$_2$CONH$_2$ | 4 SmI$_2$, THF, MeOH, H$_2$O, 6 LiNH$_2$, rt, 3 s | Ph(CH$_2$)$_3$OH (69) + Ph(CH$_2$)$_3$NH$_2$ (tr) | 42 |
| | 4 SmI$_2$, THF, H$_3$PO$_4$ (85%), rt, 4 min | Ph(CH$_2$)$_2$CHO (6) | 46 |

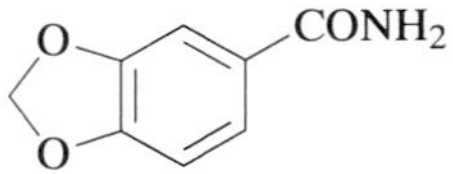

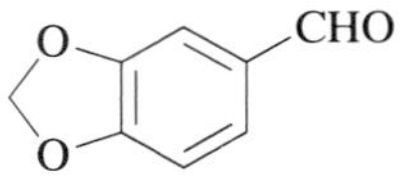

Table V. REDUCTION OF CONJUGATED CARBONYL SUBSTRATES

| Substrate | Conditions | Product(s) and Yield(s) (%) | Refs. |
|---|---|---|---|
| | A. Conjugated Aldehydes and Ketones | | |
| $C_6$ 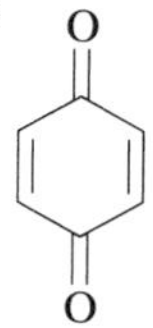 (benzoquinone) | 2 SmI$_2$, THF, MeOH, 65°, 5 min | 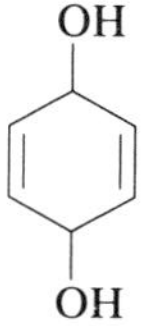 (hydroquinone, OH...OH) (93) | 49 |
| $C_7$ 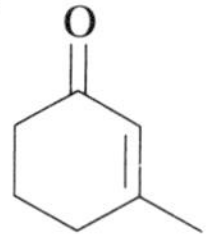 (3-methylcyclohex-2-enone) | 2 SmI$_2$, THF, MeOH, rt, 1 d | 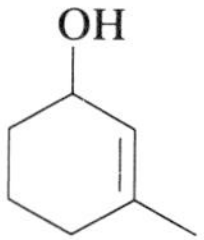 (3-methylcyclohexenol, OH) (67) 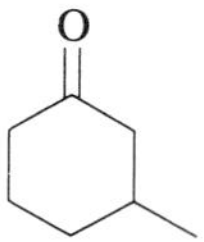 + (3-methylcyclohexanone, O) (28) | 3 |
| $C_9$ 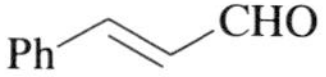 Ph–CH=CH–CHO (trans) | " | Polymers | 3 |
| $C_{10}$ 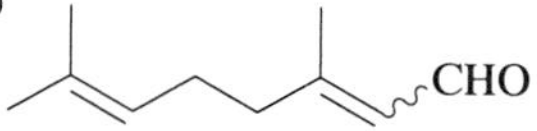 (farnesal type, CHO) | " | 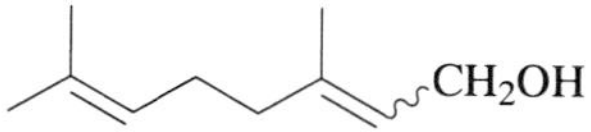 CH$_2$OH (70) 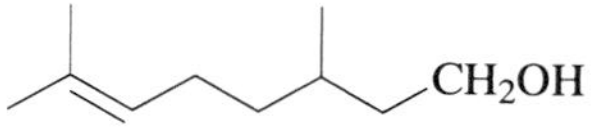 + CH$_2$OH (18) 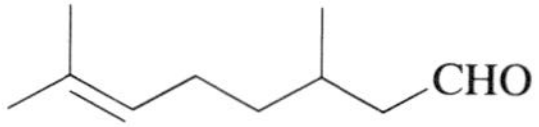 + CHO (8) | 3 |
| | B. Conjugated Esters | | |
| $C_4$ 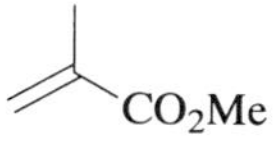 (methacrylate, CO$_2$Me) | 2 SmI$_2$, THF, HMPA, rt, 30 min | 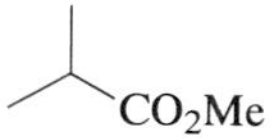 CO$_2$Me (67) | 50 |

301

Table V. REDUCTION OF CONJUGATED CARBONYL SUBSTRATES (*Continued*)

| Substrate | Conditions | Product(s) and Yield(s) (%) | Refs. |
|---|---|---|---|
| 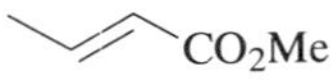 ⌇CO$_2$Me | 2 SmI$_2$, THF, HMPA, rt, 30 min |  ⌇CO$_2$Me (59) | 50 |
|  ⌇CO$_2$C$_6$H$_{11}$ | |  ⌇CO$_2$C$_6$H$_{11}$  **I** | |
| | 2.5 SmI$_2$, THF, *t*-BuOH or C$_6$H$_{11}$OH, rt, 72 h | **I** (tr) | 51 |
| | 2.5 SmI$_2$, THF, *t*-BuOH or C$_6$H$_{11}$OH, rt, DMF, 5 min | **I** (67) | 51 |
| | 2.5 SmI$_2$, THF, *t*-BuOH or C$_6$H$_{11}$OH, rt, DMA, 2 min | **I** (92) | 51 |
| | 2.5 SmI$_2$, THF, *t*-BuOH or C$_6$H$_{11}$OH, rt, TMU, 15 min | **I** (45) | 51 |
| | 2.5 SmI$_2$, THF, *t*-BuOH or C$_6$H$_{11}$OH, rt, NMP, 15 min | **I** (10) | 51 |
| | 2.5 SmI$_2$, THF, *t*-BuOH or C$_6$H$_{11}$OH, rt, HMPA, <1 min | **I** (7) | 51 |
| | 2.5 SmI$_2$, THF, *t*-BuOH or C$_6$H$_{11}$OH, rt, Me$_2$N(CH$_2$)$_3$NMe$_2$, 6 h | **I** (82) | 51 |
| | 2.5 SmI$_2$, THF, *t*-BuOH or C$_6$H$_{11}$OH, rt, Me$_2$N(CH$_2$)$_2$NMe$_2$, 24 h | **I** (61) | 51 |
| | 2.5 SmI$_2$, THF, *t*-BuOH or C$_6$H$_{11}$OH, rt, [Me$_2$N(CH$_2$)$_2$]$_2$NMe, 24 h | No reaction | 51 |

302

Table V. REDUCTION OF CONJUGATED CARBONYL SUBSTRATES (*Continued*)

| Substrate | Conditions | Product(s) and Yield(s) (%) | Refs. |
|---|---|---|---|
| (2-ethylhexyl methacrylate ester) | 2.5 SmI$_2$, THF, *t*-BuOH, DMA, rt, 15 min | 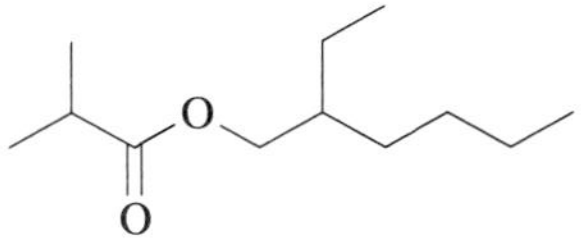 (78) | 51 |
| 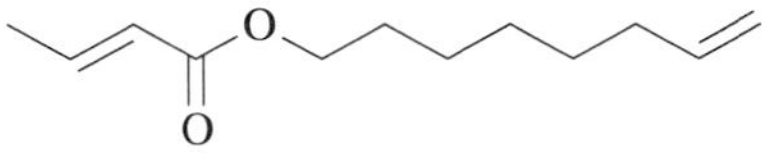 (homoallyl crotonate ester) | 2.5 SmI$_2$, THF, *t*-BuOH, DMA, rt, 2 min | 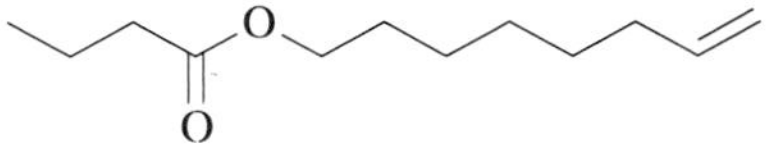 (92) | 51 |
| 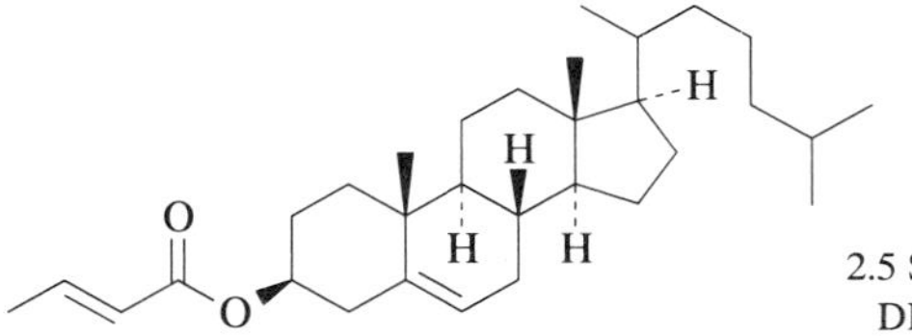 (steroid crotonate ester) | 2.5 SmI$_2$, THF, *t*-BuOH, DMA, rt, 15 min | 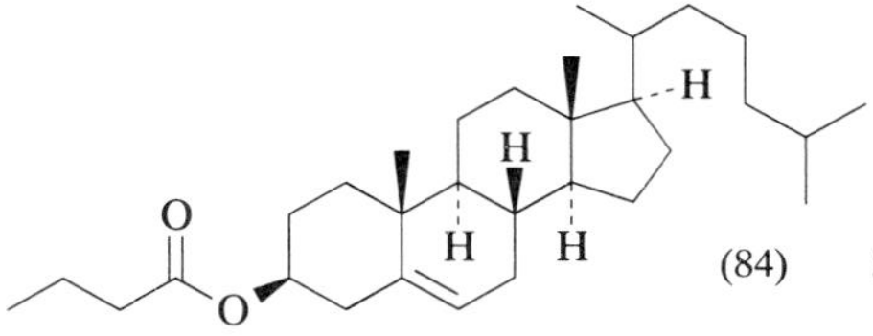 (84) | 51 |
| C$_7$ (alkynyl crotonate ester) | 2.5 SmI$_2$, THF, *t*-BuOH, DMA, rt, 2 min | (99) | 51 |
| C$_7$ (cyclohexenyl CO$_2$Me) | 2 SmI$_2$, THF, HMPA, rt, 30 min | (64) | 50 |
| C$_9$ Ph–CH=CH–CH$_2$–CO$_2$Et | 2 SmI$_2$, THF, MeOH, rt, 1 d | (85–98) | 3 |
| C$_{11}$ Ph–…–CO$_2$Et | 2.5 SmI$_2$, THF, EtOH, DMA, rt, 12 min | (96) | 51 |

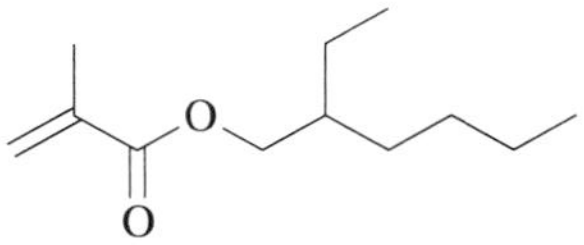

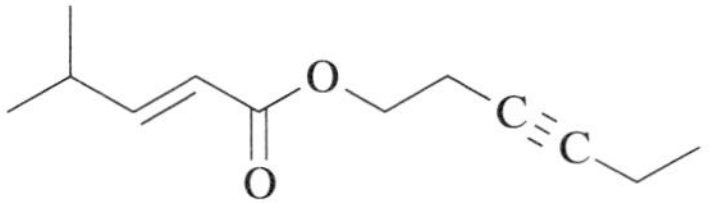

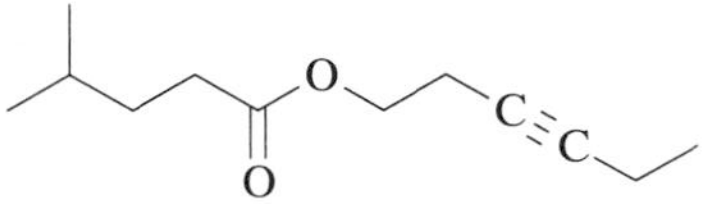

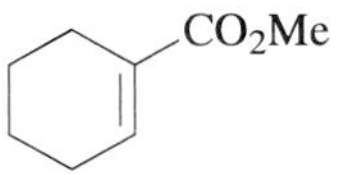

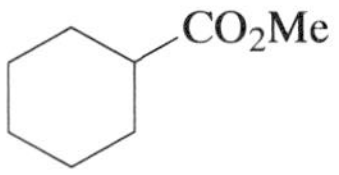

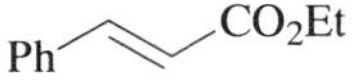

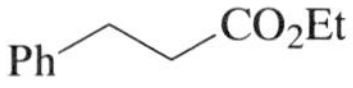

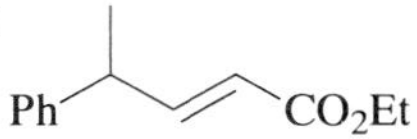

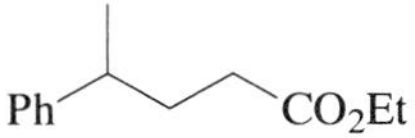

303

Table V. REDUCTION OF CONJUGATED CARBONYL SUBSTRATES (*Continued*)

| Substrate | Conditions | Product(s) and Yield(s) (%) | Refs. |
|---|---|---|---|
| C$_{12}$ (Ph...CO$_2$Me) | 2.5 SmI$_2$, THF, MeOH, DMA, rt, 10 min | (CO$_2$Me) (86) | 51 |
| 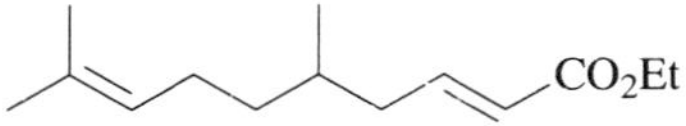 C$_{14}$ (CO$_2$Et structure, 88 : 12, *E/Z*) | 2.5 SmI$_2$, THF, EtOH, DMA, rt, 2 min | 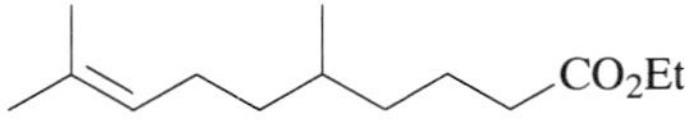 (CO$_2$Et) (97) | 51 |
| 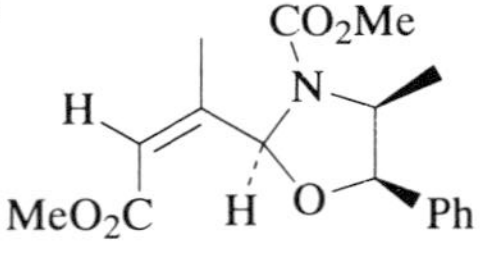 (oxazolidine CO$_2$Me, N, O, Ph, H, MeO$_2$C structure) | SmI$_2$, THF, H$_2$O, rt | 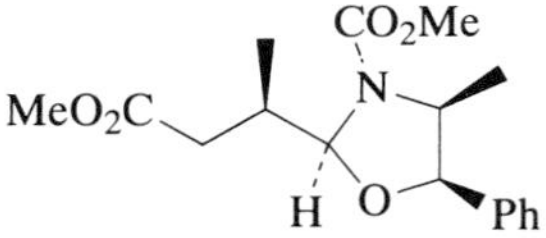 **I** 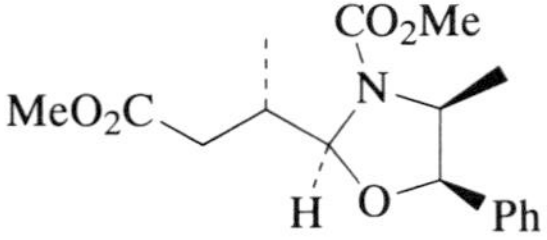 + **II** (35) **I:II** = 9 | 52 |

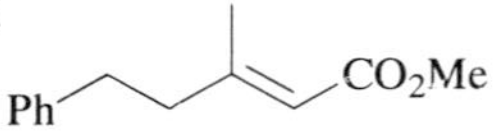

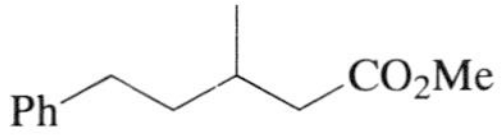

*C. Conjugated Amides*

| Substrate | Conditions | Product(s) and Yield(s) (%) | Refs. |
|---|---|---|---|
| 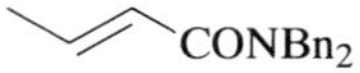 C$_4$ (CONBn$_2$) | 2.5 SmI$_2$, THF, *t*-BuOH, DMA, rt, 5 min |  (CONBn$_2$) (30) + (CONBn$_2$ / CONBn$_2$) (50) | 51 |
| 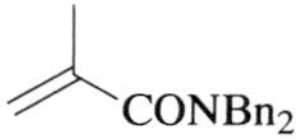 (CONBn$_2$) | 2.5 SmI$_2$, THF, *t*-BuOH, DMA, rt, 15 min | 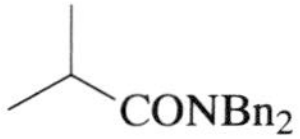 (CONBn$_2$) (87) | 51 |

304

Table V. REDUCTION OF CONJUGATED CARBONYL SUBSTRATES (*Continued*)

| Substrate | Conditions | Product(s) and Yield(s) (%) | Refs. |
|---|---|---|---|
| $C_5$ 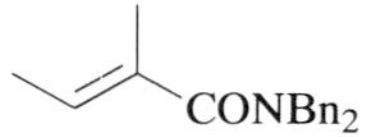  (structure, CONBn$_2$) | 2.5 SmI$_2$, THF, $t$-BuOH, DMA, rt, 3 min | 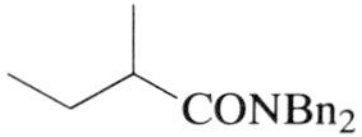 (structure, CONBn$_2$) (99) | 51 |
| $C_9$ 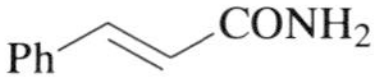  Ph (structure, CONH$_2$) | 2 SmI$_2$, THF, HMPA, rt, 1 min | 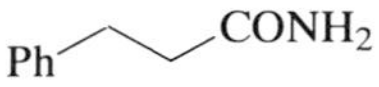 Ph (structure, CONH$_2$) (65) | 50 |

*D. Conjugated Carboxylic Acids*

| Substrate | Conditions | Product(s) and Yield(s) (%) | Refs. |
|---|---|---|---|
|  $C_4$  HO$_2$C (structure) CO$_2$H | 2 SmI$_2$, THF, HMPA, rt, 5 min | 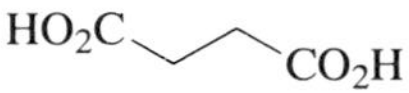 HO$_2$C (structure) CO$_2$H (99) | 50 |
| 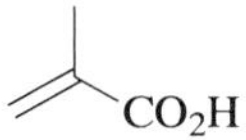 $C_5$  (structure) CO$_2$H | 2 SmI$_2$, THF, HMPA, rt, 1 min | 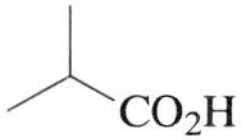 (structure) CO$_2$H (80) | 50 |
| $C_5$ 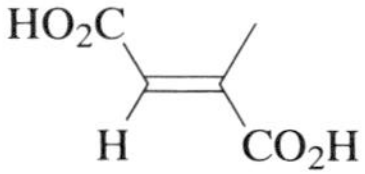  HO$_2$C / H (structure) CO$_2$H | 2 SmI$_2$, THF, HMPA, rt, 5 min | 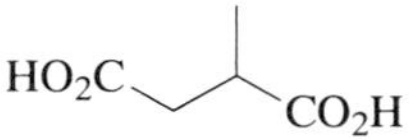 HO$_2$C (structure) CO$_2$H (88) | 50 |
| $C_6$ 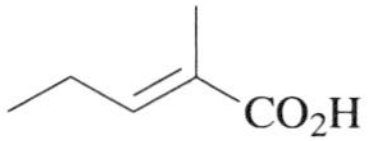  (structure) CO$_2$H | 2 SmI$_2$, THF, HMPA, rt, 30 min | 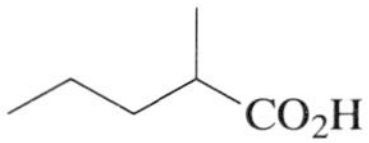 (structure) CO$_2$H (91) | 50 |
| $C_7$ 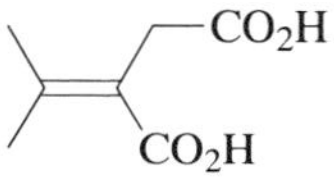  (structure) CO$_2$H / CO$_2$H | 2 SmI$_2$, THF, HMPA, rt, 30 min | 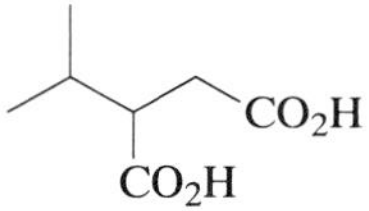 (structure) CO$_2$H / CO$_2$H (75) | 50 |

Table V. REDUCTION OF CONJUGATED CARBONYL SUBSTRATES (*Continued*)

| Substrate | Conditions | Product(s) and Yield(s) (%) | Refs. |
|---|---|---|---|
| Ph⁀CO₂H | 4 SmI₂, THF, H₂O, 8 NaOH, rt, 5 s |  Ph⁀OH **I** (39)  + Ph⁀⁀CO₂H **II** (7) | 47 |
| | 2 SmI₂, THF, MeOH, rt, 1 d | **II** (98) | 3 |
| | 2 SmI₂, THF, HMPA, rt, 1 min | **II** (95) | 50 |
| CO₂H Ph (structure) | 2 SmI₂, THF, HMPA, rt, 1 min | Ph CO₂H (60) | 50 |
| C₁₂ Ph CO₂H CO₂H (structure) | 2 SmI₂, THF, HMPA, rt, 30 min | 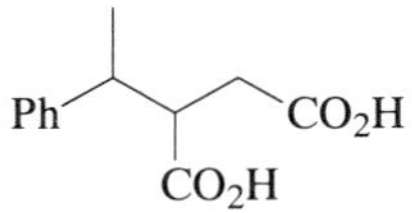 Ph CO₂H CO₂H (53) | 50 |
| *E. Conjugated Anhydrides* | | | |
| C₄ (maleic anhydride structure) | 2 SmI₂, THF, HMPA, rt, 5 min | (succinic anhydride structure) (96) | 50 |

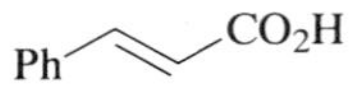

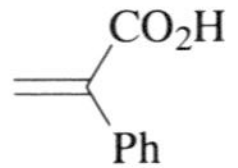

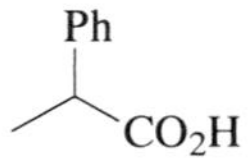

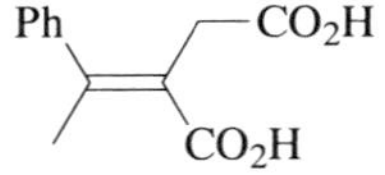

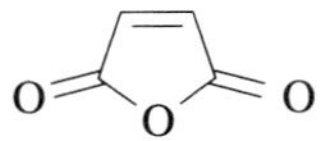

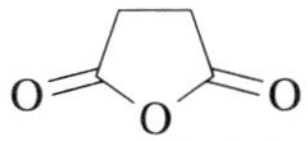

306

Table VI. REDUCTIVE CLEAVAGE OF α-HETEROSUBSTITUTED CARBONYL AND RELATED SUBSTRATES

| Substrate | Conditions | Product(s) and Yield(s) (%) | Refs. |
|---|---|---|---|
| | | A. *α-Halo Carbonyl Substrates* | |
| C$_4$ <br> (Br, CO$_2$Et structure) | 2 SmI$_2$, THF, MeOH, -78° | (CO$_2$Et structure) (98) | 53 |
| (Br, CO$_2$Et structure) | 2 Sm, 6 TMSCl, 6 NaI, MeCN, MeOH, -40°, 0.5 h | (CO$_2$Et structure) (73) | 8 |
| (O, Cl structure) | " | (O structure) (62) | 8 |
| C$_5$ (Br, O, lactone structure) | " | (lactone structure) (79) | 8 |
| C$_6$ (Cl, O cyclohexanone structure) | " | **I** (88) | 8 |
| (O, Cl, Ph structure) | 2 SmI$_2$, THF, MeOH, -78° | **I** (100) | 53 |
| 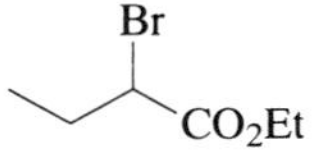 | 2 Sm, 6 TMSCl, 6 NaI, MeCN, MeOH, -40°, 0.5 h | 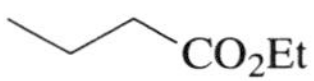 (O, Ph structure) (80) | 8 |
| 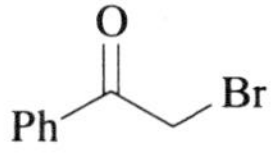 C$_8$ (O, Br, Ph structure) | 2 SmI$_2$, THF, MeOH, rt, 2 min | " (84) | 49 |

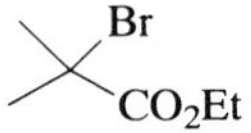

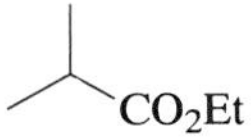

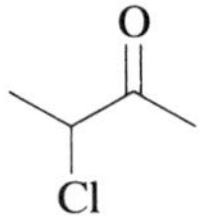

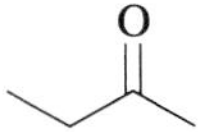

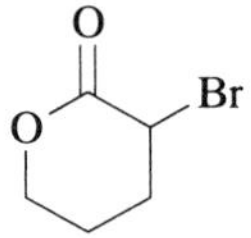

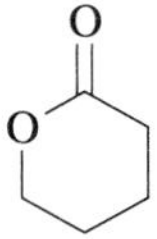

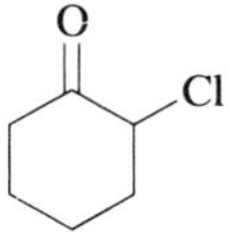

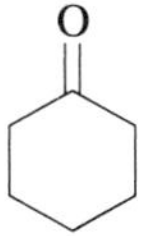

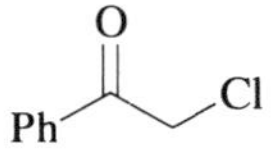

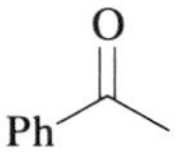

307

Table VI. REDUCTIVE CLEAVAGE OF α-HETEROSUBSTITUTED CARBONYL AND RELATED SUBSTRATES (*Continued*)

| Substrate | Conditions | Product(s) and Yield(s) (%) | Refs. |
|---|---|---|---|
| C₁₀ | 2 SmI₂, THF, MeOH, -78° | 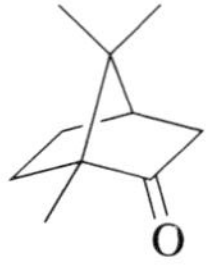 (85) | 53 |
| C₁₁ | 2.8 SmI₂, THF, MeOH, -78°, 1 h | 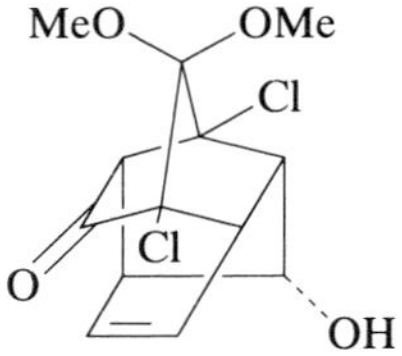 (68) 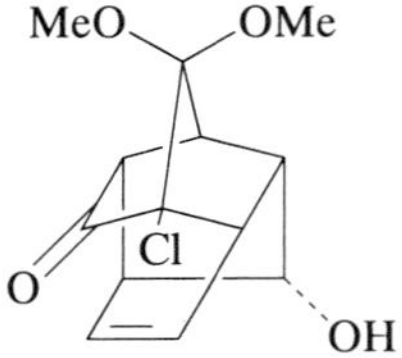 + (1) | 29 |
| | 5.5 SmI₂, THF, MeOH, -78° | 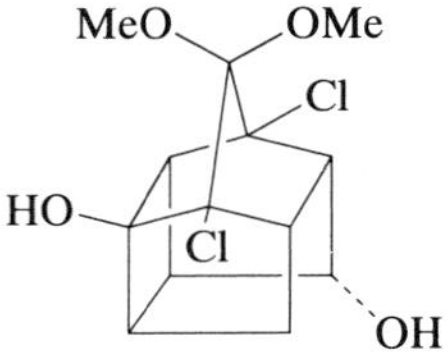 (70) | 29 |

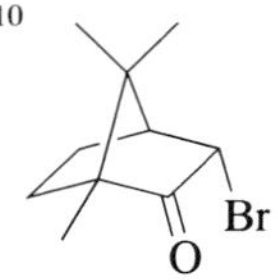

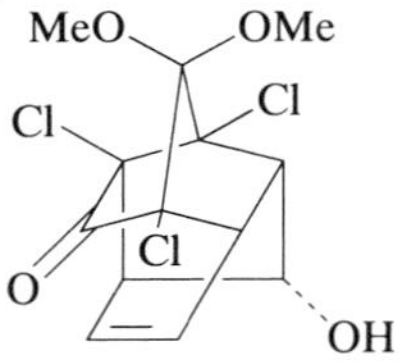

308

Table VI. REDUCTIVE CLEAVAGE OF α-HETEROSUBSTITUTED CARBONYL AND RELATED SUBSTRATES (*Continued*)

| Substrate | Conditions | Product(s) and Yield(s) (%) | Refs. |
|---|---|---|---|
| *B. α-Hydroxy and Vinylogous Hydroxy Carbonyl Substrates* | | | |
| C$_4$   EtO$_2$C—CH(OH)—CH(CO$_2$Et), HO | 3 SmI$_2$, THF, HO(CH$_2$)$_2$OH, rt, 20 min | EtO$_2$C—CH$_2$—CH(OH)—CO$_2$Et   (72) | 54 |
| 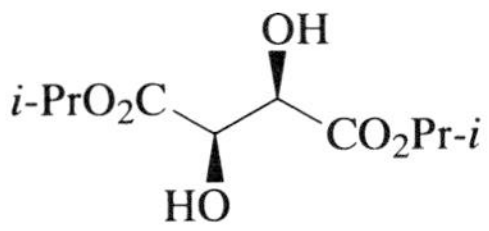 *i*-PrO$_2$C—CH(OH)—CH(CO$_2$Pr-*i*), HO | 3 SmI$_2$, THF, HO(CH$_2$)$_2$OH, rt, 1 h | 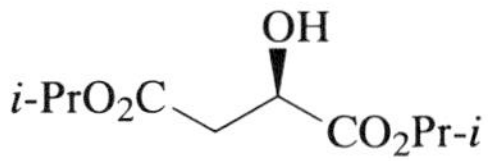 *i*-PrO$_2$C—CH$_2$—CH(OH)—CO$_2$Pr-*i*   (99) | 54 |
| 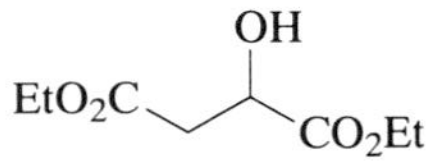 EtO$_2$C—CH(OH)—CH$_2$—CO$_2$Et | 3 SmI$_2$, THF, HMPA, pivalic acid, 20–22°, 2–4 h | 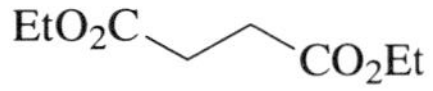 EtO$_2$C—CH$_2$—CH$_2$—CO$_2$Et   (71) | 54 |
| 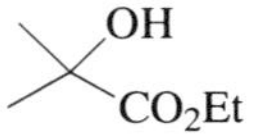 (CH$_3$)$_2$C(OH)—CO$_2$Et | " | 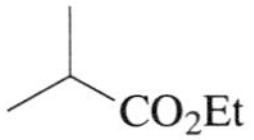 (CH$_3$)$_2$CH—CO$_2$Et   (73) | 54 |
| C$_5$ | 3 SmI$_2$, THF, HMPA, HO(CH$_2$)$_2$OH, rt, 3 h |   (90) | 55 |
| C$_6$ | 3 SmI$_2$, THF, HMPA, HO(CH$_2$)$_2$OH, rt, <1 min |   (75) | 55 |

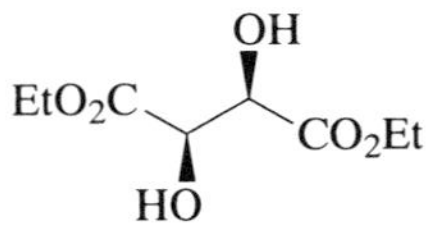

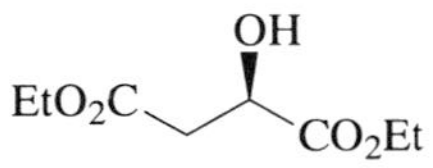

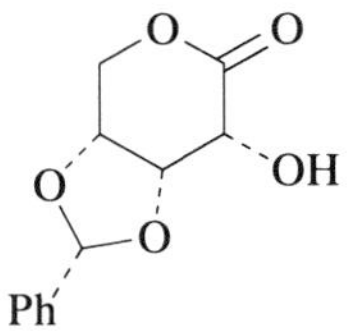

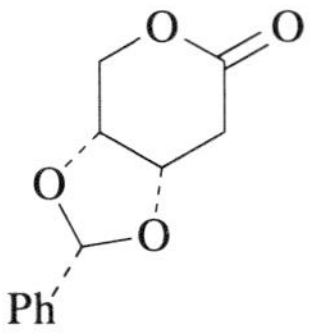

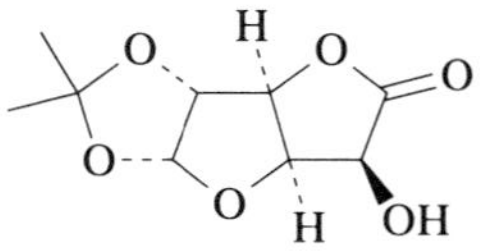

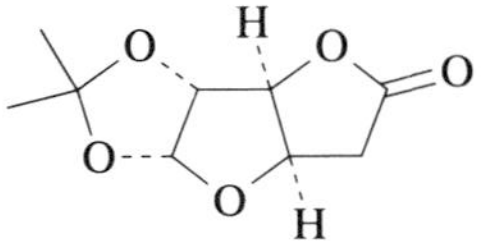

309

Table VI. REDUCTIVE CLEAVAGE OF α-HETEROSUBSTITUTED CARBONYL AND RELATED SUBSTRATES (*Continued*)

| Substrate | Conditions | Product(s) and Yield(s) (%) | Refs. |
|---|---|---|---|
| 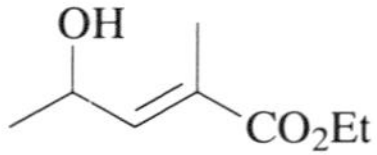 | 2 SmI$_2$, THF | 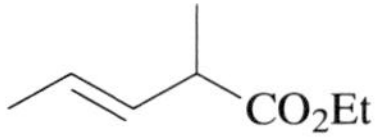 (90) | 56 |
| 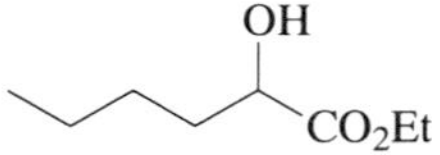 | 3 SmI$_2$, THF, HMPA, pivalic acid, 20-22°, 2-4 h |  (89) | 54 |
| 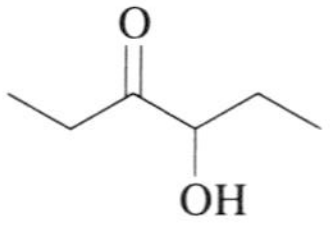 | 2 SmI$_2$, THF, Ac$_2$O, -78° | 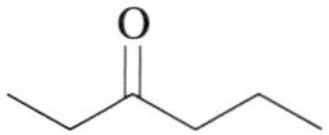 (59) | 53 |
| C$_7$ | 3 SmI$_2$, THF, HMPA, HO(CH$_2$)$_2$OH, rt, <1 min | (72) | 55 |
| 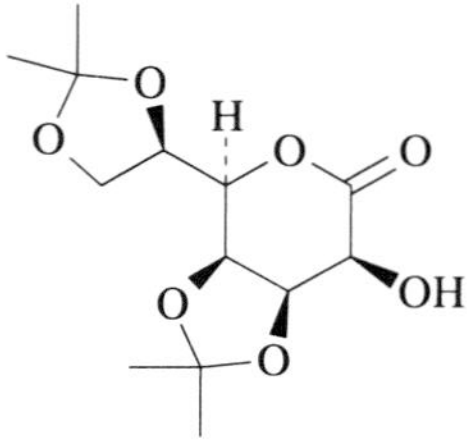 | 3 SmI$_2$, THF, HMPA, HO(CH$_2$)$_2$OH, rt, <1 min | 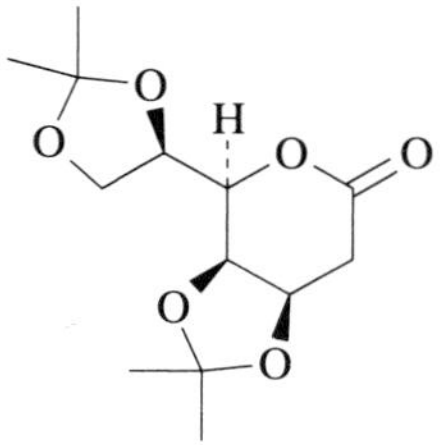 (99) | 55 |

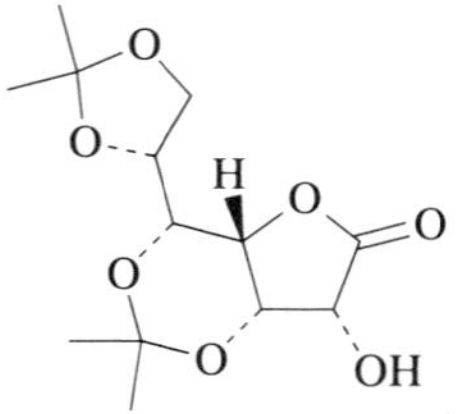

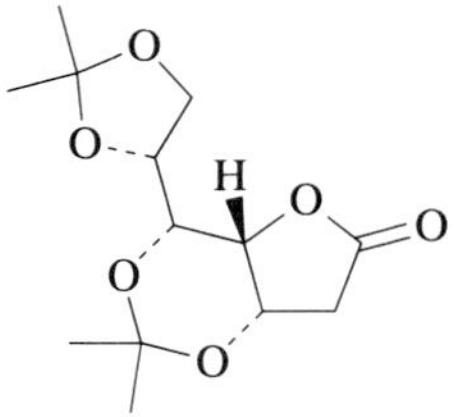

Table VI. REDUCTIVE CLEAVAGE OF α-HETEROSUBSTITUTED CARBONYL AND RELATED SUBSTRATES (*Continued*)

| Substrate | Conditions | Product(s) and Yield(s) (%) | Refs. |
|---|---|---|---|
| 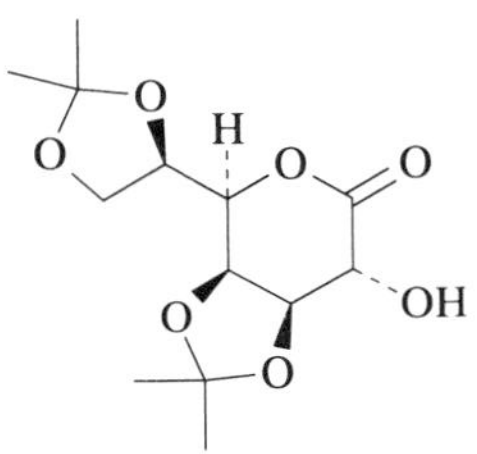 $C_8$ | 3 SmI$_2$, THF, HMPA, HO(CH$_2$)$_2$OH, rt, <1 min | 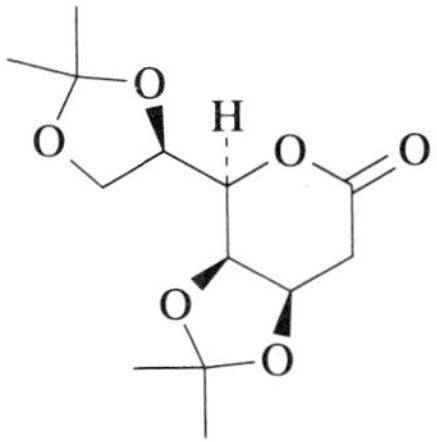 (98) | 55 |
| 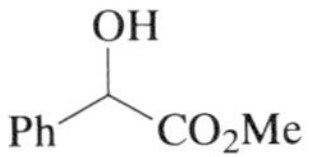 $C_{12}$ OH / Ph CO$_2$Me | 3 SmI$_2$, THF, HMPA, pivalic acid, 20-22°, 2-4 h | 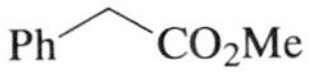 Ph CO$_2$Me (75) | 54 |
| 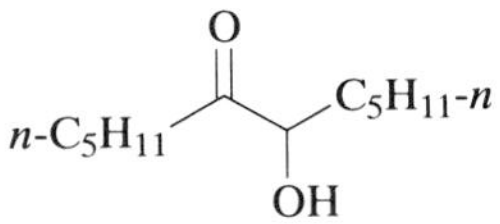 O / $n$-C$_5$H$_{11}$ C$_5$H$_{11}$-$n$ OH | 2 SmI$_2$, THF, MeOH, –78° | 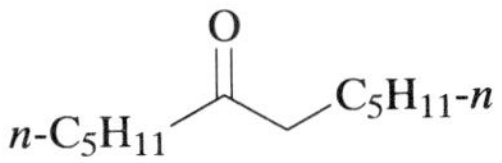 O / $n$-C$_5$H$_{11}$ C$_5$H$_{11}$-$n$ (29) | 53 |
| $C_{20}$ | 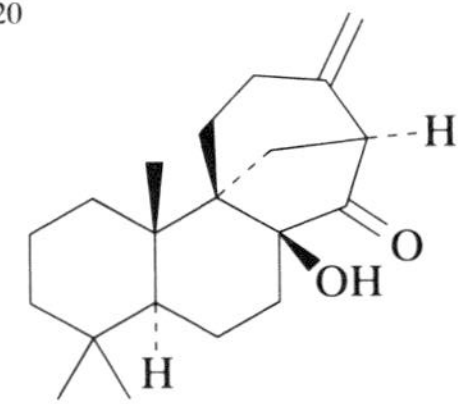 2 SmI$_2$, THF, $t$-BuOH, rt, 12 h | 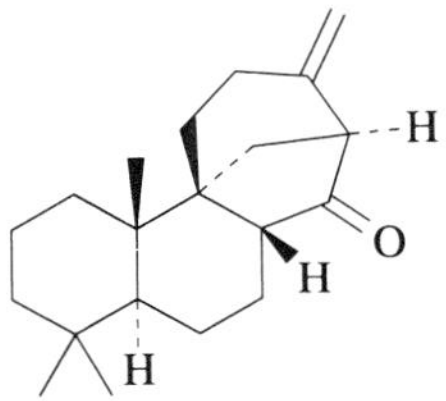 (87) | 57 |
| *C. α-Alkoxy and α-Silyloxy Carbonyl Substrates* | | | |
| $C_2$ MeOCH$_2$CO$_2$C$_8$H$_{17}$-$n$ | 2.5-3 SmI$_2$, THF, HMPA, $n$-C$_8$H$_{17}$OH, 20-22°, 3 h | MeCO$_2$C$_8$H$_{17}$-$n$ (73) | 54 |

Table VI. REDUCTIVE CLEAVAGE OF α-HETEROSUBSTITUTED CARBONYL AND RELATED SUBSTRATES (Continued)

| Substrate | Conditions | Product(s) and Yield(s) (%) | Refs. |
|---|---|---|---|

C<sub>4</sub>

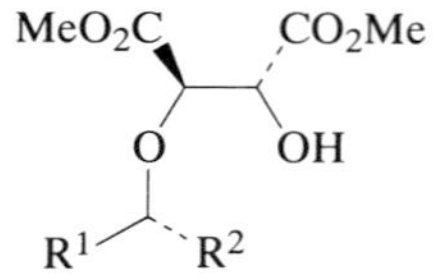

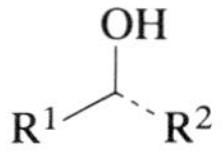

| R¹ | R² | | |
|---|---|---|---|
| n-C₉H₁₉ | Me | Sml₂, THF, HMPA, MeOH | I (—) | 58 |
| C₆H₁₁ | Me | " | I (—) | 58 |
| i-Pr | n-Bu | " | I (—) | 58 |

| | | | |
|---|---|---|---|
| C₅ | 4.5 SmI₂, THF, HO(CH₂)₂OH, rt, <1 min | 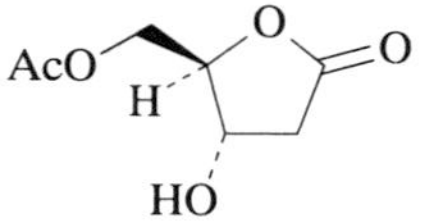 (98) | 55 |

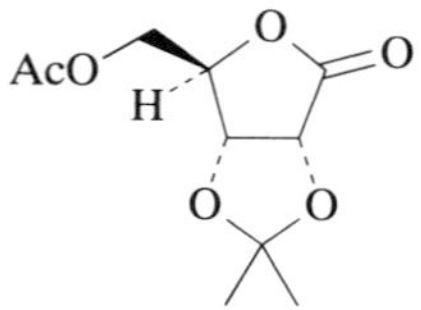

| | | | |
|---|---|---|---|
| C₆ | 4.5 SmI₂, THF, HO(CH₂)₂OH, rt, <1 min | 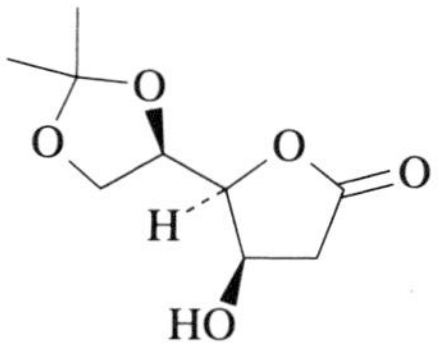 (98) | 55 |

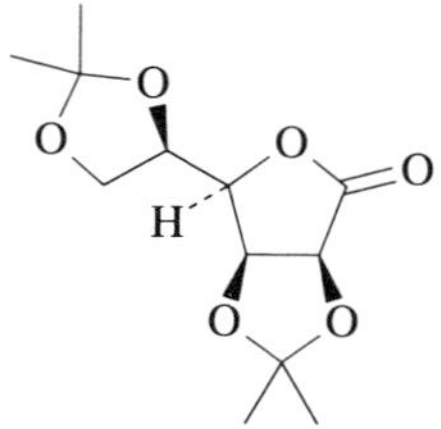

312

Table VI. REDUCTIVE CLEAVAGE OF α-HETEROSUBSTITUTED CARBONYL AND RELATED SUBSTRATES (*Continued*)

| Substrate | Conditions | Product(s) and Yield(s) (%) | Refs. |
|---|---|---|---|
| 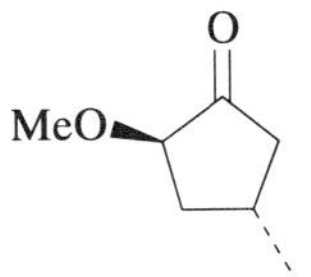 (MeO, cyclopentanone with methyl substituent) | SmI$_2$, THF | 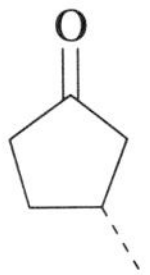 (69) | 59 |
| C$_8$ 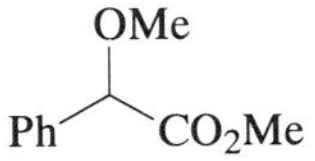 (OMe, Ph, CO$_2$Me) | 2.5-3 SmI$_2$, THF, HMPA, MeOH, 20-22°, 3 h | 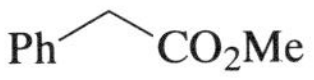 Ph⁀CO$_2$Me (89) | 54 |
| C$_9$ 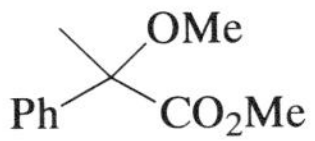 (OMe, Ph, CO$_2$Me) | 2.5-3 SmI$_2$, THF, HMPA, MeOH, 20-22°, 12 h | 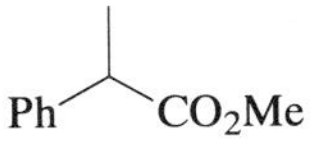 Ph⁀CO$_2$Me (95) | 54 |
| 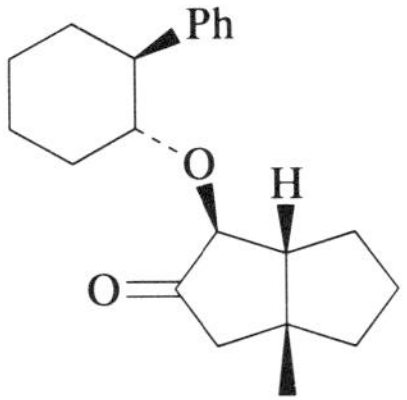 (bicyclic O-ether, Ph, cyclohexane) | SmI$_2$, THF | 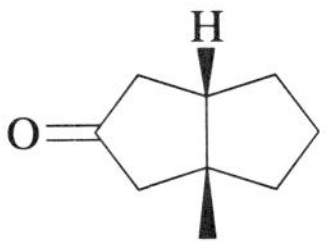 **I** (90) | 60 |
| 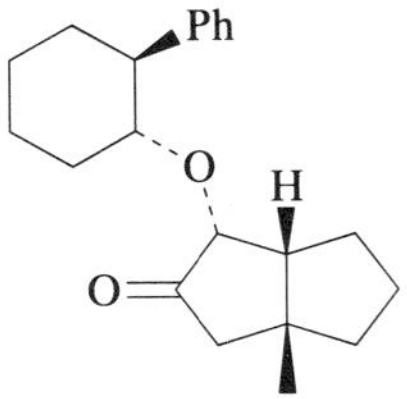 (bicyclic O-ether, Ph, cyclohexane) | SmI$_2$, THF | **I** (—) | 60 |

313

Table VI. REDUCTIVE CLEAVAGE OF α-HETEROSUBSTITUTED CARBONYL AND RELATED SUBSTRATES (*Continued*)

| Substrate | Conditions | Product(s) and Yield(s) (%) | Refs. |
|---|---|---|---|
| C₁₁ (MeO, MeO, OMEM, H, O structure) | 2 SmI₂, THF, MeOH, −78° | (OMEM, H, O structure) (92) | 61 |
| 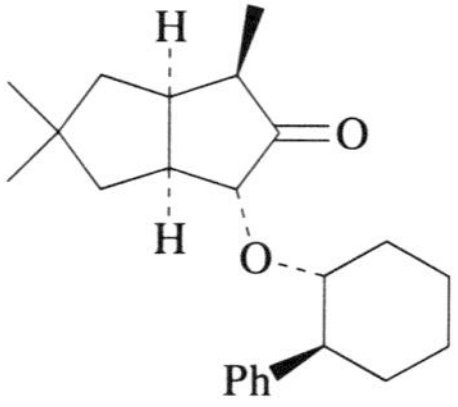 (Ph, O, cyclohexane, cyclopentanone structure) | SmI₂, THF | 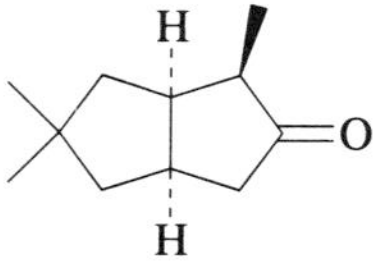 (O, H, H structure) (85) | 60 |
| C₁₂ (O, OMOM, H structure) | SmI₂, THF | **I** (O, H, HO, OMOM, H structure) + (95) **I:II** = 3  **II** (OH, H, O, OMOM, H structure) | 62 |

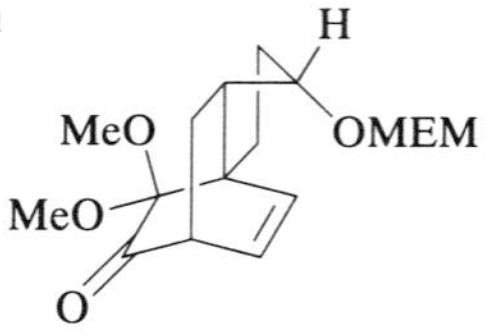

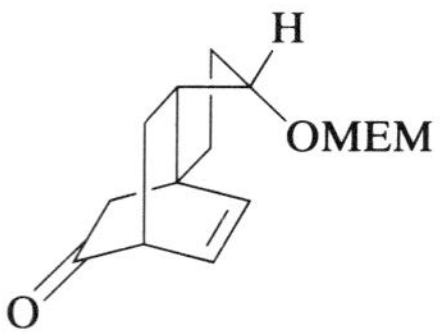

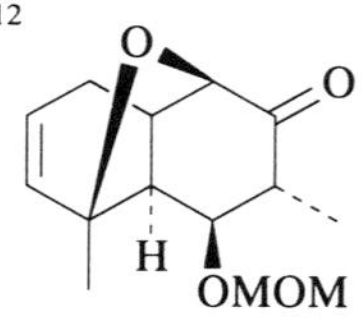

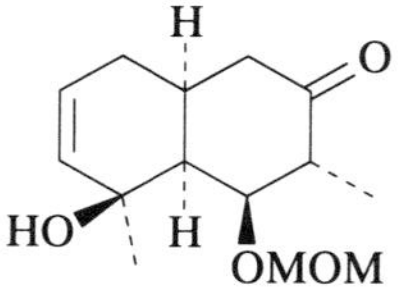

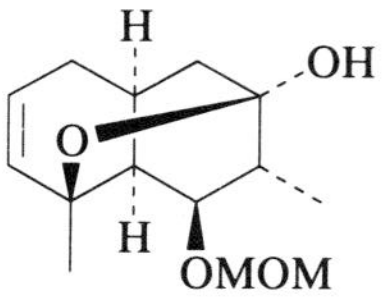

314

Table VI. REDUCTIVE CLEAVAGE OF α-HETEROSUBSTITUTED CARBONYL AND RELATED SUBSTRATES (*Continued*)

| Substrate | Conditions | Product(s) and Yield(s) (%) | Refs. |
|---|---|---|---|
| 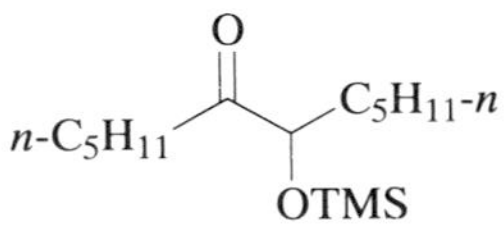 $n$-C$_5$H$_{11}$ ... C$_5$H$_{11}$-$n$, OTMS | 2 SmI$_2$, THF, MeOH, -78° | 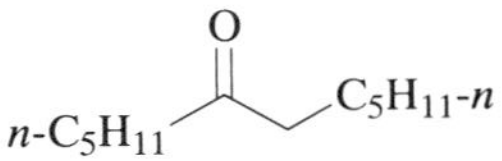 $n$-C$_5$H$_{11}$ ... C$_5$H$_{11}$-$n$ (98) | 53 |
| C$_{15}$ 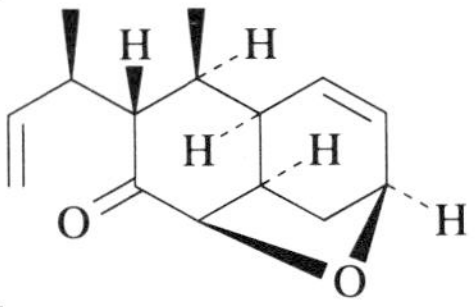 [structure] | 2.4 SmI$_2$, THF, -78°, 0.5 h | 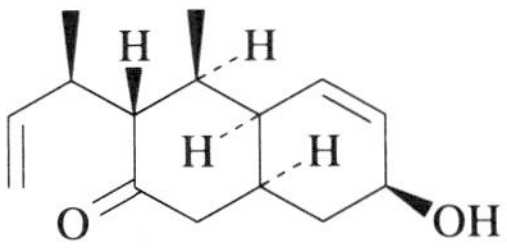 [structure] (—) | 63 |
| C$_{17}$ 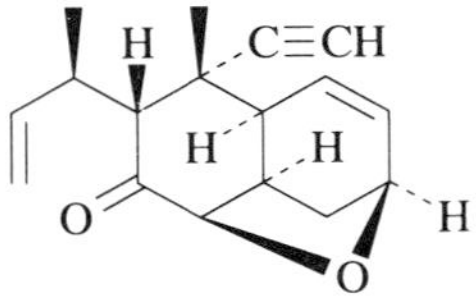 [structure] | 2.4 SmI$_2$, THF, -78° | 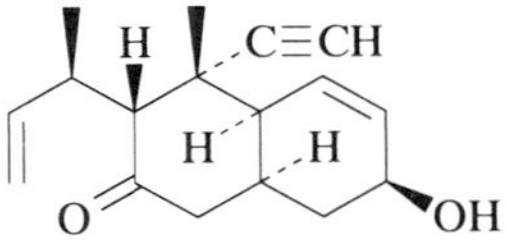 [structure] (—) | 64 |
| | *D. α-Carboalkoxy and α-Tosyloxy Carbonyl Substrates* | | |
| C$_3$ MeCH(OAc)CO$_2$Et | 2 SmI$_2$, THF, MeOH, -78° | No reaction | 53 |
| C$_4$ 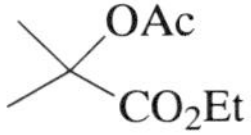 [structure with OAc, CO$_2$Et] | 2.5-3 SmI$_2$, THF, HMPA, EtOH, 20-22°, <5 min | 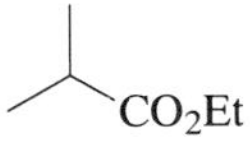 [CO$_2$Et structure] (>95) | 54 |
| C$_5$ 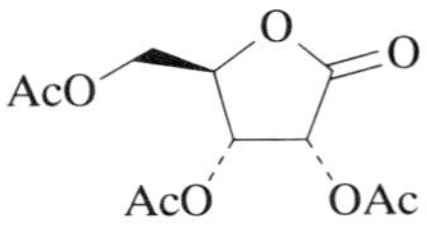 [structure with AcO, OAc] | | 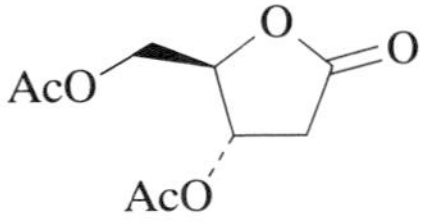 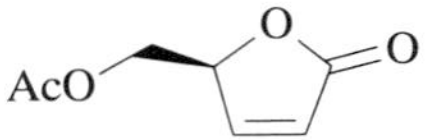 I + II | |

315

Table VI. REDUCTIVE CLEAVAGE OF α-HETEROSUBSTITUTED CARBONYL AND RELATED SUBSTRATES *(Continued)*

| Substrate | Conditions | Product(s) and Yield(s) (%) | Refs. |
|---|---|---|---|
| | 3-4 SmI₂, THF, pivalic acid, rt, 20 min | **I** + **II** (47), **I:II** = 36:64 | 65 |
| | 3-4 SmI₂, THF, AcOH, rt, 20 min | **II** (30) | 65 |
| | 3 SmI₂, THF, rt, 30 min | 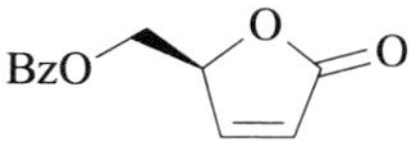 (60) | 55 |
| 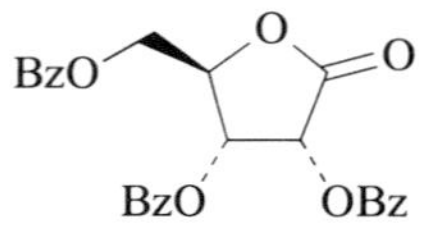 | " | " (55) | 55 |
| | 3 SmI₂, THF, HO(CH₂)₂OH, rt, <1 min | 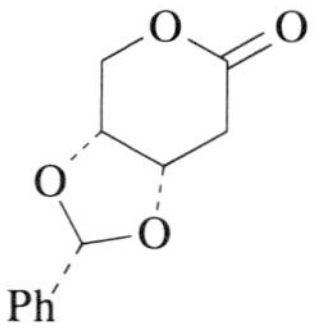 (92) | 55 |
| C₆ | " | (81) | 55 |

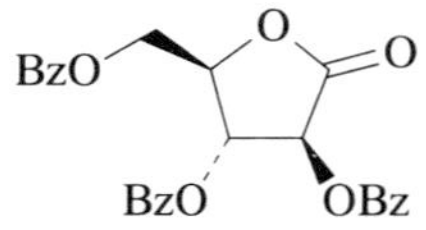

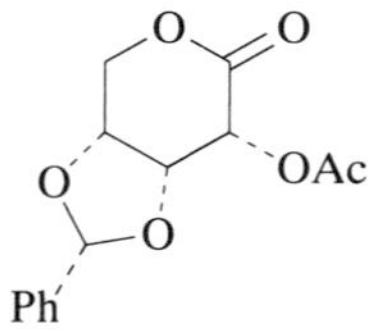

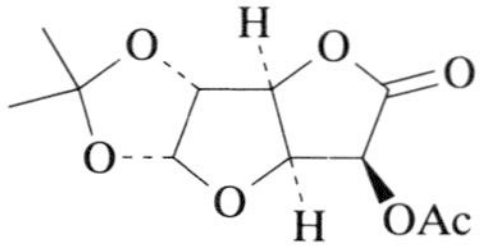

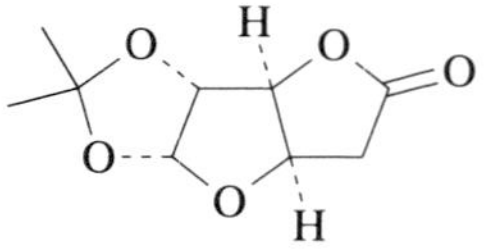

316

Table VI. REDUCTIVE CLEAVAGE OF α-HETEROSUBSTITUTED CARBONYL AND RELATED SUBSTRATES (*Continued*)

| Substrate | Conditions | Product(s) and Yield(s) (%) | Refs. |
|---|---|---|---|
| | 3-4 SmI₂, THF, pivalic acid, rt, 20 min | **I** (67) | 65 |
| | 3-4 SmI₂, THF, AcOH, rt, 20 min | **I** (60) | 65 |
| | 3-4 SmI₂, THF, pivalic acid, rt, 20 min | **I** (58) | 65 |
| | 3-4 SmI₂, THF, AcOH, rt, 20 min | **I** (60) | 65 |
| | 3 SmI₂, THF, HO(CH₂)₂OH, rt | 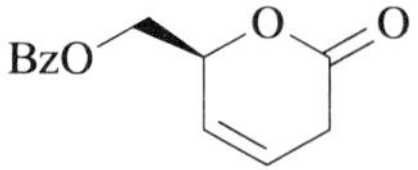 (43) 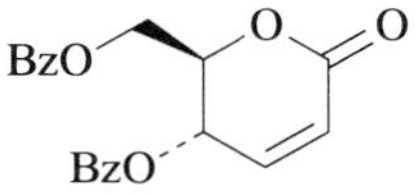 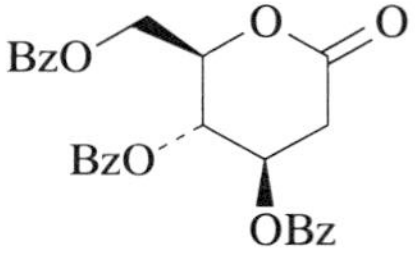 + (—) + (—) | 55 |

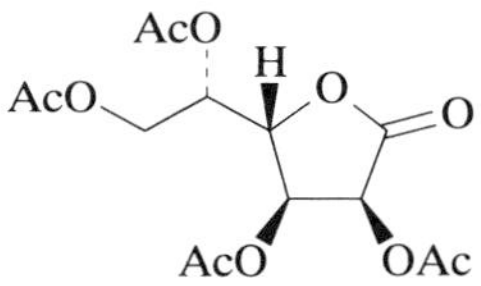

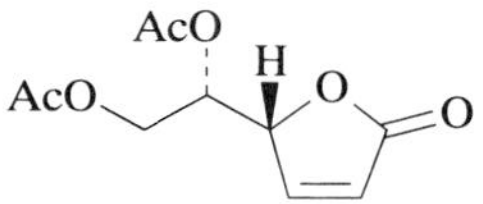

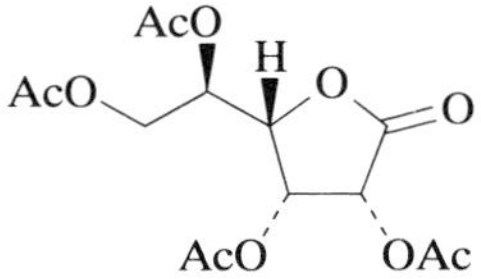

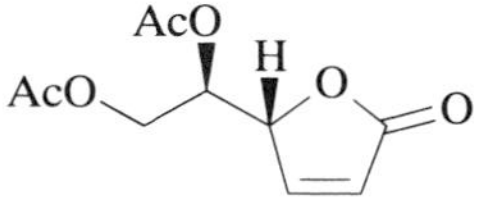

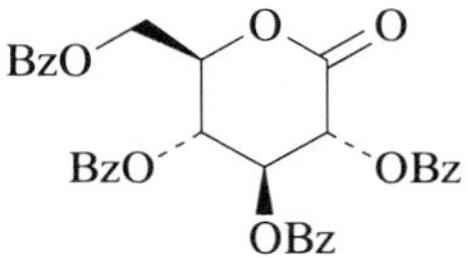

317

Table VI. REDUCTIVE CLEAVAGE OF α-HETEROSUBSTITUTED CARBONYL AND RELATED SUBSTRATES (*Continued*)

| Substrate | Conditions | Product(s) and Yield(s) (%) | Refs. |
|---|---|---|---|
| 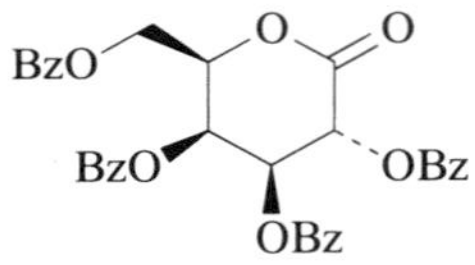 | 3 SmI$_2$, THF, rt, 30 min | 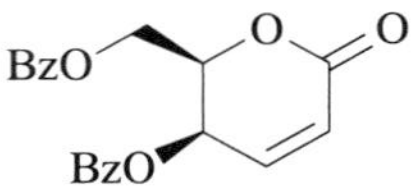 (48) | 55 |
| | 3-4 SmI$_2$, THF, pivalic acid, rt, 20 min | **I** + **II** (47), **I:II** = 1:1 | 65 |
| | 3-4 SmI$_2$, THF, AcOH, rt, 20 min | **II** (68) | 65 |
| C$_7$ | 3 SmI$_2$, THF, HMPA HO(CH$_2$)$_2$OH, rt, 1 h | (85) | 55 |

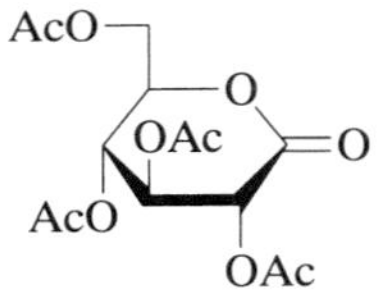

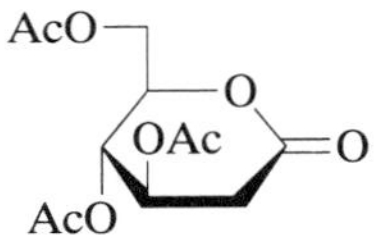

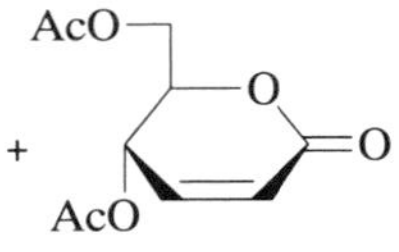

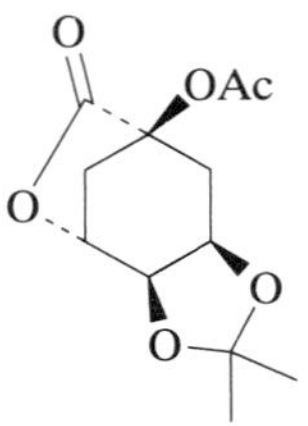

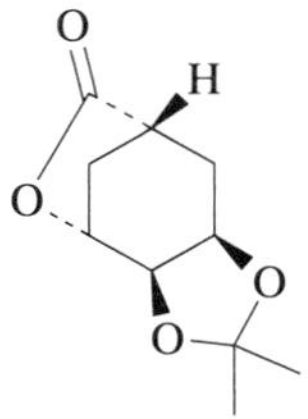

318

Table VI. REDUCTIVE CLEAVAGE OF α-HETEROSUBSTITUTED CARBONYL AND RELATED SUBSTRATES (*Continued*)

| Substrate | Conditions | Product(s) and Yield(s) (%) | Refs. |
|---|---|---|---|
| 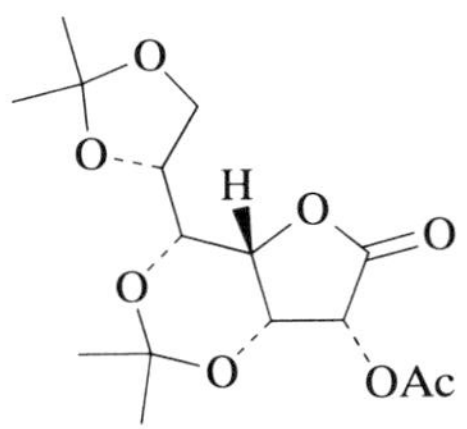 | 3 SmI₂, THF, HO(CH₂)₂OH, rt, <1 min | 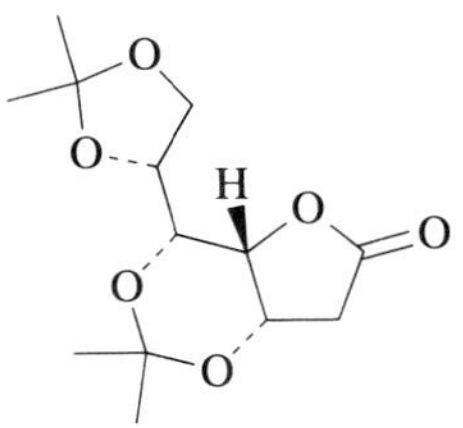 (94) | 55 |
| 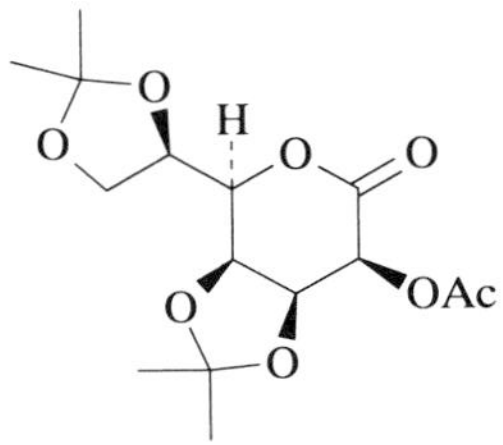 | 3 SmI₂, THF, HO(CH₂)₂OH, rt, <1 min | 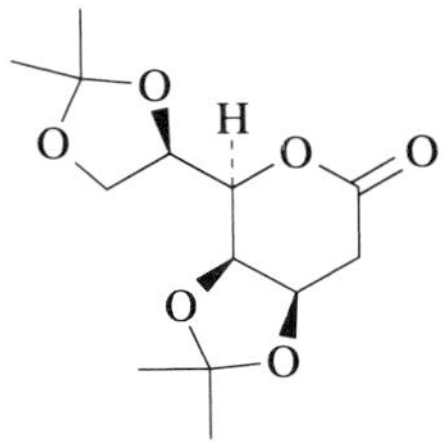 (99) | 55 |
| 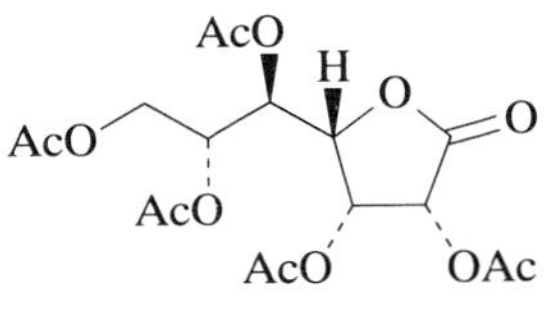 | | 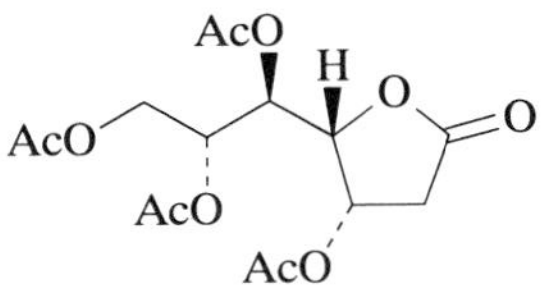 I + 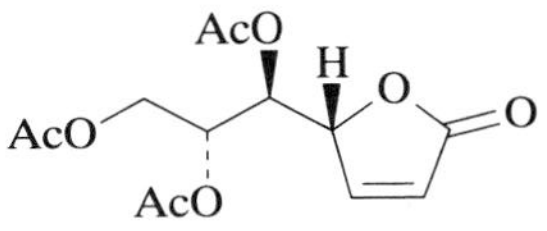 II | |

319

Table VI. REDUCTIVE CLEAVAGE OF α-HETEROSUBSTITUTED CARBONYL AND RELATED SUBSTRATES (*Continued*)

| Substrate | Conditions | Product(s) and Yield(s) (%) | Refs. |
|---|---|---|---|
| $n\text{-}C_5H_{11}CH(OAc)CO_2Me$ | 3-4 SmI$_2$, THF, pivalic acid, rt, 20 min | **I** + **II** (54), **I:II** = 42:58 | 65 |
|  | 3-4 SmI$_2$, THF, AcOH, rt, 20 min | **II** (67) | 65 |
| C$_8$ |  |  |  |
| $n\text{-}C_5H_{11}CH(OAc)CO_2Me$ | 2.5-3 SmI$_2$, THF, HMPA, MeOH, 20-22°, <5 min | $n\text{-}C_5H_{11}CH_2CO_2Me$ (>95) | 54 |
| C$_{11}$ |  |  |  |
| $PhCH(OAc)CO_2Me$ | 2.5-3 SmI$_2$, THF, HMPA, MeOH, 20-22°, 1 min | $PhCH_2CO_2Me$ (96) | 54 |
| $PhCOCH(OAc)(CH_2)_3I$ | 2 SmI$_2$, THF, MeOH, -78° | $PhCO(CH_2)_4I$ (87) | 53 |
| (tricyclic ketone with BzO, OBz substituents) | 2 SmI$_2$, THF, MeOH, -78 to 25241 | 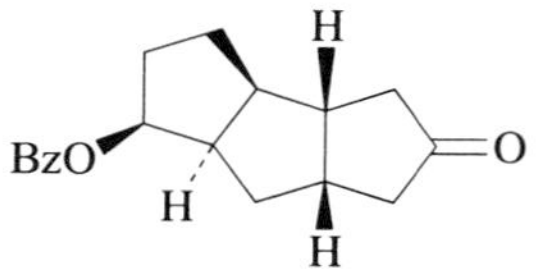 (tricyclic ketone, H, BzO) (65) | 61 |
| C$_{12}$ 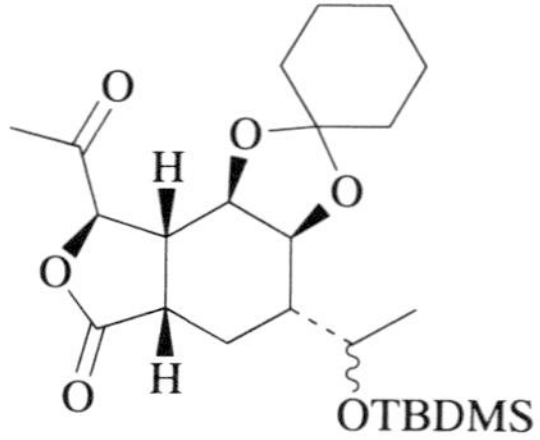 (cyclohexane spiro structure, OTBDMS) | 2 SmI$_2$, THF, FeCl$_3$ (cat.), rt, 15 min | 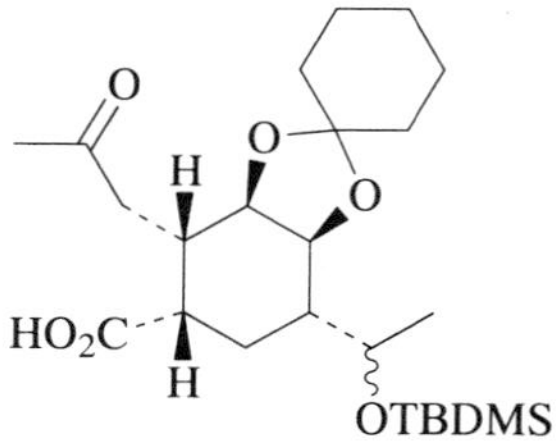 (cyclohexane spiro structure, HO$_2$C, OTBDMS) (84) | 66 |
| $n\text{-}C_5H_{11}COCH(OAc)C_5H_{11}\text{-}n$ | 2 SmI$_2$, THF, MeOH, -78° | $n\text{-}C_5H_{11}COCH_2C_5H_{11}\text{-}n$ (75) | 53 |
| $n\text{-}C_5H_{11}COCH(O_2CBn)C_5H_{11}\text{-}n$ | " | " (100) | 53 |
| $n\text{-}C_5H_{11}COCH(OTs)C_5H_{11}\text{-}n$ | " | " (94) | 53 |

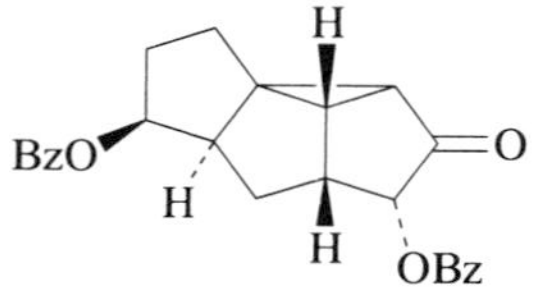

Table VI. REDUCTIVE CLEAVAGE OF α-HETEROSUBSTITUTED CARBONYL AND RELATED SUBSTRATES (*Continued*)

| Substrate | Conditions | Product(s) and Yield(s) (%) | Refs. |
|---|---|---|---|

*E. α,β-Epoxy Carbonyl Substrates*

$C_3$

|  | 2.25 SmI$_2$, THF, HMPA, DMAE, rt, 1 min | 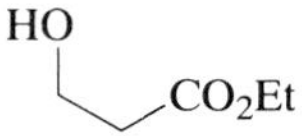 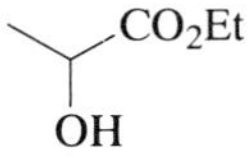 HO–CH$_2$CH$_2$CO$_2$Et  **I**  +  CH$_3$CH(OH)CO$_2$Et  **II**  (50) **I** : **II** = 17 | 67 |

$C_4$

| | | HO–CH(CH$_3$)CH$_2$CO$_2$Et  **I**  +  CH$_3$CH$_2$CH(OH)CO$_2$Et  **II** | |
| | 2.5 SmI$_2$, THF, rt, 1 h | **I + II** (5) | 67 |
| | 2.5 SmI$_2$, THF, HMPA, rt, 1 min | **I + II** (19) | 67 |
| | 2.5 SmI$_2$, THF, HMPA, *i*-PrOH, rt, 1 min | **I + II** (34), **I:II** = 10 | 67 |
| | 2.5 SmI$_2$, THF, DMAE, rt, 30 min | **I + II** (60), **I:II** = 20 | 67 |
| | 2.5 SmI$_2$, THF, HMPA, TMEDA, *i*-PrOH, rt, 1 min | **I + II** (46), **I:II** = 200 | 67 |
| | 2.5 SmI$_2$, THF, HMPA, DMAE, rt, 1 min | **I + II** (68), **I:II** = >200 | 67 |

| 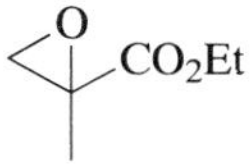 | 2.5 SmI$_2$, THF, HMPA, DMAE, rt, 1 min | 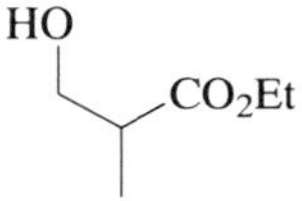 HO–CH$_2$CH(CH$_3$)CO$_2$Et  (62) | 67 |

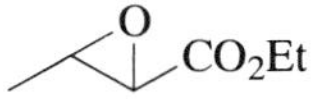

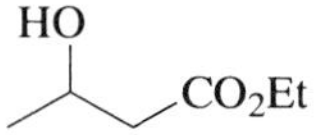

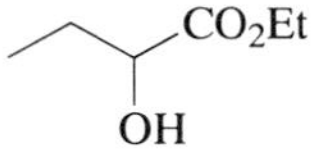

321

Table VI. REDUCTIVE CLEAVAGE OF α-HETEROSUBSTITUTED CARBONYL AND RELATED SUBSTRATES (*Continued*)

| Substrate | Conditions | Product(s) and Yield(s) (%) | Refs. |
|---|---|---|---|
| C$_5$ <br> (structure: O epoxide with CO$_2$Me and NBn$_2$) | 2 SmI$_2$, THF, <br> -78 to 0°, 2 h | (structure: OH, CO$_2$Me, NBn$_2$) (70) | 68 |
| C$_6$ <br> (structure: cyclohexanone epoxide) | 2 SmI$_2$, THF, MeOH, <br> -90° | (structure: cyclohexanone with OH) (74) | 69 |
| 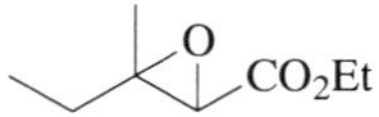 (structure: epoxide with CO$_2$Et) | 2.25 SmI$_2$, THF, HMPA, <br> DMAE, rt, 1 min | 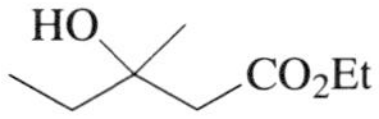 (structure: HO, CO$_2$Et) (62) | 67 |
| 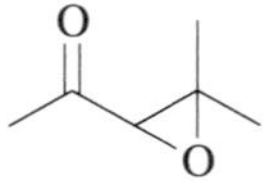 C$_7$ <br> (structure: epoxide ketone) | 2 SmI$_2$, THF, MeOH, <br> -90° | 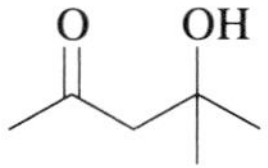 (structure: OH, O) (79) | 69 |
| 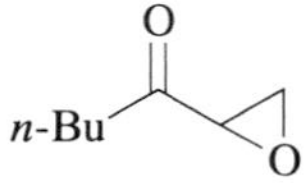 (structure: *n*-Bu epoxide ketone) | 2 SmI$_2$, THF, MeOH, <br> -90° | 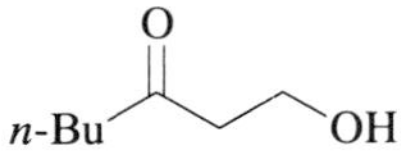 (structure: O, *n*-Bu, OH) (97) | 69 |
| C$_8$ <br> (structure: acetyl cyclohexane epoxide) | 2 SmI$_2$, THF, MeOH, <br> -90° | (structure: O, OH cyclohexane) (79) | 69 |

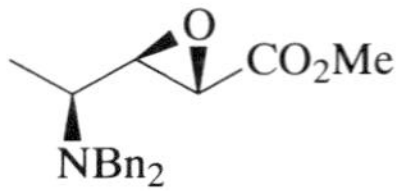

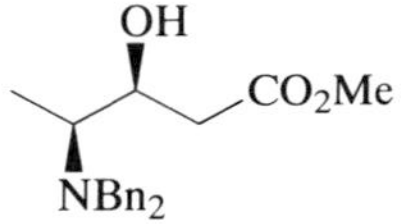

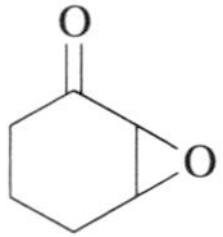

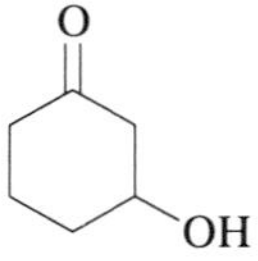

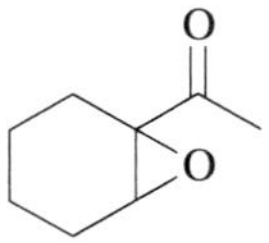

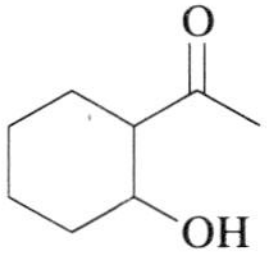

322

Table VI. REDUCTIVE CLEAVAGE OF α-HETEROSUBSTITUTED CARBONYL AND RELATED SUBSTRATES (*Continued*)

| Substrate | Conditions | Product(s) and Yield(s) (%) | Refs. |
|---|---|---|---|
| C₉ | | | |
| (structure with O, CO₂Et cyclohexane spiro epoxide) | 2.25 SmI₂, THF, HMPA, DMAE, rt, 1 min | (structure: OH, CO₂Et) (76) | 67 |
| 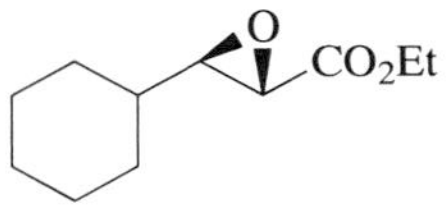 (structure CO₂Et epoxide cyclohexane, >98% ee) | 2.25 SmI₂, THF, HMPA, DMAE, rt, 1 min | 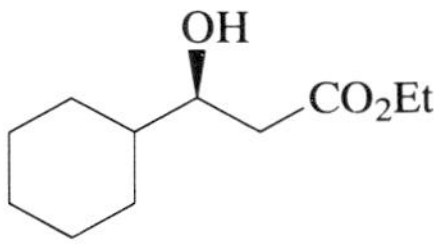 (structure: OH, CO₂Et) (—), >98% ee | 67 |
| (bicyclic epoxy ketone) | 2 SmI₂, THF, MeOH, -90° | (dimethyl cyclohexanone OH) (82) | 69 |
| (epoxy ketone with C₅H₁₁-n) | 2 SmI₂, THF, MeOH, -90° | (OH, C₅H₁₁-n ketone) (81) | 69 |
| 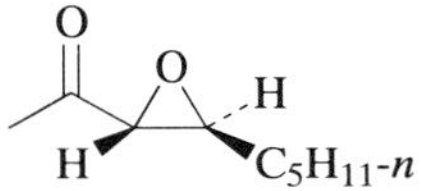 (epoxy ketone with H, C₅H₁₁-n, H, 95% ee) | 2 SmI₂, THF, MeOH, -90° | 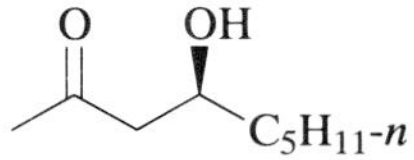 (OH, C₅H₁₁-n ketone) (94), 90% ee | 69 |
| C₁₀ (carvone epoxide) | 2 SmI₂, THF, MeOH, -90° | (epoxide ketone isopropenyl) (34) | 69 |

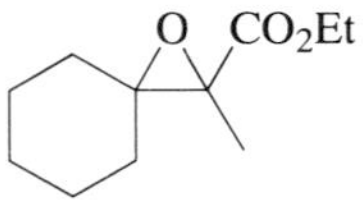

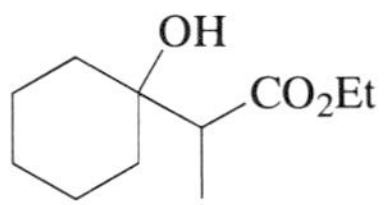

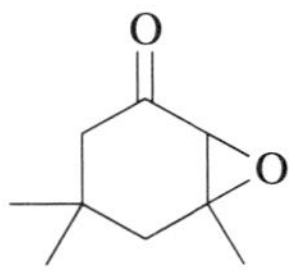

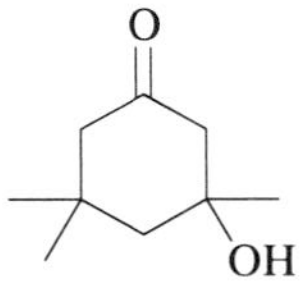

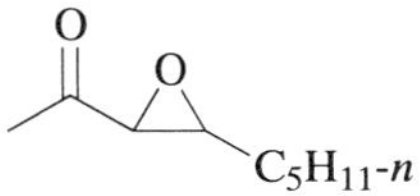

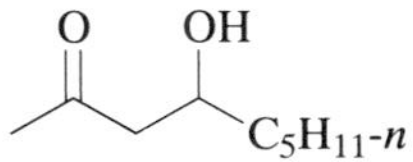

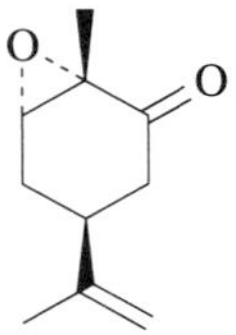

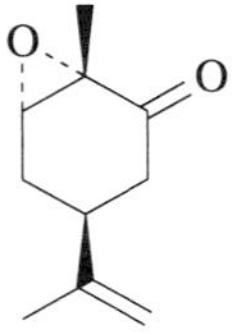

323

Table VI. REDUCTIVE CLEAVAGE OF α-HETEROSUBSTITUTED CARBONYL AND RELATED SUBSTRATES (*Continued*)

| Substrate | Conditions | Product(s) and Yield(s) (%) | Refs. |
|---|---|---|---|
| | 2.25 SmI$_2$, THF, HMPA, DMAE, rt, 1 min | (68) | 67 |
| 1.7 : 1 mixture of diastereomers | | 3.8 : 1 mixture of diastereomers | |
| 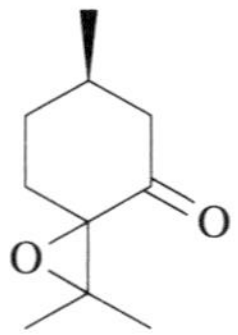 | 2 SmI$_2$, THF, MeOH, -90° | 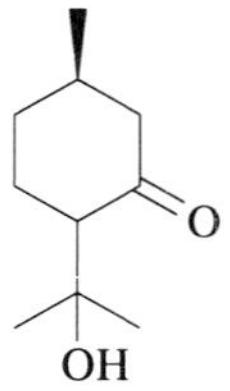 (79) | 69 |
| C$_{11}$ | 2 SmI$_2$, THF, -78 to 0°, 2 h | (70) | 68 |
| 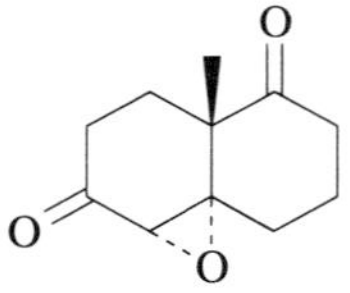 | 2 SmI$_2$, THF, MeOH, -90° | 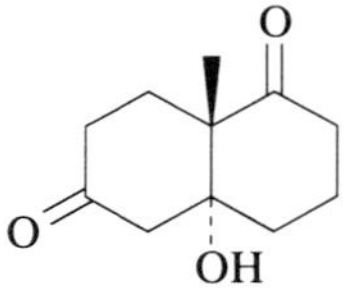 (76) | 69 |

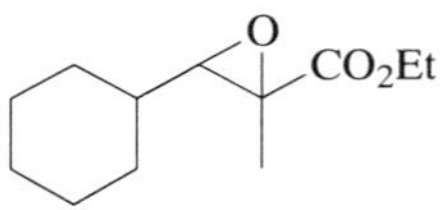

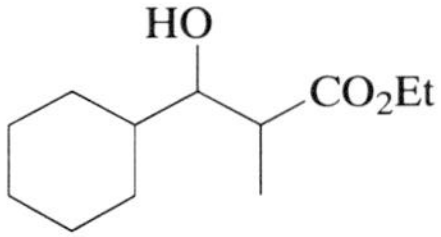

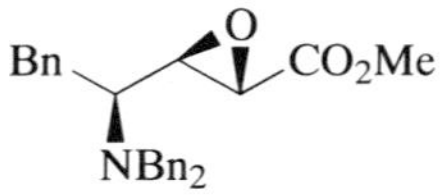

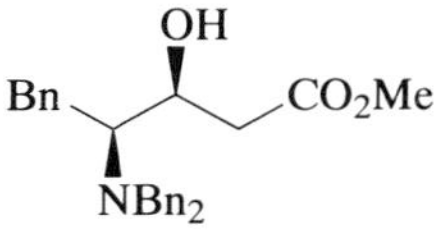

324

Table VI. REDUCTIVE CLEAVAGE OF α-HETEROSUBSTITUTED CARBONYL AND RELATED SUBSTRATES (*Continued*)

| Substrate | Conditions | Product(s) and Yield(s) (%) | Refs. |
|---|---|---|---|
| *F. Vinylogous Epoxy Carbonyl and Related Substrates* | | | |
| C₆ | | | |
| [epoxide CO₂Et structure] | 2.1 SmI₂, THF, HMPA, DMAE, rt, 1 min | [HO-CH CO₂Et structure] (73) | 67 |
| [epoxide CH₃ CO₂Et structure] | 2 SmI₂, THF, EtOH, -98° | [HO CO₂Et structure] (77) + [HO CO₂Et structure] (4) | 70 |
| C₇ | | | |
| [epoxide CO₂Et structure] | 2.1 SmI₂, THF, HMPA, DMAE, rt, 1 min | [HO CO₂Et structure] (74) | 67 |
| [epoxide CO₂Et structure] | 2.1 SmI₂, THF, HMPA, DMAE, rt, 1 min | [HO CO₂Et structure] (80) | 67 |
| [epoxide CO₂Et structure] | 2.1 SmI₂, THF, HMPA, DMAE, rt, 1 min | [HO CO₂Et structure] (90) | 67 |

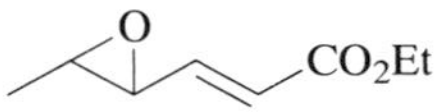

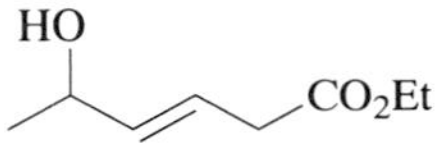

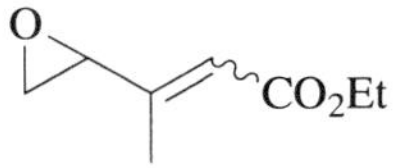

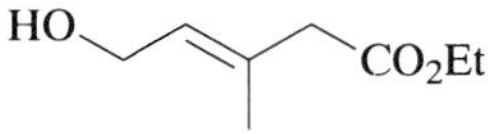

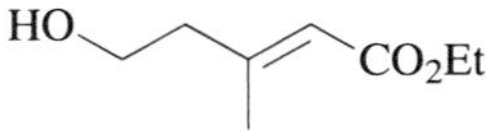

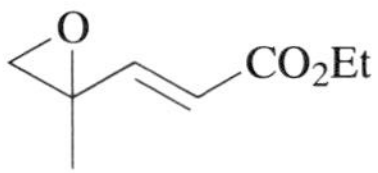

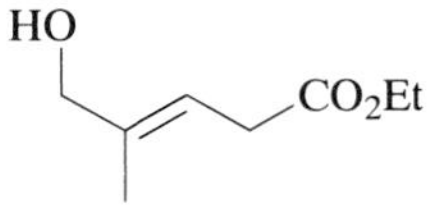

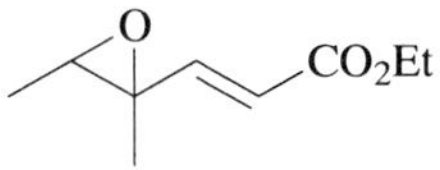

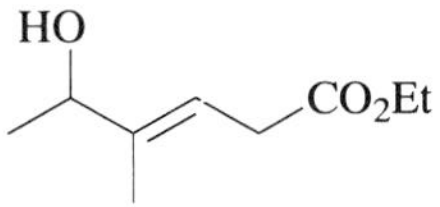

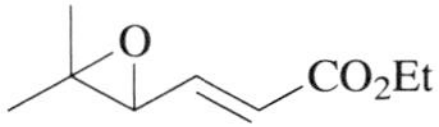

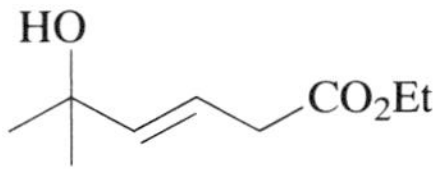

325

| Substrate | Conditions | Product(s) and Yield(s) (%) | Refs. |
|---|---|---|---|

C₈

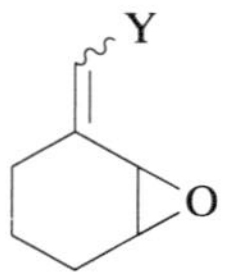

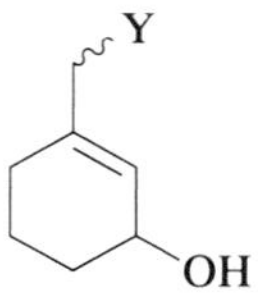

Y = CO₂Et  $\quad$ 2 SmI₂, THF, MeOH or EtOH, -90°  $\quad$ **I** (68)  $\quad$ 70

Y = CN  $\quad$ "  $\quad$ **I** (71)  $\quad$ 70

Y = COMe  $\quad$ "  $\quad$ **I** (69)  $\quad$ 70

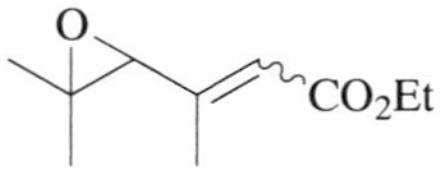

2 SmI₂, THF, EtOH, -98°  $\quad$ (75)  $\quad$ 70

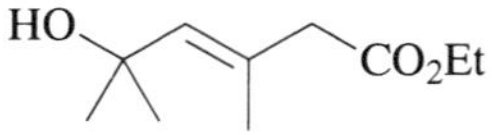

C₉  $\quad$ 2 SmI₂, THF, EtOH, -98°  $\quad$ (78) + (9)  $\quad$ 70

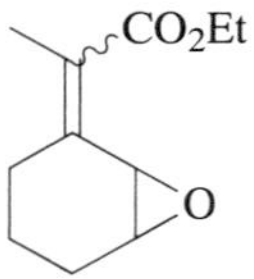

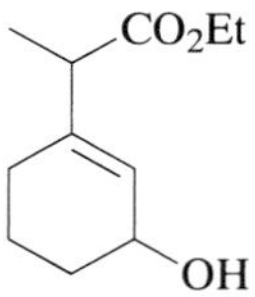

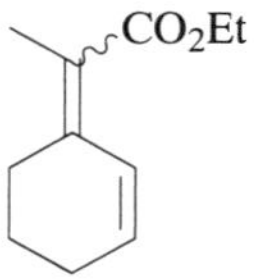

C₁₀  $\quad$ 2 SmI₂, THF, EtOH, -90°  $\quad$ (65)  $\quad$ 70

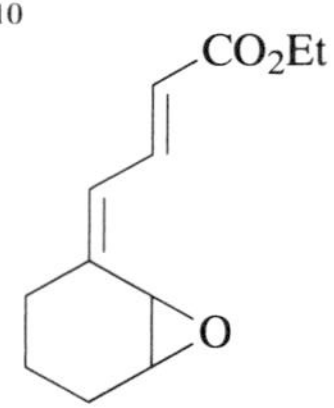

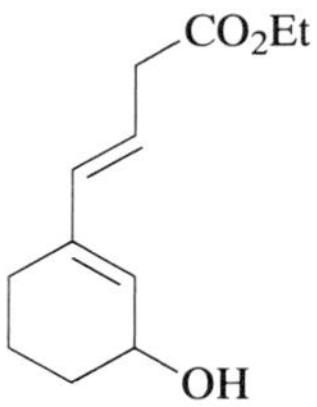

Table VI. REDUCTIVE CLEAVAGE OF α-HETEROSUBSTITUTED CARBONYL AND RELATED SUBSTRATES (*Continued*)

| Substrate | Conditions | Product(s) and Yield(s) (%) | Refs. |
|---|---|---|---|
| 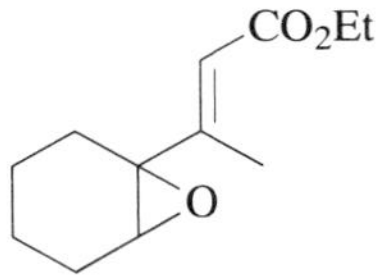 (CO$_2$Et epoxide cyclohexane) | 2 SmI$_2$, THF, EtOH, –98° | 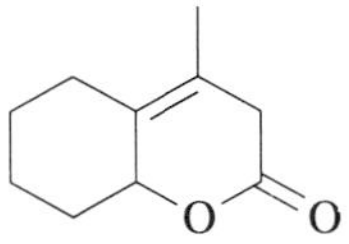 (43) 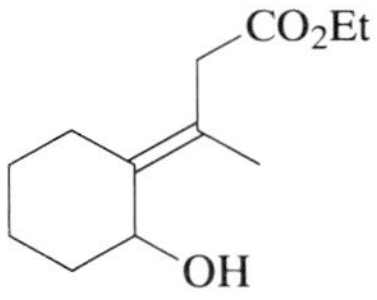 (lactone) + (44) (cyclohexane with OH, =CH–CH$_3$, CH$_2$CO$_2$Et) | 70 |
| 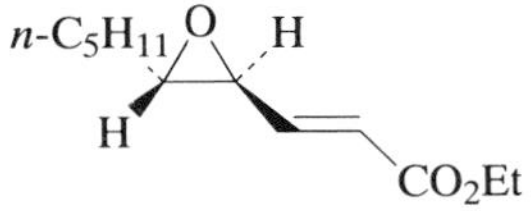 $n$-C$_5$H$_{11}$, O, H, H, CO$_2$Et — 95% ee | 2 SmI$_2$, THF, EtOH, –98° | 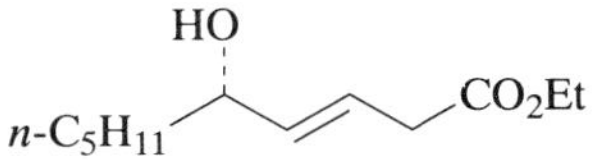 HO···, $n$-C$_5$H$_{11}$, CO$_2$Et (84), 95% ee | 70 |
| C$_{11}$ (CO$_2$Et epoxide) | 2 SmI$_2$, THF, EtOH, –98° | CO$_2$Et, OH (81) | 70 |
| O, H epoxide with vinyl-Y;  Y = COSEt | 2 SmI$_2$, THF, MeOH or EtOH, –90° | **I** (80) | 70 |
| Y = S(O$_2$)Ph | " | **I** (82) + (OH, S(O$_2$)Ph structure) (5) | 70 |

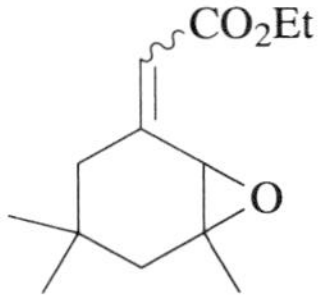

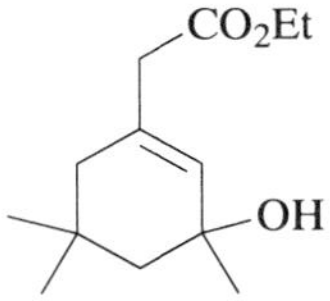

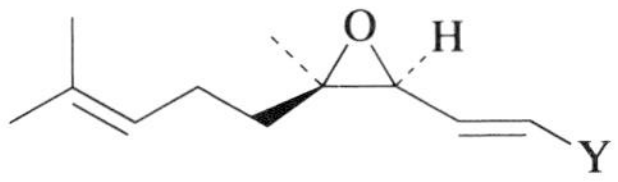

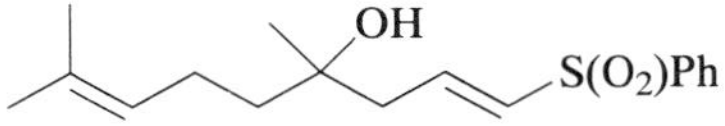

Table VI. REDUCTIVE CLEAVAGE OF α-HETEROSUBSTITUTED CARBONYL AND RELATED SUBSTRATES (Continued)

| Substrate | Conditions | Product(s) and Yield(s) (%) | Refs. |
|---|---|---|---|
| Y = PO(OEt)$_2$ | 2 SmI$_2$, THF, MeOH or EtOH, -90° | **I** (84) | 70 |
| Y = H | 2 SmI$_2$, THF, MeOH or EtOH, rt | **I** (69) 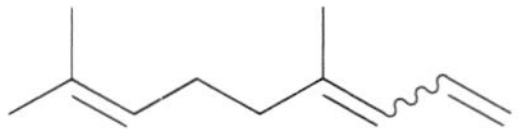 + [structure] (9) | 70 |
| Y = Me | 2 SmI$_2$, THF, MeOH or EtOH, 0° | **I** (42) 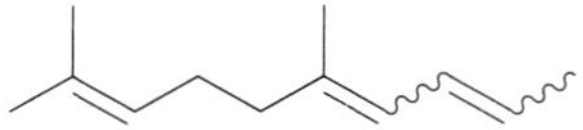 + [structure] (32) | 70 |
| Y = SPh | " | **I** 54 | 70 |
| C$_{12}$ 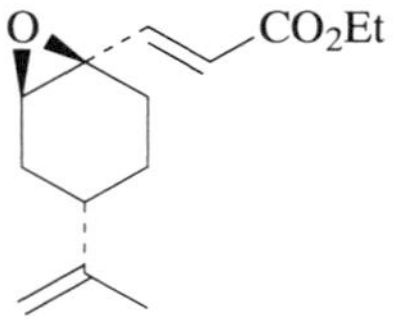 [epoxide substrate with CO$_2$Et and isopropenyl cyclohexane] | 2 SmI$_2$, THF, EtOH, -98° | 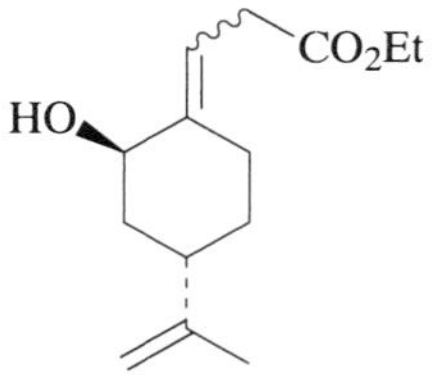 [HO-substituted cyclohexane with CO$_2$Et] (79) | 70 |
| 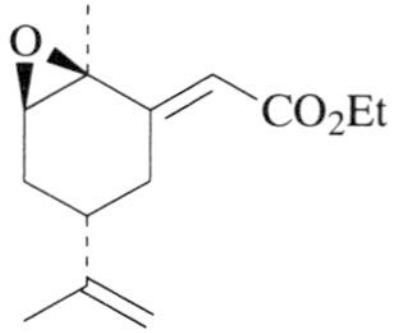 [epoxide substrate with CO$_2$Et and isopropenyl cyclohexane] | 2 SmI$_2$, THF, EtOH, -98° | 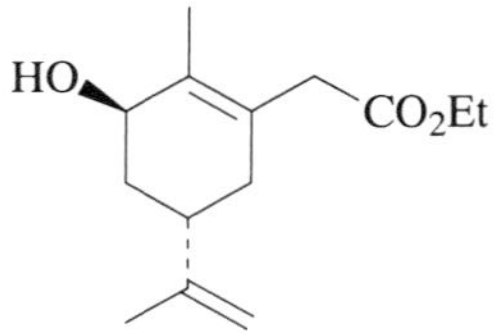 [HO-substituted cyclohexene with CO$_2$Et] (81) | 70 |

328

Table VI. REDUCTIVE CLEAVAGE OF α-HETEROSUBSTITUTED CARBONYL AND RELATED SUBSTRATES (*Continued*)

| Substrate | Conditions | Product(s) and Yield(s) (%) | Refs. |
|---|---|---|---|
| 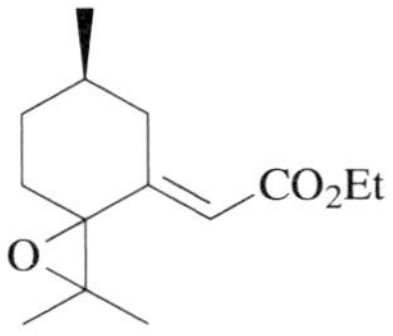 $C_6$ <br> (structure with $CO_2Et$) | 2 SmI$_2$, THF, EtOH, -98° | 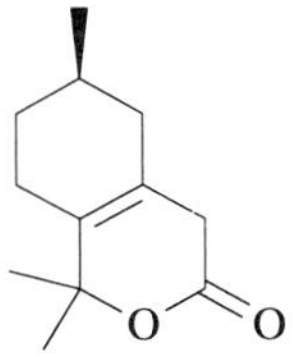 (85) | 70 |
| *G. α-Keto Sulfides, α-Keto Sulfoxides, and α-Keto Sulfones* | | | |
| 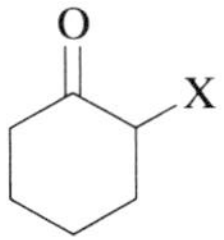 (cyclohexanone with X) | | 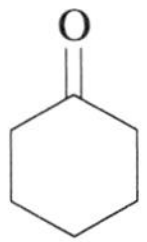 **I** | |
| X = SPh | 2 SmI$_2$, THF, MeOH, -78° | **I** (76) | 53 |
| X = S(O)Ph | " | **I** (64) | 53 |
| X = S(O$_2$)Ph | " | **I** (88) | 53 |
| *H. α-Cyano Phosphates* | | | |
| $C_6$ 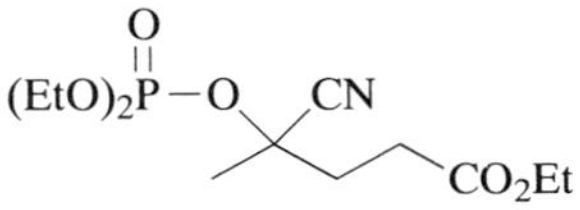 <br> (EtO)$_2$P(O)—O—C(CN)(CH$_3$)—CO$_2$Et structure | 3 SmI$_2$, THF, *t*-BuOH, rt, 2 h | 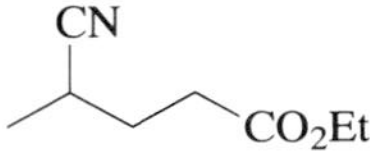 (80) <br> CN—CH(CH$_3$)—CO$_2$Et structure | 71 |

Table VI. REDUCTIVE CLEAVAGE OF α-HETEROSUBSTITUTED CARBONYL AND RELATED SUBSTRATES (*Continued*)

| Substrate | Conditions | Product(s) and Yield(s) (%) | Refs. |
|---|---|---|---|
| (EtO)₂P(O)—O— piperidine NR / NC | | | |
| R = Bn | 3 SmI₂, THF, *t*-BuOH, rt, 2 h | **I** (79) | 71 |
| R = CO₂CH=CH₂ | 3 SmI₂, THF, *t*-BuOH, rt, 1 h | **I** (92) | 71 |
| R = Ts | 3 SmI₂, THF, *t*-BuOH, rt, 15 min | **I** (92) | 72 |
| C₇ (EtO)₂P(O)—O— CN cyclohexadienone | 3 SmI₂, THF, *t*-BuOH, rt, 12 min | NC—⬡—OH (82) | 72 |
| C₈ (EtO)₂P(O)—O—CH(Ph)CN | 3 SmI₂, THF, *t*-BuOH, rt, 2 h | Ph—CH₂—CN (85) | 71 |
| pyridine—C(CH₃)(CN)—O—P(O)(OEt)₂ | 3 SmI₂, THF, *t*-BuOH, rt, 10 min | pyridine—CH(CH₃)—CN (69) | 71 |

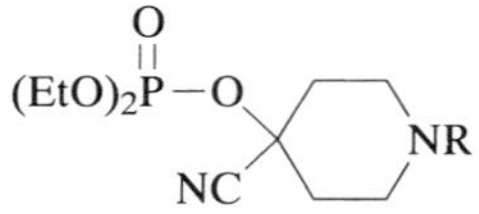

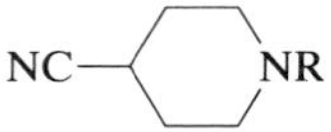

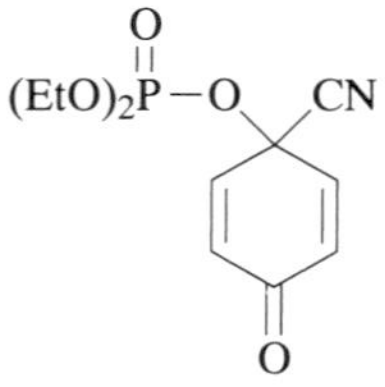

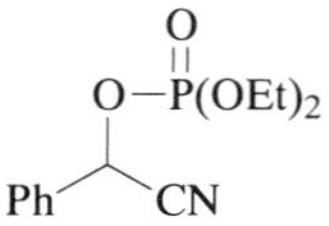

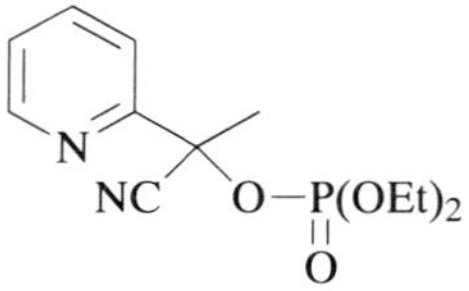

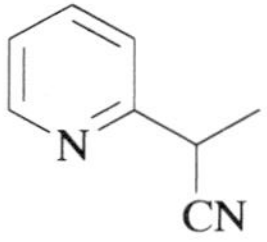

330

Table VI. REDUCTIVE CLEAVAGE OF α-HETEROSUBSTITUTED CARBONYL AND RELATED SUBSTRATES (*Continued*)

| Substrate | Conditions | Product(s) and Yield(s) (%) | Refs. |
|---|---|---|---|
| $n$-BuC≡CCHCN with O–P(OEt)$_2$ (O=P) | 3 SmI$_2$, THF, $t$-BuOH, rt, 10 min | $n$-BuC≡CCH$_2$CN (60) | 72 |
| C$_9$ $n$-BuC≡CCCN with Me and O–P(OEt)$_2$ (O=P) | 3 SmI$_2$, THF, $t$-BuOH, rt, 30 min | $n$-BuC≡CCH(Me)CN (98) | 72 |
| 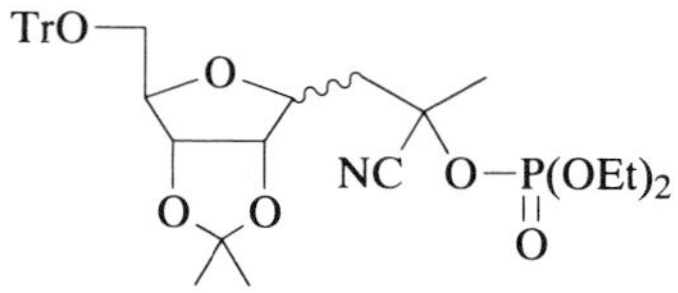 (TrO furanose isopropylidene, NC—C(Me)$_2$—O–P(OEt)$_2$) | 3 SmI$_2$, THF, $t$-BuOH, rt, 3 h | 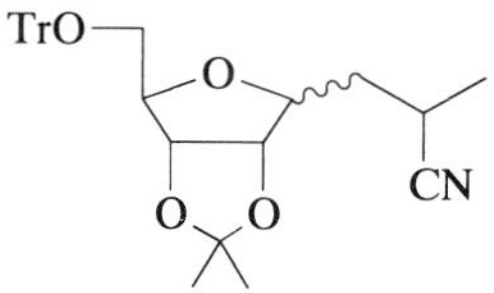 (TrO furanose isopropylidene, CH(Me)CN) (61) | 72 |
| 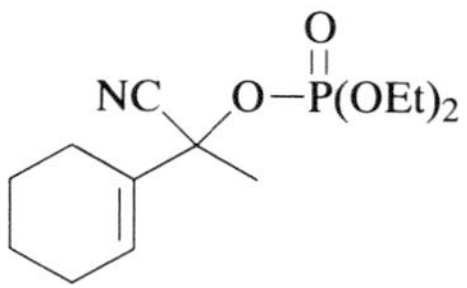 C$_{10}$ (cyclohexenyl)—C(Me)(CN)—O–P(OEt)$_2$ | 3 SmI$_2$, THF, $t$-BuOH, rt, 15 min | 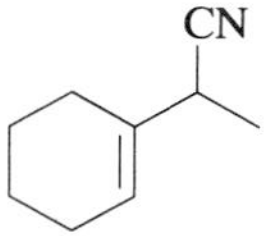 (cyclohexenyl)—CH(Me)CN (86) | 71 |
| PhC≡CCCN with Me and O–P(OEt)$_2$ (O=P) | 3 SmI$_2$, THF, $t$-BuOH, rt, 30 min | PhC≡CCH(Me)CN (85) | 72 |

331

Table VI. REDUCTIVE CLEAVAGE OF α-HETEROSUBSTITUTED CARBONYL AND RELATED SUBSTRATES (*Continued*)

| Substrate | Conditions | Product(s) and Yield(s) (%) | Refs. |
|---|---|---|---|
| 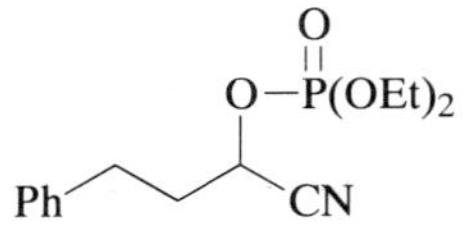 $C_{11}$ Ph–CH(O–P(=O)(OEt)$_2$)(CN)–CH$_2$CH$_2$Ph | 3 SmI$_2$, THF, $t$-BuOH, rt, 3 h |  Ph–(CH$_2$)$_3$–CN  (82) | 71 |
| 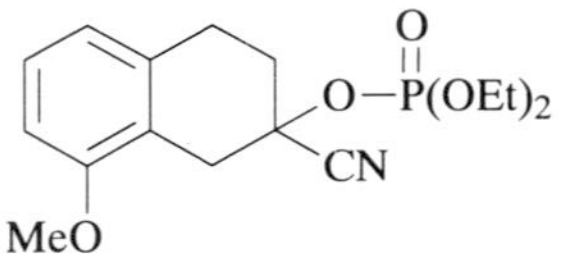 MeO-substituted tetralin, O–P(=O)(OEt)$_2$, CN | 3 SmI$_2$, THF, $t$-BuOH, rt, 4 h | 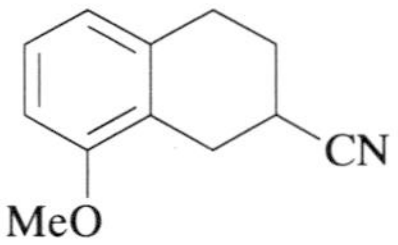 MeO-substituted tetralin, CN  (88) | 72 |
| 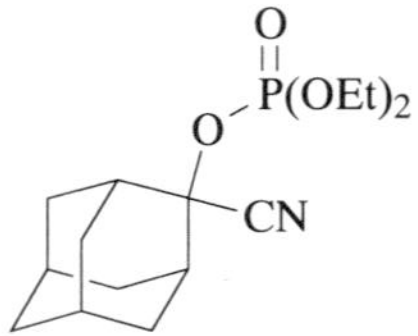 adamantyl O–P(=O)(OEt)$_2$, CN | 3 SmI$_2$, THF, $t$-BuOH, rt, 4 h | 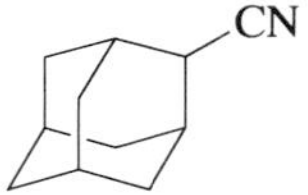 adamantyl–CN  (100) | 71 |
| 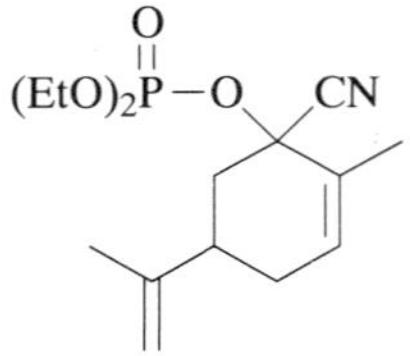 (EtO)$_2$P(=O)–O, CN, methylcyclohexene with isopropenyl | 3 SmI$_2$, THF, $t$-BuOH, rt, 30 min | 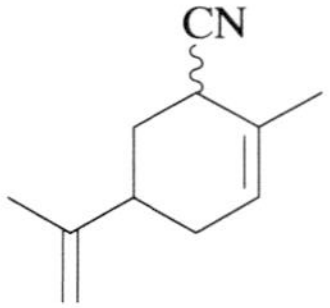 CN-substituted methylcyclohexene with isopropenyl  (86) | 71 |

332

Table VI. REDUCTIVE CLEAVAGE OF α-HETEROSUBSTITUTED CARBONYL AND RELATED SUBSTRATES (*Continued*)

| Substrate | Conditions | Product(s) and Yield(s) (%) | Refs. |
|---|---|---|---|
| 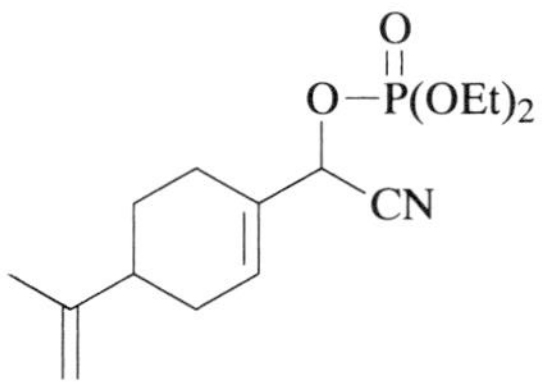 | 3 SmI$_2$, THF, *t*-BuOH, rt, 30 min | 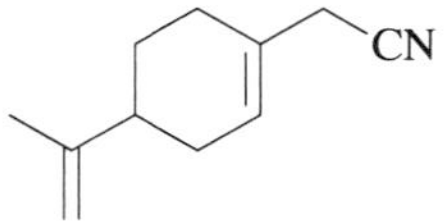 (90) | 72 |
| 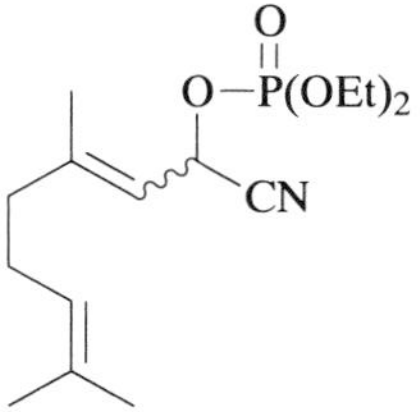 | 3 SmI$_2$, THF, *t*-BuOH, rt, 2 h | 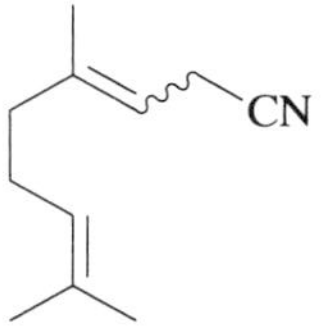 (89) | 71 |
| C$_{12}$ | 3 SmI$_2$, THF, *t*-BuOH, rt, 30 min | (92) | 72 |
| C$_{13}$ | 3 SmI$_2$, THF, *t*-BuOH, rt, 10 min | (96) | 71 |

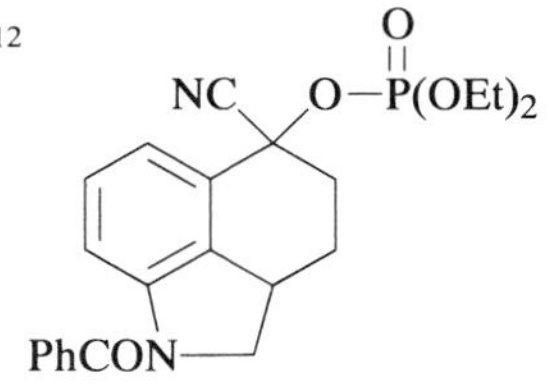

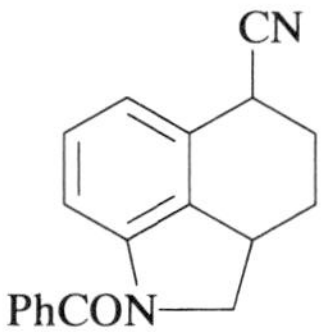

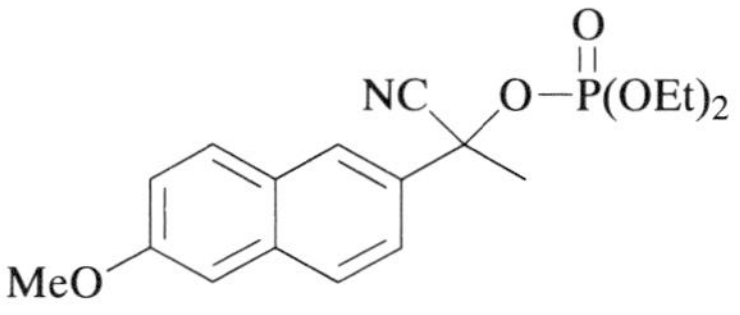

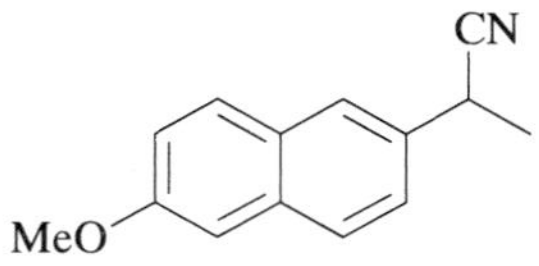

333

Table VI. REDUCTIVE CLEAVAGE OF α-HETEROSUBSTITUTED CARBONYL AND RELATED SUBSTRATES (*Continued*)

| Substrate | Conditions | Product(s) and Yield(s) (%) | Refs. |
|---|---|---|---|
| 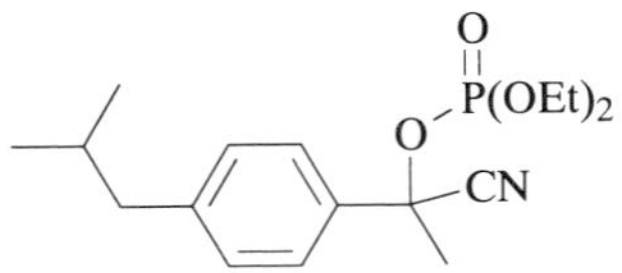 C$_{14}$ | 3 SmI$_2$, THF, *t*-BuOH, rt, 10 min | 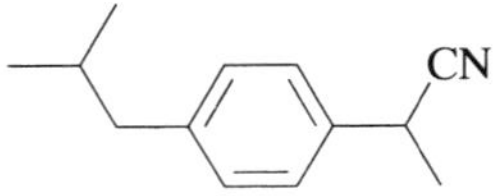 (94) | 71 |
| 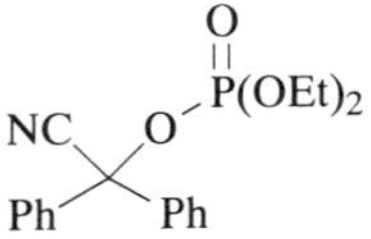 | 3 SmI$_2$, THF, *t*-BuOH, rt, 10 min | 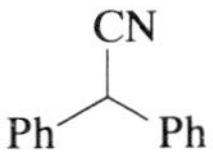 (90) | 71 |
| 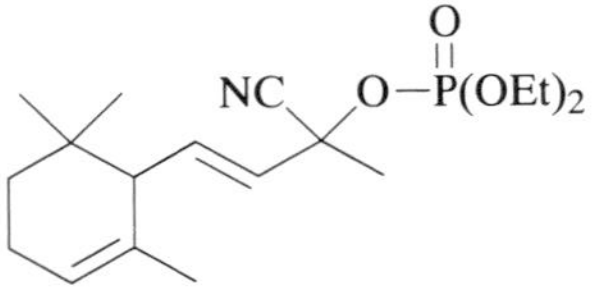 | 3 SmI$_2$, THF, *t*-BuOH, rt, 10 min | 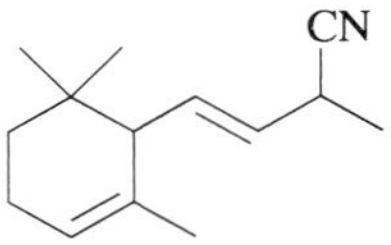 (97) | 71 |
| C$_{17}$ | 3 SmI$_2$, THF, *t*-BuOH, rt, 3 h | (97) | 71 |

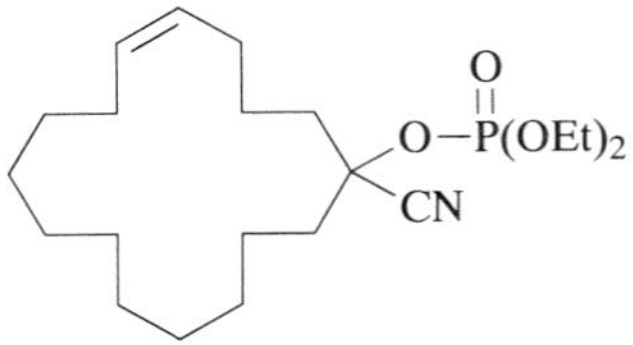

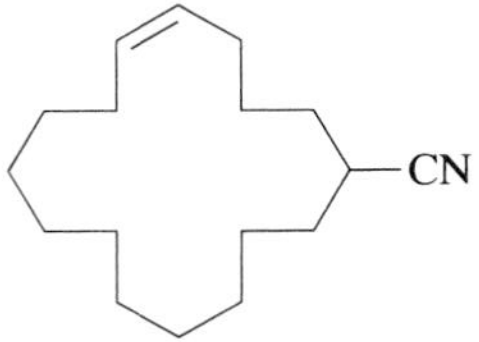

Table VI. REDUCTIVE CLEAVAGE OF α-HETEROSUBSTITUTED CARBONYL AND RELATED SUBSTRATES (*Continued*)

| Substrate | Conditions | Product(s) and Yield(s) (%) | Refs. |
|---|---|---|---|
| $C_{20}$ 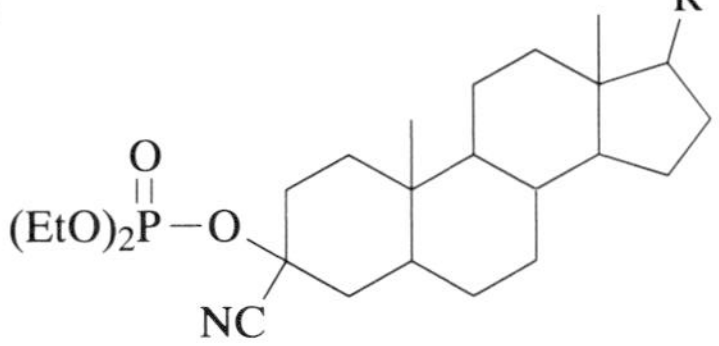 | | 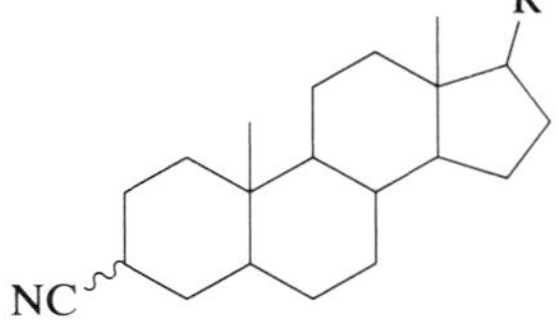 | |
| R = $C_8H_{17}$ | 3 $SmI_2$, THF, *t*-BuOH, rt, 3 h | **I** (83), 3α:3β = 3:2 | 71 |
| R = OMOM | 3 $SmI_2$, THF, *t*-BuOH, rt, 3 h | **I** (84), 3α:3β = 3:2 | 71 |
| R = OH | 3 $SmI_2$, THF, *t*-BuOH, rt, 4 h | **I** (92), 3α:3β = 4:1 | 71 |
| $C_{21}$ 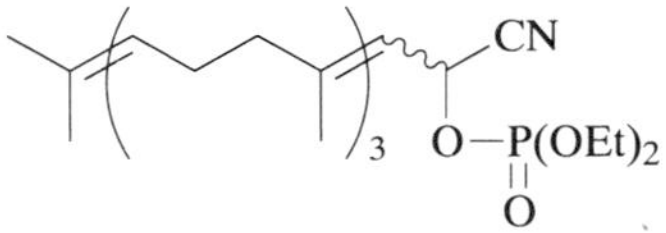 | 3 $SmI_2$, THF, *t*-BuOH, rt, 15 min | 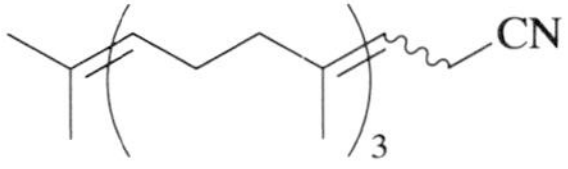 (93) | 72 |
| $C_{28}$ 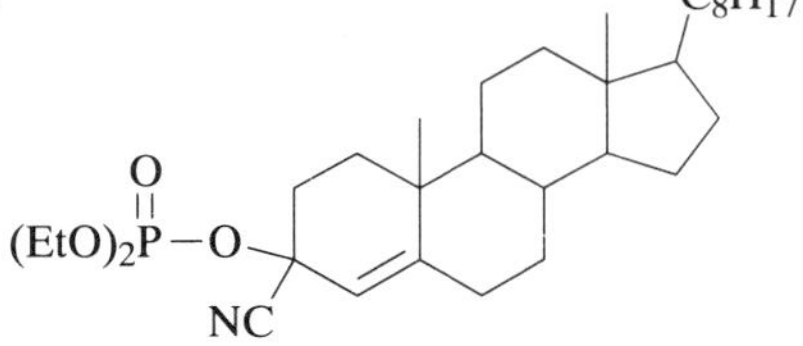 | 3 $SmI_2$, THF, *t*-BuOH, rt, 10 min | 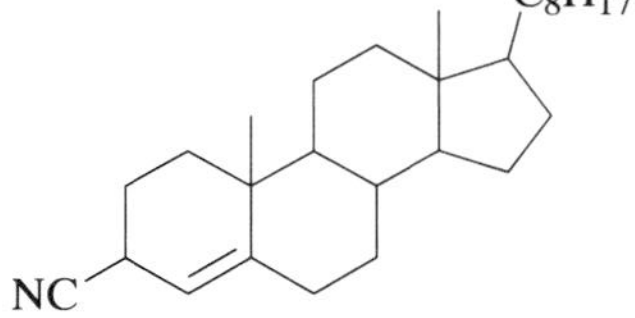 (76) | 71 |

*I. Miscellaneous Substrates*

| $C_6$ 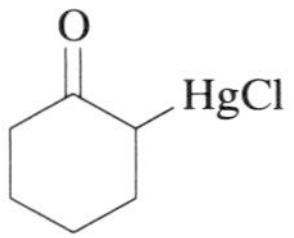 | 2 $SmI_2$, THF, MeOH, –78° | 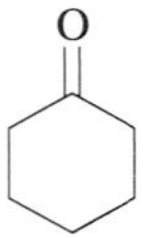 (94) | 53 |
|---|---|---|---|

Table VII. REDUCTIVE CLEAVAGE OF CYCLOPROPYL KETONES

| Substrate | Conditions | Product(s) and Yield(s) (%) | Refs. |
|---|---|---|---|
| C$_{10}$ | 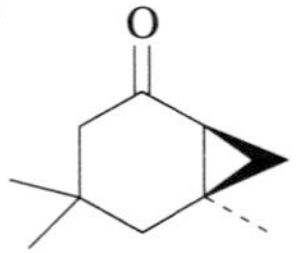 SmI$_2$, THF, DMPU, rt, 5 min | 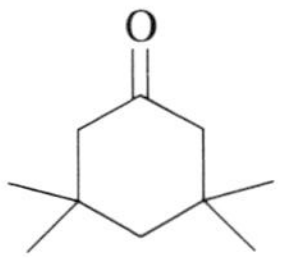 (39) | 74 |
| C$_{11}$ | 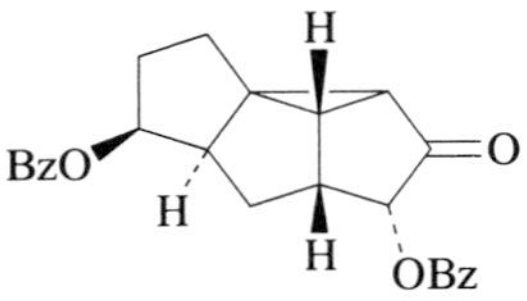 2 SmI$_2$, THF, MeOH, -78 to 25° | 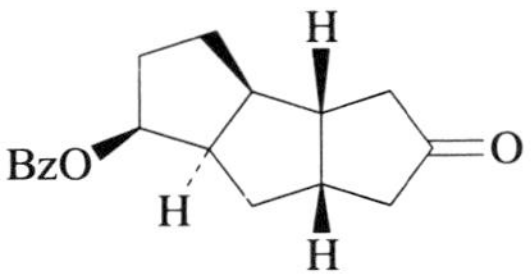 (65) | 61 |
| 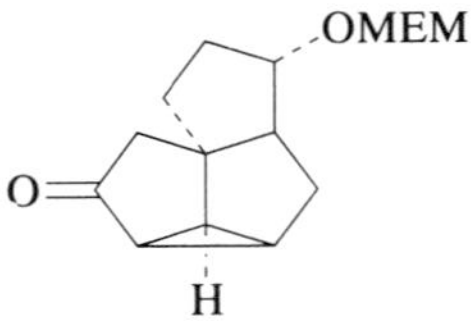 | 2 SmI$_2$, THF, MeOH, 25° | 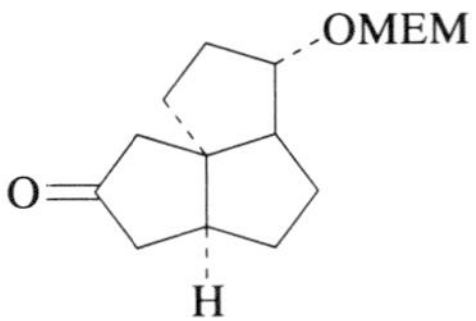 (86) | 61 |
| 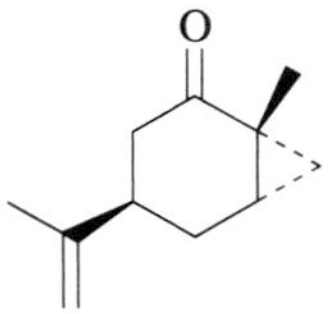 | SmI$_2$, THF, DMPU, rt, 5 min | 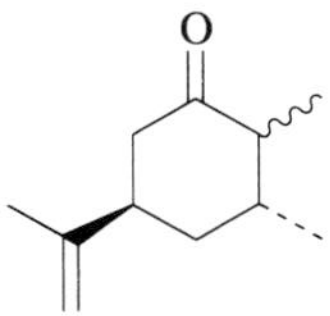 (34) | 74 |

Table VII. REDUCTIVE CLEAVAGE OF CYCLOPROPYL KETONES (*Continued*)

| Substrate | Conditions | Product(s) and Yield(s) (%) | Refs. |
|---|---|---|---|
| 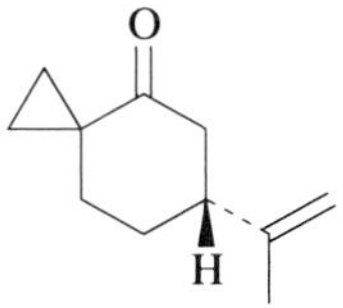 | 2 SmI$_2$, Fe(DBM)$_3$ (cat.), THF, $t$-BuOH | 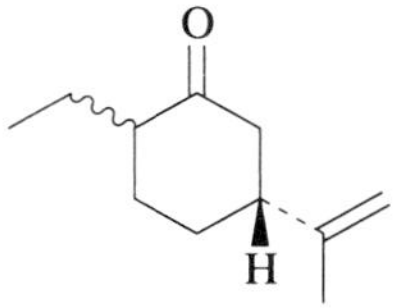 (81) | 73 |
| C$_{14}$  $n$-C$_9$H$_{19}$ | SmI$_2$, THF, DMPU, rt, 5 min | 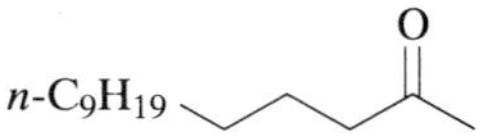 $n$-C$_9$H$_{19}$ (49) | 74 |

## Table VIII. DEOXYGENATION REACTIONS

| Substrate | Conditions | Product(s) and Yield(s) (%) | Refs. |
|---|---|---|---|
| *A. Epoxides* | | | |
| $C_6$ | 2 SmI$_2$, THF, HMPA, DMAE, rt, 2.5 h | Et⁓OAc  (88), *E:Z* = 3:1 | 75 |
| $C_8$ | 2 SmI$_2$, THF, *t*-BuOH, rt, 24 h | **I** (65) | 3 |
| | 2 SmI$_2$, THF, *t*-BuOH, 65°, 2 h | **I** (95) | 3 |
| $C_9$ | 2 SmI$_2$, THF, 65°, 4 d | **I** (92) | 3 |
| 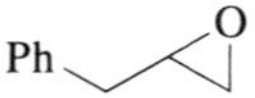 | 2 SmI$_2$, THF, *t*-BuOH, rt, 24 h |  **I** (82) 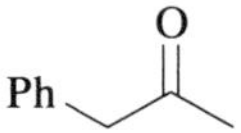 + **II** 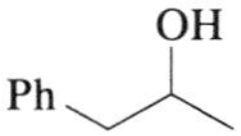 + Ph⁓OH **III**  **II + III** (18) | 3 |
| $C_{10}$ $n$-C$_8$H$_{17}$ | 4 SmI$_2$, THF, *t*-BuOH, rt, 2 d | $n$-C$_8$H$_{17}$⁓ **I** (92) + **II** + $n$-C$_8$H$_{17}$⁓OH **III**  **II + III** (6) | 3 |

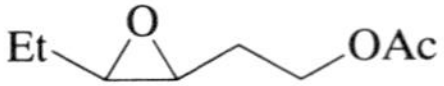

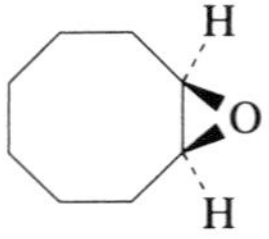

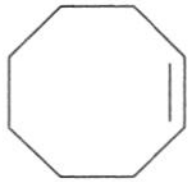

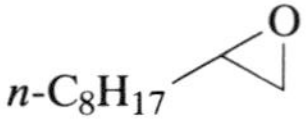

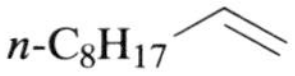

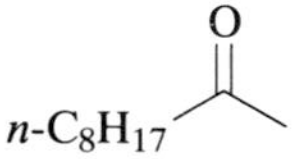

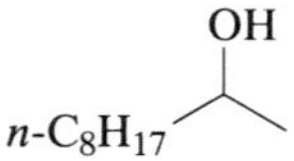

338

Table VIII. DEOXYGENATION REACTIONS (*Continued*)

| Substrate | Conditions | Product(s) and Yield(s) (%) | Refs. |
|---|---|---|---|
|  $n$-Bu, Bu-$n$ | 2 SmI$_2$, THF, $t$-BuOH, rt, 3 h | 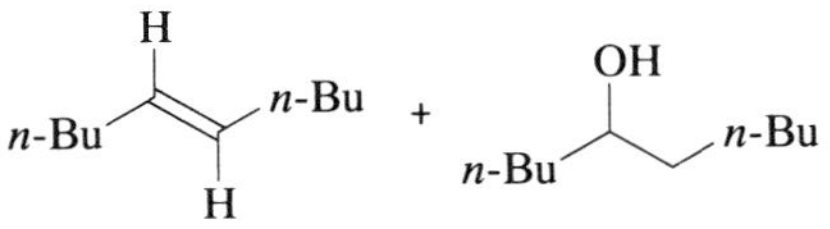 **I** (76) + (17) | 3 |
| | 4 SmI$_2$, THF, $t$-BuOH, rt, 2 d | **I** (96) | 3 |
| | 2 SmI$_2$, THF, HMPA, DMAE, rt, 24 h | **I** (95) $E{:}Z = 3{:}1$ | 75 |
| | 2 SmI$_2$, THF, HMPA, glutaric anhydride, rt, 1 h | **I** (69) $E{:}Z = 3{:}1$ | 75 |
| C$_{12}$ $n$-C$_{10}$H$_{21}$ | | $n$-C$_{10}$H$_{21}$ **I** | |
| | 1.2 SmI$_2$, THF, HMPA, rt, 2 h | **I** (37) | 75 |
| | 2 SmI$_2$, THF, DMAE, rt, 24 h | **I** (88) | 75 |
| | 2 SmI$_2$, THF, HMPA, DMAE, rt, 1.5 h | **I** (95) | 75 |
| | 2 SmI$_2$, THF, HMPA, glutaric anhydride, rt, 1 h | **I** (90) | 75 |

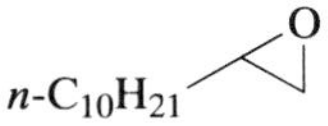

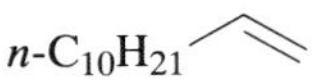

Table VIII. DEOXYGENATION REACTIONS (Continued)

| Substrate | Conditions | Product(s) and Yield(s) (%) | Refs. |
|---|---|---|---|
| $C_{14}$ | 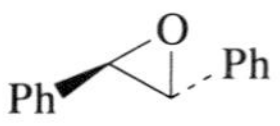 2 SmI$_2$, THF, $t$-BuOH, rt, 30 min | 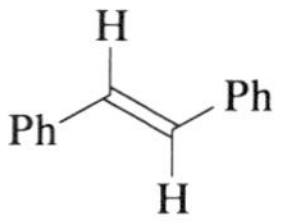 **I** (94), *trans:cis* = 2:1 | 3 |
| | 2 SmI$_2$, THF, $t$-BuOH, 65°, 5 min | **I** (98), *trans:cis* = 2:1 | 3 |
| | | | |
| *B. Sulfoxides* | | | |
| $C_7$  Ph(Me)SO | 2.2 SmI$_2$, THF, HMPA, 20°, 1 min | PhSMe  (93) | 41,75 |
| $C_8$  Bu$_2$SO | 2.2 SmI$_2$, THF, HMPA, 20°, 1 min | Bu$_2$S  (99) | 41,75 |
| $C_9$ | 2.2 SmI$_2$, THF, 20°, 24 h | "  (71) | 41,75 |
| $C_{11}$  $p$-Tol(Et)SO | 2 SmI$_2$, THF, 65°, 4 h | $p$-TolSEt  (77) | 3 |
| $C_{11}$ | 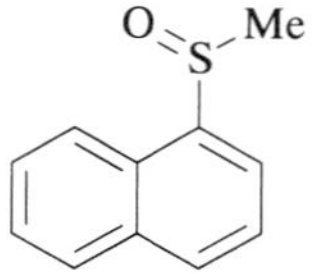 2 SmI$_2$, THF, 65°, 1 h | 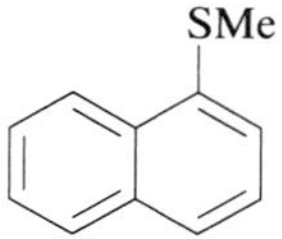  (90) | 3 |
| $C_{12}$  Ph$_2$SO | 2.2 SmI$_2$, THF, HMPA, 20°, 1 min | Ph$_2$S  (94) | 41,75 |
| | 2.5 SmI$_2$, THF, rt, 2 d | "  (95) | 3 |
| | 2 SmI$_2$, THF, 65°, 2 h | "  (90) | 3 |

## Table VIII. DEOXYGENATION REACTIONS (Continued)

| Substrate | Conditions | Product(s) and Yield(s) (%) | Refs. |
|---|---|---|---|
| | *C. Sulfones and Related Substrates* | | |
| C$_6$ 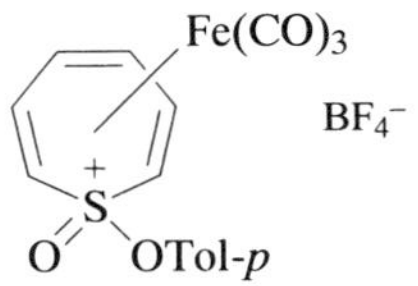 <br> Fe(CO)$_3$ BF$_4^-$ <br> S$^+$ O OTol-*p* | 15 SmI$_2$, THF, 0° | 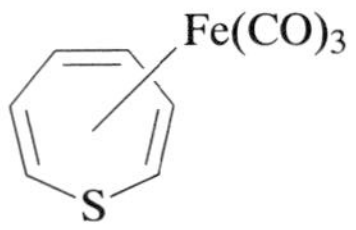 Fe(CO)$_3$ <br> S (38) | 77 |
| C$_7$ Ph(Me)SO$_2$ | 4.4 SmI$_2$, THF, HMPA, 20°, 10 min | PhSMe (99) | 41,75 |
| | 4.4 SmI$_2$, THF, 20°, 24 h | " (0) | 41,75 |
| C$_8$ Bu$_2$SO$_2$ | 4.4 SmI$_2$, THF, HMPA, 65°, 8 h | Bu$_2$S (26) | 41,75 |
| C$_{12}$ Ph$_2$SO$_2$ | 4.4 SmI$_2$, THF, HMPA, 20°, 1 min | Ph$_2$S (93) | 41,75 |
| | 4.4 SmI$_2$, THF, 20°, 24 h | " (0) | 41,75 |
| | *D. N-Oxides* | | |
| C$_4$ 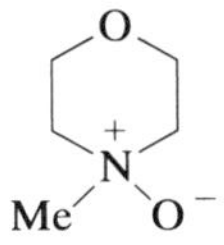 <br> O N$^+$ O$^-$ Me | 2.2 SmI$_2$, THF, HMPA, 20°, 1 min | 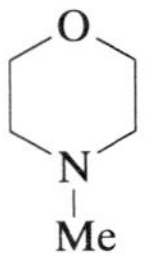 O N-Me (98) | 76 |

Table VIII. DEOXYGENATION REACTIONS (*Continued*)

| Substrate | Conditions | Product(s) and Yield(s) (%) | Refs. |
|---|---|---|---|
| 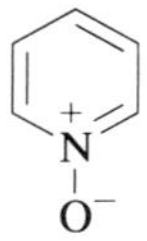 C5 | 2 SmI$_2$, THF, 65°, 10 min | 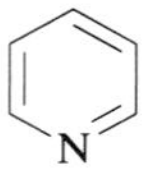 (80) | 49 |
| 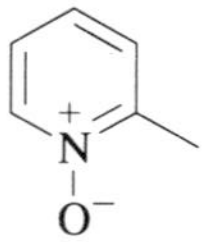 C6 | 2 SmI$_2$, THF, rt, <5 min | 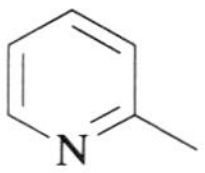 (83) | 49 |
| 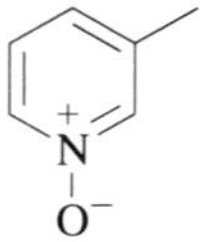 C9 | 2 SmI$_2$, THF, rt, 15-20 min | 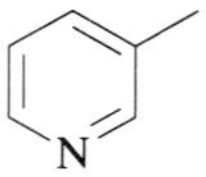 (80) | 49 |
| 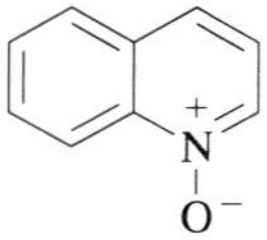 | 2.2 SmI$_2$, THF, HMPA, 20°, 1 min | 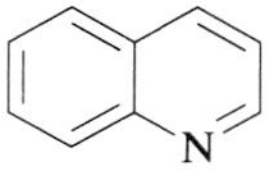 (96) | 41,75 |
| C24 (C$_8$H$_{17}$)$_3$NO | 2 SmI$_2$, THF, 65°, 10 min | (C$_8$H$_{17}$)$_3$N (79) | 49 |
| | *E. Phosphine Oxides* | | |
| C18 Ph$_3$PO | 2.2 SmI$_2$, THF, HMPA, 65°, 16 h | Ph$_3$P (75) | 41,75 |
| | 2.2 SmI$_2$, THF, 65°, 24 h | " (0) | 41,75 |

Table VIII. DEOXYGENATION REACTIONS (*Continued*)

| Substrate | Conditions | Product(s) and Yield(s) (%) | Refs. |
|---|---|---|---|
| *F. Arsine Oxides* | | | |
| $C_{18}$ | | | |
| $Ph_3AsO$ | $2 \ SmI_2$, THF, 65°, 5 min | $Ph_3As$ (82) | 49 |
| *G. Miscellaneous Deoxygenations and Desulfurizations* | | | |
| $C_4$ | | | |
| *n*-BuNCS | $SmI_2$, THF, HMPA, rt, 0.5 h | *n*-BuNC (79) | 78 |
| $C_6$ | | | |
| PhNCS | $SmI_2$, THF, HMPA, rt, 0.5 h | PhNC (83) | 78 |
| $C_7$ | | | |
| *p*-TolNCS | $SmI_2$, THF, HMPA, rt, 0.5 h | *p*-TolNC (81) | 78 |
| $C_{12}$ | | | |
| $(Bu_3Sn)_2O$ | $2.2 \ SmI_2$, THF, HMPA, 20°, 1 min | $(Bu_3Sn)_2$ (92) | 41,75 |
| | $2.2 \ SmI_2$, THF, 20° | " (0) | 41,75 |

Table IX. REDUCTION OF NITROGEN-BASED FUNCTIONAL GROUPS

| Substrate | Conditions | Product(s) and Yield(s) (%) | Refs. |
|---|---|---|---|
| | | A. *Nitro Compounds* | |
| C$_2$ | | | |
| (structure: ethyl, O, O, NO$_2$) | 4 SmI$_2$, THF, MeOH, rt, 3 min | (structure: ethoxy, O, O, NHOH) (58) | 79 |
| | 1. 6 SmI$_2$, THF, MeOH, rt, 8 h | Complex mixture | 79 |
| | 2. *p*-PhC$_6$H$_4$COCl, CH$_2$Cl$_2$, Et$_3$N, rt, 16 h | | |
| 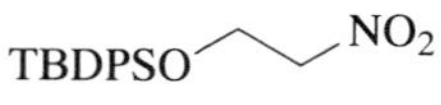 TBDPSO~~~NO$_2$ | 4 SmI$_2$, THF, MeOH, rt, 3 min | TBDPSO~~~NHOH (79) | 79 |
| | 1. 6 SmI$_2$, THF, MeOH, rt, 8 h | TBDPSO~~~NHCOC$_6$H$_4$Ph-*p* (80) | 79 |
| | 2. *p*-PhC$_6$H$_4$COCl, CH$_2$Cl$_2$, Et$_3$N, rt, 16 h | | |
| C$_3$ | | | |
| TBDPSO~~~~NO$_2$ | 4 SmI$_2$, THF, MeOH, rt, 3 min |  TBDPSO~~~~NHOH (83) | 79 |
| | 1. 6 SmI$_2$, THF, MeOH, rt, 8 h | TBDPSO~~~~NHCOC$_6$H$_4$Ph-*p* (76) | 79 |
| | 2. *p*-PhC$_6$H$_4$COCl, CH$_2$Cl$_2$, Et$_3$N, rt, 16 h | | |
| 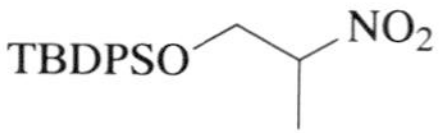 TBDPSO~~~NO$_2$ (branched) | 4 SmI$_2$, THF, MeOH, rt, 3 min | TBDPSO~~~NHOH (85) | 79 |
| | 1. 6 SmI$_2$, THF, MeOH, rt, 8 h | TBDPSO~~~NHCOC$_6$H$_4$Ph-*p* (79) | 79 |
| | 2. *p*-PhC$_6$H$_4$COCl, CH$_2$Cl$_2$, Et$_3$N, rt, 16 h | | |

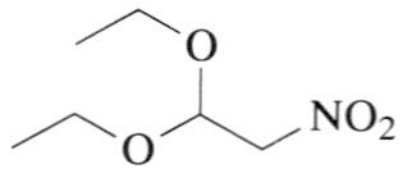

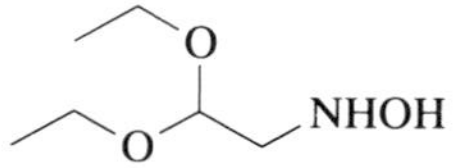

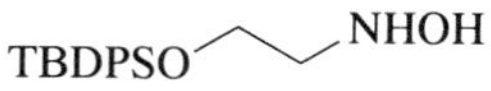

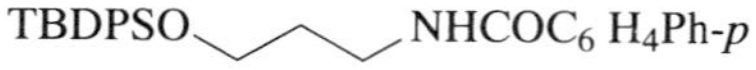

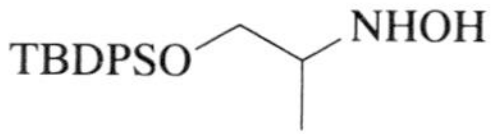

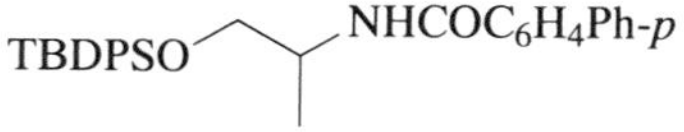

344

Table IX. REDUCTION OF NITROGEN-BASED FUNCTIONAL GROUPS (*Continued*)

| Substrate | Conditions | Product(s) and Yield(s) (%) | Refs. |
|---|---|---|---|
| **C₄** | | | |
| MeO₂C⌒⌒NO₂ | 4 SmI₂, THF, MeOH, rt, 3 min | 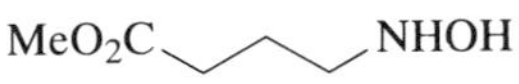 MeO₂C⌒⌒NHOH (complex mixture) | 79 |
| TBDPSO–C(CH₃)₂–CH₂NO₂ | 1. 6 SmI₂, THF, MeOH, rt, 8 h  2. *p*-PhC₆H₄COCl, CH₂Cl₂, Et₃N, rt, 16 h | TBDPSO⋯NHOH (93);  TBDPSO⋯NHCOC₆H₄Ph-*p* (93) | 79  79 |
| **C₆** | | | |
| PhNO₂ | 8 SmI₂, THF, MeOH, rt, 5 min | PhNH₂  **I** (85) | 49 |
| | " | **I** (25)  +  PhN=NPh (25) | 80 |
| *p*-O₂N–C₆H₄–NH₂ | 8 SmI₂, THF, MeOH, rt, 5 min | H₂N–C₆H₄–NH₂ (89) | 49 |
| (1,4-dioxaspiro[4.5]decane with NO₂) | 4 SmI₂, THF, MeOH, rt, 3 min | (1,4-dioxaspiro[4.5]decane with NHOH) (42) | 79 |

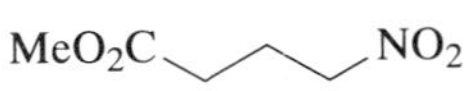

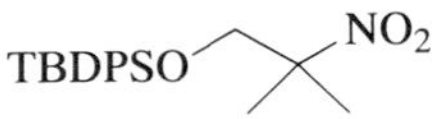

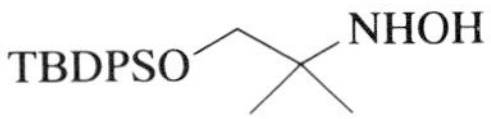

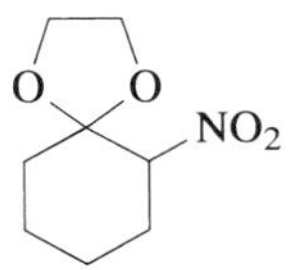

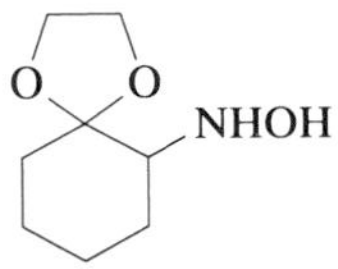

345

Table IX.  REDUCTION OF NITROGEN-BASED FUNCTIONAL GROUPS (Continued)

| Substrate | Conditions | Product(s) and Yield(s) (%) | Refs. |
|---|---|---|---|
| C7    3-NO2-C6H4-CN | 1. 6 SmI2, THF, MeOH, rt, 8 h<br>2. p-PhC6H4COCl, CH2Cl2, Et3N, rt, 16 h | Complex mixture | 79 |
| 4-NO2-C6H4-CN | 6 SmI2, THF, MeOH, 5 min | 3-NH2-C6H4-CN  (95) | 80 |
|  4-NO2-C6H4-CN | 6 SmI2, THF, MeOH, rt, 5 min |  4-NH2-C6H4-CN  (84) | 80 |
| C10   MeO-C6H4-(CH2)4-NO2 | 4 SmI2, THF, MeOH, rt, 3 min | MeO-C6H4-(CH2)4-NHOH  (72) | 79 |
| C10   MeO-C6H4-(CH2)4-NO2 | 1. 6 SmI2, THF, MeOH, rt, 8 h<br>2. p-PhC6H4COCl, CH2Cl2, Et3N, rt, 16 h | 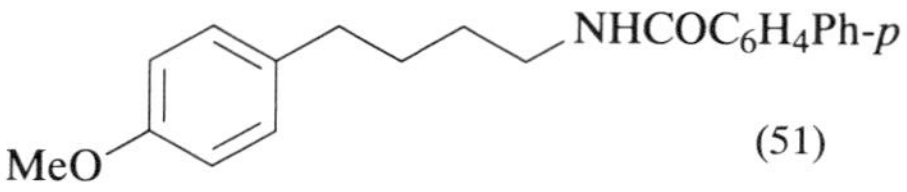 MeO-C6H4-(CH2)4-NHCOC6H4Ph-p  (51) | 79 |
| C12   (1,3-dioxolane)-CH2-CH(Ph)-C(CH3)2-NO2 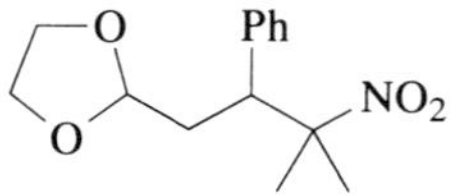 | 4 SmI2, THF, MeOH, rt, 3 min | 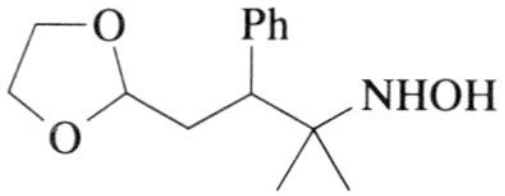 (1,3-dioxolane)-CH2-CH(Ph)-C(CH3)2-NHOH  (88) | 79 |

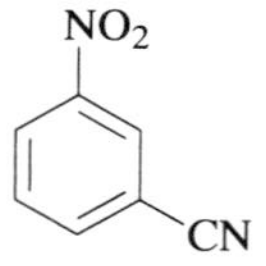

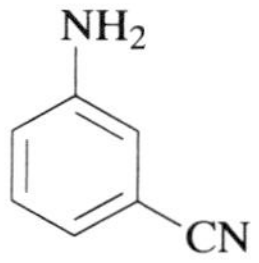

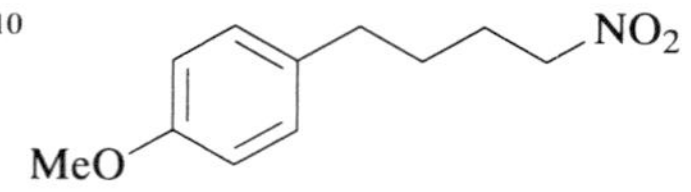

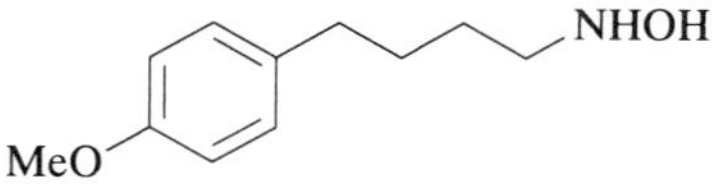

Table IX. REDUCTION OF NITROGEN-BASED FUNCTIONAL GROUPS (*Continued*)

| Substrate | Conditions | Product(s) and Yield(s) (%) | Refs. |
|---|---|---|---|
| | 1. 6 SmI$_2$, THF, MeOH, rt, 8 h<br>2. p-PhC$_6$H$_4$COCl, CH$_2$Cl$_2$, Et$_3$N, rt, 16 h | 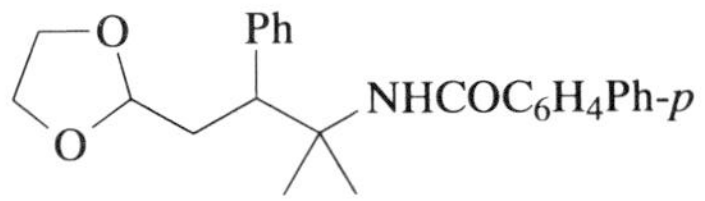 (structure) NHCOC$_6$H$_4$Ph-p (81) | 79 |
| **B. Nitriles** | | | |
| C$_7$ | | | |
| PhCN | 4 SmI$_2$, THF, MeOH, rt | No reaction | 80 |
| | 8 SmI$_2$, THF, H$_3$PO$_4$ (85%), rt, 6 min | PhCH$_2$NH$_2$ (99) | 46 |
| | 8 SmI$_2$, THF, KOH (50%), rt, 25 min | " (91) | 46 |
| o-ClC$_6$H$_4$CN | 4 SmI$_2$, THF, H$_3$PO$_4$ (85%), rt, 5 s | o-ClC$_6$H$_4$CH$_2$NH$_2$ **I** (76) | 46 |
| | 4 SmI$_2$, THF, KOH (50%), rt, 3 min | **I** (26) + PhCH$_2$NH$_2$ **II** (40) | 46 |
| | 8 SmI$_2$, THF, KOH (50%), rt, 4 min | **I** (6) + **II** (77) | 46 |
| p-ClC$_6$H$_4$CN | 4 SmI$_2$, THF, H$_3$PO$_4$ (85%), rt, 10 s | p-ClC$_6$H$_4$CH$_2$NH$_2$ **I** (83) | 46 |
| | 4 SmI$_2$, THF, KOH (50%), rt, 5 min | **I** (39) + PhCH$_2$NH$_2$ **II** (24) | 46 |
| | 8 SmI$_2$, THF, KOH (50%), rt, 3 min | **I** (12) + **II** (56) | 46 |

Table IX. REDUCTION OF NITROGEN-BASED FUNCTIONAL GROUPS (*Continued*)

| Substrate | Conditions | Product(s) and Yield(s) (%) | Refs. |
|---|---|---|---|
| $o$-HOC$_6$H$_4$CN | 4 SmI$_2$, THF, H$_3$PO$_4$ (85%), rt, 30 s | $o$-HOC$_6$H$_4$CH$_2$NH$_2$ (90) | 46 |
| | 8 SmI$_2$, THF, KOH (50%), rt, 9 min | " (88) | 46 |
| $p$-HOC$_6$H$_4$CN | 4 SmI$_2$, THF, H$_3$PO$_4$ (85%), rt, 30 s | $p$-HOC$_6$H$_4$CH$_2$NH$_2$ (72) | 46 |
| | 8 SmI$_2$, THF, KOH (50%), rt, 10 min | " (81) | 46 |
| C$_8$ | | | |
| $o$-MeC$_6$H$_4$CN | 4 SmI$_2$, THF, H$_3$PO$_4$ (85%), rt, 35 s | $o$-MeC$_6$H$_4$CH$_2$NH$_2$ **I** (52) + ($o$-MeC$_6$H$_4$CH$_2$)$_2$NH **II** (10) | 46 |
| | 4 SmI$_2$, THF, KOH (50%), rt, 2 min | **I** (58) + **II** (8) | 46 |
| | 8 SmI$_2$, THF, KOH (50%), rt, 12 min | **I** (70) + **II** (16) | 46 |
| $p$-MeC$_6$H$_4$CN | 4 SmI$_2$, THF, H$_3$PO$_4$ (85%), rt, 10 s | $p$-MeC$_6$H$_4$CH$_2$NH$_2$ **I** (55) + ($p$-MeC$_6$H$_4$CH$_2$)$_2$NH **II** (14) | 46 |
| | 4 SmI$_2$, THF, KOH (50%), rt, 6 min | **I** (64) | 46 |
| | 8 SmI$_2$, THF, KOH (50%), rt, 12 min | **I** (83) + **II** (2) | 46 |
| PhCH$_2$CN | 4 SmI$_2$, THF, H$_3$PO$_4$ (85%), rt, 50 s | Ph(CH$_2$)$_2$CN (48) | 46 |
| | 4 SmI$_2$, THF, KOH (50%), rt, 35 min | " (58) | 46 |

Table IX. REDUCTION OF NITROGEN-BASED FUNCTIONAL GROUPS (*Continued*)

| Substrate | Conditions | Product(s) and Yield(s) (%) | Refs. |
|---|---|---|---|
| $C_{11}$ 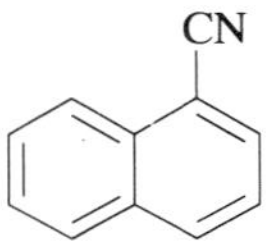 (CN-naphthalene) | 4 SmI$_2$, THF, H$_3$PO$_4$ (85%), rt, 10 s | 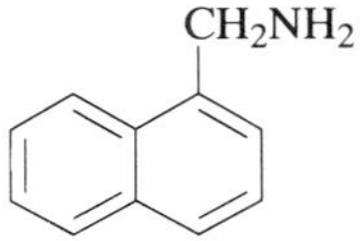 CH$_2$NH$_2$ (71) | 46 |
| | 4 SmI$_2$, THF, KOH (50%), rt, 5 min | " (83) | 46 |
| 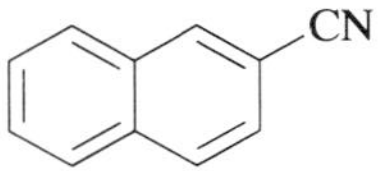 (CN-naphthalene) | 4 SmI$_2$, THF, H$_3$PO$_4$ (85%), rt, 5 s | 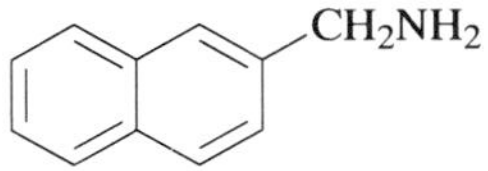 CH$_2$NH$_2$ (84) | 46 |
| | 4 SmI$_2$, THF, KOH (50%), rt, 4 min | " (86) | 46 |
| $C_{12}$ $n$-C$_{11}$H$_{23}$CN | 4 SmI$_2$, THF, rt, 2 MeOH | No reaction | 80 |
| $C_{13}$ | *C. Imines* | | |
| PhCH=NPh | 2.5 SmI$_2$, THF, MeOH, rt, 24 h | PhCH$_2$NHPh (80) | 80 |
| $C_{19}$ 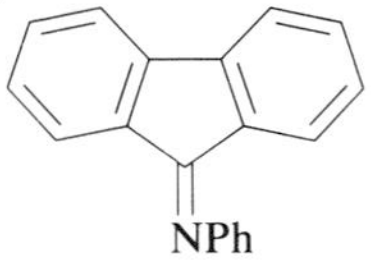 (fluorenone NPh) | 2 SmI$_2$, THF, 65°, 1 h | 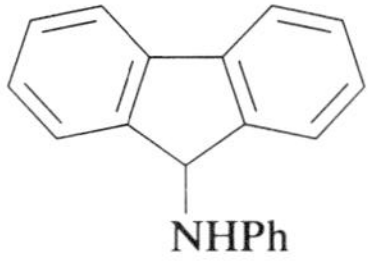 NHPh (97) | 81 |
| Ph$_2$C=NPh | 2 SmI$_2$, THF, 65°, 1 h | Ph$_2$CHNHPh (98) | 81 |

349

Table IX. REDUCTION OF NITROGEN-BASED FUNCTIONAL GROUPS (*Continued*)

| Substrate | Conditions | Product(s) and Yield(s) (%) | Refs. |
|---|---|---|---|

*D. Oximes and Related Substrates*

$C_3$

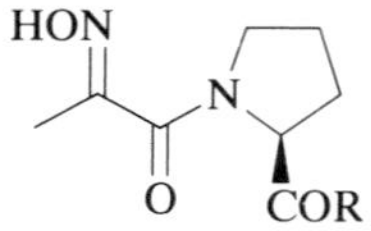

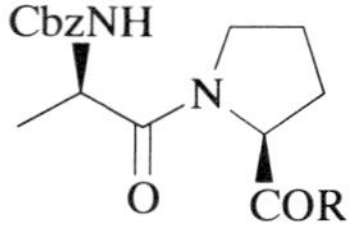

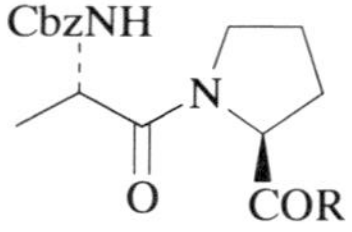

1. 5 SmI$_2$, THF, MeOH, −40°, 1 h
2. CbzCl, pyridine

**I** + **II**

R = OMe    **I** (54) + **II** (10)    82

R = OBu-*t*    **I** (60) + **II** (7)    82

R = —N⟩ (pyrrolidine)    **I** (81) + **II** (1)

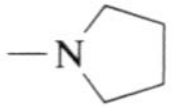

"

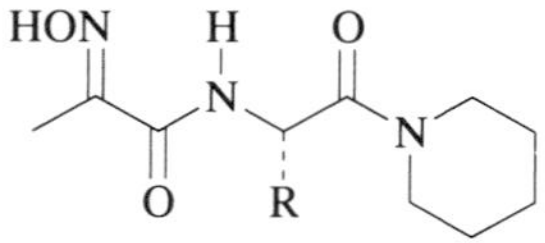

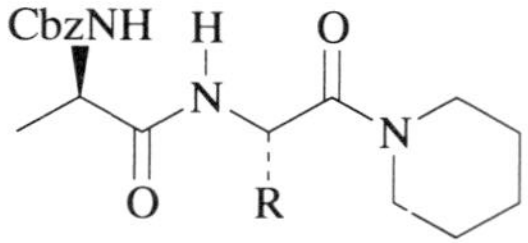

**I** + **II**

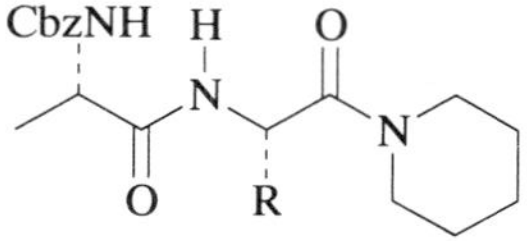

R = Me    **I** (56) + **II** (14)    82

R = *i*-Pr    **I** (69) + **II** (8)    82

R = *t*-Bu    **I** (66) + **II** (7)    82

Table IX. REDUCTION OF NITROGEN-BASED FUNCTIONAL GROUPS (*Continued*)

| Substrate | Conditions | Product(s) and Yield(s) (%) | Refs. |
|---|---|---|---|
| 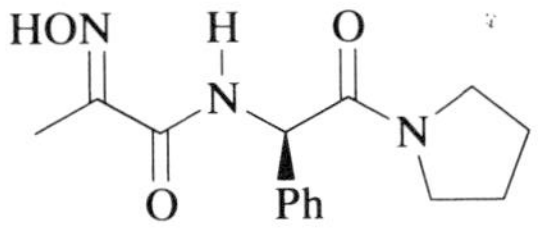 | 1. 5 SmI$_2$, THF, MeOH, -40°, 1 h<br>2. CbzCl, pyridine | 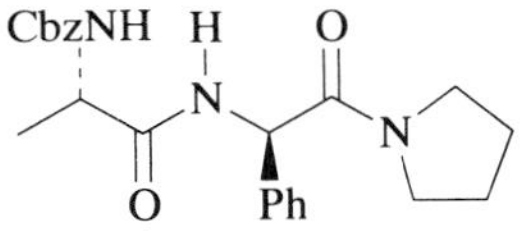 (69) 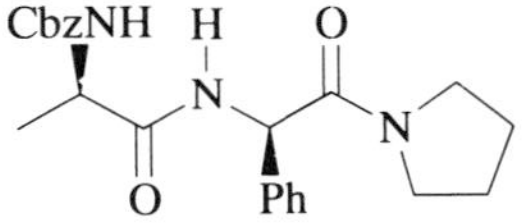 + (15) | 82 |
| 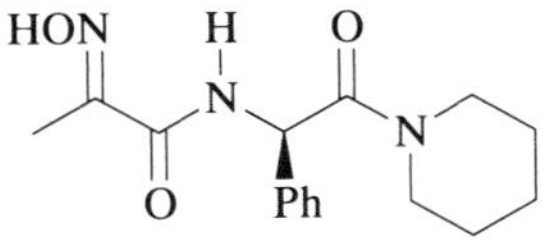 | " | 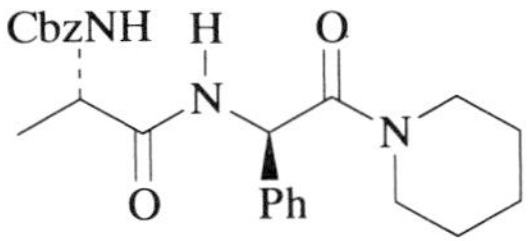 (72) + 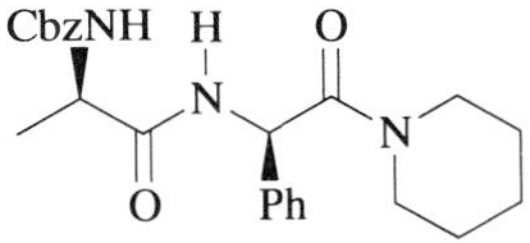 (8) | 82 |
| C$_4$ | 1. 5 SmI$_2$, THF, MeOH, -40°<br>2. CbzCl, pyridine | (66) + (2) | 82 |

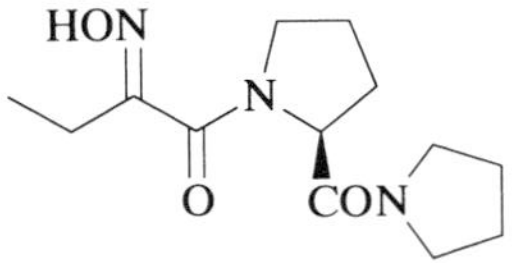

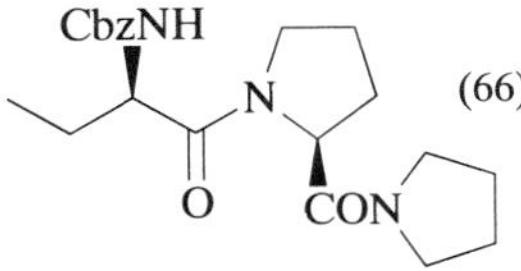

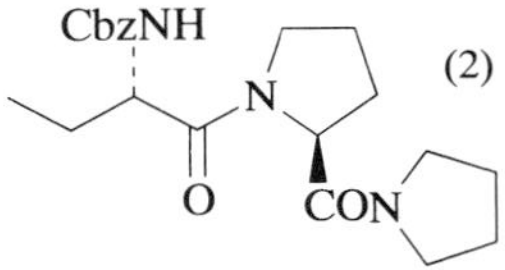

351

Table IX. REDUCTION OF NITROGEN-BASED FUNCTIONAL GROUPS (*Continued*)

| Substrate | Conditions | Product(s) and Yield(s) (%) | Refs. |
|---|---|---|---|
| C$_6$ | 1. 5 SmI$_2$, THF, MeOH, $-40°$<br>2. CbzCl, pyridine | (69) +<br> (6) | 82 |
| C$_7$ PhCH=NOH | 4 SmI$_2$, THF, MeOH,<br>8 LiNH$_2$, rt, 3 s | PhCH$_2$NH$_2$ (45) + PhCH$_2$OH (3) | 42 |
| | 4 SmI$_2$, THF, 4 MeOH,<br>rt, 5 min | PhCH(NH$_2$)CH$_2$Ph (75) +<br>PhCH$_2$NHCH$_2$Ph (25) | 80 |
| 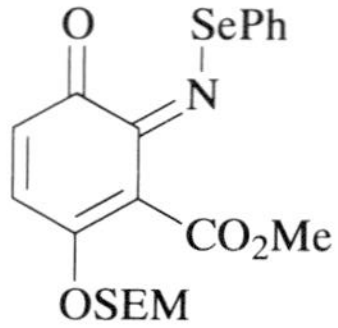 | SmI$_2$, THF, MeOH, $-10°$ | 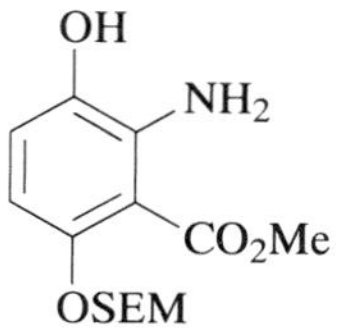 (74) | 83 |

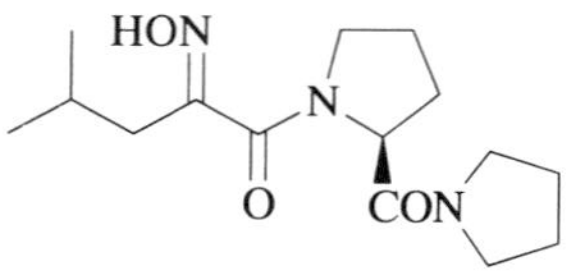

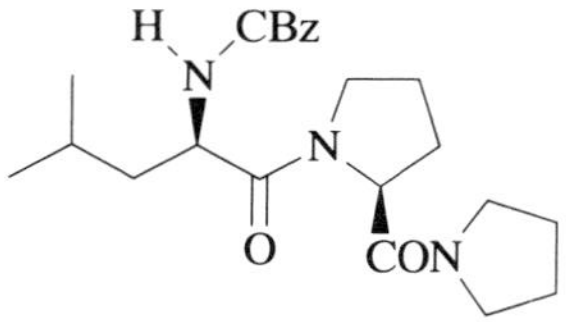

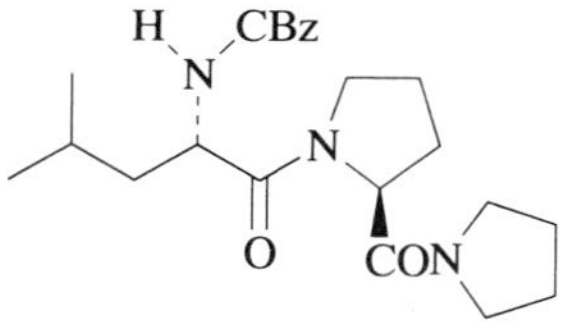

*E. Azo Compounds*

| | | | |
|---|---|---|---|
| C$_{12}$ PhN=NPh | 2 SmI$_2$, THF, MeOH,<br>rt, 5 min | PhNHNHPh (95) | 49 |

Table IX.  REDUCTION OF NITROGEN-BASED FUNCTIONAL GROUPS (*Continued*)

| Substrate | Conditions | Product(s) and Yield(s) (%) | Refs. |
|---|---|---|---|
| C$_8$ | | | |
| (structure: p-MeOC$_6$H$_4$ hydrazide with Ph) | 8 SmI$_2$, THF, MeOH, rt, 24 h | PhNH$_2$  (50) | 80 |

*F. Hydrazines and Related Substrates*

| Substrate | Conditions | Product(s) and Yield(s) (%) | Refs. |
|---|---|---|---|
| 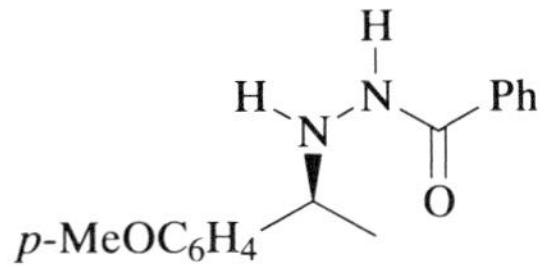 (structure: hydrazide R–C(=O), with p-MeOC$_6$H$_4$ substituent) | 2.2 SmI$_2$, THF, MeOH, rt, 0.5 h | 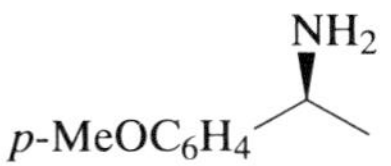 NH$_2$  (—) ; p-MeOC$_6$H$_4$ | 84 |
| | | **I**   Ph—CH(NH$_2$)— | |
| R = Ph, 89% ee | " | **I** (72), 89% ee | 84 |
| R = p-MeOC$_6$H$_4$ | " | **I** (—) | 84 |
| R = p-Me$_2$NC$_6$H$_4$ | " | **I** (—) | 84 |
| (structure: hydrazide R–C(=O), with Ph substituent) | " | NH$_2$  (—) ; Ph—CH(NH$_2$)— | 84 |

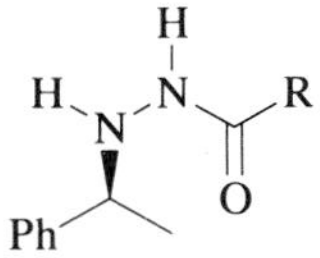

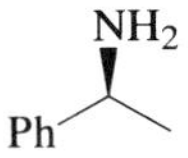

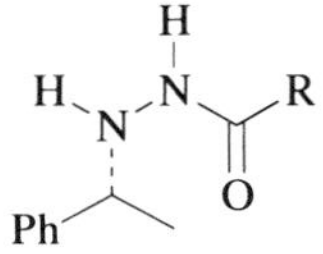

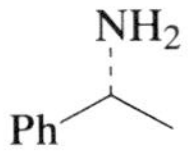

353

Table IX. REDUCTION OF NITROGEN-BASED FUNCTIONAL GROUPS (*Continued*)

| Substrate | Conditions | Product(s) and Yield(s) (%) | Refs. |
|---|---|---|---|
| C₉ | 2.2 SmI₂, THF, MeOH, rt, 0.5 h | (—) | 84 |
| 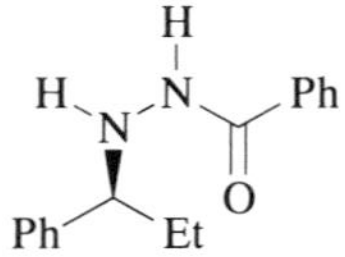 | " | 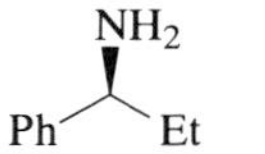 (—) | 84 |
| C₁₂ | " | (—) | 84 |
| PhNHNHPh | 2 SmI₂, THF, rt, 4 d | PhNH₂ (55) | 80 |
| 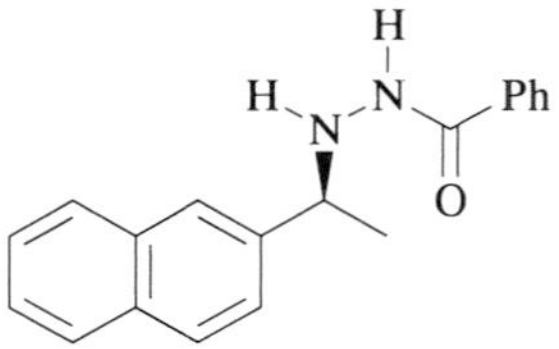 | 2.2 SmI₂, THF, MeOH, rt, 0.5 h | 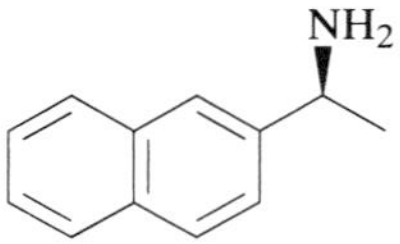 (—) | 84 |

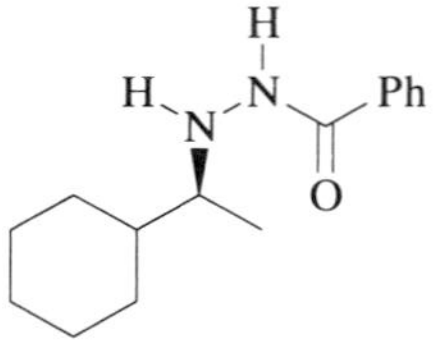

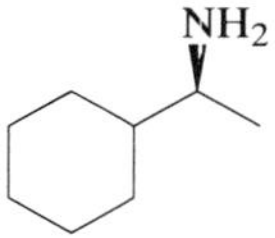

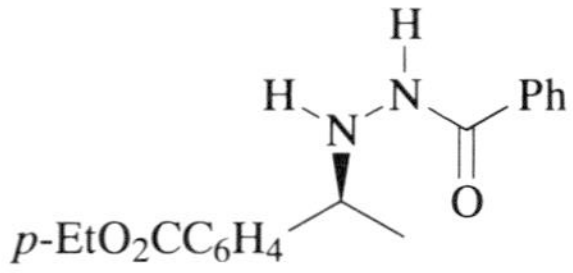

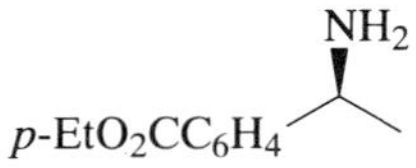

354

Table X. REDUCTION OF MISCELLANEOUS FUNCTIONAL GROUPS

| Substrate | Conditions | Product(s) and Yield(s) (%) | Refs. |
|---|---|---|---|
| | *A. Peroxides* | | |
| C7 | SmI₂, THF | 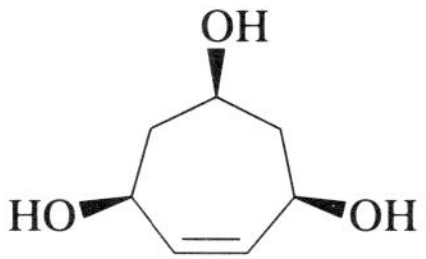 (77) | 85 |
| C13 | SmI₂, THF, 25° | 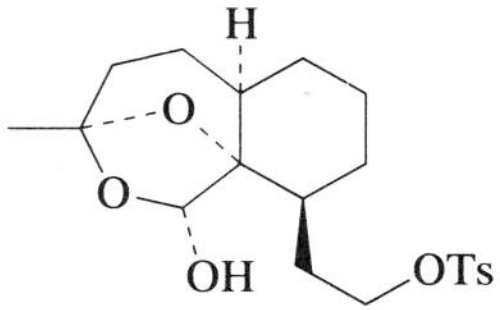 (—) | 86 |
| | *B. Isoxazoles* | | |
| C5 | 2 SmI₂, THF, rt, 1 h | 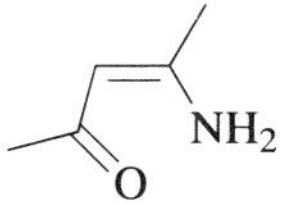 (57) | 25 |
| C6 | 2 SmI₂, THF, rt, 1 h | 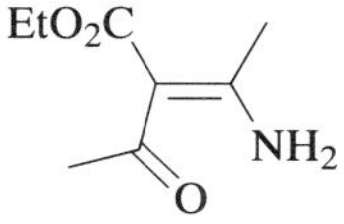 (96) | 25 |

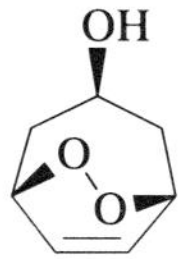

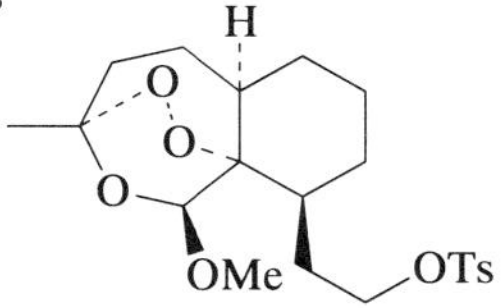

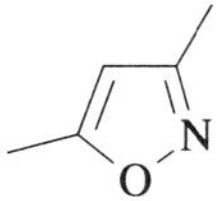

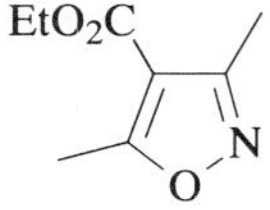

355

Table X. REDUCTION OF MISCELLANEOUS FUNCTIONAL GROUPS (*Continued*)

| Substrate | Conditions | Product(s) and Yield(s) (%) | Refs. |
|---|---|---|---|
| 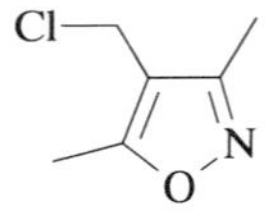 | 2 SmI$_2$, THF, 1 h | 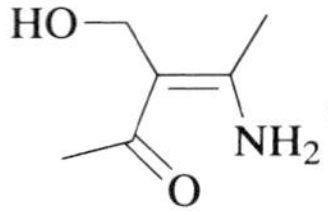 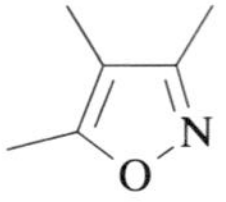 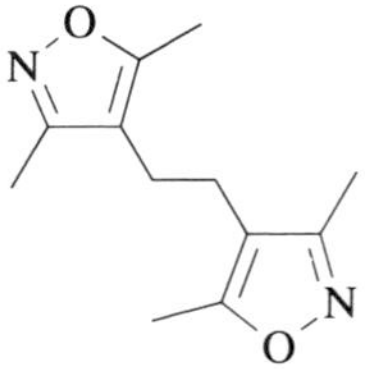 | 25 |

*C. Halo Phosphine Oxides and Sulfides*

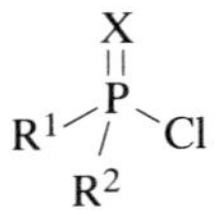

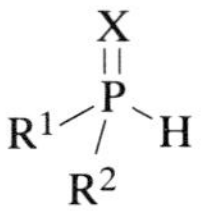

| R$^1$ | R$^2$ | X | | | |
|---|---|---|---|---|---|
| OEt | OEt | O | 2 SmI$_2$, THF, 65°, 3 h | **I** (10) | 87 |
| OEt | OEt | S | 2 SmI$_2$, THF, rt, 7 d | **I** (20) | 87 |
| Ph | OEt | O | 2 SmI$_2$, THF, rt, 7 d | **I** (0) | 87 |
| Ph | OEt | S | 2 SmI$_2$, THF, rt, 48 h | **I** (91) | 87 |
| Ph | Ph | O | 2 SmI$_2$, THF, rt, 35 min | **I** (75) + Ph$_2$P(O)P(O)Ph$_2$ (7) | 87 |
| Ph | Ph | S | 2 SmI$_2$, THF, rt, 40 min | **I** (89) | 87 |

356

Table X. REDUCTION OF MISCELLANEOUS FUNCTIONAL GROUPS (*Continued*)

| Substrate | Conditions | Product(s) and Yield(s) (%) | Refs. |
|---|---|---|---|
| 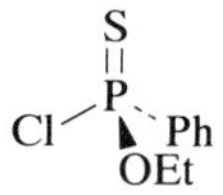 S=P(Cl)(Ph)(OEt) 90% ee | 2 SmI₂, THF, rt, 48 h | 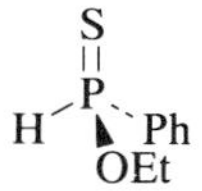 S=P(H)(Ph)(OEt) (83), 20% ee | 87 |
| 90% ee | 2 SmI₂, THF, rt, 60 h | " (83), 4.5% ee | 87 |
| 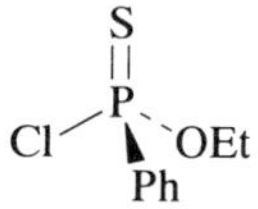 S=P(Cl)(OEt)(Ph) 88% ee | 2 SmI₂, THF, rt, 48 h | 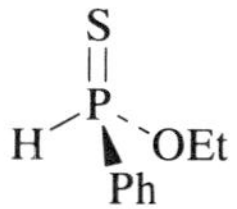 S=P(H)(OEt)(Ph) (80), 19% ee | 87 |
| 91% ee | 2 SmI₂, THF, rt, 60 h | 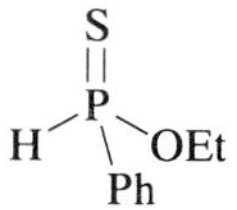 S=P(H)(OEt)(Ph) (85), racemic | 87 |

D. *Pd-Catalyzed Reduction of Allylic and Propargylic Carboxylates*

| | | | |
|---|---|---|---|
| C₈ OMEM ...OAc | 2 SmI₂, Pd(PPh₃)₄ (1 mol%), THF, *i*-PrOH, rt, 2 h | OMEM (79) | 88 |
| C₉ Ph OAc | 2 SmI₂, Pd(PPh₃)₄ (1 mol%), THF, *i*-PrOH, rt, 1 h | 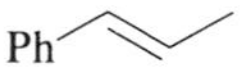 Ph **I** (20) 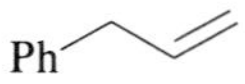 + Ph **II** (80) | 88 |
| | 2 SmI₂, PdCl₂ (1 mol%), PPh₃ (4 mol%), THF, *i*-PrOH, rt, 1 h | **I** (20) + **II** (80) | 88 |

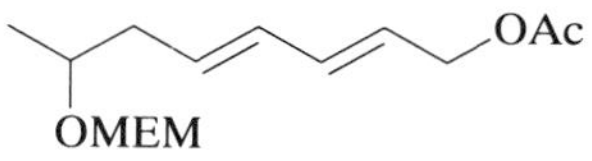

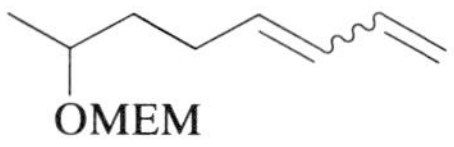

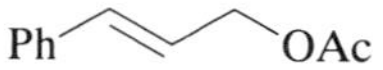

357

Table X. REDUCTION OF MISCELLANEOUS FUNCTIONAL GROUPS (*Continued*)

| Substrate | Conditions | Product(s) and Yield(s) (%) | Refs. |
|---|---|---|---|
| 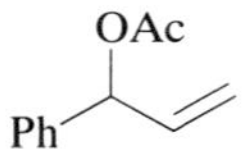 (OAc, Ph) | 2 SmI$_2$, Pd(PPh$_3$)$_4$ (1 mol%), THF, *i*-PrOH, rt, 1 h | **I** (20) + **II** (80) | 88 |
| C$_{10}$ | | | |
| 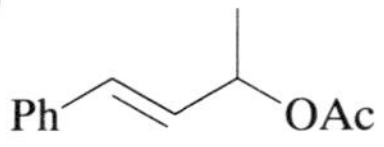 (Ph, OAc) | 2 SmI$_2$, Pd(PPh$_3$)$_4$ (1 mol%), THF, *i*-PrOH, rt, 1 h | 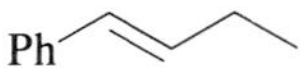 (67) 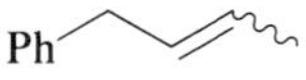 + Ph (14) | 88 |
| 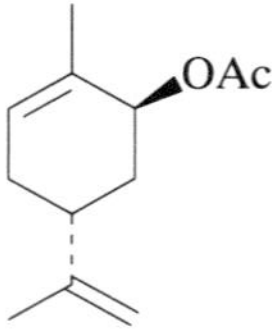 enantiomerically enriched (OAc) | 2 SmI$_2$, Pd(PPh$_3$)$_4$ (5 mol%), THF, *i*-PrOH, 65°, 8 h |  (81), racemic | 88 |
| 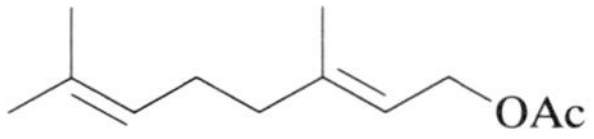 (OAc) | 2 SmI$_2$, Pd(PPh$_3$)$_4$ (5 mol%), THF, *i*-PrOH, 65°, 2 h | 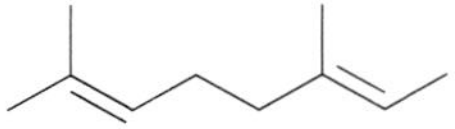 **I** (81) 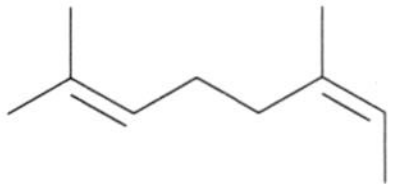 + **II** (4) 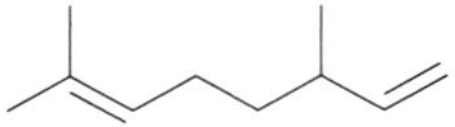 + **III** (6) | 88 |
| 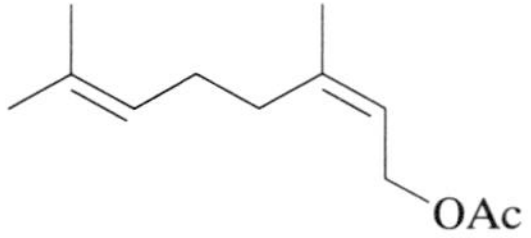 (OAc) | 2 SmI$_2$, Pd(PPh$_3$)$_4$ (5 mol%), THF, *i*-PrOH, 65°, 2 h | **I** (9) + **II** (72) + **III** (14) | 88 |

358

Table X. REDUCTION OF MISCELLANEOUS FUNCTIONAL GROUPS (*Continued*)

| Substrate | Conditions | Product(s) and Yield(s) (%) | Refs. |
|---|---|---|---|
| $C_{11}$ 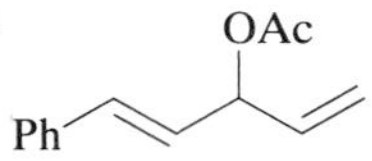 Ph, OAc, allyl substituted | 2 $SmI_2$, $Pd(PPh_3)_4$ (1 mol%), THF, $i$-PrOH, rt, 1 h | 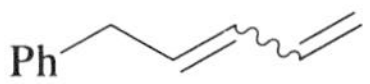 (44) 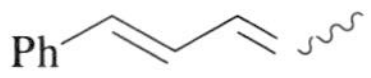 + Ph∼∼∼ (38) 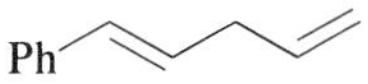 + Ph∼ (6) | 88 |
| 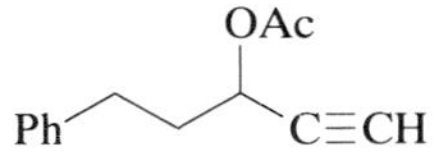 Ph, OAc, C≡CH | 2.5 $SmI_2$, $Pd(PPh_3)_4$ (5 mol%), THF, $i$-PrCHOHPr-$i$, 40°, 2 h | 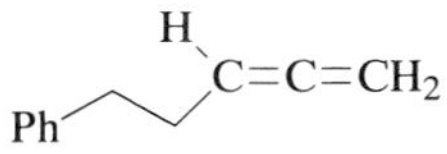 H,C=C=CH$_2$ Ph (71)  + Ph C≡CH (4) | 89 |
| 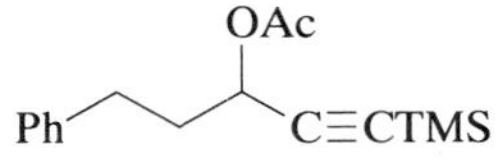 Ph, OAc, C≡CTMS | 2.5 $SmI_2$, $Pd(PPh_3)_4$ (5 mol%), THF, $i$-PrCHOHPr-$i$, 40°, 2 h | 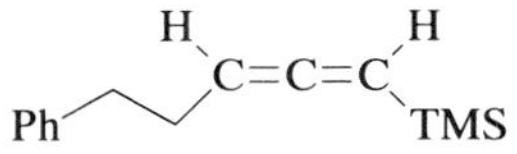 H,C=C=C(H)(TMS) Ph (54) 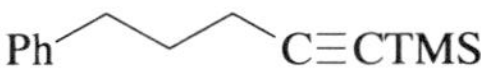 + Ph C≡CTMS (22) | 89 |
| $C_{12}$ 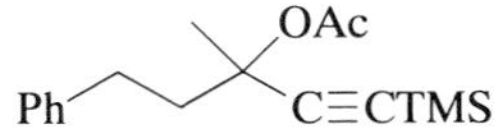 Ph, OAc, C≡CTMS | 2.5 $SmI_2$, $Pd(PPh_3)_4$ (5 mol%), THF, $i$-PrCHOHPr-$i$, 40°, 2 h | 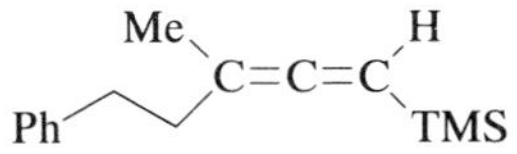 Me,C=C=C(H)(TMS) Ph (70) 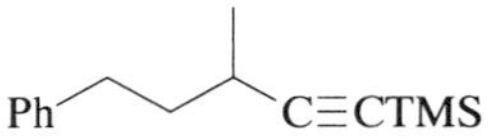 + Ph C≡CTMS (12) | 89 |
| $C_{13}$ 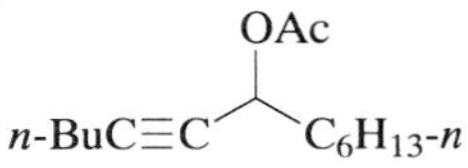 $n$-BuC≡C, OAc, $C_6H_{13}$-$n$ | 2 $SmI_2$, $Pd(PPh_3)_4$ (5 mol%), THF, $i$-PrOH, 40°, 2 h | 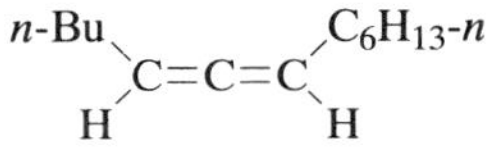 $n$-Bu,C=C=C(H)($C_6H_{13}$-$n$) (83) 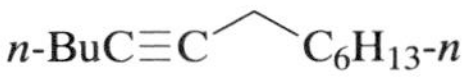 + $n$-BuC≡C—$C_6H_{13}$-$n$ (8) | 89 |

Table X. REDUCTION OF MISCELLANEOUS FUNCTIONAL GROUPS (*Continued*)

| Substrate | Conditions | Product(s) and Yield(s) (%) | Refs. |
|---|---|---|---|
| C$_{14}$ | | | |
| PhC≡C, AcO (cyclohexane) | 2.5 SmI$_2$, Pd(PPh$_3$)$_4$ (2 mol%), THF, *i*-PrOH, rt, 2 h | Ph$\diagdown$C=C=C$\diagup$H (90) | 89 |
| *n*-C$_8$H$_{17}$C≡C (γ-butyrolactone) | 1. 2.5 SmI$_2$, Pd(PPh$_3$)$_4$ (5 mol%), THF, *i*-PrCHOHPr-*i*, 40°, 2 h  2. CH$_2$N$_2$ | *n*-C$_8$H$_{17}$$\diagup$C=C=C$\diagup$H$\diagdown$H ... CO$_2$Me (66) | 89 |
| C$_{15}$ | | | |
| Ph Ph, AcO—C≡CH | 2.5 SmI$_2$, Pd(PPh$_3$)$_4$ (1 mol%), THF, *i*-PrOH, rt, 2 min | Ph$\diagdown$C=C=CH$_2$ (95) Ph$\diagup$ | 89 |
| OAc, Ph—C≡CBu-*n* | 2.5 SmI$_2$, Pd(PPh$_3$)$_4$ (5 mol%), THF, Y, 40°, 2 h | Ph$\diagdown$C=C=C$\diagup$Bu-*n* **I** + Ph—C≡CBu-*n* **II** | 89 |

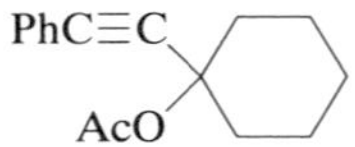

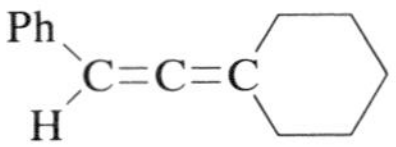

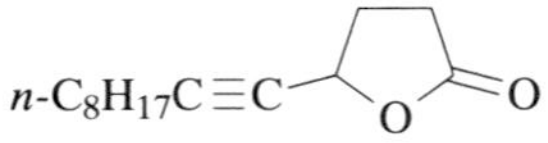

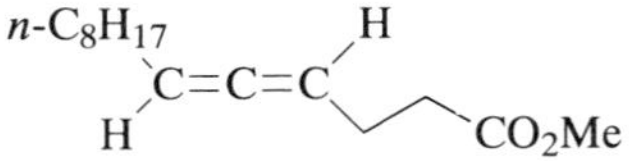

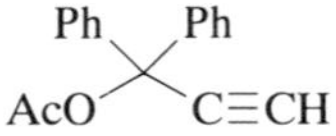

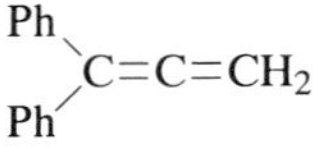

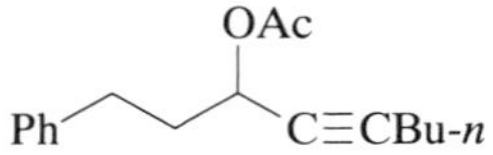

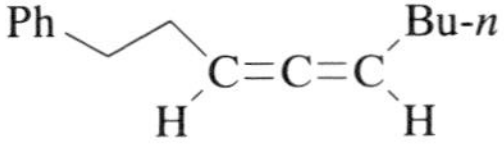

| Y | I + II | I:II |
|---|---|---|
| *i*-PrOH | (74) + (11) | 6.7:1 |
| H$_2$O | (—) | 1:1.5 |
| MeOH | (—) | 2:1 |
| *i*-PrCO$_2$H | (—) | 3:1 |
| PhOH | (—) | 9:1 |
| *t*-BuOH | (—) | 15:1 |
| Ph$_3$COH | (—) | 17:1 |
| *i*-PrCHOHPr-*i* | (—) | 20:1 |

Table X. REDUCTION OF MISCELLANEOUS FUNCTIONAL GROUPS (*Continued*)

| Substrate | Conditions | Product(s) and Yield(s) (%) | Refs. |
|---|---|---|---|
| 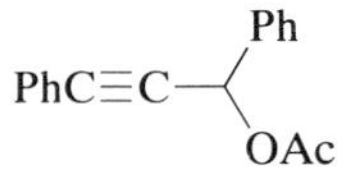 PhC≡C—CH(Ph)OAc | 2.5 SmI$_2$, Pd(PPh$_3$)$_4$ (2 mol%), THF, *i*-PrOH, rt, 0.5 h | 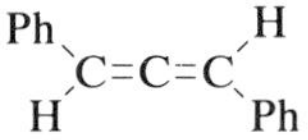 Ph\C=C=C/H\Ph with H (88)  + PhC≡C—CH$_2$Ph (5) | 89 |
| C$_{16}$   Ph—CH$_2$CH$_2$—C(Me)(OAc)—C≡CBu-*n* | 2.5 SmI$_2$, Pd(PPh$_3$)$_4$ (5 mol%), THF, *i*-PrCHOHPr-*i*, 40°, 2 h | 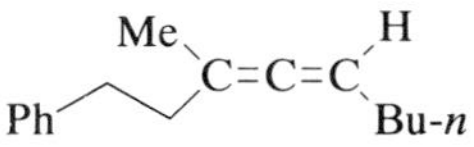 Me\C=C=C/H\Bu-*n* with Ph (96) | 89 |
| C$_{21}$  *n*-C$_{18}$H$_{37}$C≡CCH$_2$OAc | 2.5 SmI$_2$, Pd(PPh$_3$)$_4$ (5 mol%), THF, *i*-PrOH, 65°, 1 h | 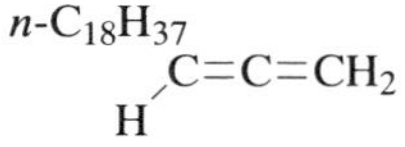 *n*-C$_{18}$H$_{37}$\C=C=CH$_2$ with H  **I** (38) + *n*-C$_{18}$H$_{37}$C≡CMe  **II** (50) | 89 |
|  | 2.5 SmI$_2$, Pd(PPh$_3$)$_4$ (5 mol%), THF, *i*-PrCHOHPr-*i*, 40°, 2 h | **I** (22) + **II** (66) | 89 |
| C$_{26}$ 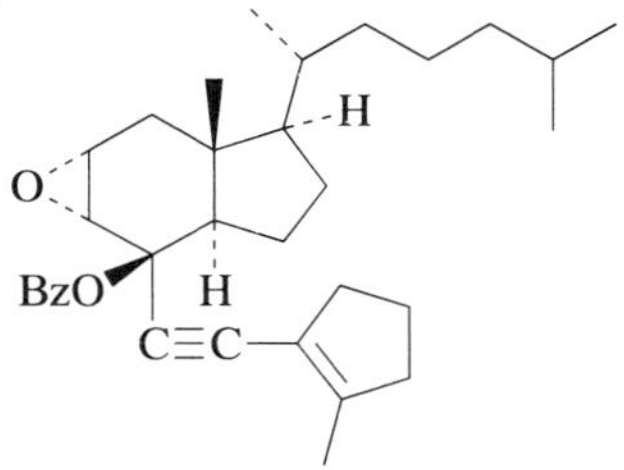 | 5 SmI$_2$, Pd(PPh$_3$)$_4$ (3 mol%), THF, rt, 1 h | 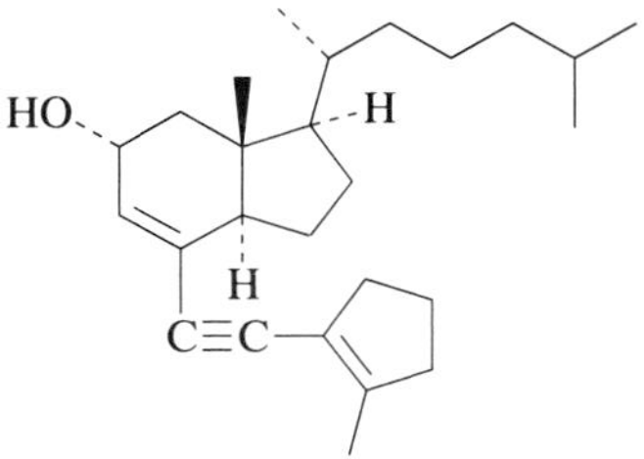 (89) | 90 |

361

Table X. REDUCTION OF MISCELLANEOUS FUNCTIONAL GROUPS (*Continued*)

| Substrate | Conditions | Product(s) and Yield(s) (%) | Refs. |
|---|---|---|---|
| *E. Transition Metal Catalyzed Reduction of Alkynes* | | | |
| $C_8$ | | | |
| PhC≡CTMS | 2 $SmI_2$, $CoCl_2$•4$PPh_3$ (3 mol%), THF, MeOH, rt, 2 h | 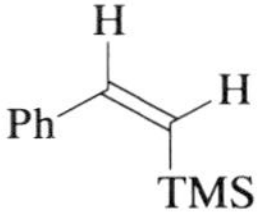 (93) + 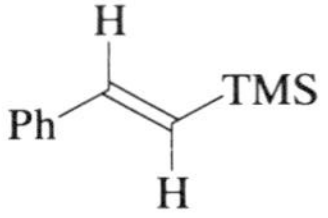 (3)<br>**I** **II** | 91 |
| | 2 $SmI_2$, $CoCl_2$•4$PPh_3$ (3 mol%), THF, *i*-PrOH, rt, 2 h | **I** (6) + **II** (86) | 91 |
| $C_{12}$ | | | |
| PhC≡CBu | 2 $SmI_2$, $CoCl_2$•4$PPh_3$ (3 mol%), THF, MeOH, rt, 3 h | 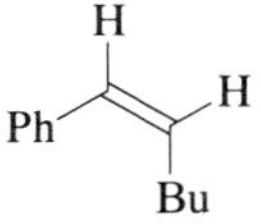 (>97) + 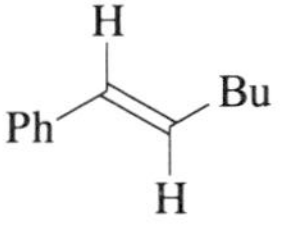 (<1) | 91 |
| $C_5H_{11}C≡CC_5H_{11}$ | 2 $SmI_2$, $CoCl_2$•4$PPh_3$ (3 mol%), THF, MeOH, rt, 5 h | 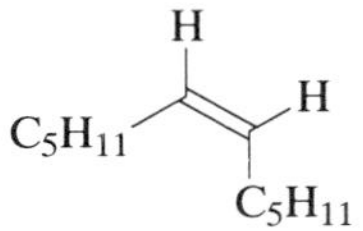 + 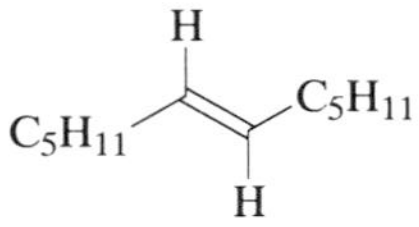<br>**I** (52) **II** (<1) | 91 |
| | 2 $SmI_2$, $CoCl_2$•4$PPh_3$ (3 mol%), THF, AcOH, rt, 5 min | **I** (88) + **II** (<1) | 91 |
| | 2 $SmI_2$, $CoCl_2$•$(H_2O)_x$ (3 mol%), THF, rt, 3 h | [**I** + **II**] (22) + 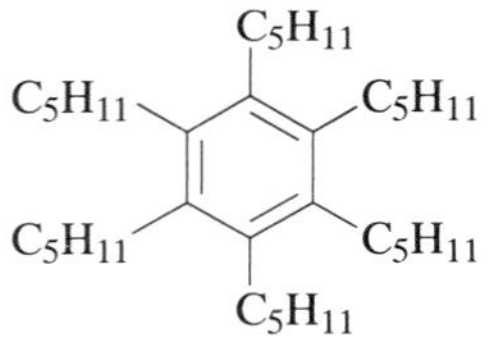 (18) | 91 |

Table X. REDUCTION OF MISCELLANEOUS FUNCTIONAL GROUPS (Continued)

| Substrate | Conditions | Product(s) and Yield(s) (%) | Refs. |
|---|---|---|---|
| C$_{13}$  BuC≡C–CH(OH)–C$_6$H$_{13}$ | 2 SmI$_2$, CoCl$_2$•4PPh$_3$ (3 mol%), THF, MeOH, rt, 15 min | (Bu)(H)C=C(H)–CH(OH)–C$_6$H$_{13}$  **I** (80)  + (Bu)(C$_6$H$_{13}$)C=C(H)–CH(OH)(H)  **II** (<1) | 91 |
| | 2 SmI$_2$, FeCl$_3$•4PPh$_3$ (3 mol%), THF, MeOH, rt, 15 min | [BuCH=C=CHC$_6$H$_{13}$ + BuC≡CC$_7$H$_{15}$] (20)  **I** (95) + **II** (<1) + Bu–CH=CH–CH=CH–C$_5$H$_{11}$ (5) | 91 |
| C$_{14}$  PhC≡CPh | 2 SmI$_2$, THF, i-PrOH, rt, 24 h | No reaction | 91 |
| | 2 SmI$_2$, THF, HMPA, i-PrOH, rt, 16 h | **I** (44) + (Ph)(H)C=C(Ph)(H) **II** (10) + Ph–CH$_2$CH$_2$–Ph **III** (44) | 91 |
| | 2 SmI$_2$, FeCl$_3$•4PPh$_3$ (3 mol%), THF, MeOH, rt, 6 h | **I** (40) + **II** (6) | 91 |

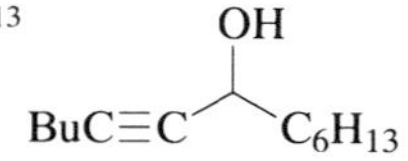

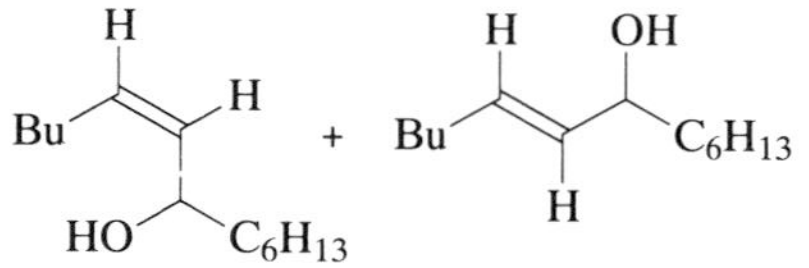

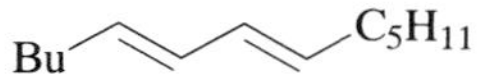

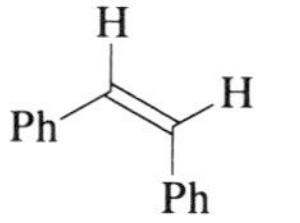

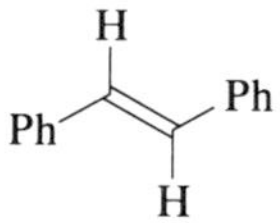

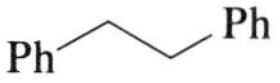

363

Table X. REDUCTION OF MISCELLANEOUS FUNCTIONAL GROUPS (*Continued*)

| Substrate | Conditions | Product(s) and Yield(s) (%) | Refs. |
|---|---|---|---|
| | 2 SmI$_2$, CoCl$_2$•(H$_2$O)$_x$ (3 mol%), THF, rt, 2 h | **I** (65) + **II** (18) + **III** (1) | 91 |
| | 2 SmI$_2$, CoCl$_2$•4PPh$_3$ (3 mol%), THF, *i*-PrOH, rt, 2 h | **I** (59) + **II** (1) | 91 |
| | 2 SmI$_2$, CoCl$_2$•4PPh$_3$ (3 mol%), THF, EtOH, rt, 1 h | **I** (77) + **II** (1) | 91 |
| | 2 SmI$_2$, CoCl$_2$•4PPh$_3$ (3 mol%), THF, MeOH, rt, 0.5 h | **I** (95) + **II** (4) | 91 |
| | 2 SmI$_2$, CoCl$_2$•4PPh$_3$ (3 mol%), THF, AcOH, rt, 0.5 h | **I** (>99) + **II** (<1) | 91 |
| | 2 SmI$_2$, CoCl$_2$•4PPh$_3$ (3 mol%), THF, *t*-BuOH, rt, 8 h | **I** (54) + **II** (2) | 91 |
| | 2 SmI$_2$, CoCl$_2$•4PPh$_3$ (3 mol%), THF, HMPA, *i*-PrOH, rt, 0.5 h | **I** (12) + **II** (70) + **III** (5) | 91 |
| | 2 SmI$_2$, NiCl$_2$•(H$_2$O)$_x$ (3 mol%), THF, rt, 2 h | **I** (59) + **II** (15) + **III** (6) | 91 |
| | 2 SmI$_2$, NiCl$_2$•4PPh$_3$ (3 mol%), THF, *i*-PrOH, rt, 2 h | **I** (62) + **II** (6) | 91 |
| | 2 SmI$_2$, NiCl$_2$•4PPh$_3$ (3 mol%), THF, HMPA, *i*-PrOH, rt, 2 h | **I** (1) + **II** (80) + **III** (8) | 91 |

## REFERENCES

[1] Jantsch, G.; Skalla, N.; Jawurek, L. *Z. Anorg. Chem.* **1931**, *201*, 207.

[2] Jantsch, G.; Skalla, N. *Z. Anorg. Chem.* **1930**, *193*, 391.

[3] Girard, P.; Namy, J. L.; Kagan, H. B. *J. Am. Chem. Soc.* **1980**, *102*, 2693.

[4] Namy, J. L.; Girard, P.; Kagan, H. B. *Nouv. J. Chem.* **1977**, *1*, 5.

[5] Namy, J. L.; Girard, P.; Kagan, H. B. *Nouv. J. Chem.* **1981**, *5*, 479.

[6] Molander, G. A.; Kenny, C. *J. Org. Chem.* **1991**, *56*, 1439.

[7] Imamoto, T.; Ono, M. *Chem. Lett.* **1987**, 501.

[8] Akane, N.; Kanagawa, Y.; Nishiyama, Y.; Ishii, Y. *Chem. Lett.* **1992**, 2431.

[9] Johnson, D. A. *J. Chem. Soc., Dalton* **1974**, 1671.

[10] Natale, N. R. *Org. Prep. Proced. Int.* **1983**, *15*, 387.

[11] Kagan, H. B.; Namy, J. L. In *Handbook on the Physics and Chemistry of the Rare Earths;* Gschneidner, K. A., Eyring, L., Eds.; Elsevier: Amsterdam, 1984; p. 525.

[12] Kagan, H. B. In *Fundamental and Technological Aspects of Organo f-Element Chemistry;* Marks, T. J., Fragalà, I. L., Eds.; Reidel: Dordrecht, 1985; p. 49.

[13] Kagan, H. B.; Namy, J. L. *Tetrahedron* **1986**, *42*, 6573.

[14] Long, J. R. In *Handbook on the Physics and Chemistry of the Rare Earths;* Gschneidner, K. A., Eyring, L., Eds.; Elsevier: Amsterdam, 1986; p. 335.

[15] Kagan, H. B. *Inorg. Chim. Acta* **1987**, *140*, 3.

[16] Kagan, H. B.; Sasaki, M.; Collin, J. *Pure Appl. Chem.* **1988**, *60*, 1725.

[17] Kagan, H. B.; Collin, J. In *Proceedings of the NATO Advanced Research Workshop on Paramagnetic Organometallic Species in Activation/Selectivity, Catalysis;* Chanon, M., Julliard, M., Poite, J. C., Eds.; Kluwer, Academic Publishers: Dordrecht, 1989; p. 131.

[18] Molander, G. A. In *The Chemistry of the Metal–Carbon Bond;* Hartley, F. R., Ed.; John Wiley & Sons: Chichester, 1989; Vol. 5, Chapter 8.

[19] Kagan, H. B. *New J. Chem.* **1990**, *14*, 453.

[20] Soderquist, J. A. *Aldrichimica Acta* **1991**, *24*, 15.

[21] Molander, G. A. In *Comprehensive Organic Synthesis*, B. M. Trost, Ed.; Pergamon Press: Oxford, 1991; Vol. 1, Chapter 1.9, p. 251.

[22] Molander, G. A. *Chem. Rev.* **1992**, *92*, 29.

[23] Kunishima, M.; Hioki, K.; Ohara, T.; Tani, S. *J. Chem. Soc., Chem. Commun.* **1992**, 219.

[24] Inanaga, J.; Ishikawa, M.; Yamaguchi, M. *Chem. Lett.* **1987**, 1485.

[25] Natale, N. R. *Tetrahedron Lett.* **1982**, *23*, 5009.

[26] Kagan, H. B.; Namy, J. L.; Girard, P. *Tetrahedron* Supplement No. 1 **1981**, *37*, 175.

[27] Honda, T.; Naito, K.; Yamane, S.; Suzuki, Y. *J. Chem. Soc., Chem. Commun.* **1992**, 1218.

[28] Beerli, R.; Brunner, E. J.; Borschberg, H. -J. *Tetrahedron Lett.* **1992**, *33*, 6449.

[29] Suri, S. C.; Hardcastle, K. I. *J. Org. Chem.* **1992**, *57*, 6357.

[30] Tabuchi, T.; Inanaga, J.; Yamaguchi, M. *Tetrahedron Lett.* **1986**, *27*, 3891.

[31] de Pouilly, P.; Vauzeilles, B.; Mallet, J. -M.; Sinaÿ, P. *C. R. Acad. Sci. Paris* **1991**, *313*, 1391.

[32] Belloch, J.; Virgili, M.; Moyano, A.; Pericàs, M. A.; Riera, A. *Tetrahedron Lett.* **1991**, *32*, 4579.

[33] Künzer, H.; Stahnke, M.; Sauer, G.; Wiechert, R. *Tetrahedron Lett.* **1991**, *32*, 1949.

[34] Crombie, L.; Rainbow, L. J. *Tetrahedron Lett.* **1988**, *29*, 6517.

[35] Evans, D. A.; Kaldor, S. W.; Jones, T. K.; Clardy, J.; Stout, T. J. *J. Am. Chem. Soc.* **1990**, *112*, 7001.

[36] Ananthanarayan, T. P.; Gallagher, T.; Magnus, P. *J. Chem. Soc., Chem. Commun.* **1982**, 709.

[37] Kende, A. S.; Mendoza, J. S. *Tetrahedron Lett.* **1990**, *31*, 7105.

[38] Molander, G. A.; Etter, J. B.; Zinke, P. W. *J. Am. Chem. Soc.* **1987**, *109*, 453.

[39] Berks, A. H. Ph.D. Thesis, University of Colorado at Boulder, 1988.

[40] Yamamoto, Y.; Matsuoka, K.; Nemoto, H. *J. Am. Chem. Soc.* **1988**, *110*, 4474.

[41] Inanaga, J. *Rev. Heteroatom Chem.* **1990**, *3*, 75.

[42] Kamochi, Y.; Kudo, T. *Tetrahedron* Lett. **1991**, *32*, 3511.

[43] Takeuchi, S.; Ohgo, Y. *Chem. Lett.* **1988**, 403.

[44] Fukuzawa, S.; Iida, M.; Nakanishi, A.; Fujinami, T.; Sakai, S. *J. Chem. Soc., Chem. Commun.* **1987**, 920.

[45] Singh, A. K.; Bakshi, R. K.; Corey, E. J. *J. Am. Chem. Soc.* **1987**, *109*, 6187.

[46] Kamochi, Y.; Kudo, T. *Tetrahedron* **1992**, *48*, 4301.

[47] Kamochi, Y.; Kudo, T. *Chem. Lett.* **1991**, 893

[48] Souppe, J.; Namy, J. -L.; Kagan, H. B. *Tetrahedron Lett.* **1984**, *25*, 2869.

[49] Zhang, Y.; Lin, R. *Synth. Commun.* **1987**, *17*, 329.

[50] Cabrera, A.; Alper, H. *Tetrahedron Lett.* **1992**, *33*, 5007.

[51] Inanaga, J.; Sakai, S.; Handa, Y.; Yamaguchi, M.; Yokoyama, Y. *Chem. Lett.* **1991**, 2117.

[52] Bernardi, A.; Carugo, O.; Pasquarello, A.; Sidjimov, A.; Poli, G. Tetrahedron **1991**, *47*, 7357.

[53] Molander, G. A.; Hahn, G. *J. Org. Chem.* **1986**, *51*, 1135.

[54] Kusuda, K.; Inanaga, J.; Yamaguchi, M. *Tetrahedron Lett.* **1989**, *30*, 2945.

[55] Hanessian, S.; Girard, C.; Chiara, J. L. *Tetrahedron Lett.* **1992**, *33*, 573.

[56] Kagan, H. B.; Namy, J. L. *Tetrahedron* **1986**, *42*, 6573.

[57] White, J. D.; Somers, T. C. *J. Am. Chem. Soc.* **1987**, *109*, 4424.

[58] Guindon, Y.; Simoneau, B.; Yoakim, C.; Gorys, V.; Lemieux, R.; Ogilvie, W. *Tetrahedron Lett.* **1991**, *32*, 5453.

[59] Smith, A. B., III; Dunlap, N. K.; Sulikowski, G. A. *Tetrahedron Lett.* **1988**, *29*, 439.

[60] Castro, J.; Sörensen, H.; Riera, A.; Morin, C.; Moyano, A.; Pericàs, M. A.; Greene, A. E. *J. Am. Chem. Soc.* **1990**, *112*, 9388.

[61] Hwang, J. -T.; Liao, C. -C. *Tetrahedron Lett.* **1991**, *32*, 6583.

[62] Evans, J. M.; Kallmerten, J. *Synlett* **1992**, 269.

[63] Pratt, D. V.; Hopkins, P. B. *J. Org. Chem.* **1988**, *53*, 5885.

[64] Pratt, D. V.; Hopkins, P. B. *Tetrahedron Lett.* **1987**, *28*, 3065.

[65] Inanaga, J.; Katsuki, J.; Yamaguchi, M. *Chem. Lett.* **1991**, 1025.

[66] White, J. D.; Nolen, E. G., Jr.; Miller, C. H. *J. Org. Chem.* **1986**, *51*, 1150.

[67] Otsubo, K.; Inanaga, J.; Yamaguchi, M. *Tetrahedron Lett.* **1987**, *28*, 4437.

[68] Reetz, M. T.; Lauterbach, E. H. *Tetrahedron Lett.* **1991**, *32*, 4477.

[69] Molander, G. A.; Hahn, G. *J. Org. Chem.* **1986**, *51*, 2596.

[70] Molander, G. A.; LaBelle, B. E.; Hahn, G. *J. Org. Chem.* **1986**, *51*, 5259.

[71] Yoneda, R.; Harusawa, S.; Kurihara, T. *Tetrahedron Lett.* **1989**, *30*, 3681.

[72] Yoneda, R.; Harusawa, S.; Kurihara, T. *J. Org. Chem.* **1991**, *56*, 1827.

[73] Molander, G. A.; McKie, J. A. *J. Org. Chem.* **1991**, *56*, 4112.

[74] Batey, R. A.; Motherwell, W. B. *Tetrahedron Lett.* **1991**, *32*, 6649.

[75] Matsukawa, M.; Tabuchi, T.; Inanaga, J.; Yamaguchi, M. *Chem. Lett.* **1987**, 2101.

[76] Handa, Y.; Inanaga, J.; Yamaguchi, M. *J. Chem. Soc., Chem. Commun.* **1989**, 298.

[77] Nishino, K.; Takagi, M.; Kawata, T.; Murata, I.; Inanaga, J.; Nakasuji, K. *J. Am. Chem. Soc.* **1991**, *113*, 5059.

[78] Liu, Y. -S.; Bei, M. -Z.; Zhou, Z. -H.; Takaki, K.; Fujiwara, Y. *Chem. Lett.* **1992**, 1143.

[79] Kende, A. S.; Mendoza, J. S. *Tetrahedron Lett.* **1991**, *32*, 1699.

[80] Souppe, J.; Danon, L.; Namy, J. L.; Kagan, H. B. *J. Organomet. Chem.* **1983**, *250*, 227.

[81] Imamoto, T.; Nishimura, S. *Chem. Lett.* **1990**, 1141.

[82] Mukaiyama, T.; Yorozu, K.; Kato, K.; Yamada, T. *Chem. Lett.* **1992**, 181.

[83] Kotecha, N. R.; Ley, S. V.; Mantegani, S. *Synlett* **1992**, 395.

[84] Burk, M. J.; Feaster, J. E. *J. Am. Chem. Soc.* **1992**, *114*, 6266.

[85] Johnson, C. R.; Senanayake, C. H. *J. Org. Chem.* **1989**, *54*, 735.

[86] Posner, G. H.; Oh, C. H. *J. Am. Chem. Soc.* **1992**, *114*, 8328.

[87] Sasaki, M.; Collin, J.; Kagan, H. B. *Tetrahedron Lett.* **1991**, *32*, 2493.

[88] Tabuchi, T.; Inanaga, -J.; Yamaguchi, M. *Tetrahedron Lett.* **1986**, *27*, 601.

[89] Tabuchi, T.; Inanaga, J.; Yamaguchi, M. *Tetrahedron Lett.* **1986**, *27*, 5237.

[90] Enas, J. D.; Shen, G. -Y.; Okamura, W. H. *J. Am. Chem. Soc.* **1991**, *113*, 3873.

[91] Inanaga, J.; Yokoyama, Y.; Baba, Y.; Yamaguchi, M. *Tetrahedron Lett.* **1991**, *32*, 5559.

[92] Walborsky, H. M.; Topolsky, M. *J. Org. Chem.* **1992**, *57*, 370.

[93] Curran, D. P.; Totleben, M. J. *J. Am. Chem. Soc.* **1992**, *114*, 6050.

[94] Curran, D. P.; Fevig, T. L.; Jasperse, C. P.; Totleben, M. J. *Synlett* **1992**, 942.

[95] Curran, D. P.; Fevig, T. L.; Totleben, M. J. *Synlett* **1990**, 773.

[96] Andriewa, C. P.; Gallardo, I.; Savéant, J. M. *J. Am. Chem. Soc.* **1989**, 111, 1620.

[97] Fevig, T. L.; Elliott, R. L.; Curran, D. P. *J. Am. Chem. Soc.* **1988**, *110*, 2565.

[98] Murakami, M.; Hayashi, M.; Ito, Y. *J. Org. Chem.* **1992**, *57*, 793.

[99] Kocienski, P. J.; Lythgoe, B.; Waterhouse, I. *J. Chem. Soc., Perkin Trans. 1* **1980**, 1045.

[100] Huffman, J. W. *Acc. Chem. Res.* **1983**, *16*, 399.

[101] Rautenstrauch, V.; Willhalm, B.; Thommen, W.; Burger, U. *Helv. Chim. Acta* **1981**, *64*, 2109.

[102] Rautenstrauch, V. *J. Chem. Soc., Chem. Commun.* **1986**, 1558.

[103] Wu, Y. -D.; Houk, K. N. *J. Am. Chem. Soc.* **1992**, *114*, 1656.

[104] House, H. O., *Modern Synthetic Reactions,* 2nd ed.; W. A. Benjamin: Menlo Park, CA, 1972.

[105] Caine, D. *Org. React.* **1976**, *23*, 1.

[106] Shono, T.; Masuda, H.; Murase, H.; Shimomura, M.; Kashimura, S. *J. Org. Chem.* **1992**, *57*, 1061.

[107] Bedenbaugh, A. O.; Bedenbaugh, J. H.; Bergin, W. A.; Adkins, J. D. *J. Am. Chem. Soc.* **1970**, *92*, 5774.

[108] Ouertani, M.; Collin, J.; Kagan, H. B. *Tetrahedron* **1985**, *41*, 3689.

[109] Kochi, J. K.; Singleton, D. M.; Andrews, L. J. *Tetrahedron* **1968**, *24*, 3503.

[110] Bertini, F.; Grasselli, P.; Zubiani, G. *J. Chem. Soc., Chem. Commun.* **1970**, 144.

[111] Kupchan, S. M.; Maruyama, M. *J. Org. Chem.* **1971**, *36*, 1187.

[112] McMurry, J. E. *J. Org. Chem.* **1975**, *40*, 2555.

[113] Gurudutt, K. N.; Ravindranth, B. *Tetrahedron Lett.* **1980**, *21*, 1173.

[114] Molander, G. A.; Harring, L. S. *J. Org. Chem.* **1989**, *54*, 3114.

[115] Collin, J.; Namy, J. L.; Bied, C.; Kagan, H. B. *Inorg. Chim. Acta* **1987**, *140*, 29.

[116] Deshayes, H.; Pete, J. -P. *J. Chem. Soc., Chem. Commun.* **1978**, 567.

[117] Barrett, A. G. M.; Godfrey, C. R. A.; Hollinshead, D. M.; Prokopiou, P. A.; Barton, D. H. R.; Boar, R. G.; Joukhadar, L.; McGhie, J. F.; Misra, S. C. *J. Chem. Soc., Perkin 1* **1981**, 1501.

[118] Larock, R. C. *Comprehensive Organic Transformations;* VCH: New York, 1989.

[119] McMurry, J. E.; Silvestri, M. G.; Fleming, M. P.; Hoz, T.; Grayston, M. W. *J. Org. Chem.* **1978**, *43*, 3249.

[120] Huheey, J. E. *Inorganic Chemistry;* Harper and Row: New York, 1983; Chapter 9.

[121] Iwase, A.; Araki, Y.; Takahashi, R. *Electrochim. Acta* **1990**, *35*, 1713.

[122] Gilbert, B.; Demarteau, V.; Duyckaerts, G. *J. Electroanal. Chem.* **1978**, *89*, 123.

[123] Nugent, J. L.; Baybarz, R. D.; Burnett, J. L. *J. Phys. Chem.* **1973**, *77*, 1528.

[124] Massaux, J.; Duyckaerts, G. *Bull. Soc. Chim. Belg.* **1975**, *84*, 6.

[125] Molander, G. A.; McKie, J. A. *J. Org. Chem.* **1992**, *57*, 3132.

[126] Hou, Z.; Kobayashi, K.; Yamazaki, H. *Chem. Lett.* **1991**, 265.

[127] Donoghue, J. T.; Fernandez, E.; McMillan, J. A.; Peter, D. A. *J. Inorg. Nucl. Chem.* **1969**, *31*, 1431.

[128] Kirshnamurty, V. N.; Soundaravajan, S. *J. Inorg. Nucl. Chem.* **1967**, *29*, 517.

[129] Moeller, T.; Vincenti, G. *J. Inorg. Nucl. Chem.* **1965**, *27*, 1477.

# CUMULATIVE CHAPTER TITLES
# BY VOLUME

*Volume 1 (1942)*

1.  **The Reformatsky Reaction**:   Ralph L. Shriner

2.  **The Arndt-Eistert Reaction**:   W. E. Bachmann and W. S. Struve

3.  **Chloromethylation of Aromatic Compounds**:   Reynold C. Fuson and C. H. McKeever

4.  **The Amination of Heterocyclic Bases by Alkali Amides**:   Marlin T. Leffler

5.  **The Bucherer Reaction**:   Nathan L. Drake

6.  **The Elbs Reaction**:   Louis F. Fieser

7.  **The Clemmensen Reduction**:   Elmore L. Martin

8.  **The Perkin Reaction and Related Reactions**:   John R. Johnson

9.  **The Acetoacetic Ester Condensation and Certain Related Reactions**: Charles R. Hauser and Boyd E. Hudson, Jr.

10.  **The Mannich Reaction**:   F. F. Blicke

11.  **The Fries Reaction**:   A. H. Blatt

12.  **The Jacobsen Reaction**:   Lee Irvin Smith

*Volume 2 (1944)*

1.  **The Claisen Rearrangement**:   D. Stanley Tarbell

2.  **The Preparation of Aliphatic Fluorine Compounds**:   Albert L. Henne

3.  **The Cannizzaro Reaction**:   T. A. Geissman

4.  **The Formation of Cyclic Ketones by Intramolecular Acylation**:   William S. Johnson

5.  **Reduction with Aluminum Alkoxides (The Meerwein-Ponndorf-Verley Reduction)**:   A. L. Wilds

6. **The Preparation of Unsymmetrical Biaryls by the Diazo Reaction and the Nitrosoacetylamine Reaction**:  Werner E. Bachmann and Roger A. Hoffman

7. **Replacement of the Aromatic Primary Amino Group by Hydrogen**: Nathan Kornblum

8. **Periodic Acid Oxidation**:  Ernest L. Jackson

9. **The Resolution of Alcohols**:  A. W. Ingersoll

10. **The Preparation of Aromatic Arsonic and Arsinic Acids by the Bart, Béchamp, and Rosenmund Reactions**:  Cliff S. Hamilton and Jack F. Morgan

*Volume 3 (1946)*

1. **The Alkylation of Aromatic Compounds by the Friedel-Crafts Method**: Charles C. Price

2. **The Willgerodt Reaction**:  Marvin Carmack and M. A. Spielman

3. **Preparation of Ketenes and Ketene Dimers**:  W. E. Hanford and John C. Sauer

4. **Direct Sulfonation of Aromatic Hydrocarbons and Their Halogen Derivatives**: C. M. Suter and Arthur W. Weston

5. **Azlactones**:  H. E. Carter

6. **Substitution and Addition Reactions of Thiocyanogen**:  John L. Wood

7. **The Hofmann Reaction**:  Everett S. Wallis and John F. Lane

8. **The Schmidt Reaction**:  Hans Wolff

9. **The Curtius Reaction**:  Peter A. S. Smith

*Volume 4 (1948)*

1. **The Diels-Alder Reaction with Maleic Anhydride**:  Milton C. Kloetzel

2. **The Diels-Alder Reaction: Ethylenic and Acetylenic Dienophiles**:  H. L. Holmes

3. **The Preparation of Amines by Reductive Alkylation**:  William S. Emerson

4. **The Acyloins**:  S. M. McElvain

5. **The Synthesis of Benzoins**:  Walter S. Ide and Johannes S. Buck

6. **Synthesis of Benzoquinones by Oxidation**:  James Cason

7. **The Rosenmund Reduction of Acid Chlorides to Aldehydes**:  Erich Mosettig and Ralph Mozingo

8. **The Wolff-Kishner Reduction**:  David Todd

*Volume 5 (1949)*

1. **The Synthesis of Acetylenes**:  Thomas L. Jacobs

2. **Cyanoethylation**:  Herman A. Bruson

3. **The Diels-Alder Reaction: Quinones and Other Cyclenones**:  Lewis W. Butz and Anton W. Rytina

4. **Preparation of Aromatic Fluorine Compounds from Diazonium Fluoborates: The Schiemann Reaction**:  Arthur Roe

5. **The Friedel and Crafts Reaction with Aliphatic Dibasic Acid Anhydrides**: Ernst Berliner

6. **The Gattermann-Koch Reaction**:  Nathan N. Crounse

7. **The Leuckart Reaction**:  Maurice L. Moore

8. **Selenium Dioxide Oxidation**:  Norman Rabjohn

9. **The Hoesch Synthesis**:  Paul E. Spoerri and Adrien S. DuBois

10. **The Darzens Glycidic Ester Condensation**:  Melvin S. Newman and Barney J. Magerlein


*Volume 6 (1951)*

1. **The Stobbe Condensation**:  William S. Johnson and Guido H. Daub

2. **The Preparation of 3,4-Dihydroisoquinolines and Related Compounds by the Bischler-Napieralski Reaction**:  Wilson M. Whaley and Tuticorin R. Govindachari

3. **The Pictet-Spengler Synthesis of Tetrahydroisoquinolines and Related Compounds**:  Wilson M. Whaley and Tuticorin R. Govindachari

4. **The Synthesis of Isoquinolines by the Pomeranz-Fritsch Reaction**: Walter J. Gensler

5. **The Oppenauer Oxidation**:  Carl Djerassi

6. **The Synthesis of Phosphonic and Phosphinic Acids**:  Gennady M. Kosolapoff

7. **The Halogen-Metal Interconversion Reaction with Organolithium Compounds**: Reuben G. Jones and Henry Gilman

8. **The Preparation of Thiazoles**:  Richard H. Wiley, D. C. England, and Lyell C. Behr

9. **The Preparation of Thiophenes and Tetrahydrothiophenes**:  Donald E. Wolf and Karl Folkers

10. **Reductions by Lithium Aluminum Hydride**:  Weldon G. Brown

*Volume 7 (1953)*

1.  **The Pechmann Reaction**:　Suresh Sethna and Ragini Phadke

2.  **The Skraup Synthesis of Quinolines**:　R. H. F. Manske and Marshall Kulka

3.  **Carbon-Carbon Alkylations with Amines and Ammonium Salts**:
    James H. Brewster and Ernest L. Eliel

4.  **The von Braun Cyanogen Bromide Reaction**:　Howard A. Hageman

5.  **Hydrogenolysis of Benzyl Groups Attached to Oxygen, Nitrogen, or Sulfur**:
    Walter H. Hartung and Robert Simonoff

6.  **The Nitrosation of Aliphatic Carbon Atoms**:　Oscar Touster

7.  **Epoxidation and Hydroxylation of Ethylenic Compounds with Organic
    Peracids**:　Daniel Swern


*Volume 8 (1954)*

1.  **Catalytic Hydrogenation of Esters to Alcohols**:　Homer Adkins

2.  **The Synthesis of Ketones from Acid Halides and Organometallic Compounds of
    Magnesium, Zinc, and Cadmium**:　David A. Shirley

3.  **The Acylation of Ketones to Form β-Diketones or β-Keto Aldehydes**:
    Charles R. Hauser, Frederic W. Swamer, and Joe T. Adams

4.  **The Sommelet Reaction**:　S. J. Angyal

5.  **The Synthesis of Aldehydes from Carboxylic Acids**:　Erich Mosettig

6.  **The Metalation Reaction with Organolithium Compounds**:　Henry Gilman and
    John W. Morton, Jr.

7.  **β-Lactones**:　Harold E. Zaugg

8.  **The Reaction of Diazomethane and Its Derivatives with Aldehydes and
    Ketones**:　C. David Gutsche


*Volume 9 (1957)*

1.  **The Cleavage of Non-enolizable Ketones with Sodium Amide**:　K. E. Hamlin and
    Arthur W. Weston

2.  **The Gattermann Synthesis of Aldehydes**:　William E. Truce

3.  **The Baeyer-Villiger Oxidation of Aldehydes and Ketones**:　C. H. Hassall

4.  **The Alkylation of Esters and Nitriles**:　Arthur C. Cope, H. L. Holmes, and
    Herbert O. House

5.  **The Reaction of Halogens with Silver Salts of Carboxylic Acids**:  C. V. Wilson

6.  **The Synthesis of β-Lactams**:  John C. Sheehan and Elias J. Corey

7.  **The Pschorr Synthesis and Related Diazonium Ring Closure Reactions**:  DeLos F. DeTar


*Volume 10 (1959)*

1.  **The Coupling of Diazonium Salts with Aliphatic Carbon Atoms**:  Stanley M. Parmerter

2.  **The Japp-Klingemann Reaction**:  Robert R. Phillips

3.  **The Michael Reaction**:  Ernst D. Bergmann, David Ginsburg, and Raphael Pappo


*Volume 11 (1960)*

1.  **The Beckmann Rearrangement**:  L. Guy Donaruma and Walter Z. Heldt

2.  **The Demjanov and Tiffeneau-Demjanov Ring Expansions**:  Peter A. S. Smith and Donald R. Baer

3.  **Arylation of Unsaturated Compounds by Diazonium Salts**:  Christian S. Rondestvedt, Jr.

4.  **The Favorskii Rearrangement of Haloketones**:  Andrew S. Kende

5.  **Olefins from Amines: The Hofmann Elimination Reaction and Amine Oxide Pyrolysis**:  Arthur C. Cope and Elmer R. Trumbull


*Volume 12 (1962)*

1.  **Cyclobutane Derivatives from Thermal Cycloaddition Reactions**:  John D. Roberts and Clay M. Sharts

2.  **The Preparation of Olefins by the Pyrolysis of Xanthates. The Chugaev Reaction**:  Harold R. Nace

3.  **The Synthesis of Aliphatic and Alicyclic Nitro Compounds**:  Nathan Kornblum

4.  **Synthesis of Peptides with Mixed Anhydrides**:  Noel F. Albertson

5.  **Desulfurization with Raney Nickel**:  George R. Pettit and Eugene E. van Tamelen


*Volume 13 (1963)*

1.  **Hydration of Olefins, Dienes, and Acetylenes via Hydroboration**:  George Zweifel and Herbert C. Brown

2. **Halocyclopropanes from Halocarbenes**:   William E. Parham and Edward E. Schweizer

3. **Free Radical Addition to Olefins to Form Carbon-Carbon Bonds**:   Cheves Walling and Earl S. Huyser

4. **Formation of Carbon-Heteroatom Bonds by Free Radical Chain Additions to Carbon-Carbon Multiple Bonds**:   F. W. Stacey and J. F. Harris, Jr.


*Volume 14 (1965)*

1. **The Chapman Rearrangement**:   J. W. Schulenberg and S. Archer

2. **α-Amidoalkylations at Carbon**:   Harold E. Zaugg and William B. Martin

3. **The Wittig Reaction**:   Adalbert Maercker


*Volume 15 (1967)*

1. **The Dieckmann Condensation**:   John P. Schaefer and Jordan J. Bloomfield

2. **The Knoevenagel Condensation**:   G. Jones


*Volume 16 (1968)*

1. **The Aldol Condensation**:   Arnold T. Nielsen and William J. Houlihan


*Volume 17 (1969)*

1. **The Synthesis of Substituted Ferrocenes and Other π-Cyclopentadienyl-Transition Metal Compounds**:   Donald E. Bublitz and Kenneth L. Rinehart, Jr.

2. **The γ-Alkylation and γ-Arylation of Dianions of β-Dicarbonyl Compounds**:   Thomas M. Harris and Constance M. Harris

3. **The Ritter Reaction**:   L. I. Krimen and Donald J. Cota


*Volume 18 (1970)*

1. **Preparation of Ketones from the Reaction of Organolithium Reagents with Carboxylic Acids**:   Margaret J. Jorgenson

2. **The Smiles and Related Rearrangements of Aromatic Systems**:   W. E. Truce, Eunice M. Kreider, and William W. Brand

3. **The Reactions of Diazoacetic Esters with Alkenes, Alkynes, Heterocyclic, and Aromatic Compounds**:   Vinod David and E. W. Warnhoff

4. **The Base-Promoted Rearrangements of Quaternary Ammonium Salts**:   Stanley H. Pine

*Volume 19 (1972)*

1. **Conjugate Addition Reactions of Organocopper Reagents**:   Gary H. Posner

2. **Formation of Carbon-Carbon Bonds via π-Allylnickel Compounds**:
Martin F. Semmelhack

3. **The Thiele-Winter Acetoxylation of Quinones**:   J. F. W. McOmie and J. M.
Blatchly

4. **Oxidative Decarboxylation of Acids by Lead Tetraacetate**:   Roger A. Sheldon and
Jay K. Kochi

*Volume 20 (1973)*

1. **Cyclopropanes from Unsaturated Compounds, Methylene Iodide, and Zinc-
Copper Couple**:   H. E. Simmons, T. L. Cairns, Susan A. Vladuchick, and
Connie M. Hoiness

2. **Sensitized Photooxygenation of Olefins**:   R. W. Denny and A. Nickon

3. **The Synthesis of 5-Hydroxyindoles by the Nenitzescu Reaction**:
George R. Allen, Jr.

4. **The Zinin Reduction of Nitroarenes**:   H. K. Porter

*Volume 21 (1974)*

1. **Fluorination with Sulfur Tetrafluoride**:   G. A. Boswell, Jr., W. C. Ripka,
R. M. Scribner, and C. W. Tullock

2. **Modern Methods to Prepare Monofluoroaliphatic Compounds**:
William A. Sheppard

*Volume 22 (1975)*

1. **The Claisen and Cope Rearrangements**:   Sara Jane Rhoads and N. Rebecca Raulins

2. **Substitution Reactions Using Organocopper Reagents**:   Gary H. Posner

3. **Clemmensen Reduction of Ketones in Anhydrous Organic Solvents**:   E. Vedejs

4. **The Reformatsky Reaction**:   Michael W. Rathke

*Volume 23 (1976)*

1. **Reduction and Related Reactions of α,β-Unsaturated Compounds with Metals in
Liquid Ammonia**:   Drury Caine

2. **The Acyloin Condensation**:   Jordan J. Bloomfield, Dennis C. Owsley, and
Janice M. Nelke

3. **Alkenes from Tosylhydrazones**:   Robert H. Shapiro

*Volume 24 (1976)*

1. **Homogeneous Hydrogenation Catalysts in Organic Synthesis:** Arthur J. Birch and David H. Williamson

2. **Ester Cleavages via S$_N$2-Type Dealkylation:** John E. McMurry

3. **Arylation of Unsaturated Compounds by Diazonium Salts (The Meerwein Arylation Reaction):** Christian S. Rondestvedt, Jr.

4. **Selenium Dioxide Oxidation:** Norman Rabjohn

*Volume 25 (1977)*

1. **The Ramberg-Bäcklund Rearrangement:** Leo A. Paquette

2. **Synthetic Applications of Phosphoryl-Stabilized Anions:** William S. Wadsworth, Jr.

3. **Hydrocyanation of Conjugated Carbonyl Compounds:** Wataru Nagata and Mitsuru Yoshioka

*Volume 26 (1979)*

1. **Heteroatom-Facilitated Lithiations:** Heinz W. Gschwend and Herman R. Rodriguez

2. **Intramolecular Reactions of Diazocarbonyl Compounds:** Steven D. Burke and Paul A. Grieco

*Volume 27 (1982)*

1. **Allylic and Benzylic Carbanions Substituted by Heteroatoms:** Jean-François Biellmann and Jean-Bernard Ducep

2. **Palladium-Catalyzed Vinylation of Organic Halides:** Richard F. Heck

*Volume 28 (1982)*

1. **The Reimer-Tiemann Reaction:** Hans Wynberg and Egbert W. Meijer

2. **The Friedländer Synthesis of Quinolines:** Chia-Chung Cheng and Shou-Jen Yan

3. **The Directed Aldol Reaction:** Teruaki Mukaiyama

*Volume 29 (1983)*

1. **Replacement of Alcoholic Hydroxy Groups by Halogens and Other Nucleophiles via Oxyphosphonium Intermediates:** Bertrand R. Castro

2. **Reductive Dehalogenation of Polyhalo Ketones with Low-Valent Metals and Related Reducing Agents**: Ryoji Noyori and Yoshihiro Hayakawa

3. **Base-Promoted Isomerizations of Epoxides**: Jack K. Crandall and Marcel Apparu

*Volume 30 (1984)*

1. **Photocyclization of Stilbenes and Related Molecules**: Frank B. Mallory and Clelia W. Mallory

2. **Olefin Synthesis via Deoxygenation of Vicinal Diols**: Eric Block

*Volume 31 (1984)*

1. **Addition and Substitution Reactions of Nitrile-Stabilized Carbanions**: Simeon Arseniyadis, Keith S. Kyler, and David S. Watt

*Volume 32 (1984)*

1. **The Intramolecular Diels-Alder Reaction**: Engelbert Ciganek

2. **Synthesis Using Alkyne-Derived Alkenyl- and Alkynylaluminum Compounds**: George Zweifel and Joseph A. Miller

*Volume 33 (1985)*

1. **Formation of Carbon-Carbon and Carbon-Heteroatom Bonds via Organoboranes and Organoborates**: Ei-Ichi Negishi and Michael J. Idacavage

2. **The Vinylcyclopropane-Cyclopentene Rearrangement**: Tomáš Hudlický, Toni M. Kutchan, and Saiyid M. Naqvi

*Volume 34 (1985)*

1. **Reductions by Metal Alkoxyaluminum Hydrides**: Jaroslav Málek

2. **Fluorination by Sulfur Tetrafluoride**: Chia-Lin J. Wang

*Volume 35 (1988)*

1. **The Beckmann Reactions: Rearrangements, Elimination-Additions, Fragmentations, and Rearrangement-Cyclizations**: Robert E. Gawley

2. **The Persulfate Oxidation of Phenols and Arylamines (The Elbs and the Boyland-Sims Oxidations)**: E. J. Behrman

3. **Fluorination with Diethylaminosulfur Trifluoride and Related Aminofluorosulfuranes**: Miloš Hudlický

*Volume 36 (1988)*

1.  **The [3 + 2] Nitrone-Olefin Cycloaddition Reaction**:   Pat N. Confalone and Edward M. Huie

2.  **Phosphorus Addition at $sp^2$ Carbon**:   Robert Engel

3.  **Reduction by Metal Alkoxyaluminum Hydrides. Part II. Carboxylic Acids and Derivatives, Nitrogen Compounds, and Sulfur Compounds**:   Jaroslav Málek

*Volume 37 (1989)*

1.  **Chiral Synthons by Ester Hydrolysis Catalyzed by Pig Liver Esterase**:   Masaji Ohno and Masami Otsuka

2.  **The Electrophilic Substitution of Allylsilanes and Vinylsilanes**:   Ian Fleming, Jacques Dunoguès, and Roger Smithers

*Volume 38 (1990)*

1.  **The Peterson Olefination Reaction**:   David J. Ager

2.  **Tandem Vicinal Difunctionalization: $\beta$-Addition to $\alpha,\beta$-Unsaturated Carbonyl Substrates Followed by $\alpha$-Functionalization**:   Marc J. Chapdelaine and Martin Hulce

3.  **The Nef Reaction**:   Harold W. Pinnick

*Volume 39 (1990)*

1.  **Lithioalkenes from Arenesulfonylhydrazones**:   A. Richard Chamberlin and Steven H. Bloom

2.  **The Polonovski Reaction**:   David Grierson

3.  **Oxidation of Alcohols to Carbonyl Compounds via Alkoxysulfonium Ylides: The Moffatt, Swern, and Related Oxidations**:   Thomas T. Tidwell

*Volume 40 (1991)*

1.  **The Pauson-Khand Cycloaddition Reaction for Synthesis of Cyclopentenones**: Neil E. Schore

2.  **Reduction with Diimide**:   Daniel J. Pasto and Richard T. Taylor

3.  **The Pummerer Reaction of Sulfinyl Compounds**:   Ottorino DeLucchi, Umberto Miotti, and Giorgio Modena

4.  **The Catalyzed Nucleophilic Addition of Aldehydes to Electrophilic Double Bonds**:   Hermann Stetter and Heinrich Kuhlmann

*Volume 41 (1992)*

1. **Divinylcyclopropane–Cycloheptadiene Rearrangement**: Thomáš Hudlický, Rulin Fan, Josephine W. Reed, and Kumar G. Gadamasetti

2. **Organocopper Reagents: Substitution, Conjugate Addition, Carbo/Metallocupration, and Other Reactions**: Bruce H. Lipshutz and Saumitra Sengupta

*Volume 42 (1992)*

1. **The Birch Reduction of Aromatic Compounds**: Peter W. Rabideau and Zbigniew Marcinow

2. **The Mitsunobu Reaction**: David L. Hughes

*Volume 43 (1993)*

1. **Carbonyl Methylenation and Alkylidenation Using Titanium-Based Reagents**: Stanley H. Pine

2. **Anion-Assisted Sigmatropic Rearrangements**: Stephen R. Wilson

3. **The Baeyer–Villiger Oxidation of Ketones and Aldehydes**: Grant R. Krow

*Volume 44 (1993)*

1. **Preparation of $\alpha,\beta$-Unsaturated Carbonyl Compounds and Nitriles by Selenoxide Elimination**: Hans J. Reich and Susan Wollowitz

2. **Enone Olefin [2 + 2] Photochemical Cycloadditions**: Michael T. Crimmins and Tracy L. Reinhold

*Volume 45 (1994)*

1. **The Nazarov Cyclization**: Karl L. Habermas, Scott E. Denmark, and Todd K. Jones

2. **Ketene [2 + 2] Cycloadditions**: John Hyatt and Peter W. Raynolds

# AUTHOR INDEX, VOLUMES 1–46

Volume number only is designated in this index.

Adams, Joe T., 8
Adkins, Homer, 8
Ager, David J., 38
Albertson, Noel F., 12
Allen, George R., Jr., 20
Angyal, S. J., 8
Apparu, Marcel, 29
Archer, S., 14
Arseniyadis, Siméon, 31

Bachmann, W. E., 1, 2
Baer, Donald R., 11
Behr, Lyell C., 6
Behrman, E. J., 35
Bergmann, Ernst D., 10
Berliner, Ernst, 5
Biellmann, Jean-François, 27
Birch, Arthur J., 24
Blatchly, J. M., 19
Blatt, A. H., 1
Blicke, F. F., 1
Block, Eric, 30
Bloom, Steven H., 39
Bloomfield, Jordan J., 15, 23
Boswell, G. A., Jr., 21
Brand, William W., 18
Brewster, James H., 7
Brown, Herbert C., 13
Brown, Weldon G., 6
Bruson, Herman Alexander, 5
Bublitz, Donald E., 17
Buck, Johannes S., 4
Burke, Steven D., 26
Butz, Lewis W., 5

Caine, Drury, 23
Cairns, Theodore L., 20
Carmack, Marvin, 3
Carter, H. E., 3
Cason, James, 4
Castro, Bertrand R., 29
Chamberlin, A. Richard, 39

Chapdelaine, Marc J., 38
Cheng, Chia-Chung, 28
Ciganek, Engelbert, 32
Confalone, Pat N., 36
Cope, Arthur C., 9, 11
Corey, Elias J., 9
Cota, Donald J., 17
Crandall, Jack K., 29
Crimmins, Michael T., 44
Crounse, Nathan N., 5

Daub, Guido H., 6
Dave, Vinod, 18
Denmark, Scott E., 45
Denny, R. W., 20
DeLucchi, Ottorino, 40
DeTar, DeLos F., 9
Djerassi, Carl, 6
Donaruma, L. Guy, 11
Drake, Nathan L., 1
DuBois, Adrien S., 5
Ducep, Jean-Bernard, 27
Dunoguès, Jacques, 37

Eliel, Ernest L., 7
Emerson, William S., 4
Engel, Robert, 36
England, D. C., 6

Fan, Rulin, 41
Fieser, Louis F., 1
Fleming, Ian, 37
Folkers, Karl, 6
Fuson, Reynold C., 1

Gadamasetti, Kumar G., 41
Gawley, Robert E., 35
Geissman, T. A., 2
Gensler, Walter J., 6
Gilman, Henry, 6, 8
Ginsburg, David, 10
Govindachari, Tuticorin R., 6

Grieco, Paul A., 26
Grierson, David, 39
Gschwend, Heinz W., 26
Gutsche, C. David, 8

Habermas, Karl L., 45
Hageman, Howard A., 7
Hamilton, Cliff S., 2
Hamlin, K. E., 9
Hanford, W. E., 3
Harris, Constance M., 17
Harris, J. F., Jr., 13
Harris, Thomas M., 17
Hartung, Walter H., 7
Hassal, C. H., 9
Hauser, Charles R., 1, 8
Hayakawa, Yoshihiro, 29
Heck, Richard F., 27
Heldt, Walter Z., 11
Henne, Albert L., 2
Hoffman, Roger A., 2
Hoiness, Connie M., 20
Holmes, H. L., 4, 9
Houlihan, William J., 16
House, Herbert O., 9
Hudlický, Miloŝ, 35
Hudlický, Tomáŝ, 33, 41
Hudson, Boyd E., Jr., 1
Hughes, David L., 42
Huie, E. M., 36
Hulce, Martin, 38
Huyser, Earl S., 13
Hyatt, John A., 45

Idacavage, Michael J., 33
Ide, Walter S., 4
Ingersoll, A. W., 2

Jackson, Ernest L., 2
Jacobs, Thomas L., 5
Johnson, John R., 1
Johnson, William S., 2, 6
Jones, G., 15
Jones, Reuben G., 6
Jones, Todd K., 45
Jorgenson, Margaret J., 18

Kende, Andrew S., 11
Kloetzel, Milton C., 4
Kobayashi, Shū, 46
Kochi, Jay K., 19
Kornblum, Nathan, 2, 12
Kosolapoff, Gennady M., 6
Kreider, Eunice M., 18

Krimen, L. I., 17
Krow, Grant R., 43
Kuhlmann, Heinrich, 40
Kulka, Marshall, 7
Kutchan, Toni M., 33
Kyler, Keith S., 31

Lane, John F., 3
Leffler, Marlin T., 1
Lipshutz, Bruce H., 41

McElvain, S. M., 4
McKeever, C. H., 1
McMurry, John E., 24
McOmie, J. F. W., 19
Maercker, Adalbert, 14
Magerlein, Barney J., 5
Málek, Jaroslav, 34, 36
Mallory, Clelia W., 30
Mallory, Frank B., 30
Manske, Richard H. F., 7
Marcinow, Zbigniew, 42
Martin, Elmore L.., 1
Martin, William B., 14
Meijer, Egbert W., 28
Mikami, Koichi, 46
Miller, Joseph A., 32
Miotti, Umberto, 40
Modena, Giorgio, 40
Molander, Gary, 46
Moore, Maurice L., 5
Morgan, Jack F., 2
Morton, John W., Jr., 8
Mosettig, Erich, 4, 8
Mozingo, Ralph, 4
Mukaiyama, Teruaki, 28, 46

Nace, Harold R., 12
Nagata, Wataru, 25
Nakai, Takeshi, 46
Naqvi, Saiyid M., 33
Negishi, Ei-Ichi, 33
Nelke, Janice M., 23
Newman, Melvin S., 5
Nickon, A., 20
Nielsen, Arnold T., 16
Noyori, Ryoji, 29

Ohno, Masaji, 37
Otsuka, Masami, 37
Owsley, Dennis C., 23

Pappo, Raphael, 10
Paquette, Leo A., 25

Parham, William E., 13
Parmerter, Stanley M., 10
Pasto, Daniel J., 40
Pettit, George R., 12
Phadke, Ragini, 7
Phillips, Robert R., 10
Pine, Stanley H., 18, 43
Pinnick, Harold W., 38
Porter, H. K., 20
Posner, Gary H., 19, 22
Price, Charles C., 3

Rabideau, Peter W., 42
Rabjohn, Norman, 5, 24
Rathke, Michael W., 22
Raulins, N. Rebecca, 22
Raynolds, Peter W., 45
Reed, Josephine W., 41
Reich, Hans J., 44
Reinhold, Tracy L., 44
Rhoads, Sara Jane, 22
Rinehart, Kenneth L., Jr., 17
Ripka, W. C., 21
Roberts, John D., 12
Rodriguez, Herman R., 26
Roe, Arthur, 5
Rondestvedt, Christian S., Jr., 11, 24
Rytina, Anton W., 5

Sauer, John C., 3
Schaefer, John P., 15
Schulenberg, J. W., 14
Schweizer, Edward E., 13
Scribner, R. M., 21
Semmelhack, Martin F., 19
Sengupta, Saumitra, 41
Sethna, Suresh, 7
Shapiro, Robert H., 23
Sharts, Clay M., 12, 21
Sheehan, John C., 9
Sheldon, Roger A., 19
Sheppard, W. A., 21
Shirley, David A., 8
Shore, Neil E., 40
Shriner, Ralph L., 1
Simmons, Howard E., 20
Simonoff, Robert, 7
Smith, Lee Irvin, 1

Smith, Peter A. S., 3, 11
Smithers, Roger, 37
Spielman, M. A., 3
Spoerri, Paul E., 5
Stacey, F. W., 13
Stetter, Hermann, 40
Struve, W. S., 1
Suter, C. M., 3
Swamer, Frederic W., 8
Swern, Daniel, 7

Tarbell, D. Stanley, 2
Taylor, Richard T., 40
Tidwell, Thomas T., 39
Todd, David, 4
Touster, Oscar, 7
Truce, William E., 9, 18
Trumbull, Elmer R., 11
Tullock, C. W., 21

van Tamelen, Eugene E., 12
Vedejs, E., 22
Vladuchick, Susan A., 20

Wadsworth, William S., Jr., 25
Walling, Cheves, 13
Wallis, Everett S., 3
Wang, Chia-Lin L., 34
Warnhoff, E. W., 18
Watt, David S., 31
Weston, Arthur W., 3, 9
Whaley, Wilson M., 6
Wilds, A. L., 2
Wiley, Richard H., 6
Williamson, David H., 24
Wilson, C. V., 9
Wilson, Stephen R., 43
Wolf, Donald E., 6
Wolff, Hans, 3
Wollowitz, Susan, 44
Wood, John L., 3
Wynberg, Hans, 28

Yan, Shou-Jen, 28
Yoshioka, Mitsuru, 25

Zaugg, Harold E., 8, 14
Zweifel, George, 13, 32

# CHAPTER AND TOPIC INDEX, VOLUMES 1–46

Many chapters contain brief discussions of reactions and comparisons of alternative synthetic methods related to the reaction that is the subject of the chapter. These related reactions and alternative methods are not usually listed in this index. In this index, the volume number is in **boldface**, the chapter number is in ordinary type.

Acetic anhydride, reaction with quinones, **19**, 3
Acetoacetic ester condensation, **1**, 9
Acetoxylation of quinones, **20**, 3
Acetylenes, synthesis of, **5**, 1; **23**, 3; **32**, 2
Acid halides:
  reactions with esters, **1**, 9
  reactions with organometallic compounds, **8**, 2
Acids, $\alpha,\beta$-unsaturated, synthesis, with alkenyl- and alkynylaluminum reagents, **32**, 2
Acrylonitrile, addition to (cyanoethylation), **5**, 2
$\alpha$-Acylamino acid mixed anhydrides, **12**, 4
$\alpha$-Acylamino acids, azlactonization of, **3**, 5
$\alpha$-Acylamino carbonyl compounds, preparation of thiazoles, **6**, 8
Acylation:
  of esters with acid chlorides, **1**, 9
  intramolecular, to form cyclic ketones, **2**, 4; **23**, 2
  of ketones to form diketones, **8**, 3
Acyl fluorides, preparation of, **21**, 1; **34**, 2; **35**, 3
Acyl hypohalites, reactions of, **9**, 5
Acyloins, **4**, 4; **15**, 1; **23**, 2
Alcohols:
  activation for displacement reactions, **42**, 335
  conversion to fluorides, **21**, 1; **34**, 2; **35**, 3
  conversion to olefins, **12**, 2
  oxidation of, **6**, 5; **39**, 3
  replacement of hydroxyl group by nucleophiles, **29**, 1
  resolution of, **2**, 9
Alcohols, preparation:
  by base-promoted isomerization of epoxides, **29**, 3
  by hydroboration, **13**, 1
  by hydroxylation of ethylenic compounds, **7**, 7
  from organoboranes, **33**, 1
  by reduction, **6**, 10; **8**, 1

Aldehydes, catalyzed addition to double bonds, **40**, 4
Aldehydes, synthesis of, **4**, 7; **5**, 10; **8**, 4, 5; **9**, 2; **33**, 1
Aldol condensation, **16**
  directed, **28**, 3
Aliphatic and alicyclic nitro compounds, synthesis of, **12**, 3
Aliphatic fluorides, **2**, 2; **21**, 1, 2; **34**, 2; **35**, 3
Alkali amides, in amination of heterocycles, **1**, 4
Alkenes, synthesis:
  with alkenyl- and alkynylaluminum reagents, **32**, 2
  from aryl and vinyl halides, **27**, 2
  from $\alpha$-halosulfones, **25**, 1
  from tosylhydrazones, **23**, 3; **39**, 1
Alkenyl- and alkynylaluminum reagents, **32**, 2
Alkenyllithiums, formation of, **39**, 1
Alkoxyaluminum hydride reductions, **34**, 1
Alkoxyphosphonium cations, nucleophilic displacements on, **29**, 1
Alkylation:
  of allylic and benzylic carbanions, **27**, 1
  with amines and ammonium salts, **7**, 3
  of aromatic compounds, **3**, 1
  of esters and nitriles, **9**, 4
  $\gamma$-, of dianions of $\beta$-dicarbonyl compounds, **17**, 2
  of metallic acetylides, **5**, 1
  of nitrile-stabilized carbonions, **31**
  with organopalladium complexes, **27**, 2
Alkylidenation by titanium-based reagents, **43**, 1
Alkylidenesuccinic acids, preparation and reactions of, **6**, 1
Alkylidene triphenylphosphoranes, preparation and reactions of, **14**, 3
Allenylsilanes, electrophilic substitution reactions of, **37**, 2

Allylic alcohols, synthesis:
  with alkenyl- and alkynylaluminum
    reagents, **32**, 2
  from epoxides, **29**, 3
  by Wittig rearrangement, **46**, 2
Allylic and benzylic carbanions,
    heteroatom-substituted, **27**, 1
Allylic hydroperoxides, in
    photooxygenations, **20**, 2
π-Allylnickel complexes, **19**, 2
Allylphenols, preparation by Claisen
    rearrangement, **2**, 1; **22**, 1
Allylsilanes, electrophilic substitution
    reactions of, **37**, 2
Aluminum alkoxides:
  in Meerwein–Ponndorf–Verley reduction,
    **2**, 5
  in Oppenauer oxidation, **6**, 5
Amide formation by oxime rearrangement,
    **35**, 1
α-Amidoalkylations at carbon, **14**, 2
Amination:
  of heterocyclic bases by alkali amides, **1**, 4
  of hydroxy compounds by Bucherer
    reaction, **1**, 5
Amine oxides:
  Polonovski reaction of, **39**, 2
  pyrolysis of, **11**, 5
Amines:
  preparation from organoboranes, **33**, 1
  preparation by reductive alkylation, **4**, 3;
    **5**, 7
  preparation by Zinin reduction, **20**, 4
  reactions with cyanogen bromide, **7**, 4
Aminophenols from anilines, **35**, 2
Anhydrides of aliphatic dibasic acids,
    Friedel–Crafts reaction with, **5**, 5
Anion-assisted sigmatropic rearrangements,
    **43**, 2
Anthracene homologs, synthesis of, **1**, 6
Anti-Markownikoff hydration of olefins, **13**, 1
π-Arenechromium tricarbonyls, reaction with
    nitrile-stabilized carbanions, **31**
Arndt–Eistert reaction, **1**, 2
Aromatic aldehydes, preparation of, **5**, 6; **28**,
    1
Aromatic compounds, chloromethylation of,
    **1**, 3
Aromatic fluorides, preparation of, **5**, 4
Aromatic hydrocarbons, synthesis of, **1**, 6;
    **30**, 1
Arsinic acids, **2**, 10
Arsonic acids, **2**, 10
Arylacetic acids, synthesis of, **1**, 2; **22**, 4

β-Arylacrylic acids, synthesis of, **1**, 8
Arylamines, preparation and reactions of, **1**, 5
Arylation:
  by aryl halides, **27**, 2
  by diazonium salts, **11**, 3; **24**, 3
  γ-, of dianions of β-dicarbonyl compounds,
    **17**, 2
  of nitrile-stabilized carbanions, **31**
  of olefins, **11**, 3; **24**, 3; **27**, 2
Arylglyoxals, condensation with aromatic
    hydrocarbons, **4**, 5
Arylsulfonic acids, preparation of, **3**, 4
Aryl thiocyanates, **3**, 6
Asymmetric Reactions, **46**, 1; **46**, 2
Azaphenanthrenes, synthesis by
    photocyclization, **30**, 1
Azides, preparation and rearrangement of, **3**,
    9
Azlactones, **3**, 5

Baeyer–Villiger reaction, **9**, 3; **43**, 3
Bamford–Stevens reaction, **23**, 3
Bart reaction, **2**, 10
Béchamp reaction, **2**, 10
Beckmann rearrangement, **11**, 1; **35**, 1
Benzils, reduction of, **4**, 5
Benzoin condensation, **4**, 5
Benzoquinones:
  acetoxylation of, **19**, 3
  in Nenitzescu reaction, **20**, 3
  synthesis of, **4**, 6
Benzylamines, from Sommelet–Hauser
    rearrangement, **18**, 4
Benzylic carbanions, **27**, 1
Biaryls, synthesis of, **2**, 6
Bicyclobutanes, from cyclopropenes, **18**, 3
Birch reaction, **23**, 1; **42**, 1
Bischler–Napieralski reaction, **6**, 2
Bis(chloromethyl) ether, **1**, 3; **19**, *warning*
Boranes, **33**, 1
Boyland–Sims Oxidation, **35**, 2
Bucherer reaction, **1**, 5

Cannizzaro reaction, **2**, 3
Carbanions:
  heteroatom-substituted, **27**, 1
  nitrile-stabilized, **31**
Carbenes, **13**, 2; **26**, 2; **28**, 1
Carbohydrates, deoxy, preparation of, **30**, 2
Carbo/metallocupration, **41**, 2
Carbon alkylations with amines and
    ammonium salts, **7**, 3
Carbon–carbon bond formation:
  by acetoacetic ester condensation, **1**, 9

by acyloin condensation, **23**, 2
by aldol condensation, **16**; **28**, 3
by alkylation with amines and ammonium
    salts, **7**, 3
by γ-alkylation and arylation, **17**, 2
by allylic and benzylic carbanions, **27**, 1
by amidoalkylation, **14**, 2
by Cannizzaro reaction, **2**, 3
by Claisen rearrangement, **2**, 1; **22**, 1
by Cope rearrangement, **22**, 1
by cyclopropanation reaction, **13**, 2; **20**, 1
by Darzens condensations, **5**, 10
by diazonium salt coupling, **10**, 1; **11**, 3;
    **24**, 3
by Dieckmann condensation, **15**, 1
by Diels–Alder reaction, **4**, 1, 2; **5**, 3; **32**, 1
by free radical additions to olefins, **13**, 3
by Friedel–Crafts reaction, **3**, 1; **5**, 5
by Knoevenagel condensation, **15**, 2
by Mannich reaction, **1**, 10; **7**, 3
by Michael addition, **10**, 3
by nitrile-stabilized carbanions, **31**
by organoboranes and organoborates, **33**, 1
by organocopper reagents, **19**, 1; **38**, 2; **41**, 2
by organopalladium complexes, **27**, 2
by organozinc reagents, **20**, 1
by rearrangement of α-halo sulfones, **25**, 1
by Reformatsky reaction, **1**, 1; **28**, 4
by vinylcyclopropane-cyclopentene
    rearrangement, **33**, 2
Carbon–halogen bond formation, by
    replacement of hydroxyl groups, **29**, 1
Carbon–heteroatom bond formation: by
    free radical chain additions to carbon–
    carbon multiple bonds, **13**, 4
by organoboranes and organoborates, **33**, 1
Carbon–phosphorus bond formation, **36**, 2
Carbonyl alkylidenation, **43**, 1
α-Carbonyl carbenes and carbenoids,
    intramolecular additions and insertions
    of, **26**, 2
Carbonyl compounds, α, β-unsaturated:
    formation by selenoxide elimination, **44**, 1
    vicinal difunctionalization, **38**, 2
Carbonyl compounds, from nitro compounds,
    **38**, 3
Carboxylic acid derivatives, conversion to
    fluorides, **21**, 1; **34**, 2; **35**, 3
    reduction of, **36**, 3
Carboxylic acids:
    preparation from organoboranes, **33**, 1
    reaction with organolithium reagents, **18**, 1
    reduction of, **36**, 3
Catalytic homogeneous hydrogenation, **24**, 1

Catalytic hydrogenation of esters to alcohols,
    **8**, 1
Chapman rearrangement, **14**, 1; **18**, 2
Chloromethylation of aromatic compounds,
    **2**, 3; **9**, *warning*
Cholanthrenes, synthesis of, **1**, 6
Chugaev reaction, **12**, 2
Claisen condensation, **1**, 8
Claisen rearrangement, **2**, 1; **22**, 1
Cleavage:
    of benzyl–oxygen, benzyl–nitrogen, and
        benzyl–sulfur bonds, **7**, 5
    of carbon–carbon bonds by periodic acid,
        **2**, 8
    of esters via $S_N2$-type dealkylation, **24**, 2
    of non-enolizable ketones with sodium
        amide, **9**, 1
    in sensitized photooxidation, **20**, 2
Clemmensen reaction, **1**, 7; **22**, 3
Cobalt–carbon monoxide complexes to
    prepare cyclopentenones, **40**, 1
Condensation:
    acetoacetic ester, **1**, 9
    acyloin, **4**, 4; **23**, 2
    aldol, **16**
    benzoin, **4**, 5
    Claisen, **1**, 8
    Darzens, **5**, 10; **31**
    Dieckmann, **1**, 9; **6**, 9; **15**, 1
    directed aldol, **28**, 3
    Knoevenagel, **1**, 8; **15**, 2
    Stobbe, **6**, 1
    Thorpe–Ziegler, **15**, 1; **31**
Conjugate addition:
    of hydrogen cyanide, **25**, 3
    of organocopper reagents, **19**, 1; **41**; 2
Cope rearrangement, **22**, 1; **41**, 1; **43**, 2
Copper–catalyzed:
    decomposition of α-diazocarbonyl
        compounds, **26**, 2
    reactions of Grignard reagents, **19**, 1; **41**, 2
Copper–Grignard complexes, conjugate
    additions of, **19**, 1; **41**, 2
Corey–Winter reaction, **30**, 2
Coumarins, preparation of, **7**, 1; **20**, 3
Coupling:
    of allylic and benzylic carbanions, **27**, 1
    of π-allyl ligands, **19**, 2
    of diazonium salts with aliphatic
        compounds, **10**, 1, 2
Cuprate reagents, **19**, 1; **38**, 2; **41**, 2
Curtius rearrangement, **3**, 7, 9
Cyanide catalysis, aldehyde addition to double
    bonds, **40**, 4

Cyanoethylation, **5**, 2
Cyanogen bromide, reactions with tertiary amines, **7**, 4
Cyclic ketones, formation by intramolecular acylation, **2**, 4; **23**, 2
Cyclization:
  with alkenyl- and alkynylaluminum reagents, **32**, 2
  of alkyl dihalides, **19**, 2
  of aryl-substituted aliphatic acids, acid chlorides, and anhydrides, **2**, 4; **23**, 2
  of $\alpha$-carbonyl carbenes and carbenoids, **26**, 2
  of diesters and dinitriles, **15**, 1
  Fischer indole, **10**, 2
  intramolecular by acylation, **2**, 4
  intramolecular by acyloin condensation, **4**, 4
  intramolecular by Diels–Alder reaction, **32**, 1
  Nazarov, **45**, 1
  of stilbenes, **30**, 1
Cycloaddition reactions, **4**, 1, 2; **5**, 3; **12**, 1; **29**, 2; **32**, 1; **36**, 1; **40**, 1; **45**, 2
Cyclobutanes, preparation:
  from nitrile-stabilized carbanions, **31**
  by thermal cycloaddition reactions, **12**, 1
Cycloheptadienes, from divinylcyclopropanes, **41**, 1
$\pi$-Cyclopentadienyl transition metal carbonyls, **17**, 1
Cyclopentenone:
  annulation, **45**, 1
  synthesis, **40**, 1; **45**, 1
Cyclopropane carboxylates, from diazoacetic esters, **18**, 3
Cyclopropanes:
  from $\alpha$-diazocarbonyl compounds, **26**, 2
  from nitrile-stabilized carbanions, **31**
  from tosylhydrazones, **23**, 3
  from unsaturated compounds, methylene iodide, and zinc–copper couple, **20**, 1
Cyclopropenes, preparation of, **18**, 3

Darzens glycidic ester condensation, **5**, 10; **31**
DAST, **34**, 2; **35**, 3
Deamination of aromatic primary amines, **2**, 7
Debenzylation, **7**, 5; **18**, 4
Decarboxylation of acids, **9**, 5; **19**, 4
Dehalogenation:
  of $\alpha$-haloacyl halides, **3**, 3
  reductive, of polyhaloketones, **29**, 2
Dehydrogenation:
  in preparation of ketenes, **3**, 3
  in synthesis of acetylenes, **5**, 1
Demjanov reaction, **11**, 2
Deoxygenation of vicinal diols, **30**, 2

Desoxybenzoins, conversion to benzoins, **4**, 5
Desulfurization:
  of $\alpha$-(alkylthio)nitriles, **31**
  in olefin synthesis, **30**, 2
  with Raney nickel, **12**, 5
Diazoacetic esters, reactions with alkenes, alkynes, heterocyclic and aromatic compounds, **18**, 3; **26**, 2
$\alpha$-Diazocarbonyl compounds, insertion and addition reactions, **26**, 2
Diazomethane:
  in Arndt–Eistert reaction, **1**, 2
  reactions with aldehydes and ketones, **8**, 8
Diazonium fluoroborates, preparation and decomposition, **5**, 4
Diazonium ring closure reactions, **9**, 7
Diazonium salts:
  coupling with aliphatic compounds, **10**, 1, 2
  in deamination of aromatic primary amines, **2**, 7
  in Meerwein arylation reaction, **11**, 3; **24**, 3
  in synthesis of biaryls and aryl quinones, **2**, 6
Dieckmann condensation, **1**, 9; **15**, 1
  for preparation of tetrahydrothiophenes, **6**, 9
Diels–Alder reaction:
  with acetylenic and olefinic dienophiles, **4**, 2
  with cyclenones and quinones, **5**, 3
  intramolecular, **32**, 1
  with maleic anhydride, **4**, 1
Dienes, synthesis with alkenyl- and alkynylaluminum reagents, **32**, 2
Diimide, **40**, 2
3,4-Dihydroisoquinolines, preparation of, **6**, 2
Diketones:
  pyrolysis of diaryl, **1**, 6
  reduction by acid in organic solvents, **22**, 3
  synthesis by acylation of ketones, **8**, 3
  synthesis by alkylation of $\beta$-diketone dianions, **17**, 2
Dimethyl sulfoxide, in oxidation reactions, **39**, 3
Diols:
  deoxygenation of, **30**, 2
  oxidation of, **2**, 8
Dioxetanes, **20**, 2
Divinyl-aziridines, -cyclopropanes, -oxiranes, and -thiiranes, rearrangements of, **41**, 1
Doebner reaction, **1**, 8

Eastwood reaction, **30**, 2
Elbs reaction, **1**, 6; **35**, 2

Electrophilic substitution reactions of allyl-
and vinylsilanes, **37**, 2
Enamines, reaction with quinones, **20**, 3
Ene reaction, in photosensitized oxygenation,
**20**, 2
Enolates, in directed aldol reactions, **28**, 3;
**46**, 1
Enone cycloadditions, **44**, 2
Enynes, synthesis with alkenyl- and
alkynylaluminum reagents, **32**, 2
Enzymatic resolution, **37**, 1
Epoxidation with organic peracids, **7**, 7
Epoxide isomerizations, **29**, 3
Esters:
acylation with acid chlorides, **1**, 9
alkylation of, **9**, 4
alkylidenation of, **43**, 1
cleavage via $S_N2$-type dealkylation, **24**, 2
dimerization, **23**, 2
glycidic, synthesis of, **5**, 10
hydrolysis catalyzed by pig liver esterase,
**37**, 1
$\beta$-hydroxy, synthesis of, **1**, 1; **22**, 4
$\beta$-keto, synthesis of, **15**, 1
reaction with organolithium reagents, **18**, 1
reduction of, **8**, 1
synthesis from diazoacetic esters, **18**, 3
synthesis by Mitsunobu reaction, **42**, 2
$\alpha,\beta$-unsaturated, synthesis with alkenyl-
and alkynylaluminum reagents, **32**, 2
Exhaustive methylation, Hofmann, **11**, 5

Favorskii rearrangement, **11**, 4
Ferrocenes, **17**, 1
Fischer indole cyclization, **10**, 2
Fluorination of aliphatic compounds, **2**, 2;
**21**, 1, 2; **34**, 2; **35**, 3
Fluorination by DAST, **35**, 3
Fluorination by sulfur tetrafluoride, **21**, 1;
**34**, 2
Formylation:
of alkylphenols, **28**, 1
of aromatic hydrocarbons, **5**, 6
Free radical additions:
to olefins and acetylenes to form carbon–
heteroatom bonds, **13**, 4
to olefins to form carbon–carbon bonds,
**13**, 3
Friedel–Crafts reaction, **2**, 4; **3**, 1; **5**, 15;
**18**, 1; **31**
Friedländer synthesis of quinolines, **28**, 2
Fries reaction, **1**, 11
$\alpha$-Functionalized sulfides by Pummerer
reaction, **40**, 3

Gattermann aldehyde synthesis, **9**, 2
Gattermann–Koch reaction, **5**, 6
Germanes, addition to olefins and acetylenes,
**13**, 4
Glycidic esters, synthesis and reactions of,
**5**, 10
Gomberg–Bachmann reaction, **2**, 6; **9**, 7
Grundmann synthesis of aldehydes, **8**, 5

Halides, displacement reactions of, **22**, 2; **27**, 2
Halides, preparation:
from alcohols, **34**, 2
alkenyl, synthesis with alkenyl- and
alkynylaluminum reagents, **32**, 2
by chloromethylation, **1**, 3
from organoboranes, **33**, 1
from primary and secondary alcohols, **29**, 1
Haller–Bauer reaction, **9**, 1
Halocarbenes, preparation and reaction of,
**13**, 2
Halocyclopropanes, reactions of, **13**, 2
Halogenated benzenes, in Jacobsen reaction,
**1**, 12
Halogen–metal interconversion reactions, **6**, 7
$\alpha$-Haloketones, rearrangement of, **11**, 4
$\alpha$-Halosulfones, synthesis and reactions of,
**25**, 1
Helicenes, synthesis by photocyclization, **30**, 1
Heterocyclic aromatic systems, lithiation of,
**26**, 1
Heterocyclic bases, amination of, **1**, 4
Heterocyclic compounds, synthesis:
by acyloin condensation, **23**, 2
by allylic and benzylic carbanions, **27**, 1
by intramolecular Diels–Alder reaction, **32**, 1
by phosphoryl-stabilized anions, **25**, 2
by Ritter reaction, **17**, 3
*see also* Azlactones, **3**, 5; Isoquinolines,
synthesis of, **6**, 2, 3, 4; $\beta$-Lactams,
synthesis of, **9**, 6; Quinolines, **7**, 2; **28**, 2;
Thiazoles, preparation of, **6**, 8;
Thiophenes, preparation of, **6**, 9
Hoesch reaction, **5**, 9
Hofmann elimination reaction, **11**, 5; **18**, 4
Hofmann exhaustive methylation, **11**, 5
Hofmann reaction of amides, **3**, 7, 9
Homogeneous hydrogenation catalysts, **24**, 1
Hunsdiecker reaction, **9**, 5; **19**, 4
Hydration of olefins, dienes, and acetylenes,
**13**, 1
Hydrazoic acid, reactions and generation of, **3**,
8
Hydroboration, **13**, 1

Hydrocyanation of conjugated carbonyl compounds, **25**, 3

Hydrogenation of esters:
  with copper chromite and Raney nickel, **8**, 1
  by homogeneous hydrogenation catalysts, **24**, 1

Hydrogenolysis of benzyl groups attached to oxygen, nitrogen, and sulfur, **7**, 5

Hydrogenolytic desulfurization, **12**, 5

Hydrohalogenation, **13**, 4

Hydroxyaldehydes, **28**, 1

5-Hydroxyindoles, synthesis of, **20**, 3

α-Hydroxyketones, synthesis of, **23**, 2

Hydroxylation of ethylenic compounds with organic peracids, **7**, 7

Hydroxynitriles, synthesis of, **31**


Imidates, rearrangement of, **14**, 1

Iminium ions, **39**, 2

Indoles, by Nenitzescu reaction, **20**, 3

Intramolecular cyclic rearrangement, **2**, 1; **18**, 2; **22**, 1

Intramolecular cyclization:
  by acylation, **2**, 4
  by acyloin condensation, **4**, 4
  of α-carbonyl carbenes and carbenoids, **26**, 2
  by Diels–Alder reaction, **32**, 1

Isoquinolines, synthesis of, **6**, 2, 3, 4; **20**, 3


Jacobsen reaction, **1**, 12

Japp–Klingemann reaction, **10**, 2


Ketene cycloadditions, **45**, 2

Ketenes and ketene dimers, preparation of, **3**, 3; **45**, 2

Ketones:
  acylation of, **8**, 3
  alkylidenation of, **43**, 1
  Baeyer–Villiger oxidation of, **9**, 3; **43**, 3
  cleavage of non-enolizable, **9**, 1
  comparison of synthetic methods, **18**, 1
  conversion to amides, **3**, 8; **11**, 1
  conversion to fluorides, **34**, 2; **35**, 3
  cyclic, preparation of, **2**, 4; **23**, 2
  cyclization of divinyl ketones, **45**, 1
  preparation from acid chlorides and organometallic compounds, **8**, 2; **18**, 1
  preparation from organoboranes, **33**, 1
  preparation from α,β-unsaturated carbonyl compounds and metals in liquid ammonia, **23**, 1
  reaction with diazomethane, **8**, 8
  reduction to aliphatic compounds, **4**, 8

reduction by alkoxyaluminum hydrides, **34**, 1

reduction in anhydrous organic solvents, **22**, 3

synthesis from organolithium reagents and carboxylic acids, **18**, 1

synthesis by oxidation of alcohols, **6**, 5; **39**, 3

Kindler modification of Willgerodt reaction, **3**, 2

Knoevenagel condensation, **1**, 8; **15**, 2

Koch–Haaf reaction, **17**, 3

Kornblum oxidation, **39**, 3

Kostanek synthesis of chromanes, flavones, and isoflavones, **8**, 3


β-Lactams, synthesis of, **9**, 6; **26**, 2

β-Lactones, synthesis and reactions of, **8**, 7

Lead tetraacetate, in oxidative decarboxylation of acids, **19**, 4

Leuckart reaction, **5**, 7

Lithiation:
  of allylic and benzylic systems, **27**, 1
  of halogen–metal interconversion, **6**, 7
  of heterocyclic and olefinic compounds, **26**, 1

Lithioorganocuprates, **19**, 1; **22**, 2; **41**, 2

Lithium aluminum hydride reductions, **6**, 10

Lossen rearrangement, **3**, 7, 9


Mannich reaction, **1**, 10; **7**, 3

Meerwein arylation reaction, **11**, 3; **24**, 3

Meerwein–Ponndorf–Verley reduction, **2**, 5

Metal alkoxyaluminum hydrides, **34**, 1; **36**, 3

Metalations with organolithium compounds, **8**, 6; **26**, 1; **27**, 1

Methylenation of carbonyl groups, **43**, 1

Methylene-transfer reactions, **18**, 3; **20**, 1

Michael reaction, **10**, 3; **15**, 1, 2; **19**, 1; **20**, 3; **46**, 1

Mitsunobu reaction, **42**, 2

Moffatt oxidation, **39**, 3


Nazarov cyclization, **45**, 1

Nef reaction, **38**, 3

Nenitzescu reaction, **20**, 3

Nitriles:
  formation from oximes, **35**, 2
  preparation from organoboranes, **33**, 1
  α,β-unsaturated:
    by elimination of selenoxides, **44**, 1
    synthesis with alkenyl- and alkynylaluminum reagents, **32**, 2

Nitrile-stabilized carbanions:
    alkylation of, **31**
    arylation of, **31**
Nitroamines, **20**, 4
Nitro compounds, conversion to carbonyl
    compounds, **38**, 3
Nitro compounds, preparation of, **12**, 3
Nitrogen compounds, reduction of, **36**, 3
Nitrone–olefin cycloadditions, **36**, 1
Nitrosation, **2**, 6; **7**, 6

Olefins:
    arylation of, **11**, 3; **24**, 3; **27**, 2
    cycloadditions, **44**, 2
    cyclopropanes from, **20**, 1
    as dienophiles, **4**, 1, 2
    epoxidation and hydroxylation of, **7**, 7
    free-radical additions to, **13**, 3, 4
    hydroboration of, **13**, 1
    hydrogenation with homogeneous catalysts,
        **24**, 1
    reactions with diazoacetic esters, **18**, 3
    reactions with nitrones, **36**, 1
    reduction by alkoxyaluminum hydrides,
        **34**, 1
Olefins, synthesis:
    with alkenyl- and alkynylaluminum
        reagents, **32**, 2
    from amines, **11**, 5
    by Bamford–Stevens reaction, **23**, 3
    by Claisen and Cope rearrangements, **22**, 1
    by dehydrocyanation of nitriles, **31**
    by deoxygenation of vicinal diols, **30**, 2
    by palladium-catalyzed vinylation, **27**, 2
    from phosphoryl-stabilized anions, **25**, 2
    by pyrolysis of xanthates, **12**, 2
    from silicon-stabilized anions, **38**, 1
    by Wittig reaction, **14**, 3
Olefin reduction by diimide, **40**, 2
Oligomerization of 1,3-dienes, **19**, 2
Oppenauer oxidation, **6**, 5
Organoboranes:
    formation of carbon–carbon and carbon–
        heteroatom bonds from, **33**, 1
    isomerization and oxidation of, **13**, 1
    reaction with anions of $\alpha$-chloronitriles, **31**
Organo-heteroatom bonds to germanium,
    phosphorus, silicon, and sulfur,
    preparation by free-radical additions, **13**, 4
Organometallic compounds:
    of aluminum, **25**, 3
    of copper, **19**, 1; **22**, 2; **38**, 2; **41**, 2
    of lithium, **6**, 7; **8**, 6; **18**, 1; **27**, 1
    of magnesium, zinc, and cadmium, **8**, 2;
        **18**, 1; **19**, 1; **20**, 1

of palladium, **27**, 2
of zinc, **1**, 1; **22**, 4
Oxidation:
    of alcohols and polyhydroxy compounds,
        **6**, 5; **39**, 3
    of aldehydes and ketones, Baeyer–Villiger
        reaction, **9**, 3; **43**, 3
    of amines, phenols, aminophenols,
        diamines, hydroquinones, and
        halophenols, **4**, 6; **35**, 2
    of $\alpha$-glycols, $\alpha$-amino alcohols, and
        polyhydroxy compounds by periodic acid,
        **2**, 8
    of organoboranes, **13**, 1
    with peracids, **7**, 7
    by photooxygenation, **20**, 2
    with selenium dioxide, **5**, 8; **24**, 4
Oxidative decarboxylation, **19**, 4
Oximes, formation by nitrosation, **7**, 6

Palladium-catalyzed vinylic substitution, **27**, 2
Pauson–Khand reaction to prepare
    cyclopentenones, **40**, 1
Pechmann reaction, **7**, 1
Peptides, synthesis of, **3**, 5; **12**, 4
Peracids, epoxidation and hydroxylation with,
    **7**, 7
    in Baeyer–Villiger oxidation, **9**, 3; **43**, 3
Periodic acid oxidation, **2**, 8
Perkin reaction, **1**, 8
Persulfate oxidation, **35**, 2
Peterson olefination, **38**, 1
Phenanthrenes, synthesis by photocyclization,
    **30**, 1
Phenols, dihydric from phenols, **35**, 2
Phosphinic acids, synthesis of, **6**, 6
Phosphonic acids, synthesis of, **6**, 6
Phosphonium salts:
    halide synthesis, use in, **29**, 1
    preparation and reactions of, **14**, 3
Phosphorus compounds, addition to carbonyl
    group, **6**, 6; **14**, 3; **25**, 2; **36**, 2
    addition reactions at imine carbon, **36**, 2
Phosphoryl-stabilized anions, **25**, 2
Photochemical additions, **44**, 2
Photocyclization of stilbenes, **30**, 1
Photooxygenation of olefins, **20**, 2
Photosensitizers, **20**, 2
Pictet–Spengler reaction, **6**, 3
Pig liver esterase, **37**, 1
Polonovski reaction, **39**, 2
Polyalkylbenzenes, in Jacobsen reaction, **1**, 12
Polycyclic aromatic compounds, synthesis by
    photocyclization of stilbenes, **30**, 1

Polyhalo ketones, reductive dehalogenation of, **29**, 2
Pomeranz–Fritsch reaction, **6**, 4
Prévost reaction, **9**, 5
Pschorr synthesis, **2**, 6; **9**, 7
Pummerer reaction, **40**, 3
Pyrazolines, intermediates in diazoacetic ester reactions, **18**, 3
Pyrolysis:
  of amine oxides, phosphates, and acyl derivatives, **11**, 5
  of ketones and diketones, **1**, 6
  for preparation of ketenes, **3**, 3
  of xanthates, **12**, 2
$\pi$-Pyrrolylmanganese tricarbonyl, **17**, 1

Quaternary ammonium salts, rearrangements of, **18**, 4
Quinolines:
  preparation by Friedländer synthesis, **28**, 2
  by Skraup synthesis, **7**, 2
Quinones:
  acetoxylation of, **19**, 3
  diene additions to, **5**, 3
  synthesis of, **4**, 6
  in synthesis of 5-hydroxyindoles, **20**, 3

Ramberg–Bäcklund rearrangement, **25**, 1
Rearrangement:
  Beckmann, **11**, 1
  Chapman, **14**, 1; **18**, 2
  Claisen, **2**, 1; **22**, 1
  Cope, **22**, 1; **41**, 1; **43**, 2
  Curtius, **3**, 7, 9
  divinylcyclopropanes, **41**, 1
  Favorskii, **11**, 4
  Lossen, **3**, 7, 9
  Ramberg–Bäcklund, **25**, 1
  sigmatropic, **43**, 2
  Smiles, **18**, 2
  Sommelet–Hauser, **18**, 4
  Stevens, **18**, 4
  [2,3] Wittig, **46**, 2
  vinylcyclopropane-cyclopentene, **33**, 2
Reduction:
  of acid chlorides to aldehydes, **4**, 7; **8**, 5
  of aromatic compounds, **42**, 1
  of benzils, **4**, 5
  by Clemmensen reaction, **1**, 7; **22**, 3
  desulfurization, **12**, 5
  with diimide, **40**, 2
  by dissolving metal, **42**, 1
  by homogeneous hydrogenation catalysts, **24**, 1

by hydrogenation of esters with copper chromite and Raney nickel, **8**, 1
hydrogenolysis of benzyl groups, **7**, 5
by lithium aluminum hydride, **6**, 10
by Meerwein–Ponndorf–Verley reaction, **2**, 5
by metal alkoxyaluminum hydrides, **34**, 1; **36**, 3
of mono- and polynitroarenes, **20**, 4
of olefins by diimide, **40**, 2
of $\alpha,\beta$-unsaturated carbonyl compounds, **23**, 1
by samarium(II) iodide, **46**, 3
by Wolff–Kishner reaction, **4**, 8
Reductive alkylation, preparation of amines, **4**, 3; **5**, 7
Reductive dehalogenation of polyhalo ketones with low-valent metals, **29**, 2
Reductive desulfurization of thiol esters, **8**, 5
Reformatsky reaction, **1**, 1; **22**, 4
Reimer–Tiemann reaction, **13**, 2; **28**, 1
Resolution of alcohols, **2**, 9
Ritter reaction, **17**, 3
Rosenmund reaction for preparation of arsonic acids, **2**, 10
Rosenmund reduction, **4**, 7

Samarium(II) iodide, **46**, 3
Sandmeyer reaction, **2**, 7
Schiemann reaction, **5**, 4
Schmidt reaction, **3**, 8, 9
Selenium dioxide oxidation, **5**, 8; **24**, 4
Selenoxide elimination, **44**, 1
Shapiro reaction, **23**, 3; **39**, 1
Sigmatropic rearrangements, **43**, 2
Silanes:
  addition to olefins and acetylenes, **13**, 4
  electrophilic substitution reactions, **37**, 2
Silyl carbanions, **38**, 1
Simmons–Smith reaction, **20**, 1
Simonini reaction, **9**, 5
Singlet oxygen, **20**, 2
Skraup synthesis, **7**, 2; **28**, 2
Smiles rearrangement, **18**, 2
Sommelet–Hauser rearrangement, **18**, 4
Sommelet reaction, **8**, 4
Stevens rearrangement, **18**, 4
Stetter reaction of aldehydes with olefins, **40**, 4
Stilbenes, photocyclization of, **30**, 1
Stobbe condensation, **6**, 1
Substitution reactions using organocopper reagents, **22**, 2; **41**, 2
Sulfide reduction of nitroarenes, **20**, 4
Sulfonation of aromatic hydrocarbons and aryl halides, **3**, 4

Sulfoxides in the Pummerer reaction, **40**, 3
Sulfur compounds, reduction of, **36**, 3
Swern oxidation, **39**, 3

Tetrahydroisoquinolines, synthesis of, **6**, 3
Tetrahydrothiophenes, preparation of, **6**, 9
Thiazoles, preparation of, **6**, 8
Thiele–Winter acetoxylation of quinones, **19**, 3
Thiocarbonates, synthesis of, **17**, 3
Thiocyanation of aromatic amines, phenols, and polynuclear hydrocarbons, **3**, 6
Thiocyanogen, substitution and addition reactions of, **3**, 6
Thiophenes, preparation of, **6**, 9
Thorpe–Ziegler condensation, **15**, 1; **31**
Tiemann reaction, **3**, 9
Tiffeneau–Demjanov reaction, **11**, 2
Tin(II) Enolates, **46**, 1
Tipson–Cohen reaction, **30**, 2
Tosylhydrazones, **23**, 3; **39**, 1

Ullmann reaction:
    in synthesis of diphenylamines, **14**, 1
    in synthesis of unsymmetrical biaryls, **2**, 6

Vinylcyclopropanes, rearrangement to cyclopentenes, **33**, 2
Vinyllithiums, from sulfonylhydrazones, **39**, 1
Vinylsilanes, electrophilic substitution reactions of, **37**, 2
Vinyl substitution, catalyzed by palladium complexes, **27**, 2
von Braun cyanogen bromide reaction, **7**, 4

Willgerodt reaction, **3**, 2
Wittig reaction, **14**, 3; **31**
[2,3]-Wittig rearrangement, **46**, 2
Wolff–Kishner reaction, **4**, 8

Xanthates, preparation and pyrolysis of, **12**, 2

Ylides:
    in Stevens rearrangement, **18**, 4
    in Wittig reaction, structure and properties, **14**, 3

Zinc–copper couples, **20**, 1
Zinin reduction of nitroarenes, **20**, 4